Contents

A Changing levels

The transition from GCSE to AS Level is demanding but can be achieved in a year if you take a positive attitude towards change. The main differences between GCSE and AS Level are the **quantity** and **quality** of information required and the level of **interpretation** and **analysis** expected. These can usefully be divided into **knowledge** and **understanding**.

- **Knowledge** This involves:
  - learning terminology, definitions, processes and examples
  - knowledge about real world events and recall of specific information.

  The main difference between GCSE and AS Level is that of quantity and depth – many of the case studies will remain the same, with only detail added.

  This skill responds well to hard work and effort and should be in place by the end of the first year in the sixth form.

- **Understanding** This is the analysis and evaluation of the material tested through response to questions. It involves:
  - an understanding of why the material has been learned and what its implications are
  - using information in a flexible manner and working out answers from data supplied by the examiners.

  The main difference between GCSE and AS Level is the amount of individual thought involved and the level of problem-solving. This requires thought and questioning as well as practice in written expression including structured and data response questions.

What the examiners say

At AS Level the examiners are looking for a combination of knowledge and understanding. Examination Boards publish details of how candidates coped with the papers, and these give a very clear insight into what the examiners expect. To gain high marks these comments also clearly tell you what to do and what to avoid. When divided into strengths and weaknesses a clear pattern emerges in both the Physical and Human sections.

Typical examiner comments:

- weak skills, particularly in interpretation of response material
- processes are not understood
- answers lack focus on the wording of the questions
- answers are too dependent on recall and do not answer the questions set
- there are a number of misconceptions about basic processes
- answers have disappointingly weak geographical language
- there is an inability to link models with current reality.

Summary of strengths and weaknesses:

- the strengths tend to be case studies and descriptions (knowledge)
- question interpretation is a problem in all papers
- the use of terminology is generally weak
- the understanding and use of processes is generally weak
- questions are often not answered
- understanding is the main problem area.

The examiners' comments are based on real scripts and are relevant to you.

- The strength of answers lies in case studies and factual material.
- The criticisms are consistent and focus on three main areas:
  1 the use of terminology
  2 the understanding of processes
  3 the ability to interpret and respond to the set question.

Use the material in the following chapters wisely and **you will succeed**.

Note: The sample answers given in this book are not necessarily the *best* answers. They simply demonstrate particular points. You could usefully discuss how good you think they are!

ESSENTIAL *AS* Ge

Simon Ross • John Morgan • Ri

Volume Editor: Simon Ro

Essential AS Geography Website

The activities in this book are supported by a website, the address of which is:

HYPERLINK
http://www.nelsonthornes.com/secondary/geography/essential_intro.htm

The user should be aware that URLs or web addresses change quite often. Although every effort will be made to provide accurate addresses, it is inevitable that some will change. We will maintain the site to ensure this frustration is an infrequent event.

Remember to bookmark this address within your browser. You may even wish to create a separate folder in which to deposit downloaded materials. We trust that the information and activities we have provided will enable you to make the most of your course in AS level Geography.

2

ailable from the British

es, Charlbury, Oxford

Printing Ltd

B   What are notes?

Notes are not books and they are not essays. They are your own record of the terminology, definitions, processes and examples required to gain success at AS/A2 Level. You will use your notes on a regular basis when completing written work and revising for mocks and examinations. It is well worth giving your notes considerable thought, attention and care. Good notes make revision and learning easier, and produce more effective tasks – they are essential for success.

- Notes should be a precise summary of the required material, with an organisation that allows rapid recall and the possibility of adding more detail at a later date. There is no point in copying out the book (as you already have it), and there is no point in including large sections of 'padding' text.

- The idea is to isolate what really matters, and to use abbreviations to reduce the length whenever possible. Some teachers dictate notes whilst others expect students to construct their own.

- The material generally has to be reworked to a more compact form well before examinations and revision.

When using a textbook, all of the information seems important and it is often difficult to know what to note and what to discard. The following sections should help you in this task. Remember that you can always add detail later when you are more familiar with the topic.

Why note?

- Rapid source of material for written work.
- Forces you to think about and select main points.
- Revision for mocks.
- Revision for AS examination.
- Revision for A2 synoptic paper.

How do I start to note?

Note-taking with only a partial understanding of a topic is a problem. It is difficult to know what is relevant, what is important, what to include and what to discard. **PQRST** (Figure 1.1) is an approach that helps to overcome this problem and also helps to increase the speed and accuracy of the memory process. The procedure may seem complicated but, when mastered, it is a very efficient way to note and learn material.

What to do	Why
P Preview Skim through the whole section, chapter or topic. Identify the main divisions using the headings. Look at the diagrams to help establish the general content. Try to divide the topic into major sections, for example three main ideas or three main scales.	This helps to establish the type of material the topic covers before you start to add detail. It opens up categories, allowing notes to be subdivided into sections. Establishing simple divisions helps understanding and learning.
Q Question Before starting to read in detail, think about why this material has been included. Think about the range of material, the sequence of the topics and how this information relates to your existing knowledge.	This increases the significance of the material. It also links the new material to your existing knowledge, making learning and remembering the material more effective.
R Read Take each of the sections in turn and read them fully. Identify and note the key definitions, terms and processes. Try to simplify the diagrams and include these with annotations in your notes.	This adds the detail and knowledge required to reach AS and A2 standards.
S Self-recitation At the end of each section run through the main terms and ideas either mentally or saying them aloud. Rehearse the main points.	Rehearsal of small chunks of information increases memory and hence learning.
T Test At the end of each section, after rehearsal, test yourself on your ability to recall the main items. If you cannot remember the data then repeat the RST stages.	This gives a feeling of achievement and reassurance that you are adding to your knowledge. You know that the material has been stored and can use this to understand the subsequent sections.

1.1   *The PQRST method of noting and learning*

Where do I find information?

Notes should be constructed from a variety of sources so that all aspects of the topics are fully covered. Some material is available through **primary sources**, or personal experience (including fieldwork, holidays and visits), but most will come from **secondary sources**, including textbooks, journals, the internet and television. By selecting relevant material from these, it is possible to build up notes that are *better* than any of the individual sources.

1 Primary information

Events directly experienced by yourself provide powerful memories and form a real strength in geography. These combine sight, sound, smell and significance and are useful in adding to an understanding of the topic. How useful these experiences are depends on your level of observation at the time. Even if this was weak in the past it would help accelerate your current progress by increasing your awareness of what is happening around you. Try to relate the theories and concepts covered in lessons and reading to your observations. See if you can make sense of what you see.

Potential sources of primary information include:

- holidays – coasts, National Parks, agriculture, tourism and development
- shopping – retail parks, new roads, central business district and suburban centres
- travel to city centre – inner city change, de-industrialisation and congestion
- travel between cities – agricultural land use, rural settlement changes, industry
- travel abroad – differences in settlement and economic patterns.

2 Secondary information

For most students this will be the main source of data (Figure 1.2), and over a one-year course you should make an attempt to explore and use as wide a range of sources as possible. You should make sure that you know which are the main topics covered by your Board at AS level and, if you intend to complete the second year, at A2 level. Material in newspapers and articles is only available for a short length of time and should be gathered on a regular basis. Your active involvement is required to collect this information and you should aim to start from the moment you decide to follow AS Geography.

Type	Advantages	Disadvantages	Source
Textbooks, CD-ROMs	Selected material to meet AS Level requirements • definitions • terminology • case examples • 'all in one' – simple to use • available when needed	examples may be out of date limited in discussion and relevance to *you*	• Stanley Thornes website • libraries • references at end of chapters/book • teacher recommendation
Dictionaries	• good detailed definitions with a breakdown of topics into parts • linkage to other relevant terms and concepts	• rather boring to use • lacking in illustration and examples	
Television, newspapers	• very up-to-date information • a vivid visual and verbal impression • can be related to current events and changes	• rapid presentation at unknown times • lacks geographical explanation • difficult to record or remember detail	Try to watch the daily news and look out for relevant current affairs programmes
Specific programmes	• powerful combination of visual and verbal information • often with graphics or maps to help understanding • good in terms of management and the human impact	• often weak in terms of process • made for the general public, and concentrate on human rather than geographical interest • limited use of terminology	Look out for relevant series
Journals	• up-to-date information • topics generally selected for relevance to current AS topics	• can be dense and over-detailed • difficult to find when needed • can be poorly related to AS questions	Stanley Thornes website
Internet	• fun to use • can discover 'oddities' and unusual information • up-to-date information • can down-load and print results • available when needed	• very time-consuming • difficult to assess reliability of the information • some of the better data is on restricted or pay sites • need internet access	Stanley Thornes website

1.2 *Main sources of secondary information*

How should notes be organised?

Notes will become your main source of information and, to be useful and efficient, careful organisation is very important. Notes consisting of page after page of text are difficult to use, make it hard to find specific material, do not help provide ideas, and are difficult to remember. The organisation of notes must be one that suits you as well as the demands of the subject.

There are several methods that you can use to organise and arrange the material in your notes. You can 'mix and match', using the technique that is most appropriate for the material being covered. Consider the following techniques.

1 Layout

The division of topics into sections is an important part of learning and this can be achieved through the use of space or layout on the page (Figure 1.3).

- Use clear headings to identify the main sections of the topic.
- Use colour or size of writing to make key divisions stand out, e.g. underline key definitions in red.
- Use indentation to allow the topic to be approached at different levels: main ideas, main sections, detailed sections.
- Underline key terms, using a distinctive colour, to allow rapid retrieval from notes.
- Fit diagrams into adjacent spaces.

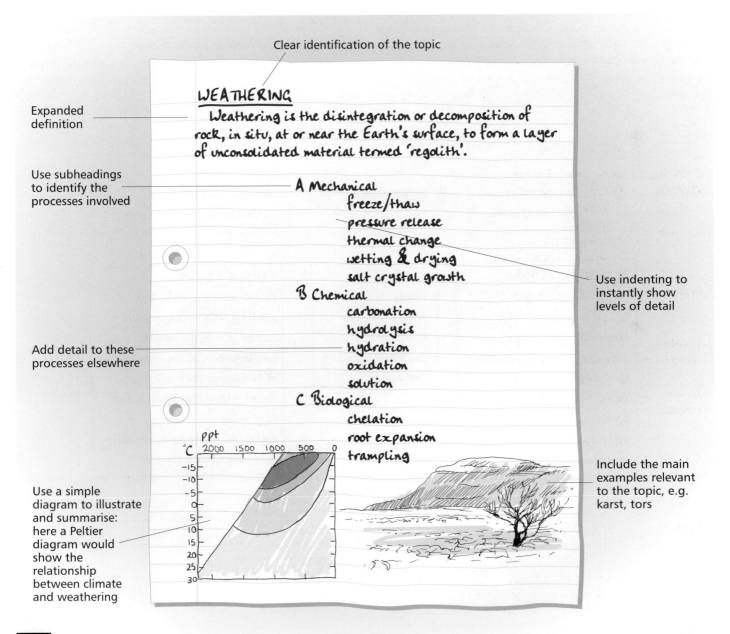

1.3   *Example: create a useful layout*

2 Diagrams

Use simple line drawings with data labels or annotations. These should show the main details but be sufficiently simple to reproduce under examination conditions.

- Diagrams help in seeing how the parts of a topic fit together.
- They give a picture that is easier to recall.
- In the Preview stage of PQRST, see if there is a diagram that helps give a picture of the whole topic.
- Diagrams can reduce the length of notes, but make sure that they are fully labelled or annotated when data must be learned.

Systems diagrams

A systems diagram simplifies often complex processes to a series of stores (boxes) and flows (arrows) (Figure 1.4). Systems are central to most topics in Geography and are useful as they show change, processes and factors in one diagram. There are two main types of system:

- **Open systems** – energy or material is lost and gained through interaction with areas outside the system.
- **Closed systems** – energy or material circulates within the system without losses or gains from external sources.

Advantages:

- shows the main components or sections (the boxes)
- shows the main processes or flows (the arrows)
- shows the main divisions of a topic and the sequence of noting
- provides a simple summary of the whole topic
- allows the addition of factors or even examples with data
- is directly related to examination questions.

Application:

- use at the start of a topic to show the main components and processes
- use when the topic involves change
- use system diagrams to work out the influence of other factors (this is common in examination questions).

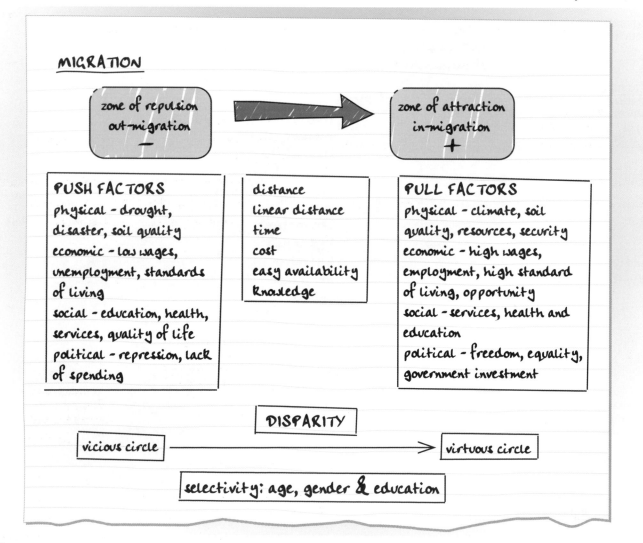

1.4   *Example: a systems diagram*

Sketch diagrams

Sketch diagrams involve drawing a simple outline (cross-section, map or profile) with labels added to identify the main features, and annotations to explain basic points (Figure 1.5). These should be simple enough to reproduce under examination conditions. Unless you are a real 'artist', aim for two-dimensional rather than three-dimensional diagrams.

Advantages:

- simple illustrations show the main features
- annotations can use the main terminology
- increases the use of visual recall
- shows how parts of the topic fit together
- direct use in examinations.

Application:

- particularly useful for physical topics
- use when shape is important, as in cross-sections
- the ability to sketch is required in examinations.

THORNTON FORCE

stepped bed due to horizontal bedding planes of varying resistance

resistant massive limestone forming the lip of the fall – resistant to erosion

undercutting by fluvial erosion: hydraulic action, corrasion and solution

plunge pool created by hydraulic action, corrasion and weathering processes

angular debris resulting from mass movement of the waterfall face – rounded load carried downstream by the river

0   10 metres

1.5 *Example: a sketch diagram*

Sketch maps

A sketch map is a simple map showing the distribution of the main features (Figure 1.6). These should be accurate enough to represent the actual shape but simple enough to draw quickly in examinations.

Advantages:

- shows where features lie in relation to each other as a pattern
- are easily recalled
- can be used as a structure for annotated data.

Application:

- use in essays and examinations to help describe patterns
- use to compare examples with theory (models)
- a major component of case study answers

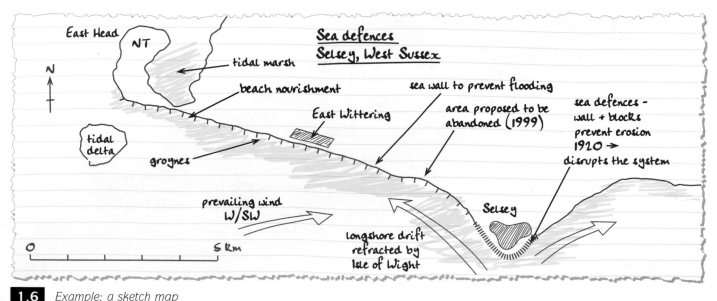

East Head
NT
tidal marsh
beach nourishment
East Wittering
tidal delta
groynes
prevailing wind W/SW
longshore drift refracted by Isle of Wight
Selsey

Sea defences Selsey, West Sussex

sea wall to prevent flooding
area proposed to be abandoned (1999)
sea defences – wall + blocks prevent erosion 1920 → disrupts the system

0   5 km

1.6 *Example: a sketch map*

Case studies

Case study diagrams contain more detail than sketch maps and can be used to provide material for a wider range of topics. Attempt to include both physical and human aspects, as this helps develop the understanding required in synoptic questions (Figure 1.7).

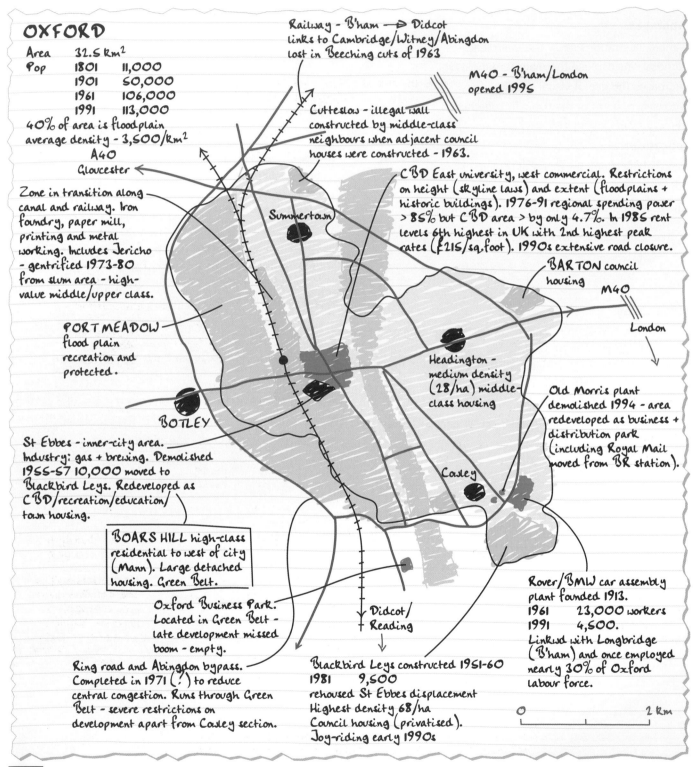

OXFORD

Area	32.5 km²
Pop	1801	11,000
	1901	50,000
	1961	106,000
	1991	113,000

40% of area is floodplain
average density - 3,500/km²

A40
Gloucester ←

Zone in transition along canal and railway. Iron foundry, paper mill, printing and metal working. Includes Jericho - gentrified 1973-80 from slum area - high-value middle/upper class.

PORT MEADOW flood plain recreation and protected.

BOTLEY

St Ebbes - inner-city area. Industry: gas + brewing. Demolished 1955-57 10,000 moved to Blackbird Leys. Redeveloped as CBD/recreation/education/town housing.

BOARS HILL high-class residential to west of city (Mann). Large detached housing. Green Belt.

Oxford Business Park. Located in Green Belt - late development missed boom - empty.

Ring road and Abingdon bypass. Completed in 1971 (?) to reduce central congestion. Runs through Green Belt - severe restrictions on development apart from Cowley section.

Railway - B'ham → Didcot links to Cambridge/Witney/Abingdon lost in Beeching cuts of 1963

M40 - B'ham/London opened 1995

Cutteslow - illegal wall constructed by middle-class neighbours when adjacent council houses were constructed - 1963.

Summertown

CBD East university, west commercial. Restrictions on height (skyline laws) and extent (floodplains + historic buildings). 1976-91 regional spending power > 85% but CBD area > by only 4.7%. In 1985 rent levels 6th highest in UK with 2nd highest peak rates (£215/sq.foot). 1990s extensive road closure.

BARTON council housing

M40

London

Headington - medium density (28/ha) middle-class housing

Old Morris plant demolished 1994 - area redeveloped as business + distribution park (including Royal Mail moved from BR station).

Cowley

Didcot/Reading

Blackbird Leys constructed 1951-60
1981   9,500
rehoused St Ebbes displacement
Highest density 68/ha
Council housing (privatised).
Joy-riding early 1990s

Rover/BMW car assembly plant founded 1913.
1961   23,000 workers
1991   4,500.
Linked with Longbridge (B'ham) and once employed nearly 30% of Oxford labour force.

0   2 km

1.7 *Example: a case study diagram*

Advantages:

- can include considerable information, through the use of annotations, and can replace note text
- provides information for a number of topics
- includes factual information, theory and visual detail
- allows the noting of various aspects on one sheet, with the topic seen as a whole
- provides a powerful visual recall.

Application:

- use in notes to increase knowledge of specific areas
- use in answers requiring reference to 'one case study/example'
- use to test your knowledge of major examples.

Give basic information about the case study. Include location, scale and any other relevant numerical information.

Outline the area showing the main features. Keep this simple, as it will be drawn in an examination under pressure (time).

Add the relevant names and terminology.

Use colour to identify the different zones or schemes. This helps you to recall information from the diagram.

Include as much numerical information as possible, as it can be used selectively in a wide range of questions.

If more detail is required on a specific zone (for example the CBD or inner city), make a new diagram on a separate sheet.

Try to identify and locate a range of information – physical as well as human. This gives the case study greater flexibility and allows its use in a greater variety of questions.

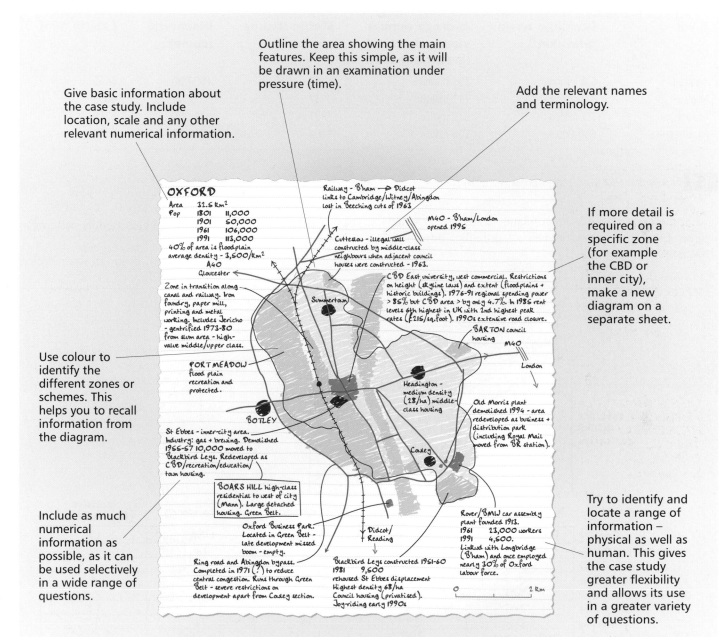

Test
Test your knowledge and memory by attempting to redraw the diagram and adding detail. Re-learn any detail that was omitted.

3 Tables

Tables allow large amounts of data to be noted in an organised manner. These are particularly useful for noting numerical data or when a number of variables are being included for each example (Figure 1.8).

Advantages:

- simple to use and to find data
- allows immediate comparison

- compresses data
- frequently used in exam questions to present data.

Application:

- more useful in noting and learning than for use in examination answers
- can be used to identify trends and patterns, including the relationship between variables
- provide a useful way of learning data.

River	Drainage basin area (km²)	Average annual suspended load (metric tonnes)	Metric tonnes per km²	Discharge (cumecs)	Length (km)
Amazon	5.8 million	363 million	63	180 000	6440
Nile	3.0 million	111 million	37	3000	6695
Mississippi	3.2 million	312 million	97	17 000	6210
Brahmaputra	666 000	726 million	1100	12 000	2700

1.8 *Example: data on some major world rivers*

It is sometimes possible to combine tables and diagrams (Figure 1.9). This has a number of potential advantages including:

- shows the general pattern in both numerical and graphical forms

- links specific data to trends and helps identify patterns in the data
- tables and diagrams are often used in examination questions
- combines theory and examples.

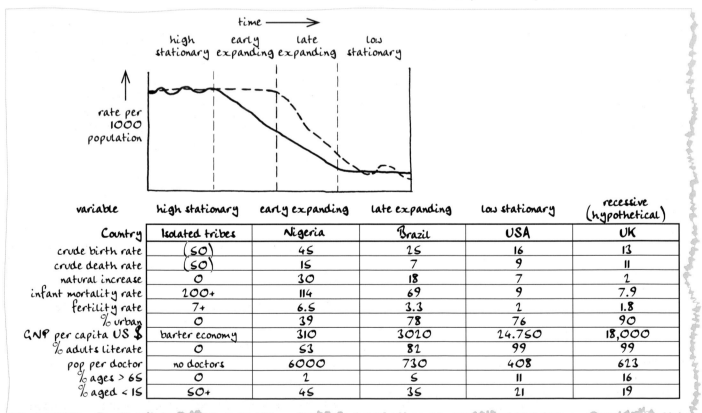

variable	high stationary	early expanding	late expanding	low stationary	recessive (hypothetical)
Country	Isolated tribes	Nigeria	Brazil	USA	UK
crude birth rate	(50)	45	25	16	13
crude death rate	(50)	15	7	9	11
natural increase	0	30	18	7	2
infant mortality rate	200+	114	69	9	7.9
fertility rate	7+	6.5	3.3	2	1.8
% urban	0	39	78	76	90
GNP per capita US $	barter economy	310	3020	14.750	18,000
% adults literate	0	53	82	99	99
pop per doctor	no doctors	6000	730	408	623
% ages > 65	0	2	5	11	16
% aged < 15	50+	45	35	21	19

1.9 *Example: combining a diagram and a table*

4 Bullets

This technique can be used to identify and separate points when noting or answering questions. Each bullet identifies a key idea or aspect and should be followed by an expansion or explanation. Bullets allow the listing of ideas and can be used to reduce length in both notes and data response answers (Figure 1.10).

River long profile

The long profile of a river is a section taken from source to mouth showing the height of the bed above sea level. The long profile is generally concave in shape and is influenced by the discharge of the river, which determines the amount of erosion; the rock type the river flows over, which determines the resistance of the surface; and the length of time the river has been following its course. The long profile may be modified by changes in sea level or land level which cause rejuvenation or drowning of the lower course. Many rivers in the United Kingdom have long profiles that also reflect past climates and processes including glaciation and periglaciation.

1.10 *Example: use of bullets*

Whilst bullet points can work well, you must make sure that sufficient explanation and discussion is included to gain full marks. This technique should *not* be used in essays or in structured questions.

5 Abbreviations

Your notes are your own, and as long as you understand abbreviations, try to use them.

- Abbreviations help to reduce the length of notes and to focus more attention on new key terms that require learning (Figure 1.11). Commonly used abbreviations include:

  = (equals)   > (greater than)   < (less than)
  ↑ (increasing)   ↓ (decreasing)

  but you can also develop your own, for example:

  E (energy)   R (range)   T (threshold)
  or F (function).

- Some terms will be used frequently in notes and the use of abbreviations does seem to help memory and recall.

- **Warning:** some abbreviations are acceptable under examination conditions (DALR or SALR in weather) whilst others are not (ZoT for zone of transition). If the abbreviation is not used in a textbook it is generally not acceptable in an examination. When in doubt, in an examination it is best to write the term in full.

A bulleted list:

- cuts down the length of notes
- focuses on specific themes
- helps to develop a data response style of writing
- makes it easier to find material
- allows new material to be added.

The use of bullets to reduce the length of text in notes is illustrated in the following example. The key ideas are taken from the text, and most unnecessary English is omitted.

River long profile

- section showing height above sea level
- concave – increasing energy with increasing discharge
- lithology and structure determine resistance
- changes in sea level – eustasy and isostasy – rejuvenation.

Market areas can be defined as the area and the population around a central place. The market area contains the threshold population within the range of the central place. When the range is less than the threshold the functions will become unprofitable and will be forced to close. When range is greater than the threshold then the functions will be profitable and competitors will establish themselves.

MA = T in R
uneconomic – R < T
profitable – R > T

1.11 *Example: abbreviation*

If your abbreviation does not immediately make sense it will force you to work it out, and this will help memory and understanding.

Summary of organisation techniques

- Think through a topic before you start to note.
- Try to divide the topic into sections.
- When possible, use visual or diagrammatic noting.
- Avoid long sections of text.
- Leave space to add more material later if it is needed.
- Develop your own style.
- Take a pride in your notes.

What should I note?

When deciding what to include and what to exclude from your notes, try to think about how you will use the material. What is it for?

You should be selective and ruthless, only including material that will be learned or material that is necessary to understand the topic. You can always return to the book source later if you do not fully understand your notes. The main items to aim for – the building blocks of geography – are:

- definitions
- terminology
- processes
- examples/case studies.

1 Definitions

Geography has its own 'language', which is used to describe everyday processes and events in a more technical and precise manner. This is the working language for AS/A2 Level and you need to take time and care to learn definitions thoroughly (Figure 1.12). Definitions are important because:

- data response questions often require definitions
- they make sure that you are writing about the correct topic
- they provide a useful way to start an essay or a paragraph
- they open up ideas and very often help in finding an answer to the question; a specialised dictionary should be used whenever possible
- a knowledge of the exact meaning of a term increases confidence and relevance
- the use of definitions gives a positive impression and helps persuade the examiner that you have worked hard and that you are trying to be concise and precise.

1.12 *Example: definition of weathering*

A minimum

> Weathering is the breakdown of rock in situ at or near the Earth's surface.

Extended version

> Weathering is the disintegration or decomposition of rock, in situ, at or near the Earth's surface, to form a layer of unconsolidated material termed 'regolith'. This occurs through a combination of mechanical, chemical and biological processes and is influenced by climate, the lithology and structure of the rock, and human activities.

2 Terminology

This is the language of the subject and must be learned in order to gain a high grade at AS Level. There are more terms than at GCSE and until they are understood, using textbooks and answering examination questions is difficult. In order to learn the terms efficiently, try to:

- underline key terms in your notes
- look for terms in **bold** type or in *italics* in your books
- make up your own lists for learning and revision (Figure 1.13)
- use the terms when writing – terminology is particularly important in data response answers where space is limited.

1.13 *Example: terminology for mass movement*

mass movement/mass wasting	stress force
cohesive force	frictional force
maximum slope angle	regolith
lubrication	rotational slip
solifluction	soil creep
talus creep	repose slope
concave/convex/rectilinear	slope decline
slope retreat	slope stability/instability.

3 Processes

Processes are actions that bring about change. These are the core of most topics in geography and must be thoroughly understood. To help you find out about and understand processes, note the following:

- A process is an action that causes change.
- Processes involve some form of energy and can often be expressed as equations or as a system (a balance between inputs and outputs over time).
- You should refer to the main factors that influence the process (Figure 1.14).
- When possible, include details of the energy and the quantity and quality of material involved.

1.14 *Example: mass movement (MM)*

> Occurs when stress force (SF) is greater than the forces of cohesion (CF) and friction (FF).
>
> MM = SF > CF + FF
>
> stable slope = SF < CF + FF
>
> maximum slope angle = SF = CF + FF
>
> SF slope angle (gradient) + weight (regolith, water, buildings). Gravity.
>
> CF Vegetation, roots, cementing < by vibration (earthquakes) and degree of weathering.
>
> FF lubrication (rainfall) + particle shape (< by clay).

Think of processes as **causing change** (Figure 1.15).

1.15   *Flow diagram: process*

Factors influence the process; they may speed it up, slow it down, or change the way in which the process works (Figure 1.16).

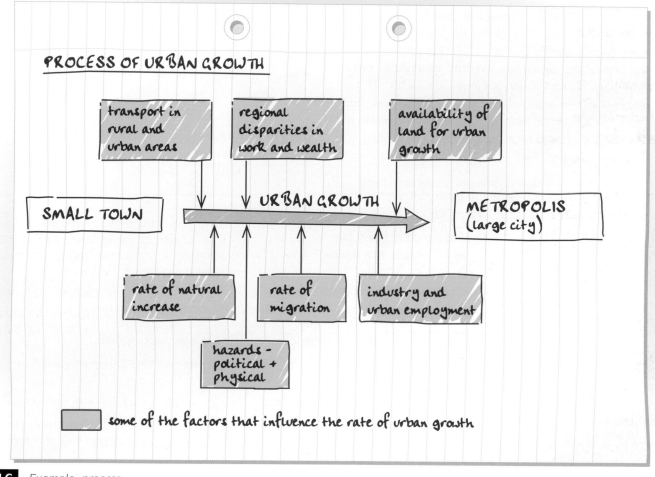

1.16   *Example: process*

4 Examples and case studies

Examples are a basic requirement for examinations. They are used to illustrate and to demonstrate an understanding of processes and theories. A knowledge of the physical and human environments is essential for AS success, and a good knowledge of examples gives confidence and allows flexibility. Consider the following points:

- It is better to start with one detailed example than to note numerous examples in a superficial manner.
- It helps to select examples from a limited number of countries/regions, as this reduces the amount of background knowledge required.
- All key topics should be supported by at least one in-depth example.

- Examples should be detailed and include as much relevant information as possible. Hypothetical examples should be avoided if possible.
- Examples require research and the use of a number of sources.
- It is generally possible to include theory with examples by asking, 'What does this example show?'.

Three scales of examples can be identified.

Basic examples

This may be a reference to a country, place or time without significant expansion. This is the 'e.g.' approach, and it should be used with extreme caution and care. Whilst this approach shows some knowledge, it is of limited application because:

- it lacks the information required to link the example to the processes, factors, causes or consequences as specified in the question
- it does not help to explain why the example is being included
- it reduces written answers to a string of examples, and annoys the examiner who is looking for evidence of understanding

- if an example is worth giving you must explain why it is included.

The use of basic examples in the following sentences add little at AS Level and would be given little or no credit:

- A good example of a mass movement is Aberfan in 1966.
- Many developing countries have a high rate of natural increase, e.g. India.

Detailed examples

This involves an attempt to develop the basic example through the addition of factors, causes, consequences and implications (Figure 1.17). Research from a number of sources may be necessary. Detailed examples should:

- contain sufficient detail to allow flexible use in a range of contexts
- be able to illustrate processes, hazards and the human response
- be large enough to form the core of a paragraph.
- enable conclusions to be drawn from the example.

1.17 *A detailed example: Aberfan 1966*

- a rapid flow from a valley-side spoil heap
- slope angle 25 degrees
- spoil tip on spring line in permeable sedimentary rocks
- water build-up caused by period of heavy rain
- saturated debris flowed 800 metres as a sludge
- 116 schoolchildren killed
- one of worst natural disasters in UK (total deaths 144)
- resulted in removal of all valley-side slag heaps in South Wales, with a major effect on the environment
- 25% of other South Wales colliery spoil heaps were found to be in a dangerous condition

Case studies

These are major examples, with a higher level of detail and a wider range of material than the detailed examples. Try to consider the following points:

- For each topic one detailed case study is usually sufficient.
- Reference should be made to a range of factors and material including physical, economic, social and political (**PESP**).
- Research will be required from a number of sources and possibly be built up over time as new material is discovered.
- Case studies allow you to answer questions requiring reference to *one* area or region that you have studied.

- Choose examples that fit the questions; make sure that the data covers the required aspects of the topic, for example management.
- Select a case study with plenty of data available. Do not worry about boring the examiner, for example by choosing the Mississippi. This is a river basin about which ample data is available in textbooks, articles, newspapers, on television and even the internet.
- Case studies are efficient as they allow a considerable range of data to be learned through *one* set of background information. This reduces the amount of material to learn whilst increasing understanding.

1.18   *Examples of case studies for selected AS/A2 topics*

Syllabus topic	Potential case studies	Topics to include	Sources available
Population	developed country (UK) developing country (Brazil)	distribution demography migration impact of change	examples in textbooks specialised textbooks newspapers (UK) internet (Brazil) *Geofile*
Settlement	major developed city (London) major developing city (São Paulo)	urban structure urban growth de-industrialisation rural–urban fringe transport	examples in textbooks specialised books *Geofile* internet newspapers
Economic development	developed country (UK) developing country (Brazil)	resource base regional disparities regional policies national change	textbooks internet (Brazil) newspapers television
Tectonics	one major volcanic/earthquake event (Montserrat, Turkey)	process (plate margins) effects consequences management issues	textbooks internet (USA) newspapers *Geofile*
Ecosystems	one major biome (tropical rainforest or boreal forest)	climate vegetation soil land use management	textbooks internet (tropical rainforest)
Rivers	one major drainage basin; one local river (Mississippi or Thames)	climate geology channels regime management flooding	textbooks newspapers news internet (USA)
Coasts	one extended stretch of coastline (Dorset or Sussex)	climate geology landforms issues management	textbooks *Geofile* newspapers news internet (weak)

See the Stanley Thornes website for other case study examples.

C How to succeed in exams

What are the different styles of question?

During noting and research you will start to build up an understanding of the topics. From an early stage of the course you will be asked to use the information to answer examination questions, and this involves a new set of skills, including **interpretation** and **understanding**.

The examination questions you will be asked to answer take three main forms:

- data response questions
- structured questions
- essay questions.

1 Data response questions

This style is frequently used at GCSE and should be familiar. The initial questions are a response to data provided (text, a table, diagram, map or photograph) and you are required to write an answer in a limited space provided. These questions are usually set in booklets and the whole script is returned to the examiners. The example in Figure 1.19 shows some of the approaches used in a successful answer.

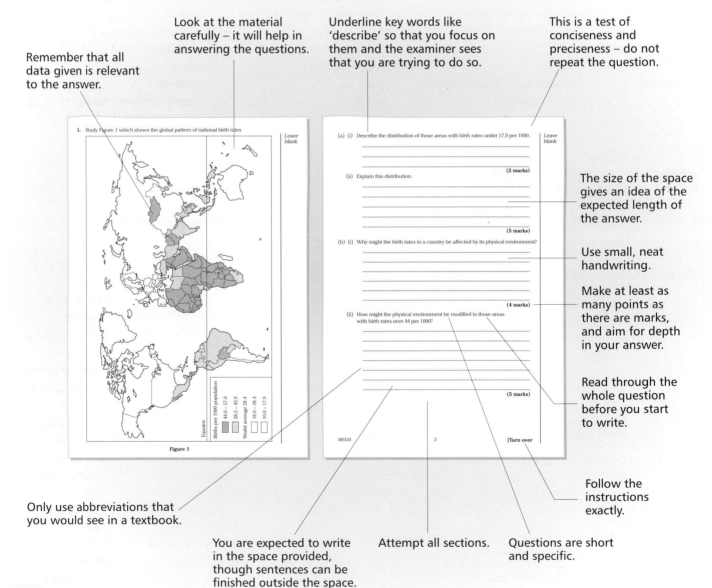

Remember that all data given is relevant to the answer.

Look at the material carefully – it will help in answering the questions.

Underline key words like 'describe' so that you focus on them and the examiner sees that you are trying to do so.

This is a test of conciseness and preciseness – do not repeat the question.

The size of the space gives an idea of the expected length of the answer.

Use small, neat handwriting.

Make at least as many points as there are marks, and aim for depth in your answer.

Read through the whole question before you start to write.

Follow the instructions exactly.

Only use abbreviations that you would see in a textbook.

You are expected to write in the space provided, though sentences can be finished outside the space.

Attempt all sections.

Questions are short and specific.

1.19 *Example: a data response question*

2 Structured questions

Structured questions are similar to data response in that a series of sub-questions are asked, each with a limited mark. Structured questions may also require reference to one or more supplied materials including maps, diagrams and tables. The main difference is that structured questions do not have space limitations and the answers are written in a separate booklet. An example of a structured question is shown in Figure 1.20.

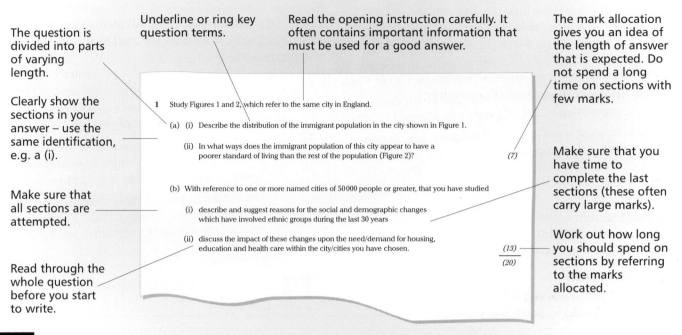

The question is divided into parts of varying length.

Clearly show the sections in your answer – use the same identification, e.g. a (i).

Make sure that all sections are attempted.

Read through the whole question before you start to write.

Underline or ring key question terms.

Read the opening instruction carefully. It often contains important information that must be used for a good answer.

The mark allocation gives you an idea of the length of answer that is expected. Do not spend a long time on sections with few marks.

Make sure that you have time to complete the last sections (these often carry large marks).

Work out how long you should spend on sections by referring to the marks allocated.

1   Study Figures 1 and 2, which refer to the same city in England.

(a) (i) Describe the distribution of the immigrant population in the city shown in Figure 1.

(ii) In what ways does the immigrant population of this city appear to have a poorer standard of living than the rest of the population (Figure 2)?   (7)

(b) With reference to one or more named cities of 50 000 people or greater, that you have studied

(i) describe and suggest reasons for the social and demographic changes which have involved ethnic groups during the last 30 years

(ii) discuss the impact of these changes upon the need/demand for housing, education and health care within the city/cities you have chosen.   (13)
                                                                                                                    (20)

1.20   *Example: a structured question*

3 Essay questions

Essay questions are not usually set at GCSE, and they can pose new problems for AS/A2 students. Essay questions tend to be brief, though they can be divided into two sections with mark allocations. They require you to select processes, theories and examples that can be used to answer the issues raised by the question. The main approaches to use are included in Figure 1.21.

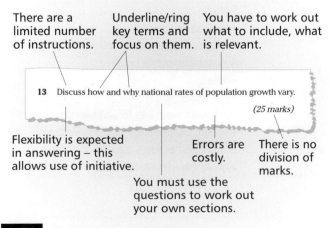

There are a limited number of instructions.

Underline/ring key terms and focus on them.

You have to work out what to include, what is relevant.

13   Discuss how and why national rates of population growth vary.

(25 marks)

Flexibility is expected in answering – this allows use of initiative.

Errors are costly.

There is no division of marks.

You must use the questions to work out your own sections.

1.21   *Example: essay question*

Summary of question types

Question style	Advantages	Disadvantages
Data response	some data given clear indication of length clear indication of time some parts easy numerous questions	writing in space provided need for terminology emphasis on understanding data
Structured questions	a number of questions some data often given reduced need to plan mark allocation given clear division of knowledge and application some parts easy	hard to complete in the time difficult to allocate knowledge difficult to know how much to write
Essay	rapid to read one clear instruction simple to time	interpretation planning relevance errors are costly

How do I interpret a question?

All questions provide **instructions**. These are to help you give a relevant answer. Use the question words and you will give an effective answer. Ignore them and you will receive low marks.

Regardless of the type, questions pose a problem that *you* must solve. It is rare for questions to neatly fit the textbook or note sequences, and you must *expect* to have to fit your knowledge to the question; this requires a plan. The most important thing to remember is that you must work from the question and not be tempted to write 'all you know' about a topic. To develop an effective question answering technique you must learn to think about the types of question asked and to respond to key words. Practice is essential in developing this technique, as you will learn through experimenting and through making mistakes.

Within every question there are instruction words telling you what is required or what to do. There are numerous terms, and each demands a specific response. It is crucial that you focus on these and respond in an appropriate manner. Figure 1.22 shows the type of instruction given in a structured question. Colours are used to identify the type of instruction being given.

1 Study Figures 1 and 2, which refer to **the same city in England**.

    (a) (i) **Describe** the distribution of **the immigrant population** in **the city shown in Figure 1**.

        (ii) **In what ways** does **the immigrant population of this city appear** to have a poorer standard of living than the rest of the population (Figure 2)?

                                        (7)

    (b) **With reference to one or more named cities of 50 000 people or greater**, that you have studied:

      (i) **describe and suggest reasons** for **the social and demographic changes** which have involved ethnic groups during the last 30 years;

      (ii) **discuss** the **impact of these changes** upon the need/demand for housing, education and health care within the city/cities you have chosen.

                                          (13)

                                          (20)

In this example four types of instruction or information are given (identified by colour). The following sections examine how you can use this information to your advantage.

1.22 *Instructions in a structured question*

Question terms

Question terms shown in **red** in Figure 1.22 tell you what to do or how you should respond, e.g. **Describe**

Knowledge		Understanding	
Question term	**What you should do**	**Question term**	**What you should do**
describe	Identify and put into words the main features of the topic. Establish targets to be used in the rest of the question. Identify trends and patterns. Identify types, classify, divide up into categories. Do not use terms like: *as, because of, reasons for*.	*explain*	You should give reasons why. Refer to processes and factors. Try to say why the pattern occurs. Use words like: *due to, caused by, because of*.
compare	What aspects do they have in common? What are the similarities?	*discuss* (a topic)	Means both *describe* and *explain*. Make observations on the subject.
contrast	What are the differences? What aspects make them separate?	*discuss* (a statement)	Evaluate. Agree and disagree. Put forward arguments for and arguments against. A balanced answer is expected.
examine	Both describe and explain – a general instruction allowing a range of observations.	*to what extent?*	Evaluate. Agree and disagree. Put forward arguments for and against. What are the advantages and disadvantages?
how?	Describe. What methods are involved?	*comment on*	Describe and explain – a general instruction allowing any relevant observation.
in what ways?	Describe. Identify a pattern.		

1.23 *Common question terms*

Sometimes a single question may have more than one instruction, as in the common instruction to 'describe and explain'. To cope with this, remember that each question word demands a different response, and the two instructions should be separated. The sequence used in the question should always be followed; describe fully before you start to explain.

During homework, If you are not sure what the question terms mean, or what you should do in response, then ask your teacher. In an examination it is sensible to look at alternative questions where you know what you should write. The failure to respond to question terms in the required manner is probably the major problem at AS and A2 Levels (see examiner comments of page 4).

The subject

This identifies the subject or topic of the question. The instruction identifies the area of information the examiners want you to focus on. In Figure 1.22, each sub-question has a subject: in b(i) this is **the social and demographic changes**

- These terms tell you the topic or the part of the topic you should include.
- They indicate the specific information you should include.
- Try to define these terms – state clearly what they mean.
- If you include other information it will be irrelevant.

Subject term	You should try to...
A topic: *immigrant population*	Define the term at the start of the paragraph. Try to break the topic down into constituent parts. Give as much detail as possible to give ideas. This is why you have learned definitions – the better the definition, the better the answer.
a specific area or example ... *with reference to an area you have studied ...* *... a named example ...*	Your answer will be example specific. Clearly identify the example – location, size and main features. If applicable, draw a map or diagram to show the example. Divide the example up into parts or areas. **These questions cannot be well answered without a good example.**

Subject qualifications

The major topic covered by the subject is frequently narrowed down by asking for specific elements or aspects.

You must focus on these to give relevance to your answer and to allow you to answer the question in time. In Figure 1.22, the section b(ii) asks you to consider the **need/demand for housing, education and health care within the city/cities you have chosen** To respond to this you should divide your answer into sections on housing, education and health care. Remember:

- These terms sharpen the subject.
- They provide precise instruction about the aspects of the subject required.
- Always follow the sequence given.
- Ring or underline these terms in the question to increase focus.
- These help to structure your answer.
- Try to think why they have been included: how do they influence your topic?

Question qualifiers	What you should do
Double subjects: *velocity and discharge of a river ...* *location and size of settlement ...* *economic and social aspects of migration ...*	These subjects have two components. Separate the two and distinguish between them (define what they mean). Divide your answer into two sections and focus on each. Use the terms in the sequence given in the question. If recognised, these terms help give a sharper idea of what the examiners require. **Use them fully.**
Specified aspects: *... housing, education and health ...* *consequences ...* *impact ...* *results of ...*	These are specific requests and must be focused on. Examine each in turn, in the order given. Try to include some comment on each. **A general and vague answer is not effective.** Refer to the outcome, what happens because of... Try to be balanced: consider advantages and disadvantages – do not be pessimistic and full of doom. Try to find a range of ideas (think Physical, Economic, Social and Political – **PESP**).
Scale: *... of at least 500 000* *... local area*	Make sure that your example fits the scale given in the question. An inappropriate example will not earn high marks.
Status: *developed ... developing*	These qualifiers must be used. Define what is involved – how does the level of development influence the topic?
Time: *... over the last 30 years ...*	Take care to limit your discussion to the relevant time scale. Ask why – what change occurred in this topic 30 years ago? (technology, or routes?)

Mark allocation

The maximum mark the examiner can award for each question, or part of a question, is usually clearly shown. This provides valuable information about what you should do:

- Structured questions and data response are divided into clear sections with marks.

- In structured questions the marks tell you how long the answer should be.

- On data response the marks suggest how much time to spend on the section.

- The marks also give some idea of the number of points you should include.

- Use the marks wisely to write enough within the allowed time to gain the maximum.

Marks out of 25	How long to spend	How much to write	
2	3 mins		one short paragraph
3	6 mins		one medium paragraph
5	9 mins		half a side
7	12 mins		two medium paragraphs
8	14 mins		two large paragraphs
10	18 mins		up to one side

1.24 *A rough guide to time allocation*

The importance of planning

Essay questions, or large sections of structured questions (greater than 4 marks), require planning before you start to write the answer. Note that:

- your answer should be relevant, and respond to the given question

- it is unlikely that the answer will be the same as your book or note sequence

- making a plan helps you recall and select the material required

- it gives you a logical sequence

- it **focuses** on the question

- it helps you avoid the temptation to write 'all you know'.

When constructing a plan, work from the question terms, following the instructions. Try to allocate ideas to form a skeleton for the paragraph (Figure 1.25).

Describe and explain how the **structure of the population** may vary from **place to place**.   *(25 marks)*

subject	• population structure • scale – country/region/city
describe	• age groups • gender • ethnic groups
explain	• levels of development/wealth • migration • employment
conclusion	

- Find the question terms.
  Use these to divide up the question.
  Allocate space to the terms.
  Think of the task as two small essays.

- Find the subject.
  Start with a clear and concise definition of the subject.
  Define the terms that qualify the subject.

- Allocate ideas to the sections to form paragraph themes.
  Keep these brief so that the theme is clear.
  Try to include a range of material.

1.25 *Planning your answer*

How to approach a structured question

Structured questions are already partially planned, as they are divided into sections. However, the larger sections (above 4 marks) still require a plan of action, as they usually contain a number of aspects that must be focused on to gain full marks (Figure 1.26).

(a) **What is meant by** a 'hierarchy' of settlements? *(5 marks)*
(b) **What criteria** would you use to establish such a **hierarchy**? *(10 marks)*
(c) **Describe** and **justify** the **fieldwork methods** you would use to establish such a **hierarchy**. *(10 marks)*

what is meant	define **hierarchy**
what criteria	population size; functions (order); retail turnover
Describe	urban-based survey; village survey (Bracey)
justify	range (market area); order

- Use the terms and sections to divide up the question.
- Make the answer into a series of tasks.
- Always follow the sequence given in the question.
- Identify the subject.
- Start with a definition – here it is requested.
- Use the marks to help work out the number of ideas and paragraphs.
- Subdivide the work as much as possible.
- Two question terms are used: divide the section into two parts, then focus on each term.

  Answering a structured question

How to approach a data response question

Your answers and the data response questions are returned to the examiner. This allows you to show the examiner that you are thinking and that you are trying to give a relevant answer to the question by modifying the paper (Figure 1.27).

During the first term of your AS/A2 course, work on question interpretation and plans is demanding and time-consuming. With practice and application it will become both easier and more automatic. It involves learning a new way of looking at questions and putting the instructions before your knowledge of the topic. This will allow you to use your data effectively and gain high marks with the minimum of risk.

Remember ...

- Do not panic about the answer – look for the question terms.
- Use the terms to divide your answer up into manageable chunks.
- Underline the question words – particularly on data response papers that will be seen by the examiner.
- Show question words in plans, and use them.

Ring or underline the key question terms so that the examiner can see you are trying to focus on the question.

Use *all* the data given in the stimulus material.

Use bullets or underlining to identify the reason or idea. Make it clear what the point is.

Use small but legible writing. Use all the space provided.

Briefly expand the idea, giving an explanation or giving detail.

Try to give a range of ideas or reasons. 4 marks mean that a number of ideas are required, not just one.

Watch out for unexpected terms like this. Candidates expect human factors, but this question asks how physical factors affect social processes.

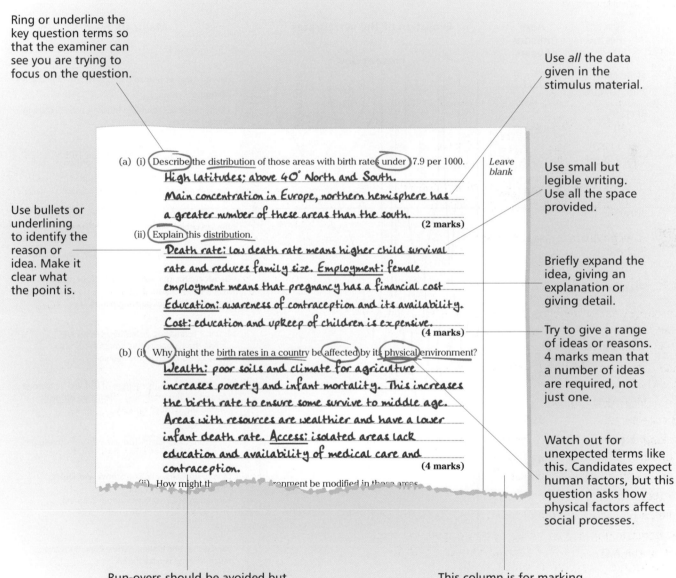

(a) (i) Describe the distribution of those areas with birth rates under 7.9 per 1000.   Leave blank

High latitudes; above 40° North and South.
Main concentration in Europe, northern hemisphere has a greater number of these areas than the south.
(2 marks)

(ii) Explain this distribution.

Death rate: low death rate means higher child survival rate and reduces family size. Employment: female employment means that pregnancy has a financial cost Education: awareness of contraception and its availability. Cost: education and upkeep of children is expensive.
(4 marks)

(b) (i) Why might the birth rates in a country be affected by its physical environment?

Wealth: poor soils and climate for agriculture increases poverty and infant mortality. This increases the birth rate to ensure some survive to middle age. Areas with resources are wealthier and have a lower infant death rate. Access: isolated areas lack education and availability of medical care and contraception.
(4 marks)

(ii) How might th... ...ronment be modified in those areas

Run-overs should be avoided but are generally acceptable provided a new sentence has not been started. Ideally, answers should be within the space provided.

This column is for marking. Never write into it as this annoys examiners and leads to marking errors.

1.27 *Answering a data response question*

A The theory of plate tectonics

What's the story . . . ?

In the opening chapter of this section we study the forces responsible for the major features of the Earth's surface, including the mountain ranges, volcanic islands and deep-sea trenches. In later chapters we examine the characteristics of rocks and study the processes that act upon them to form our natural landscapes.

The Earth is some 4600 million years old. The last 570 million years have been well documented by geologists because it is during this most recent stage in Earth's history that life has become abundant. Geologists have produced a **geological time scale** and have divided it into a number of distinct **geological periods** (Figure 2.1). You will come across references to particular periods throughout your studies, and you should take time to familiarise yourself with their names.

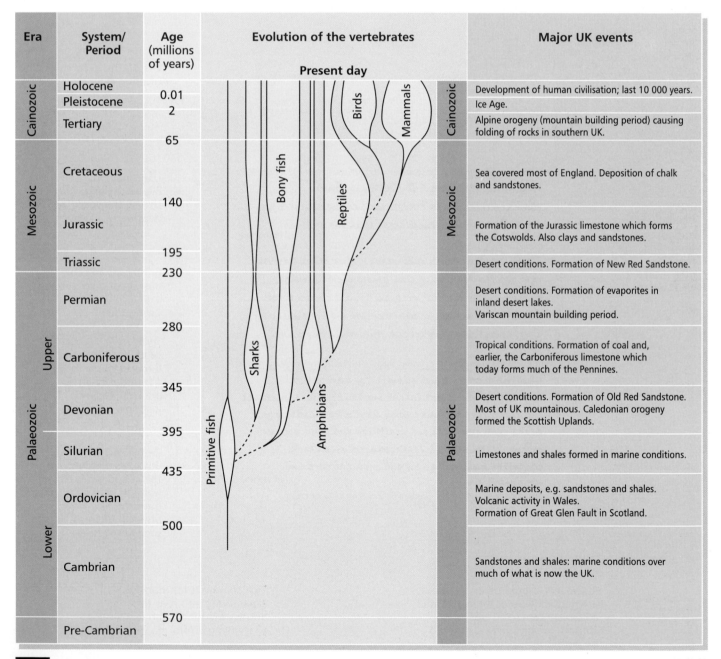

2.1 *Geological time scale*

The shape of the Earth is close to a perfect sphere. If you were to slice it through the middle, you would discover four distinct layers. At the centre are the inner and outer **cores** which are largely composed of nickel and iron. These are encircled by a very thick, dense layer called the **mantle**. The outermost layer, the **crust**, is extremely thin relative to the other layers but it is by far the most important as far as human activity is concerned.

The Earth's crust is made up of three types of rock: **igneous**, **sedimentary** and **metamorphic** (Figure 2.2). Once formed, these rocks are often subjected to powerful earth movements which may lead to **folding** (crumpling) and **faulting** (vertical or horizontal displacements). We will look at the effects of folding and faulting in Chapter 2F.

What are the major relief features of the Earth?

There are a number of major surface relief features. Figure 2.3 shows these features in a great deal of detail. Figure 2.4 is a simplified version. As you read through this chapter, locate the various features on the two maps.

Rock type	Formation	Characteristics	Examples
Igneous	Formed by the cooling of molten magma either underground (**intrusive**), or on the ground surface (**extrusive**) by volcanic activity.	Igneous rocks are composed of interlocking crystals (they are said to be **crystalline**). They are generally tough rocks and are resistant to erosion.	**Basalt**, **andesite** and **rhyolite** are examples of extrusive lavas. **Granite**, **gabbro** and **dolerite** are intrusive rocks.
Sedimentary	Formed by the compaction and cementation of sediments; usually deposited in the sea. Also includes organic material (e.g. coal) and rocks precipitated from solutions (e.g. limestone).	Sedimentary rocks usually form layers called **beds**. They often contain fossils. Whilst some rocks can be tough (e.g. **limestone**), most are weaker than igneous and metamorphic rocks.	Common sedimentary rocks include **sandstone**, **limestone**, **shale** and **mudstone**. The rock **chalk** is a form of limestone.
Metamorphic	Formed by the alteration of pre-existing igneous, sedimentary or metamorphic rocks by heat and/or pressure.	Metamorphic rocks are also crystalline. They often exhibit layering (*not* beds) called **cleavage** (as with the rock **slate**), and banding. Metamorphic rocks tend to be very tough and resistant to erosion.	**Slate** is one of the most common metamorphic rocks. Other examples include **gneiss** (pronounced 'nice') and **schist**.

2.2   Crustal rock types

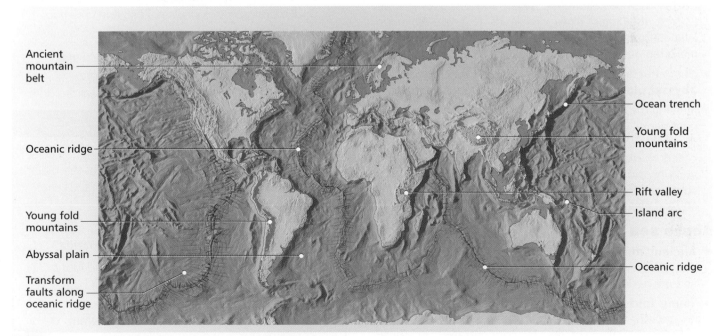

2.3   The physical world

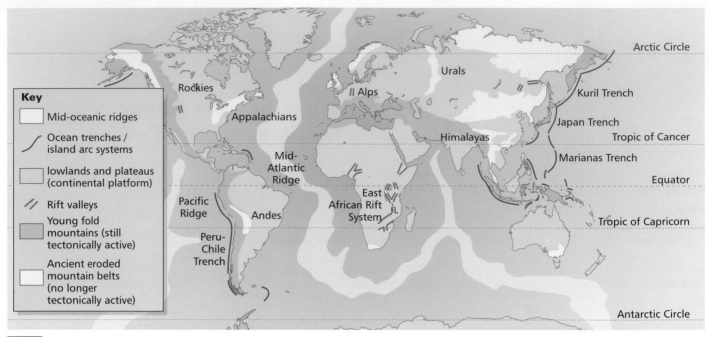

2.4 *Major features of the Earth's surface*

Below sea level

- **Oceanic ridges** extend up to 3000 m above the sea floor along the centre of the oceanic basins. The Mid-Atlantic Ridge is perhaps the best known of these underwater mountain chains. In places it breaks the surface to form islands, for example Iceland.

- There are deep **ocean trenches**, such as those encircling the Pacific Ocean. Over 10 000 m deep, they are far deeper than the equivalent height above sea level of Mt Everest (8848 m). Lines of islands (e.g. the Solomon Islands, to the north-east of Australia) called **island arcs** are often associated with ocean trenches.

- Huge tracts of deep (3000–5000 m) but largely flat **abyssal plain** make up the bulk of the ocean floor.

- Marking the edge of each continent, and strictly speaking continental rather than oceanic, are two distinct slopes. The first is a very gentle slope (approximately 75 km wide) called the **continental shelf**. This becomes a slightly steeper slope (called the **continental slope**) that stretches for about 100 km before merging into the abyssal plain.

Above sea level

- **Ancient mountain belts**, such as the Appalachians in the USA, have been severely eroded over millions of years and are no longer tectonically active.

- **Young fold mountains**, are still tectonically active and are growing by several centimetres every year. Volcanic and earthquake activity is concentrated in these mountain belts.

- Deep valleys called **rift valleys** occur in the middle of some continents, for example the East African Rift Valley.

NOTING ACTIVITIES

1 Make a copy of Figure 2.4 on a blank world outline. Use the information in Figure 2.3 together with the text above to label the major relief features of the Earth. Refer to an atlas to discover some examples of the features you have plotted.

2 Write brief and simple definitions of each major relief feature. Give an example of each one.

STRUCTURED QUESTION 1

Figure 2.5 is a cross-section through the Earth's crust at the Tropic of Capricorn. It shows some of the major relief features between Africa and the Pacific Ocean.

a What are the major relief features at A, B, C, D and E? (5)

b What is the width of the Mid-Atlantic Ridge? (1)

c Describe the shape of the Mid-Atlantic Ridge. (2)

d The ocean trench and the Andes mountains are two of the most dramatic relief features on Figure 2.5. Use the scales on the diagram to compare these two features. (2)

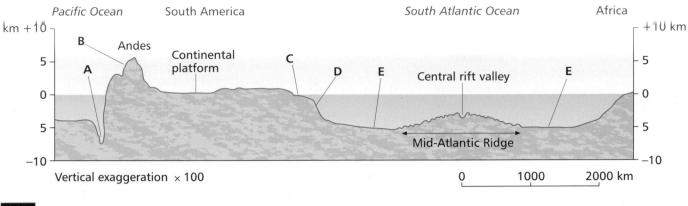

2.5   *Cross-section along the Tropic of Capricorn*

A world of extremes . . .		
Highest mountain	Mt Everest (Nepal/China)	8848 m
Largest island	Greenland	2 175 597 km²
Longest river	Nile (Africa)	6695 km
Largest continent	Asia	43 608 000 km²
Largest ocean	Pacific	165 384 000 km²
Largest lake	Caspian Sea	371 795 km²
Largest desert	Sahara	9 000 000 km²
Largest depression	Dead Sea	395 m below sea level
Greatest waterfall	Angel Falls (Venezuela)	drop of 979 m

What is the theory of plate tectonics?

Look back to Figures 2.3 and 2.4 and notice that the major relief features of the Earth form several distinct linear belts. See how some features occur close to one another, for example ocean trenches and young fold mountains. These clear patterns puzzled early scientists: how could they be explained? Today, it is the theory of **plate tectonics** that enables scientists to account for them.

Look at Figure 2.6. It describes in detail the outermost 500 km of the Earth. You can see the familiar terms 'crust' and

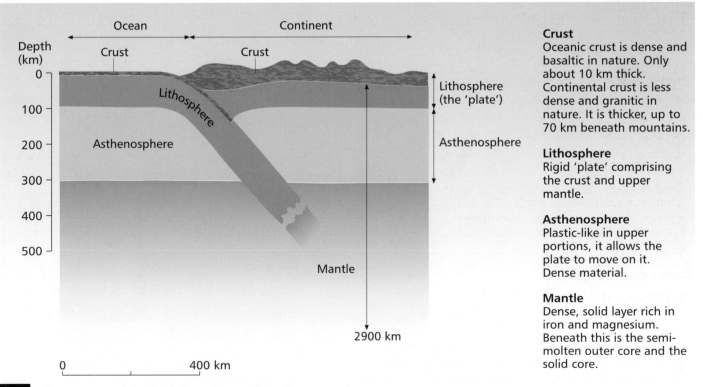

Crust
Oceanic crust is dense and basaltic in nature. Only about 10 km thick. Continental crust is less dense and granitic in nature. It is thicker, up to 70 km beneath mountains.

Lithosphere
Rigid 'plate' comprising the crust and upper mantle.

Asthenosphere
Plastic-like in upper portions, it allows the plate to move on it. Dense material.

Mantle
Dense, solid layer rich in iron and magnesium. Beneath this is the semi-molten outer core and the solid core.

2.6   *The structure of the Earth from the mantle to the crust*

27

'mantle', but recent research has enabled scientists to identify a rather more complex picture. It is now generally accepted that the crust and the upper part of the mantle form a single layer called the **lithosphere**. This 100 km thick layer comprises a number of rigid slabs called **plates**. Each plate rests on a 'mobile' layer called the **asthenosphere** – without such a layer, the plates would be unable to move. The plates move at an average of about 50 mm per year (roughly equivalent to the speed of growth of a fingernail), but this varies considerably from one part of a plate to another.

How do we know where the plates are?

The extent of the plates and the precise position of their margins has been plotted using the following evidence:

- The location of volcanoes and earthquakes (Figure 2.7).
- Studies of surface heat flow, where high values suggest rising magma.

- The location of the ocean trenches where the plates are believed to descend to be destroyed.
- The presence of gravity anomalies. The force of gravity measured at the surface largely depends on altitude and on the gravitational pull of the underlying rock. The greater the density of the underlying rock, the greater the pull of gravity. Therefore, if dense rock material is being formed at the surface, as happens at a constructive margin, there will be a higher gravity reading than on either side, thus creating a **gravity anomaly**.
- The presence of active faults on the ground surface indicate that large-scale movement is occurring causing the ground to split apart.

The outcome of all this research is the map of the plates (Figure 2.8). You should take time to familiarise yourself with the names of the major plates, their directions of movement and the features that correspond with their margins. There is some evidence that other plate margins may exist, for example through the rift valley system in East Africa.

Although broadly accepted by scientists, the theory of plate tectonics is an evolving one. There are still a number of uncertainties and questions that need answering.

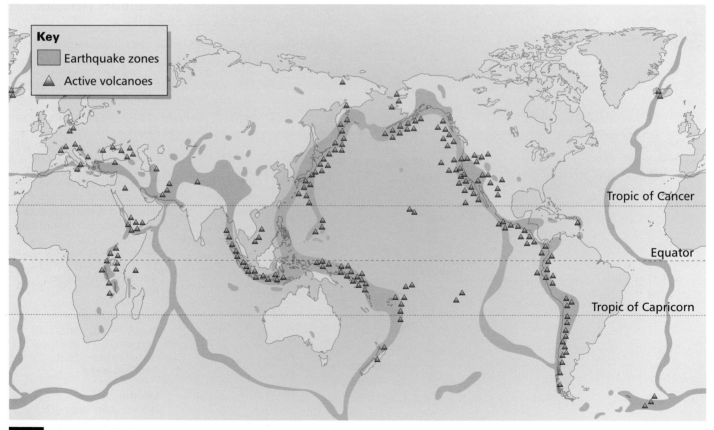

Key
Earthquake zones
▲ Active volcanoes

Tropic of Cancer

Equator

Tropic of Capricorn

2.7 *World distribution of earthquakes and active volcanoes*

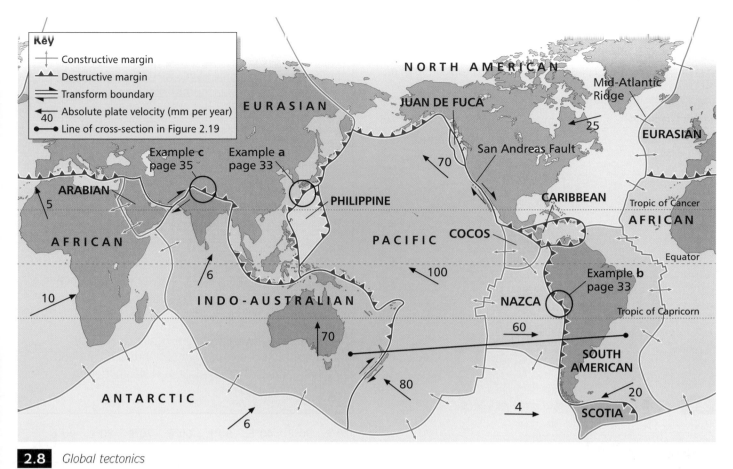

Key

- ┼─ Constructive margin
- ▲▲ Destructive margin
- ═══ Transform boundary
- ◄── 40 Absolute plate velocity (mm per year)
- ●──● Line of cross-section in Figure 2.19

NORTH AMERICAN

EURASIAN

JUAN DE FUCA

Mid-Atlantic Ridge

25

EURASIAN

Example c page 35

Example a page 33

San Andreas Fault

70

ARABIAN

5

PHILIPPINE

CARIBBEAN

Tropic of Cancer

AFRICAN

AFRICAN

COCOS

PACIFIC

6

10

INDO-AUSTRALIAN

100

NAZCA

Example b page 33

Equator

Tropic of Capricorn

70

60

80

SOUTH AMERICAN

ANTARCTIC

6

4

20

SCOTIA

2.8 *Global tectonics*

A revolution on paper

Based on an extract from 'Living through the revolution' by Tim Burt in Geography Review, November 1998

As soon as maps of the Atlantic Ocean were available, people noticed that the continents on either side fitted together remarkably well. However, no one thought continents could move about, so the jigsaw fit attracted no serious attention. The question was left until the early years of this century when two people independently made the suggestion that there had once been just one continental mass and that this had broken up during geological time and floated apart.

One of them, a German scientist called Alfred Wegener, provided plenty of evidence that a single continent had once existed. For example, it was clear that, in the southern hemisphere, there had been a period of glaciation in the late Carboniferous period (about 290 million years ago). Glacial deposits are found in South America, Antarctica, and India. This can only be explained if these land masses (now thousands of kilometres apart) were once part of a single continent with a single ice cap covering the area. A further line of evidence was the fossil reptile Mesosaurus, found in both South Africa and South America, now too far apart for any land animal to swim between them.

Wegener's ideas were first published in 1912. His ideas were, however, more or less completely rejected in the 1920s and 1930s.

Wegener's theory became known as continental drift. The lack of a mechanism prevented his ideas being accepted at the time. More recent research, however, has established the presence of huge convection currents, which can be used to explain the movement apart of the continents for which Wegener found evidence many years ago.

NOTING ACTIVITIES

1 What was the early evidence that suggested that the continents used to form a single land mass?

2 What additional evidence did Wegener cite that led to his theory of continental drift?

What causes the plates to move?

The cause of plate movement is far from certain. Many scientists believe that large-scale **convection** (heat) **currents** circulating beneath the lithosphere are responsible for plate movement. These currents are probably generated at the mantle/core boundary and spread up through the mantle itself (see Figure 2.9). Some scientists, however, feel that the convection currents are much more shallow affairs circulating within the asthenosphere only, i.e. down to a depth of about 300 km.

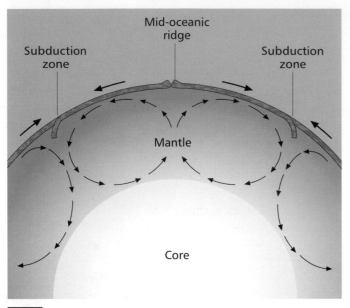

2.9 *Convection currents and plate movement*

NOTING ACTIVITIES

1   Carefully define the following terms:
    - lithosphere
    - plate
    - asthenosphere.

2   What is a gravity anomaly? How can such a characteristic be used to identify a plate margin?

3   Apart from gravity anomalies, what other evidence have scientists used to help them plot plate margins?

4   Draw a map to show the major global plates. Name each plate and mark on its direction and rate of movement.

5   Describe, with the aid of a diagram, how convection currents may be responsible for plate movement.

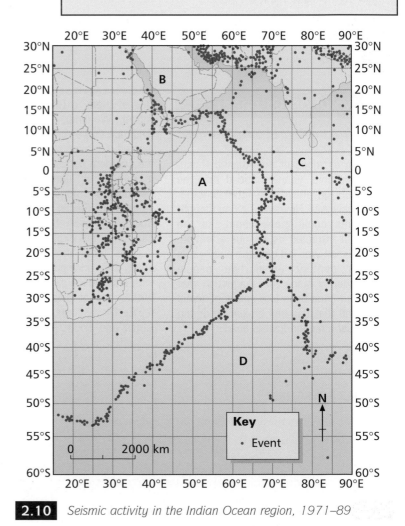

2.10 *Seismic activity in the Indian Ocean region, 1971–89*

B Tectonic activity

It is the movement of the plates relative to one another that accounts for the existence of the major relief features of the Earth, for example the mountain chains and the deep ocean trenches (Figure 2.3 page 25). When the plates move, massive forces are released resulting in earthquakes and volcanic activity. These forces, and the features that they form, are concentrated along the active margins of the plates.

What happens at plate margins?

Look back to Figure 2.8 (page 29) to discover that there are three types of plate margin:

1 **Constructive** or divergent margins, where two plates move away from each other.

2 **Destructive** or convergent margins, where two plates move together.

3 **Conservative** or transform margins, where two plates move alongside each other.

Understanding the processes operating at plate margins enables us to explain how the major relief features of the world were formed and why the majority of earthquakes and volcanoes occur where they do.

1 Constructive margins

Constructive margins occur where the lithosphere splits apart allowing fresh molten **magma** to escape to the surface. Once on the surface, the magma, now called **lava**, cools to form new crustal material.

One of the best examples of a constructive margin is the **Mid-Atlantic ridge** (Figure 2.11). Notice that the ridge itself is a very broad feature, over 1000 km wide. The centre of the ridge is marked by a **central rift valley** and it is through here that the magma rises to form new rock. The ridge is offset right along its length by many **transform faults** which cause the rocks to slip sideways. These faults are caused by the enormous stresses that build up along the plate margin.

Over millions of years, as new rock is constantly being formed at the plate margin, the plates grow steadily outwards. This is called **sea floor spreading**. There are two important pieces of evidence that support this process:

- **Fossil magnetism** The rock that forms at a constructive margin is called **basalt**. It is rich in iron and,

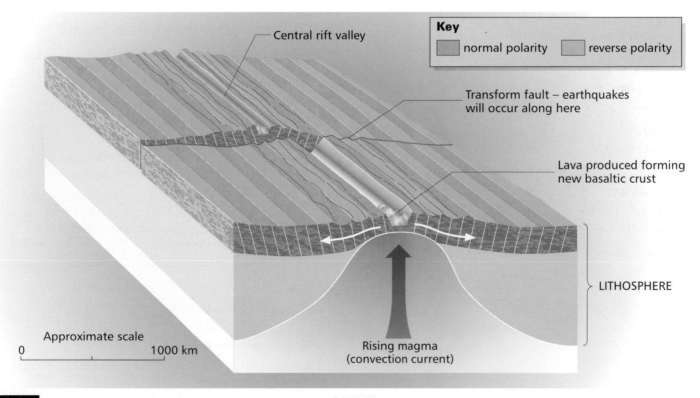

Key
normal polarity   reverse polarity

Central rift valley

Transform fault – earthquakes will occur along here

Lava produced forming new basaltic crust

LITHOSPHERE

Approximate scale

0   1000 km

Rising magma (convection current)

2.11 *Processes in operation at the Mid-Atlantic Ridge* in the same way that iron filings react to a bar magnet in a physics experiment, the iron in basalt reacts to the Earth's magnetic field. On cooling, the basalt records information about the Earth's magnetic field as a fossil record. This is called **palaeomagnetism** ('palaeo' means 'fossil'). Over geological time, the Earth's polarity has periodically 'switched'. This means that magnetic north becomes south, and vice versa. Scientists studying rocks from the sea floor have found a pattern of this switching polarity (Figure 2.11).

The pattern is symmetrical on either side of the central rift, supporting the notion that new plate is being formed here and is then spreading outwards on either side.

- **Age of rocks** Studies of rocks on the sea floor and, in Iceland, where the Mid-Atlantic ridge breaks the surface, have shown that they increase in age away from the centre of the ridge. This supports the notion that new rock is forming at the ridge and then spreading outwards.

NOTING ACTIVITIES

1 What are the three major relief features that are associated with a constructive margin?

2 With the aid of a simple diagram, describe the processes operating at a constructive plate margin.

3 What is meant by palaeomagnetism and how does it support the idea of sea floor spreading?

4 a Why is Iceland an ideal outdoor laboratory for studying the processes at work along a constructive margin?

   b How have the ages of the rocks been used to support the idea of sea floor spreading?

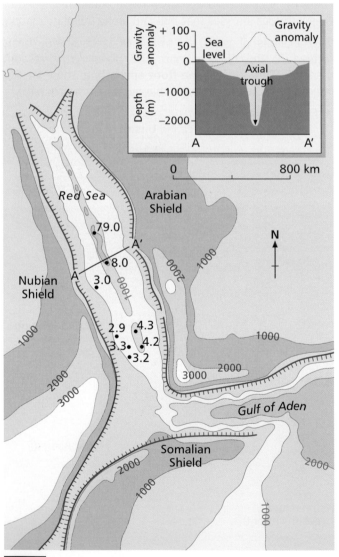

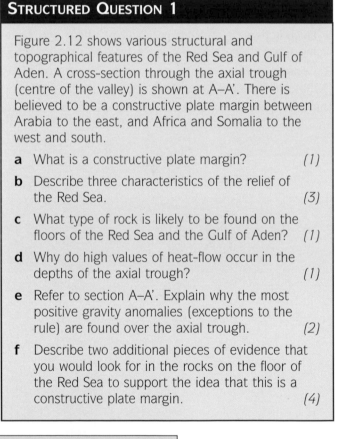

STRUCTURED QUESTION 1

Figure 2.12 shows various structural and topographical features of the Red Sea and Gulf of Aden. A cross-section through the axial trough (centre of the valley) is shown at A–A'. There is believed to be a constructive plate margin between Arabia to the east, and Africa and Somalia to the west and south.

a What is a constructive plate margin? *(1)*

b Describe three characteristics of the relief of the Red Sea. *(3)*

c What type of rock is likely to be found on the floors of the Red Sea and the Gulf of Aden? *(1)*

d Why do high values of heat-flow occur in the depths of the axial trough? *(1)*

e Refer to section A–A'. Explain why the most positive gravity anomalies (exceptions to the rule) are found over the axial trough. *(2)*

f Describe two additional pieces of evidence that you would look for in the rocks on the floor of the Red Sea to support the idea that this is a constructive plate margin. *(4)*

Key

⌇ Contours at 1000 m intervals

3.2
• Heat flow values
(microcals cm^{-2} per sec^{-1})

⌢⌢⌢ Fault scarp

 The Red Sea constructive plate margin

2 Destructive margins

At a destructive plate margin, two plates are moving towards one another. Where they meet, one plate dives or **subducts** below the other. The cause of this subduction is thought to be a difference in density between the two plates, with the heavier one subducting below the lighter one.

Enormous pressures build up at a destructive margin. They are released periodically in the form of powerful earthquakes. Volcanoes too are common at destructive margins. The subducting plate is heated by friction and intense pressures, until eventually it melts to form fresh magma. This rises to the surface, often producing cataclysmic volcanic eruptions.

It is possible to identify three types of destructive plate margin. We shall study each type, with reference to three case study examples.

a Ocean-ocean margin: the Ryukyu Islands

The western Pacific Ocean, between Japan to the north and Papua New Guinea to the south is, tectonically speaking, an extremely complex area. Look back to Figure 2.8 (page 29) to discover that there are several plates subducting beneath one another. As Figures 2.3 and 2.4 show (pages 25–26), several relief features, such as ocean trenches, are associated with the plate margins in this area.

The Ryukyu Islands, just to the south of the Japanese island of Kyushu (Figure 2.13), are a direct result of the destructive plate activity in the western Pacific. The Philippine Plate to the east is diving beneath the Eurasian Plate to the west. As it subducts, tremendous stresses are released, causing earthquakes. Notice that the points where the earthquakes occur (the **foci**) plot the angle of the subducting plate as it enters the asthenosphere.

At about 100 km below the surface, the subducting plate begins to melt in the so-called **Wadati–Benioff Zone** and magma escapes to the surface to form volcanoes. After several eruptions, these volcanoes break the ocean surface to form islands; a series of islands, such as the Ryukyu Islands, form an **island arc**.

b Ocean-continent margin: the Andes

The Andes mountains stretch for some 9000 km along the entire western side of South America (see Figure 2.4 page 26), reaching heights well in excess of 6000 m. The highest peak is Aconcagua at 6960 m (Mt Everest in the Himalayas is 8848 m). The Andes, with its many active volcanoes and glaciers, forms some of the most spectacular mountain scenery in the world (see Figure 2.14).

Look back to the plate map (Figure 2.8 page 29) to discover that the Andes mountains lie close to a plate margin, where the dense oceanic Nazca Plate is subducting beneath the South American Plate (Figure 2.15).

There is plenty of evidence suggesting the presence of a destructive plate margin here:

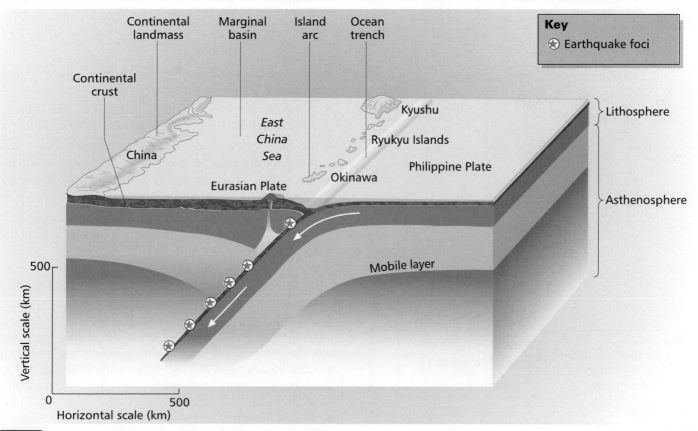

2.13 *Cross-section through the ocean trench off the Ryukyu Islands*

2.14 *Part of the Andes mountain range, in Peru*

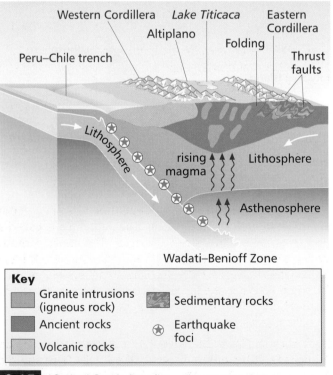

Key

Granite intrusions (igneous rock)	Sedimentary rocks
Ancient rocks	⊛ Earthquake foci
Volcanic rocks	

2.15 *Central South America – the present day*

- there is a deep ocean trench (the Chilean Trench) running parallel to the South American coast
- there are negative gravity anomalies at the trench, suggesting that dense material is being subducted
- the rocks of the Andes have been severely folded and faulted, suggesting the presence of great tectonic forces
- vast bodies of igneous rock attributed to melting plate make up the roots of the Andes
- active volcanoes are found in the Andes; these erupt a viscous lava that forms the rock andesite, hence the name 'Andes'
- frequent earthquakes rock the mountains and their inclined foci suggest the presence of a subducting plate beneath the Andes.

The melting of the subducting Nazca Plate has led to the creation of huge igneous intrusions which have increased the thickness of the crust. Surface volcanoes and their associated lava flows and ash deposits have also added to the vertical extent of the Andes. The horizontal pressures associated with the colliding plates have in turn led to a great deal of vertical movement as rocks have been forced to buckle and **fault** – this is particularly obvious in the Eastern Cordilleras.

c Continent-continent margin: the Himalayas and Tibetan Plateau

Continental crust is less dense than oceanic crust and so, when two continents meet at a destructive plate margin, there is a slow collision rather than any marked subduction. This results in intense folding, faulting and uplift and leads to the formation of mountains (a mountain building period is called an **orogeny**).

In the case of the Himalayas (see Figure 2.8 page 29) the Indo-Australian Plate to the south is colliding with the Eurasian Plate to the north (Figure 2.16). The rocks have become severely **folded** (buckled) and faulted to form the immense mountain range that incorporates Mt Everest, the world's highest mountain.

As there is very little – if any – subduction at this plate margin, there are few earthquakes and no volcanoes.

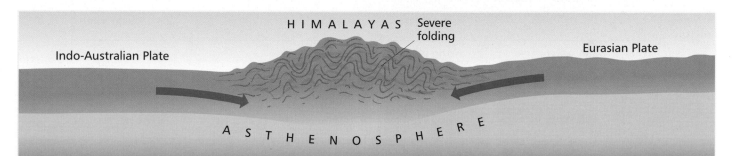

| **2.16** | *The formation of the Himalayas* |

NOTING ACTIVITIES

1 What is meant by a destructive margin?

2 Define the following important terms:

- subduction
- Wadati–Benioff Zone
- earthquake focus (foci)
- fault
- fold
- orogeny.

3 Explain how the following features and events are associated with destructive margins:

- earthquakes
- volcanoes
- ocean trenches
- island arcs.

4 With the aid of a simple diagram, describe the processes and landforms associated with the ocean-ocean margin to the south of Japan.

5 What is the evidence that the Andes mountains have formed at a destructive plate margin? Include a summary diagram to describe the evidence.

6 **a** Describe the formation of the Himalayas.

 b Why are there few earthquakes and no volcanoes in the Himalayas?

STRUCTURED QUESTION 2

Figure 2.17 shows ocean depths and the foci of earthquakes occurring during 1965 in the area of the volcanic Tonga Islands ridge in the south-west Pacific. The plate margin shown is a destructive one.

a What is meant by the 'focus' of an earthquake? *(1)*

b What tectonic feature is indicated by the letter A? *(1)*

c In which direction is the Indo-Australian Plate moving? *(1)*

d **(i)** Describe three characteristics of the pattern of earthquake foci. *(3)*

 (ii) Account for the pattern of earthquake foci. *(2)*

e Explain the formation of the volcanic Tonga Islands ridge. *(3)*

3 Conservative margins

In some places, the relative movement between two plates is horizontal rather than vertical. Such a margin is called a conservative or **transform** margin. The most famous example of a conservative margin is the San Andreas Fault in California (see Figure 2.18). The whole fault system extends for some 1200 km and it has been estimated that there has been about 1000 km of lateral movement in the last 25 million years.

Look back to Figure 2.8 (page 29) to discover that the San Andreas Fault lies at the margin of the Pacific Plate and the North American Plate. Although both plates are moving in roughly the same direction, the Pacific Plate is moving faster

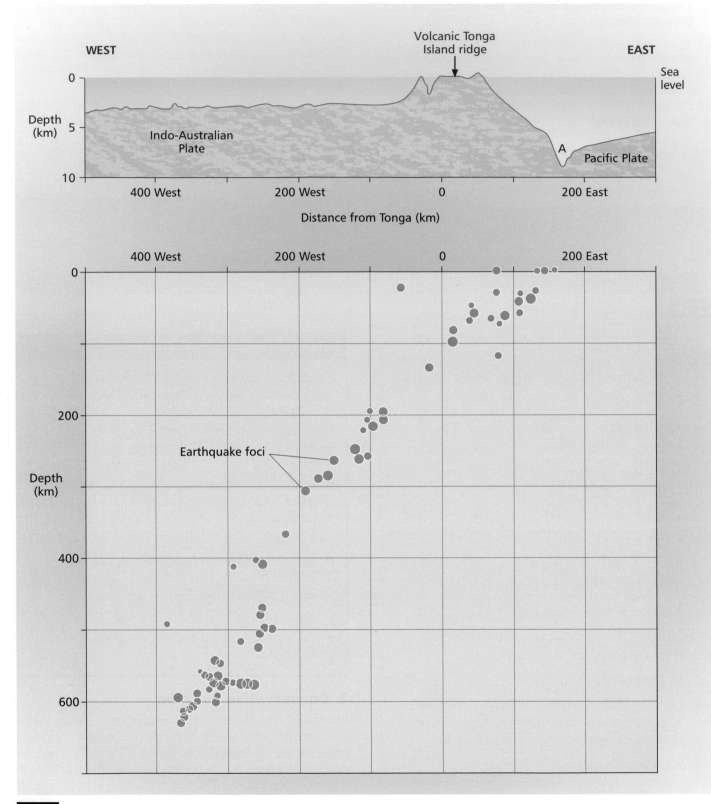

2.17 *Tectonic activity in the south-west Pacific, 1965*

so that there is relative displacement along the margin (Figure 2.18). Slippage along the San Andreas Fault has been responsible for many serious earthquakes, including the devastating 1906 San Francisco earthquake which killed over 700 people. However, as no subduction is occurring at this plate margin, there are no active volcanoes.

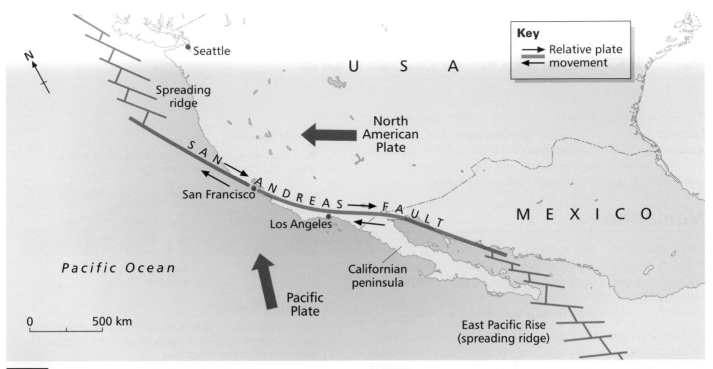

The San Andreas Fault: a conservative plate margin

NOTING ACTIVITIES

1. Use a simple diagram to define a conservative plate margin.

2. How has movement along the San Andreas Fault been responsible for shaping the landscape of western California?

3. Explain why no volcanoes exist along conservative plate margins.

4. Figure 2.19 is a cross-section through the lithosphere from New Zealand to the South Atlantic Ocean (see Figure 2.8 page 29).

   a Make a copy of the cross-section and, using earlier maps and diagrams, label the following:

   Nazca Plate

   Indo-Australian Plate

   African Plate

   South American Plate

   Pacific Plate

   South America

   East Pacific Rise (ridge)

   Pacific Ocean

   Andes young fold mountains

   Kermadec Trench (off New Zealand)

   Chile–Peru Trench

   Wadati–Benioff Zone

   subduction

   rising magma

   earthquake foci.

   b Locate a number of the earthquake foci on your diagram.

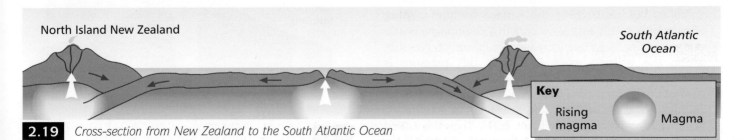

2.19   *Cross-section from New Zealand to the South Atlantic Ocean*

C Tectonic hazards

The enormous forces and high temperatures associated with tectonic activity can trigger **earthquakes** and **volcanic eruptions**. These natural events can be both creative in forming spectacular features of the landscape, and destructive in generating tragic natural disasters.

Volcanoes

When molten magma escapes to the surface it forms a volcano. Most volcanoes are formed when magma escapes through a single hole or vent. These are called **central vent volcanoes**. If, however, the magma escapes along a crack, a line of volcanoes called **fissure volcanoes** is formed.

The shape of a volcano depends on the chemical composition of the magma and the explosiveness of the eruption (Figure 2.20). Deep below an active volcano lies a vast **magma chamber**. A magma chamber is like a cauldron containing many different ingredients, rather like a cake mix. However, it is the relative amounts of just two of the ingredients that determines the shape of the overlying volcano:

- The element silica affects the magma's resistance to flow, or its **viscosity**. The greater the proportion of silica, the more viscous the magma and the greater its resistance to flow.

- Superheated water (i.e. water above its normal boiling point due to it being at a greater pressure than normal atmospheric pressure) combines with expanding gases to act like the fizz in a bottle of champagne. It is the characteristics of these liquids and gases, collectively called **volatiles**, that determine the explosiveness of an eruption.

Where are volcanoes found?

Almost all volcanoes are found at the margins of the tectonic plates (see Figure 2.8 page 29). There is a particular concentration of volcanoes around the edge of the Pacific Ocean – the so-called 'ring of fire'.

In general terms there is a relationship between the type of plate margin and the volcano that occurs there:

- **Destructive** margin volcanoes are formed from magma that is rich in silica. As the magma is so viscous, it often solidifies before reaching the surface to form a **plug** (Figure 2.20c). This acts like a cork in a champagne bottle and often leads to an increased build-up of pressure within the volcano itself. Eventually this pressure is released in the form of a violent explosion. Volcanoes at destructive margins tend to be conical in shape and, if cut through in cross-section, often reveal an alternating layering of ash and lava (Figure 2.21). They are called **composite** volcanoes. The layers of ash and lava

a Gentle (basaltic)

Very fluid magma enables gas bubbles to expand freely and the eruption is very gentle.

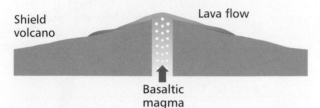

b Explosive (andesitic)

Viscous andesitic magma allows some expansion of gas bubbles but there is still sufficient pressure to cause an explosive eruption.

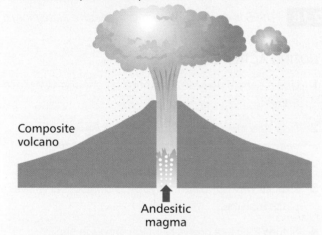

c Violent (rhyolitic)

Extremely viscous rhyolitic magma prevents gas bubbles expanding. As the magma is so thick, a solid plug may form in the vent causing the eventual cataclysmic eruption to be lateral rather than vertical. If the volcano blows its top off, a huge depression called a caldera is formed.

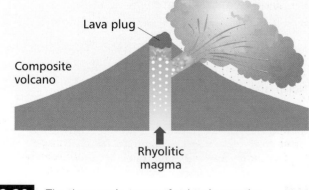

2.20 *The three main types of volcanic eruption*

represent individual eruptive events that build up on one another to form a record of a volcano's history, in much the same way that the rings in a tree-trunk record its age.

- **Constructive** margin volcanoes, for example those found at the Mid-Atlantic Ridge, are fed by silica-poor magma. As the magma rises to the surface, the greater fluidity enables gas bubbles to expand on the way up, so preventing the sudden surface explosive activity that is associated with thicker magmas. Eruptions tend to be fiery but generally far less explosive than those at destructive margins, and their consequences tend to be less catastrophic. As the lava is of a low viscosity, it tends to flow long distances before cooling. This results in a much broader **shield** volcano (Figure 2.22), which has very gentle slopes and can extend over a vast area.

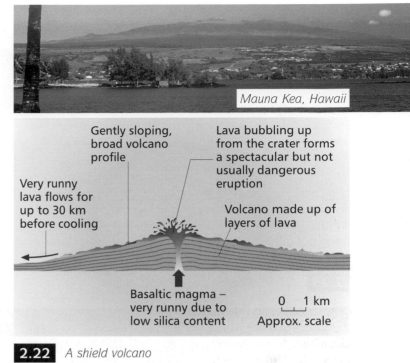

Mauna Kea, Hawaii

Gently sloping, broad volcano profile

Lava bubbling up from the crater forms a spectacular but not usually dangerous eruption

Very runny lava flows for up to 30 km before cooling

Volcano made up of layers of lava

Basaltic magma – very runny due to low silica content

0   1 km
Approx. scale

2.22   *A shield volcano*

Why are some volcanoes found in the middle of plates?

The volcanic islands of Hawaii in the Pacific Ocean (see Figure 2.24) are exceptions to the general rule. They are not located at a plate margin.

The Hawaiian Islands have formed over a source of magma called a **hot spot** (Figure 2.23). Hot spots are irregularly distributed around the world and they are thought to result from rising plumes of hot mantle material originating far below the lithosphere. Over geological time, hot spots are thought to remain more or less stationary. So, as the Pacific Plate has steadily moved over it, a whole series of islands have been formed.

The volcanoes themselves are very similar to those formed at constructive plate margins. They tend to be shield volcanoes and are formed by non-violent eruptions.

Taal volcano, Philippines

Key

Lava layer

Pyroclastic layer – rocks and ash

Cloud of ash and rocks

Crater of main cone

Dyke – igneous rock layer cutting across beds of rock

New lava flow

Parasitic cone – formed by an offshoot of magma

Approx. scale
0   0.25 km

Rising magma – very thick due to high silica content

Sill – igneous rock layer running between beds of rock

2.21   *A composite volcano*

NOTING ACTIVITIES

1   What is the difference between central vent volcanoes and fissure volcanoes?

2   How do the relative amounts of silica and volatiles affect the nature of volcanic eruptions?

3   With the aid of diagrams, compare composite and shield volcanoes. Refer to the shape of the volcanoes, the nature of their eruptions and their relationship with plate margins.

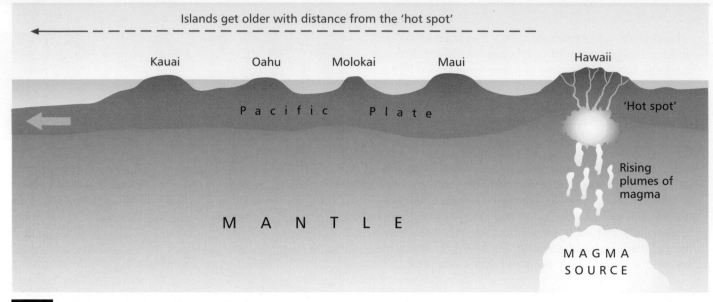

Islands get older with distance from the 'hot spot'

2.23 *The Pacific 'hot spot' beneath Hawaii*

STRUCTURED QUESTION

Figure 2.24 shows the Hawaiian Islands in the Pacific Ocean.

a What is a 'hot spot'? *(1)*

b Stating your evidence from Figure 2.24, identify which Hawaiian Island is currently lying above a 'hot spot'. *(2)*

c Calculate the average rate of movement of the Pacific Plate. Express your answer in kilometres per 1 million years, and show your working. *(2)*

d **(i)** In which direction is the Pacific Plate moving? *(1)*

(ii) What evidence is there to support your answer to part **(i)**? *(1)*

e Notice on Figure 2.24 that the Hawaiian Islands vary in their size and in their height above sea level. Suggest *two* possible reasons why the island of Kauai is much smaller and reaches a lower altitude than the island of Hawaii. *(4)*

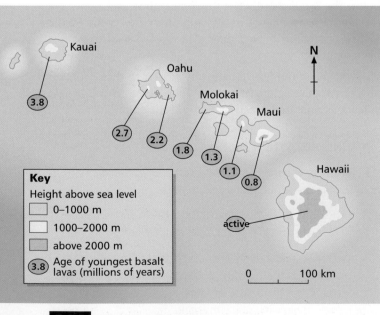

Key

Height above sea level

☐ 0–1000 m

☐ 1000–2000 m

☐ above 2000 m

(3.8) Age of youngest basalt lavas (millions of years)

0   100 km

2.24 *The Hawaiian Islands*

What is the volcanic hazard?

A volcanic eruption is an extraordinarily powerful event and, if it occurs in a populated area, there are several ways in which it can represent a hazard to people:

- **Lava flows** – **lava** is magma that flows on the ground surface. It is most commonly associated with the gentler, constructive margin eruptions and does not usually pose much of a threat to life. This is because the cooling outer crust of the lava slows down the rate of flow to walking pace. However, lava flows will destroy property and farmland. After several eruptions, the lava can build up to form a significant landform and, if it is extruded from a fissure, it can form a **lava plateau** (see page 69).

- **Ash** – one of the most common sights during an eruption is the black cloud of ash that billows high up into the atmosphere. When the ash falls back to Earth, it can cause massive loss of life as building roofs collapse under its sheer weight. Air, thick with ash, caused the asphyxiation of many people in the town of Pompeii when Vesuvius erupted in AD 79.

a   Mt Pinatubo eruption

b   Mudflows from Mt Pinatubo

2.25   *Volcanic hazards*

- **Pyroclastic flows** – violent eruptions often produce a deadly cocktail of burning gases and hot rocks known as a **pyroclastic flow**. Such flows can reach temperatures of 800°C and can travel at speeds of over 200 km/hour as they roll down a volcano's flanks. In 1991, an eruption of the Japanese volcano Mt Unzen sent a pyroclastic flow down the mountainside which destroyed all in its path. Fortunately, over 6000 people had just been evacuated, but as it was, 34 were killed.

- **Lahars** – when ash combines with water (precipitation or ice-melt), a thick, hot slurry or **lahar** may be formed. Lahars are one of the most deadly products of an eruption because they flow very fast down river valleys, demolishing property and bridges, and literally burying people alive in what amounts to a quick-setting cement. In 1985, an eruption of the volcano Nevado del Ruiz in Colombia generated a lahar that destroyed the town of Armero, and most of its population (22 000 people).

- **Tsunamis** – volcanic eruptions (and earthquakes) can send shockwaves through the world's oceans to form huge waves called **tsunamis**. When these waves, often up to 10 m high, hit a low-lying coastal region, tremendous damage can result. In 1883, the eruption of Krakatoa triggered huge waves that battered the coasts of Java and Sumatra, drowning over 30 000 people.

Although it is true to say that the majority of the most damaging eruptions have involved destructive margin volcanoes, it would be misleading to suggest that they are necessarily any more hazardous to humans than constructive margin volcanoes.

Constructive margin volcanoes erupt more frequently, so the hazard is more 'real' to people. Frequent flows of lava and falls of ash render large areas sterile – such areas cannot realistically be farmed or settled and often have very low population densities. Destructive margin volcanoes, on the other hand, erupt far less frequently. There can often be many hundreds of years between eruptions, so the volcanic hazard becomes less 'real' to people. Memories are short and people think that the risk is minimal.

Volcanic activity is not all bad news. It can bring considerable benefits to people:

- the heat can be used to generate geothermal electricity and, in Iceland, the hot water is used directly to heat buildings and greenhouses, enabling the country to be self-sufficient in most foodstuffs despite its high latitude

- lava weathers to produce fertile soils, encouraging farming

- volcanic rocks provide physically strong building materials.

NOTING ACTIVITIES

1 Lava flows may look dramatic but they are often not as life-threatening as they may seem. Why?

2 The most devastating hazards associated with volcanic activity are pyroclastic flows and lahars. Try to explain why this is so.

3 How important is the passage of time as far as people's perceptions of the volcanic hazard is concerned?

4 Suggest some of the positive aspects of volcanic activity as far as landscape development and human activities are concerned.

CASE STUDY

Montserrat, 1995–97

The island of Montserrat (Figure 2.26) lies within the Caribbean in the same island arc as the island of Martinique, on which Mt Pelée erupted in a spectacular pyroclastic flow in 1902. In July 1995, after a dormant period lasting 350 years, the first eruption on Montserrat occurred. By the following April, most of the population of Plymouth, the capital, had been evacuated. Although several minor eruptions had taken place, no major incidents occurred until 25 June 1997. By then, a series of early warnings had allowed a total evacuation of the hazard area early on the day of the big eruption, when a series of pyroclastic flows began from the northern flank of the Soufrière Hills volcano. During the following three days, the death toll rose to 23 and a further 30 people were rescued from the hazard area. Meanwhile, further activity was reported from the volcano, with another series of earthquakes and pyroclastic flows.

Several warning signs were given that the volcano was about to erupt. Up to 100 small earthquakes of various sizes were noted between 13 and 27 May. These suddenly started again on 22 June, lasting until 25 June, just before the main eruptions took place. The earthquakes gained in intensity and were accompanied in their later stages by rockfalls. Rapid degradation of the north face of the volcano had begun in May 1997, and was accompanied by dome growth in the summit area during mid-June. Regular monitoring of the northern crater walls indicated that deformation was taking place in early March, and by May the site was considered too dangerous to visit. Just before the main eruption, earthquake activity was almost continuous, reaching six quakes per minute. Suddenly, at 12.55 pm, a dense, dark ash cloud rose vertically from the north flank of the volcano, reaching 10 000 m within minutes. Pyroclastic eruptions such as these, in which rocks, dust and gases escape at temperatures exceeding 500°C, are amongst the most dangerous volcanic hazards, as they move extremely fast and make escape difficult.

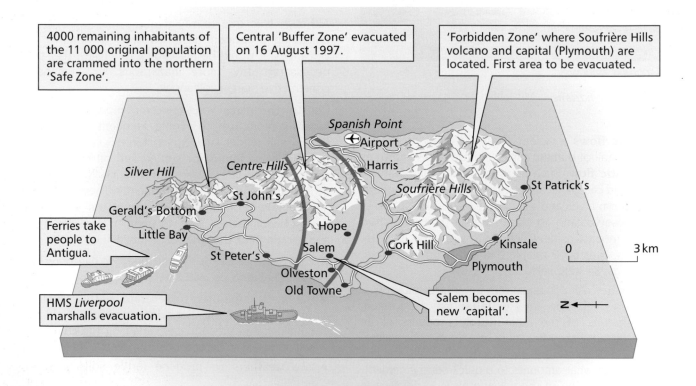

2.26 *Montserrat*

1 **a** On what type of plate margin do the island of Montserrat and its Caribbean neighbours lie?

   b Describe the nature of the eruption that would be expected on this type of plate margin.

2 What were the volcanic hazards associated with the eruptions of the Soufrière Hills volcano 1995–97?

3 Try to suggest some short-term and some longer-term effects of the eruptions on the people of the island.

Earthquakes

An earthquake is a sudden and intense shaking motion that usually lasts for a few seconds. It is caused by the 'snapping' of rocks as enormous stresses, which may have built up over many years, are suddenly released. The point where the 'snapping' occurs is called the **focus** (Figure 2.27). The point on the ground surface immediately above the focus is called the **epicentre** (Figure 2.27).

The vast majority of earthquakes occur at plate margins because it is here that sufficient stresses build up. They can occur at all plate margins but the most powerful are usually associated with destructive margins. As with volcanoes, earthquakes can occur mid-plate. These earthquakes are often triggered by human activity, for example subsidence associated with deep underground mining, or the abstraction of underground water leading to sudden pressure changes.

What is the earthquake hazard?

When an earthquake occurs, it send shockwaves through the crust which can be picked up and recorded by a **seismograph**. If an earthquake occurs in a populated area, the shaking can cause buildings and bridges to collapse, rails to buckle and piped services to rupture (Figure 2.28). Whilst some loss of life is often inevitable, the scale of disaster will depend on several factors:

- The strength of the earthquake – the magnitude of an earthquake is measured using the Richter scale. With each point on the scale, there is a ×30 increase in energy released. Most serious earthquakes have a magnitude in excess of 5.5.

- Depth of focus – generally, the shallower the focus, the more damaging the resulting shockwaves.

- Nature of the bedrock – solid rock can withstand shaking far better than weak sands and clays, which can turn to jelly (liquefaction) and cause buildings to topple over.

- Building design and competence – it is possible to construct buildings to withstand shaking. However, in countries where there are no enforced building regulations, corners are cut to keep costs down and shoddily-built structures simply collapse.

- Population distribution – nearly half the world's largest cities are located in earthquake-prone areas (Figure 2.29) close to plate margins. Many are at the coast where the threat of tsunamis is greatest.

- Wealth of a country – wealthy countries have good communications and they can afford to stockpile emergency supplies of water, food, medicines and forms of shelter (tents). These countries, such as the USA, tend to suffer far less loss of life than the poorer countries of the world, such as China and Afghanistan.

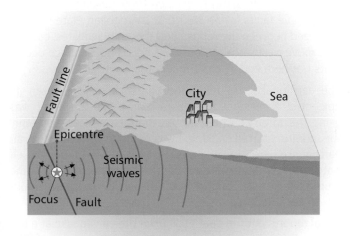

2.27 *Earthquake terminology*

2.28 *Earthquake damage, Kobe 1995*

Algiers	Caracas	Jakarta	Los Angeles	Rome	Teheran
Ankara	Casablanca	Kabul	Managua	San Francisco	Tianjin
Athens	Chongqing	Kanpur	Manila	Santiago	Tokyo
Bangkok	Davao	Kobe	Mexico City	Seoul	Tripoli
Beijing	Dhaka	Kuala Lumpur	Milan	Shanghai	Turin
Bogotà	Guatemala City	Kunming	Nanjing	Shenyang	Wuhan
Bucharest	Harbin	Lahore	Naples	Singapore	Xi'an
Cairo/Alexandria	Havana	Lanzhou	Osaka	Surabaya	Yangon (Rangoon)
Calcutta	Hong Kong	Lima	Pyongyang	Taipei	Yokohama
Canton	Istanbul	Lisbon	Rangoon	Tashkent	

2.29 *Some large cities at risk from earthquakes*

NOTING ACTIVITIES

1 Write your own definitions of:

- earthquake

- focus

- epicentre.

2 Study Figure 2.28. An earthquake causes both short-term and long-term problems.

a What short-term problems are illustrated in the photograph?

b Suggest some long-term problems that might affect the area.

3 Study Figure 2.29.

a Plot the location of the cities listed onto a world outline.

b Now add the plate margins to your map, using Figure 2.8 page 29.

c Why are so many of the world's largest cities located at plate margins?

4 Why *do* people in LEDCs often suffer greater hardship following an earthquake than those in MEDCs?

CASE STUDY

The 1999 Turkish earthquake

On 17 August 1999, at 3.00 am, western Turkey was hit by a massive earthquake measuring 7.4 on the Richter scale. Its epicentre was close to the heavily populated and industrial town of Izmit (Figure 2.30), which was devastated by the earthquake. Many nearby towns, including Gölcük and

2.30 *The Turkish earthquake, 1999*

Istanbul (Figure 2.30), were also badly damaged, and many people were killed.

The earthquake was caused by slippage along the North Anatolian transform fault (Figure 2.31). Over 35 earthquakes have occurred along this fault since 1900, making it one of the most active faults in the world. The Anatolian Plate is being squeezed between the African and the Arabian Plates, causing it to move westwards. As pressure builds up, slippage occurs at its northern end where it meets the Eurasian Plate.

The earthquake was felt hundreds of kilometres from its epicentre. A resident in Ankara, some 200 km away, was using the Internet when the earthquake struck. 'Suddenly the monitor started shaking. I was going to write "there is an earthquake here", but the electricity went off. I threw myself onto the bed. My room was shaking and there were some crunching noises coming from the walls. It lasted about 20–25 seconds.'

Initial reports put the death toll from the earthquake at a few hundred but, as dawn broke, the sheer scale of the destruction became clear (Figure 2.32). Turkey was in the throes of a major disaster.

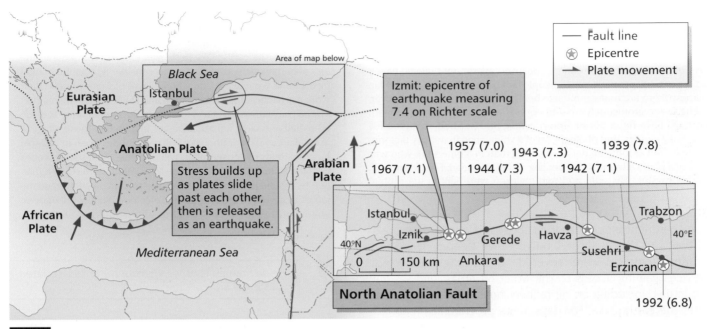

 2.31 *What caused the earthquake?*

2.32 *Aftermath of the 1999 Turkish earthquake*

A reporter for the *Daily Telegraph* described the scene in Izmit the morning after the earthquake:

> Scores of people gathered round a collapsed apartment building which was home to 100 people. Only 15 people had been pulled out alive so far. 'We know there are many more like her,' said Gurkan Acar, a taxi driver, pointing to the grotesquely twisted corpse of a young woman beneath what must have been her bedroom ceiling and her bed. A stuffed toy panda, baby clothes, and photographs of a young man in military uniform, were scattered among the debris. A young man in a blue shirt, who was clutching a photo album, kept repeating 'If they are dead, I will kill myself. I will kill myself.' His wife and two children were still in the rubble.

The apartment block described above was one of thousands of buildings that collapsed, burying their sleeping residents. Communications were severed, preventing emergency services gaining access to the worst-hit areas, and several industries, including an oil refinery near Izmit, were ablaze after the earthquake. For days, news programmes reported the progress of the emergency teams, armed with sniffer-dogs and heat-seeking equipment, as they painstakingly pulled individuals out of collapsed buildings.

The final death toll of the earthquake has been put at 14 500 with over 25 000 injured. Some 20 000 buildings were completely destroyed, and many thousands were damaged.

The sheer scale of the disaster can be attributed to a number of factors. The earthquake was extremely powerful and its focus, at a depth of only 17 km, was very close to the surface. The epicentre was near to a number of heavily populated towns. People were asleep in their houses with no chance to get out into the relative safety of an open area. Despite the existence of building codes, many buildings had been shoddily built, using poor-quality cement and insufficient supports to enable them to withstand shaking. Corners had been cut by unscrupulous builders trying to keep the costs down. Scientific reports had suggested the likelihood of an earthquake in the area, and scientists had even advised against the building of the oil refinery at Izmit.

The authorities in some towns were very slow to respond and there was little local provision of emergency supplies. Some officials were even discovered to have inflated their local provincial death toll in an attempt to receive more aid. Necati Ozfatura, a columnist on the daily *Turkiye* newspaper, wrote the following:

> Centralisation is the biggest reason for the disaster. The local government's authority is weak. Turkish search and rescue teams are not as effective as the foreign teams, so the public had to help themselves. Our people have sent
> so much food to help the victims whereas the bureaucracy has done nothing but send bread which happened to be rubbish.

The long-term cost of the earthquake will be immense. The following extract was written by Savas Akat, a columnist on the daily *Sabah* newspaper:

> First of all, the human loss is the biggest loss. What is the cost of a human life? 1000 dollars? 10 000 dollars? 100 000 dollars? How will we be able to determine the losses of those who have lost loved ones? The earthquake destroyed urban areas and thus the loss of human life was high. We can consider it to be the human investment; we call the educational level of these people, their work experience, their information and skills 'human resources'. They all have a social value. I think the educational value of those who died and were injured is higher than that of those who live in rural areas. The earthquake will cause a decrease in national income. Rates of consumption will decrease until society copes with the shock. Tourists will come to Istanbul more seldom than before. Buildings will need to be reconstructed and, if we add the infrastructural damage to roads, the electricity supply, telephones, water, etc. then we are talking about billions of dollars.

See Stanley Thornes website for links to related Internet sites.

NOTING ACTIVITIES

1 Describe, with the aid of a sketch map, the cause of the Turkish earthquake.

2 What were the short-term effects of the earthquake?

3 a What do you think the journalist Savas Akas means by the 'human resource'?

 b In what ways will the loss of the 'human resource' affect Turkey's ability to re-build?

 c Suggest other long-term effects of the earthquake.

4 To what extent was the Turkish earthquake a purely 'natural' disaster?

How can tectonic hazards be reduced?

There is absolutely nothing that can be done to prevent a volcanic eruption occurring and, although attempts have been made to lubricate faults with oil, there is also nothing that can be done to stop an earthquake. Tectonic events have to be 'lived with' rather than prevented.

Tectonically active areas are carefully monitored by scientists in the hope that a volcanic eruption or earthquake can be predicted. Volcanoes do have a tendency to 'warm up' prior to an eruption, giving the following clues to watching scientists:

- rising heat in the form of infrared radiation can be measured using satellites

- ice and snow on the summit may melt as magma starts to rise
- a series of earthquakes centred on the volcano often precedes an eruption
- physical bulges occur in the volcano as it swells to accommodate the rising magma
- changes in water pressure and gas emissions may occur prior to an eruption.

When an eruption seems imminent, people can be warned and evacuated, saving countless lives. Damage to property and land, however, is not so easily prevented.

Scientists are also keen to monitor earthquakes in the hope that one day it might be possible to predict an earthquake. Figure 2.33 shows some of the monitoring equipment currently available to scientists. The highly sensitive equipment is designed to detect the slightest changes in

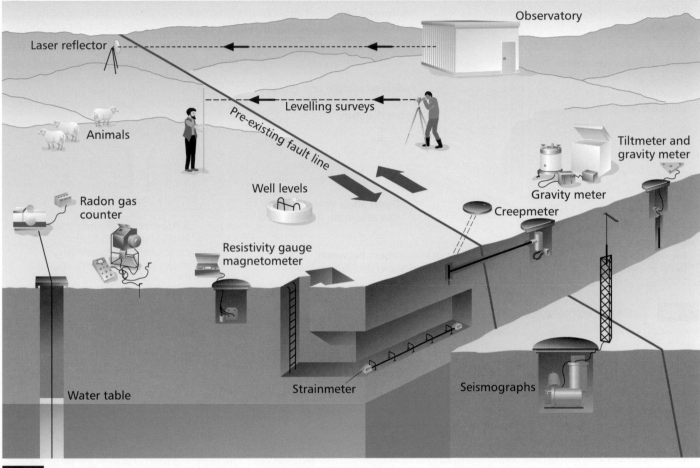

2.33 *Monitoring methods that may help to predict an earthquake*

stress, gas emissions, ground movement and electrical activity. Some people believe that animal intuition may be a far cheaper way of predicting an earthquake (Figure 2.34). However, unlike volcanoes, earthquakes do not give reliable warning signals. Whilst some physical and chemical changes have come to light retrospectively, no reliable means of prediction exists today. Even if it did, what real value would there be in having a few seconds' or, at most, a few minutes' warning?

The only realistic means of reducing the earthquake hazard is to build structures designed to cope with shaking (Figure 2.35) and to make sure that there is the wherewithal (administrative structures and supplies) available to cope with the aftermath of an earthquake.

2.34 *Can animals predict earthquakes?*

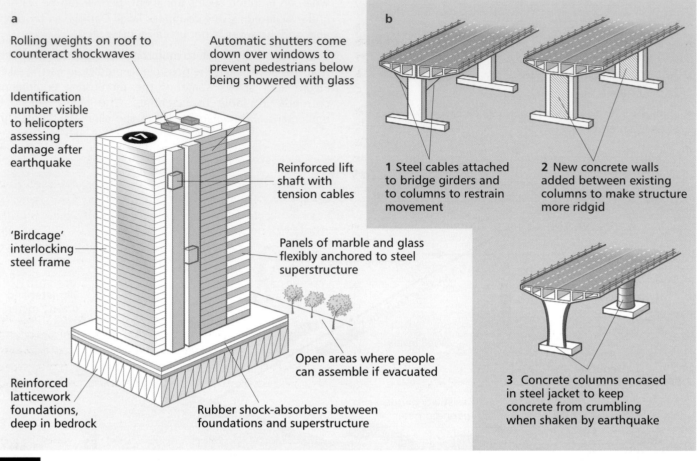

a

Rolling weights on roof to counteract shockwaves

Automatic shutters come down over windows to prevent pedestrians below being showered with glass

Identification number visible to helicopters assessing damage after earthquake

Reinforced lift shaft with tension cables

'Birdcage' interlocking steel frame

Panels of marble and glass flexibly anchored to steel superstructure

Reinforced latticework foundations, deep in bedrock

Rubber shock-absorbers between foundations and superstructure

Open areas where people can assemble if evacuated

b

1 Steel cables attached to bridge girders and to columns to restrain movement

2 New concrete walls added between existing columns to make structure more ridgid

3 Concrete columns encased in steel jacket to keep concrete from crumbling when shaken by earthquake

2.35 *Earthquake-proofing: **a** buildings **b** bridges*

NOTING ACTIVITIES

1 What are the warning signs of an impending volcanic eruption?

2 If volcanic eruptions can be predicted reasonably successfully, why is there often considerable loss of life?

3 Study Figure 2.33.

   a Choose four methods of earthquake monitoring shown in the diagram. For each one, attempt to describe what it is measuring.

   b Is there any point in spending huge sums of money trying to predict earthquakes? Suggest some arguments on both sides of the debate.

4 Study Figure 2.35.

   a Identify two measures aimed at reducing the direct effect of shockwaves on buildings.

   b Suggest the purpose of the building's 'birdcage' interlocking steel frame.

   c Why should heavy objects and panels of marble and glass be firmly secured to a building's superstructure?

   d How can bridges be made to withstand an earthquake more effectively?

   e What sort of provisions should be available as emergency supplies following an earthquake?

D Weathering

What is weathering?

Weathering is the decay or decomposition of rocks *in situ* (in its original position) resulting from physical or chemical actions. Most forms of weathering, as the term suggests, involve elements of the weather, such as water, frost and oxygen. However, weathering can also result from the activity of plants and animals, such as the physical effects of tree roots or the action of organic acids.

It is the lack of movement that distinguishes weathering from erosion. **Erosion** is the picking up and transportation of rock material by agents such as rivers, glaciers and the sea. Although weathering and erosion are, in theory, quite different, in many situations they are almost impossible to separate. Working together, weathering and erosion cause landscapes to be worn down: this is called **denudation**.

Weathering and erosion depend on one another, and one could not exist very effectively without the other. Weathering breaks down a rock surface to produce a pile of loose debris called **regolith**. Erosion then removes the broken-down rock fragments, often using the weathered fragments as 'tools' for erosion. This exposes a fresh rock surface to the processes of weathering, and so the two processes continue.

It is wrong to think that weathering only happens in the 'natural' countryside, for it is extremely effective in towns and cities. Figure 2.36 shows a limestone gargoyle from St Paul's Cathedral in London which has become severely disfigured by chemical weathering. You only have to look at the gravestones in your local churchyard to see that it is often hard to read inscriptions engraved over one hundred years ago.

There are three types of weathering. **Physical weathering** involves the disintegration of rocks into smaller fragments but without any chemical change taking place. In contrast, **chemical weathering** does involve a chemical change, causing rocks to decompose. **Biological weathering** involves the effects (both physical and chemical) of plants and animals on rocks.

Weathering processes cause rocks to be broken down in a number of ways, as Figure 2.37 illustrates. When individual grains in a rock are weathered the rock may crumble to form a pile of separate particles. This is referred to as **granular disintegration**. Weathering may, on the other hand, cause the outer layer of rock to peel away, a process called

2.36 *A weathered gargoyle*

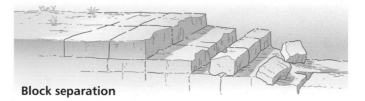

Granular disintegration

Exfoliation

Block separation

Shattering

2.37 *The breakdown of rocks*

exfoliation. The process of **block separation** involves a well-bedded and jointed rock breaking up into distinct blocks, whereas **shattering** causes rocks to break apart in a random way, producing angular rock fragments. When a chemical change occurs, minerals within a rock may be converted into a clay or may even be dissolved (a process called **solution**) in water.

Physical weathering

There are five main processes of physical weathering.

1 **Frost-shattering** This involves the disintegration of rock resulting from stresses caused by freezing water. Water may collect in joints or pores and, on freezing, it will expand by approximately 9 per cent by volume, causing considerable stresses within a rock. When temperatures rise above freezing, the ice thaws and the stresses are released. If many cycles of freezing and thawing take place, the joints and pores may become enlarged and the rock will eventually shatter (Figure 2.38). The shattered, angular fragments of rock which collect at the foot of a frost-weathered slope are called **scree**.

Scientists believe that the following factors encourage frost-shattering:

- rapid freezing with a minimum temperature of –5°C
- frequent cycles of freeze–thaw; some studies have suggested that diurnal (daily) cycles, such as occur in Alpine or Arctic coastal environments, lead to more shattering than seasonal cycles typical of true Arctic environments

- high degree of porosity or density of cracks in a rock
- presence of large amounts of water.

2 **Insolation weathering** This is the expansion and contraction of rock particles resulting from extreme variations in temperature. It is particularly significant in desert environments where diurnal temperature ranges may be in the order of 40–50°C. Rock is a poor conductor of heat, so only the outer skin responds to temperature changes and subsequent cycles of expansion and contraction. As a result, the outer 'skin' may flake off – this is **exfoliation**. Rocks that contain minerals of different colours can be affected by insolation weathering in a slightly different way. The darker minerals will absorb more heat than the lighter ones and will expand, causing stresses within the rock. After many cycles, the rock may begin to show the effects of granular disintegration.

3 **Pressure release** Although not strictly speaking a form of weathering, it does occur *in situ* and does involve the disintegration of rocks. When overlying rocks are removed rapidly by erosion, the release of the overlying pressure may cause these rocks to expand. The expansion may lead to the formation of cracks which will subsequently be exploited by weathering. The result is that whole sheets of rock, several centimetres thick, may become detached to give a rounded appearance to a rocky landscape. Figure 2.39 shows the effect of exfoliation on an exposure of granite in Yosemite National Park, USA. The cause of pressure release here was severe erosion by ice which removed huge thicknesses of rock in a relatively short period of time.

2.38 *A frost-shattered boulder, Iceland*

2.39 *Exfoliation dome in Yosemite National Park*

4 Salt weathering This type of weathering is particularly effective in dry environments. The high temperatures, high rates of evaporation and low rainfall often result in significant concentrations of salt lying on, or just below, the ground surface. The growth of salt crystals acts rather like the ice crystals in frost-shattering, causing stresses to be created within rocks and man-made structures. In parts of the Middle East, for example, salt weathering has caused damage to roads, runways and building foundations. Salt weathering is also common at the coast in Britain where it causes flaking of paint on buildings: home-owners often complain of the need to frequently redecorate the outside of their homes. Roads, roadside buildings and bridge supports suffer from the effect of rock salt being spread on roads to prevent icing in the winter (Figure 2.40).

5 Hydration Hydration involves the expansion of minerals or salts resulting from the absorption of water. Such expansion is common in certain clay minerals which are capable of absorbing large quantities of water into their crystal structures. When this happens, they exert stresses within the rock which may eventually cause it to break apart. The wetting and drying of a mass of clay – for example an exposure at the coast – can be very effective in its breakdown and subsequent erosion.

Salt is the hidden menace inside
by Nick Rufford

SALT attack has caught Britain's highway engineers on the hop, because bridge designers failed to anticipate the rapid growth in the use of de-icing salt in winter.

Each year, in their battle to keep roads open in winter, councils in England spread 2.4 million tonnes of salt. The salt causes an insidious and destructive form of corrosion. Chloride eats away at vital steel reinforcing in concrete bridges. The corrosion is caused by surface water or traffic spray that has become contaminated with de-icing salt. The water seeps through cracks in the bridge concrete.

Once the water reaches the steel, the bridge rapidly deteriorates. To make matters worse, frequently the rusting metal expands, splitting away the surrounding concrete. In some concrete bridges, which rely for their strength on steel cables under stress, the rusting may cause the cables to break. These are the bridges most in danger of collapsing.

What particularly worries engineers is that there is often no visible indication that the bridge is under attack. Cables are in ducts, buried in the concrete of the bridge.

Other forms of attack, including corrosion of steel bearings which support bridge roadways and corrosion of steel box girders, have also caused problems, particularly on the Midlands Links route connecting the M1 with the M5 and M6.

In a recent report, the Building Research Establishment, a government-funded research body, warned that salt attack could cause the failure of all the tendons in one bridge duct without external signs.

But despite the danger signals, Britain has been slow to learn from the experience of the United States, where more than 230 000 concrete bridges – half the national stock – have been classified as 'structurally and functionally obsolete'. The number is increasing at the rate of 3500 each year and the cost of repair or replacement is put at $50 billion.

In Britain, the dilemma for the Department of Transport now is how to carry out the necessary repairs without causing chaos on the motorways. Replacing the concrete supports on motorway or trunk road bridges involves lane closures and often traffic diversion.

It is sometimes almost as cheap to replace the whole bridge. A typical motorway bridge now costs about £250 000, but repairing and rebuilding an existing one costs more because of the added cost of demolition and traffic disruption.

Some firms are offering a new system of risk prevention imported from the USA. It involves passing a weak electric current through the steelwork of the bridge to repel the salt. But it cannot be used to save bridges that are already in advanced decay.

NOTING ACTIVITIES

1 What is the source of the salt that is causing all the problems, and what was its intended purpose?

2 Describe some of the effects of the salt on bridge structures.

3 Try to assess the financial implications at the time of the salt weathering.

4 Are there any solutions to the problem?

2.40   *An article in the* Sunday Times, *26 April 1987*

NOTING ACTIVITIES

1 Define the following terms:
  a weathering
  b erosion
  c denudation
  d regolith.

2 For each of the processes of physical weathering:
  • describe how the process works
  • suggest the circumstances under which the process will operate most effectively (consider climate, rock characteristics, etc.)
  • identify the products and effects of the process
  • suggest any implications for human activities.

Chemical weathering

1 **Solution** This is quite simply the dissolving of minerals in water. Some minerals such as halite (rock salt) dissolve very readily whereas others such as quartz dissolve at an exceedingly slow rate. Rocks that are made of calcium carbonate (limestones) will be easily dissolved by rainwater that has absorbed carbon dioxide from the atmosphere to become a weak carbonic acid. This specific form of solution is called **carbonation**. It is particularly effective when humic acids from the soil have also been incorporated into the rainwater.

2 **Hydrolysis** Hydrolysis is often associated with the process of hydration (see page 51), because a chemical change often occurs when a mineral absorbs water. The mineral feldspar, a major constituent of the rock granite (see page 69), is particularly vulnerable to hydrolysis. Weakly acidic rainwater causes the feldspar to be converted into a white powdery clay called kaolin (china clay) with the result that the whole rock gradually breaks apart.

3 **Oxidation** Oxygen dissolved in water can react with certain minerals, particularly iron which is converted to iron oxide. This chemical conversion weakens the mineral bonding and makes it more vulnerable to other weathering processes. A characteristic red or yellow staining results from this type of weathering, commonly seen on the insides of bridges or canal tunnels.

4 **Chelation** Chelation is the effect of organic acids on rock. These acids are derived either from the decomposition of **humus** (rotted vegetation) in the soil, or by direct secretions from organisms such as lichen. Chelation is thought to be very important in promoting the effects of hydrolysis and carbonation because weathering of rock under soil seems to be more active than where bare rock is exposed to the elements.

Biological weathering

Living organisms can lead to both physical and chemical weathering. Plant roots, particularly of trees, will seek out joints and bedding planes in rocks and will gradually prise rocks apart (Figure 2.41). Trees extract water from soils

2.41 *A tree root prising apart joints in rock*

which can lead to shrinkage, particularly during times of drought. This in turn can cause buildings to subside or foundations to become cracked.

The presence of organic matter can produce humic acids which promote chemical weathering, particularly of limestones. At the coast, crustaceans (e.g. piddocks) actually bore holes in rocks, and the secretions of shellfish can increase the rate of chemical weathering. Even bird droppings can have an expensive effect on car paintwork!

> ## NOTING ACTIVITIES
>
> 1 Hydrolysis and hydration are often confused with one another. How are they different?
>
> 2 What is chelation? Discuss the importance of organic acids in promoting chemical weathering.
>
> 3 With the aid of a simple sketch, describe how tree roots can prise apart rocks.
>
> 4 'Weathering processes rarely occur completely separately from one another on a particular rock outcrop or building.' Comment on this statement.

What factors affect weathering?

The main factors affecting the nature and rate of weathering are climate, rock characteristics, and vegetation.

Climate

Study Figure 2.42, which describes the relationship between climate (temperature and rainfall) and weathering processes. Notice that chemical weathering tends to be most intense in wet and warm climates where high temperatures promote chemical reactions, and the heavy rainfall provides the moisture necessary for the processes to operate. Deep weathering profiles are common in such tropical climates and the characteristic red soils reflect active oxidation. Physical weathering is particularly active in cold climates where frost-shattering dominates. In desert environments, although physical weathering might be expected to dominate due to the absence of water, chemical weathering is important, particularly oxidation.

It is, however, important to bear in mind a couple of points:

- Climates vary with altitude as well as latitude – so, for example, a mountain range situated in the tropics may have a succession of climates from temperate to arctic as height increases.

- Climates change over time – therefore particular weathering processes and features should not necessarily be linked to present-day conditions. For example, the

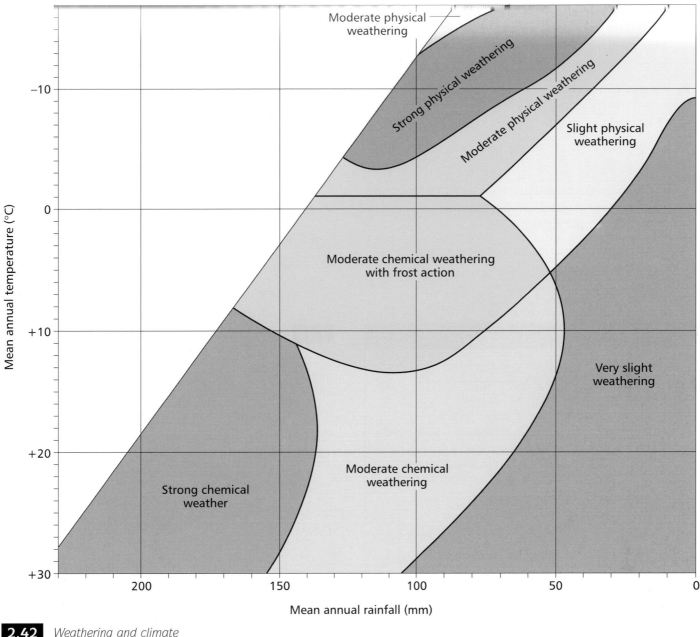

2.42   *Weathering and climate*

deep weathering profiles that exist in parts of Scotland are unlikely to have been formed under present conditions and are more likely to have been formed during a much warmer inter-glacial period.

Rock characteristics

The rate of chemical weathering is affected by the chemical composition of a rock. Some minerals are more prone to chemical change than others. For example, in granite, both the minerals feldspar and mica are chemically vulnerable, whereas quartz weathers extremely slowly. However, the fact that some of the constituents of granite are prone to weathering means that the whole rock is likely to break apart. The presence of iron minerals and salts will affect oxidation and hydration respectively, and rocks rich in calcium carbonate (i.e. limestones) will be affected by carbonation. Joints and bedding planes promote weathering as they enable water to penetrate deep into rocks.

Vegetation

The presence of vegetation also promotes weathering, in that organic acids speed up hydrolysis, and plant roots may prise apart jointed rocks. Some forms of vegetation, such as moss, cling to rock surfaces, holding water against them like a wet sponge thereby encouraging chemical weathering. However, this same vegetation might, at the same time, protect a rock surface from temperature extremes, thereby reducing the effects of physical weathering.

NOTING ACTIVITIES

1 'Climate is generally thought to be the most important factor affecting weathering.' Do you agree with this statement, and if so why?

2 Study Figure 2.42. Attempt to answer the following questions to help you understand the diagram.

a Why do you think the top left-hand corner of the graph is blank?

b Describe the temperature and rainfall characteristics of the zone labelled 'Very slight weathering'. Try to explain why it is placed here.

c Where would you place 'frost-shattering' on the graph, and why?

d Why is 'Mean annual temperature' not an ideal measure when attempting to locate 'frost-shattering' on this diagram?

e Where would you place 'insolation weathering', and why?

f Where would you place 'biological weathering', and why?

3 How do rock characteristics affect the following weathering processes:

- frost-shattering
- hydrolysis
- insolation weathering
- prising apart of rocks by plant roots?

STRUCTURED QUESTION 1

Weathering processes and rates change throughout the year. Refer to Figure 2.43, which shows weathering processes in a small sandstone depression in Poland. The climate is temperate continental with cold winters (January monthly mean −8°C) and warm summers (July monthly mean 17°C). Precipitation (approximately 1100 mm per annum) is concentrated in late spring and autumn.

a (i) Why is chemical weathering at a minimum in high summer? *(1)*

(ii) In which season would you expect to find the greatest intensity of frost weathering, and why? *(2)*

The sandstone is composed mainly of quartz grains, although there are small quantities of feldspar. The grains are held together with a clayey, ferruginous (containing iron) cement. Mosses and lichens are found in the wetter places.

b Explain how moss and lichen might affect the rate of weathering in the depression. *(1)*

c (i) Name one specific weathering process (apart from frost weathering) which probably operates in this depression. *(1)*

(ii) Explain how this process will operate in the depression. *(2)*

2.43 *Weathering processes in a sandstone depression in Poland*

Early spring, with frosts and thaws

Early summer, with alternate short spells of wet and dry weather

High summer, with long periods of dry weather

Autumn, with prolonged rainfall

Winter, with long periods of frosty weather

	Water		Ice		Sediment	Coarse

Fine

Precipitation

Washing in and out

Deflation (wind action)

Evaporation

0 200 400 600 800 1000 mm
Vertical and horizontal scale

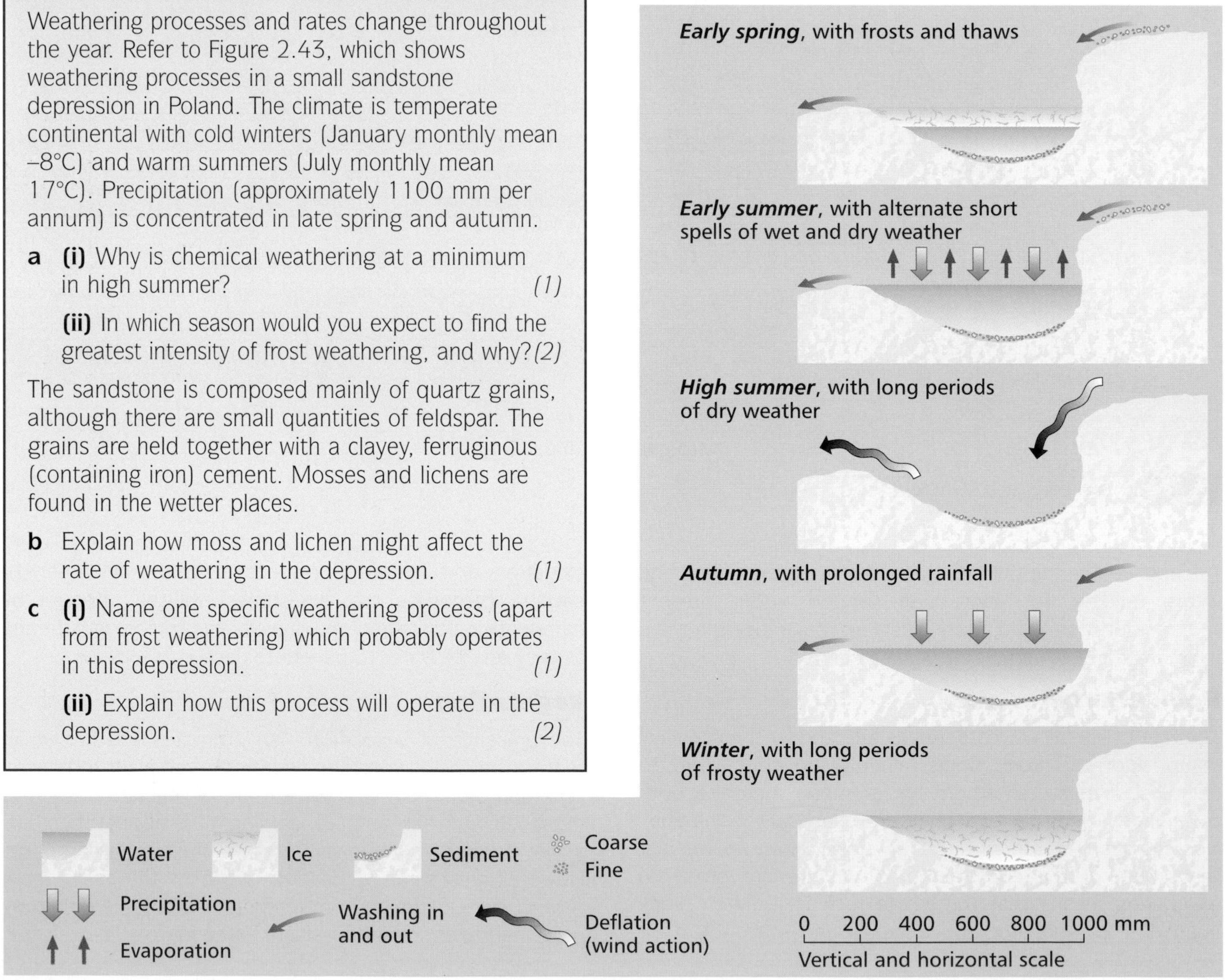

STRUCTURED QUESTION 2

Figure 2.44 shows the way in which precipitation, temperature, evaporation, leaf fall and the depth of weathering vary with latitude.

a **(i)** What type of weathering would you expect to dominate in the tropical forest zone? *(1)*

(ii) Give three reasons why this type of weathering is dominant. *(3)*

b Describe how the depth of the regolith varies with latitude. *(2)*

c Suggest two reasons for the shallow depth of the regolith in the following zones:

• tundra zone

• desert and semi-desert. *(4)*

d The arrow marked A represents the approximate latitude of the British Isles. Explain why the diagram may not adequately represent weathering conditions in the British Isles. *(2)*

e How can weathering processes affect human activities? *(4)*

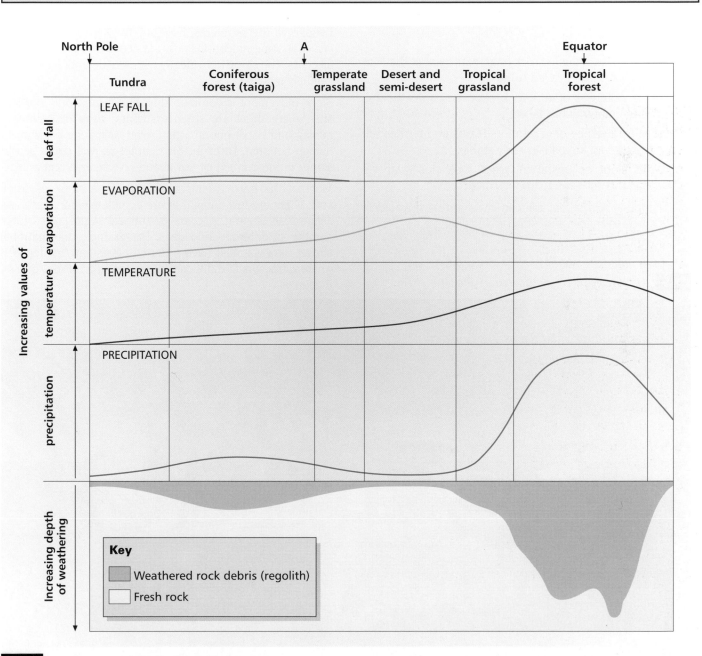

2.44   *Latitudinal variations in weathering*

E Slopes

Why are slopes important?

Figure 2.45 shows one of the most active and dangerous sections of road in the world. It is part of the Karakoram Highway which runs between China and Pakistan and which is subject to frequent severe landslides. Every year landslides and rockfalls close the road for several days until the debris can be cleared. Occasionally whole sections of the road collapse downhill.

There is plenty of evidence in the photograph that this slope is currently active:

- There are a lot of angular rock fragments lying at the foot of the slope – this is called **talus**.
- There are outcrops of bare rock, exposed to weathering, from where material has fallen.
- There is an absence of vegetation. Plants and bushes will only colonise an area if it is fairly stable.

Why is this slope so unstable? Again, look closely at the photograph and spot the following clues:

- The slope is very steep. You can imagine that it wouldn't take much to trigger a rockfall or landslide.

- The road construction may well have added to the problem by cutting into the base of the slope and, in effect, making it more unstable.
- The absence of vegetation means that the slope receives the full force of any rainfall that occurs. In addition, the lack of roots means that the soil and rock fragments are not being held together.
- Traffic vibrations may well have caused some slope collapse.

Despite all the problems, however, the Karakoram Highway is a vital communications link and needs to be kept open at all costs. The slope here clearly needs to be understood and carefully managed.

Slope movement is common throughout the world, particularly where there are steep gradients, weak rocks, heavy rainfall and basal undercutting (river or marine erosion, or human actions). Later in this chapter we will come across slopes in the context of fluvial and coastal environments.

Most of the Earth's land surface forms a slope of some sort. There are few areas that are completely horizontal. As Figure 2.46 illustrates, there are critical slope angles for a number of different land uses. For example, international airport runways must have a slope of less than 0.5 degrees, whereas tractors find difficulty in operating on slopes of over 8 degrees.

2.45 *Landslides on the Karakoram Highway*

Slope angle	Land use limitation
<1°	Few obstacles to land use except poor drainage and flood risk. Limit for international airport runways.
1°	Mainline railways and large motor vehicles affected. Limit for local airport runways. Beyond this angle some influence felt on large-scale agriculture. Flood risk still present.
2°	Ideally the maximum for major roads and railways. Large-scale agricultural machinery and irrigation hindered. Some influence on building development. Flood risk replaced by threat of soil erosion.
4°	Development of housing and roads difficult.
5°	Real problems for large-scale mechanised agriculture. Contour farming advisable. Maximum for railways and for large-scale industry.
8°	Limit for large-scale site development. Problems for wheeled vehicles including tractors. Ploughing impossible without contour terraces.
25°	Mostly forestry and pasture land. Transport possible only with special vehicles.
35°	Extreme limit for caterpillar vehicles. No possibility of agriculture or building. Even forestry now limited by difficulty of mechanised extraction.
55°	No further economic utilisation apart from mountaineering, though land still 'useful', for example as water catchment or for aesthetic appeal.

2.46 *Critical slope angles for various land uses*

NOTING ACTIVITIES

1 With reference to Figures 2.45 and 2.46, comment on the importance of slopes to human activity.

2 Make a list of the features that you might see if a slope was currently active.

3 What factors make a slope vulnerable to collapse?

What are the major slope processes?

There are several different processes that operate on a slope:

- Weathering will directly affect bare rock outcrops, causing rock fragments to break away and collect at the base of a slope to form deposits of **scree**.

- Rainwater will act upon a slope, possibly flowing over it to form small channels called **rills** or larger **gullies**. Under extreme conditions, water may flow as a sheet (**sheetflow**) down a slope.

- Mass movement processes (Figure 2.47), such as landslides and mudflows, are the most significant because they can involve the downslope movement of huge quantities of material and can completely alter a slope's shape. The processes of mass movement vary enormously in their rate of movement, the mechanisms involved and their need for water. Figure 2.48 is a triangular graph used to show all three factors. The position of each type of mass movement is relative: no absolute figures are used on the graph.

NOTING ACTIVITIES

1 Study Figure 2.47.

   a What is the main difference between a landslide and a rotational slip?

   b How is the presence of water often very important in triggering a landslide?

   c What is a lahar?

   d Define the toe, the head and the main scarp features associated with a rotational slip.

   e What is the difference between solifluction and gelifluction?

   f Draw a simple diagram to show how soil particles may creep downhill as a result of expansion and contraction.

   g Apart from terracettes, what other features would you look for as evidence of soil creep?

2 Study Figure 2.48.

   a Which type of mass movement is the fastest?

   b Which type is the wettest?

   c What are the moisture and speed characteristics of soil creep?

   d Which of mudflows and landslides are the most rapid?

   e What is the typical rate of movement of the process of solifluction?

   f Study Figure 2.47. Make a copy of Figure 2.48, and suggest the best position for rockfalls and rotational slips.

a Rockfall

This rapid movement usually occurs on the steepest slopes. Individual rock fragments, or occasionally whole slabs of rock, suddenly become detached and fall to the base of the slope. They may be detached by gradual processes such as freeze–thaw or by sudden and dramatic events such as earthquakes. The angular debris collects at the base of the slope to form **scree** or **talus**.

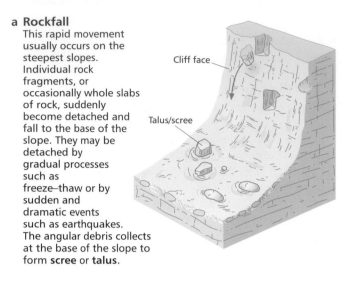

b Landslide

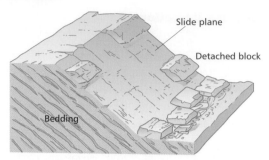

The key thing about a landslide is that movement takes place along a flat or planar **slide plane**. Landslides commonly occur along bedding planes, particularly when the underlying bed is impermeable leading to a high moisture content which will lubricate the slide surface. Landslides are very rapid and can cause huge damage and loss of life.

c Rotational slip

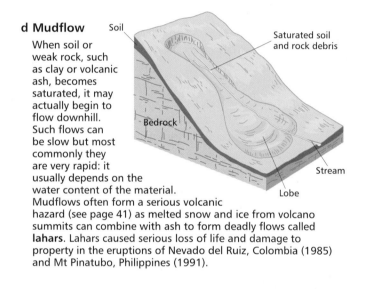

A rotational slip, or landslip, differs from a landslide in that the slide plane is concave in shape, causing more of a rotational movement. These are probably the most common form of visible mass movement in the UK. They can be seen along the coast (e.g. Norfolk; Folkestone, Kent; and Lulworth, Dorset) and on the verges of main roads. They usually occur in weak rock (e.g. clay) or in soil that has become saturated and, in response to gravity, simply collapses.

d Mudflow

When soil or weak rock, such as clay or volcanic ash, becomes saturated, it may actually begin to flow downhill. Such flows can be slow but most commonly they are very rapid: it usually depends on the water content of the material.

Mudflows often form a serious volcanic hazard (see page 41) as melted snow and ice from volcano summits can combine with ash to form deadly flows called **lahars**. Lahars caused serious loss of life and damage to property in the eruptions of Nevado del Ruiz, Colombia (1985) and Mt Pinatubo, Philippines (1991).

e Solifluction

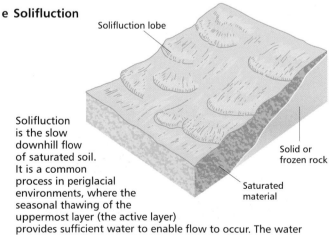

Solifluction is the slow downhill flow of saturated soil. It is a common process in periglacial environments, where the seasonal thawing of the uppermost layer (the active layer) provides sufficient water to enable flow to occur. The water reduces the effects of cohesion and friction, thus promoting movement. The term **gelifluction** refers to solifluction that takes place on top of frozen ground.

f Soil creep

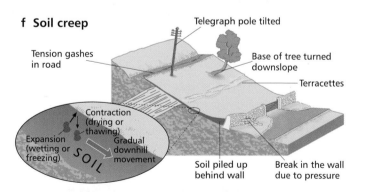

Soil creep involves a heave process whereby individual particles rise and fall in response to expansion and contraction due to wetting and drying or freezing and thawing. It is a very slow process. Soil creep is very widespread in the UK, particularly on clay slopes, as clay is vulnerable to the effects of wetting and drying. You may well have seen **terracettes** on grassy slopes: these are thought to result from soil creep.

2.47 *The main types of mass movement*

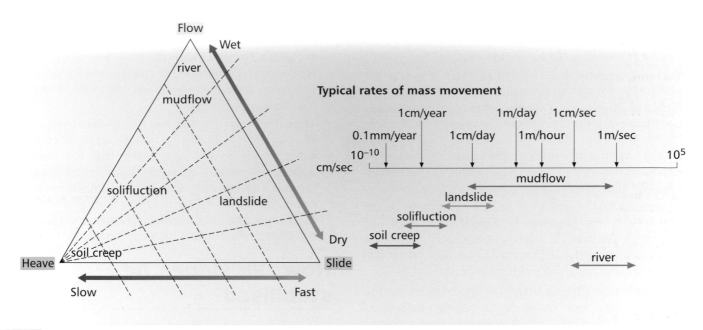

2.48 *Classification of mass movement*

What factors affect slope processes?

Figure 2.49 shows the **slope system**. Notice that it is an example of an **open** system because there are inputs from outside (e.g. heat and precipitation) and outputs (e.g. water and weathered rock) into other systems. This figure illustrates some of the many factors that affect slope processes:

- **Rock type** Generally speaking, the tougher a rock the more able it is to support a steep slope. Igneous and metamorphic rocks are extremely strong and are capable of supporting near-vertical slopes, whereas sands and gravels can only support very gentle slopes.
- **Geological structure** Rock slabs may become detached along bedding planes or joints, promoting rockfalls and landslides.

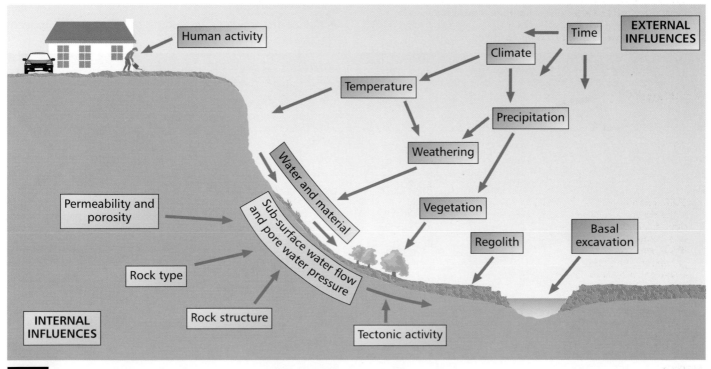

2.49 *The slope system*

- **Permeability and porosity** These characteristics affect what happens to water on a slope. An impermeable rock will be liable to surface water flow, and deep gullies may form. (See page 64 for definitions.)

- **Tectonic activity** Some of the steepest slopes in the world are found in tectonically active areas. The Himalayan region has a large number of dramatic slopes (see Figure 2.45) resulting from the gradual uplift of the mountains (the collision of the Indian and European Plates) combining with rapid rates of erosion. Earthquakes may also trigger slope failure.

- **Climate** The climate of an area will affect the type of weathering that operates on a slope. It will also govern the nature, and presence or absence, of water and vegetation. Most of the processes operating on a slope (e.g. mudflows and soil creep) are dependent on aspects of the climate, particularly precipitation.

- **Weathering** Weathering affects the upper slopes, particularly any bare rock outcrops. In general, mechanical weathering, particularly frost-shattering, will lead to a more jagged, angular, bare rock surface whereas chemical weathering, with its tendency to dissolve and produce fine clays, will tend to produce more rounded slopes.

- **Vegetation** If a slope is forested or covered in bushes and grass it is less likely to be active. This is because it will protect a slope from the direct effects of rainfall and help bind together particles of rock and soil.

- **Basal excavation** Basal excavation can take the form of a river undercutting a slope or the sea cutting a notch in a cliffline. Human activity such as road construction can have the same effect. Basal excavation can lead to a steepening of a slope, so making it unstable.

- **Human activity** People are capable of altering slopes directly by mining and quarrying, constructing roads and housing estates, and terracing land for farming. Slopes can also be altered indirectly when, for example, forests are cut down for firewood or to make way for agriculture. This deforestation will encourage more surface runoff and soil erosion may occur.

- **Time** The length of time that a slope has been exposed to weathering is bound to be an important controlling factor. Newly formed landscapes that are steep and unvegetated are actively weathered and eroded until they assume a shape that is in balance with their environmental conditions. Of course, if the environment changes (global warming, for example) the balance may be upset and the slope profile will be forced to adapt.

How can slopes be stabilised?

Occasionally a slope collapse can have a significant effect on human activities. Mountainsides can collapse after the heavy rains associated with tropical cyclones (Figure 2.50), and cliff failure at the coast is common. There are several ways that a slope can be made more stable:

- plant vegetation to bind the soil together and intercept rainfall
- improve drainage to prevent the slope becoming saturated and to stop lines of weakness, for example bedding planes, becoming lubricated
- use wire nets and metal stakes to hold a slope together (Figure 2.51)
- reduce the gradient by adding material to the base of a slope.

2.50 *The collapse of a hillside following heavy rain*

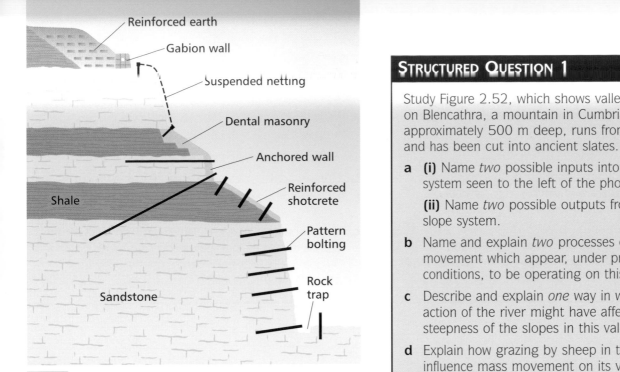

Reinforced earth

Gabion wall

Suspended netting

Dental masonry

Anchored wall

Reinforced shotcrete

Shale

Pattern bolting

Rock trap

Sandstone

2.51 *Methods of slope stabilisation*

STRUCTURED QUESTION 1

Study Figure 2.52, which shows valley-side slopes on Blencathra, a mountain in Cumbria. The valley is approximately 500 m deep, runs from west to east and has been cut into ancient slates.

a **(i)** Name *two* possible inputs into the slope system seen to the left of the photograph. *(2)*

(ii) Name *two* possible outputs from this slope system. *(2)*

b Name and explain *two* processes of mass movement which appear, under present climatic conditions, to be operating on this slope. *(4)*

c Describe and explain *one* way in which the action of the river might have affected the steepness of the slopes in this valley. *(2)*

d Explain how grazing by sheep in this valley might influence mass movement on its valley sides. *(2)*

NOTING ACTIVITIES

1 Study Figure 2.49.

a Suggest why the slope system can be described as being an 'open' system.

b Attempt to make a list of three inputs and three outputs.

c How is a diagram like this helpful in understanding how slopes work?

2 Some of the slopes in Figure 2.45 used to be forested. What effects do you think deforestation has had?

3 Apart from deforestation, how can human activity make slopes more active and unstable?

4 Suggest ways that a slope can be made more stable and less likely to collapse.

2.52 *Valley-side slopes on Blencathra, a mountain in Cumbria*

STRUCTURED QUESTION 2

Refer to Figure 2.53. The Hope landslide occurred in the Cascade Range of British Columbia in January 1965. The area is composed mainly of schists which dip in a south-westerly direction. The landslide buried a 3 km section of highway to a maximum depth of 75 m. The landslide is more correctly called a debris avalanche. A **debris avalanche** may be defined as a rapid flow of large masses of rock fragments downslope.

a The majority of debris avalanches are due to natural factors. These natural factors can be long-term favourable conditions and sudden trigger mechanisms. For the Hope landslide, suggest and explain:

- one long-term favourable condition (2)
- one trigger mechanism. (2)

b Human activity can also cause debris avalanches. Describe and explain any one such human impact. (2)

c The Hope landslide travelled over 1 km away from the base of the slope before coming to rest. Suggest and explain one possible mechanism to account for this mobility. (2)

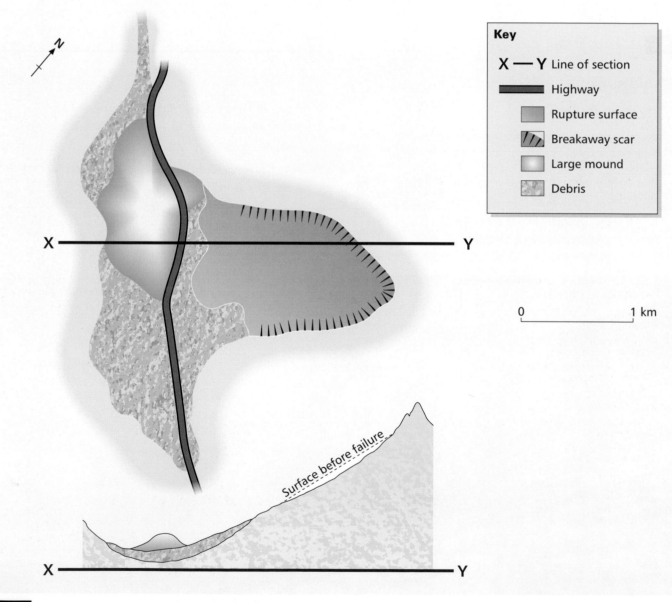

Key

X — Y Line of section

Highway

Rupture surface

Breakaway scar

Large mound

Debris

0   1 km

2.53 *The Hope landslide, British Columbia*

F Rock characteristics and landscape

The landscape of the UK is extremely varied. You do not have to travel very far in any direction in order to pass through very different landscapes: bleak moorlands, deep valleys, gently rolling hills, or vast, flat plains. One of the main reasons for this diversity is the nature of the rocks themselves.

Look at Figure 2.54, which shows the geology of the British Isles. Notice that there are a great many different colours on the map. Each colour represents a different geological time period or a different rock type. Look back to Figure 2.1 on page 24 to remind yourself about the different geological time periods. It is the features represented by the many colours in Figure 2.54 that account for our tremendously diverse natural landscape.

2.54 *Geological map of Britain and Ireland*

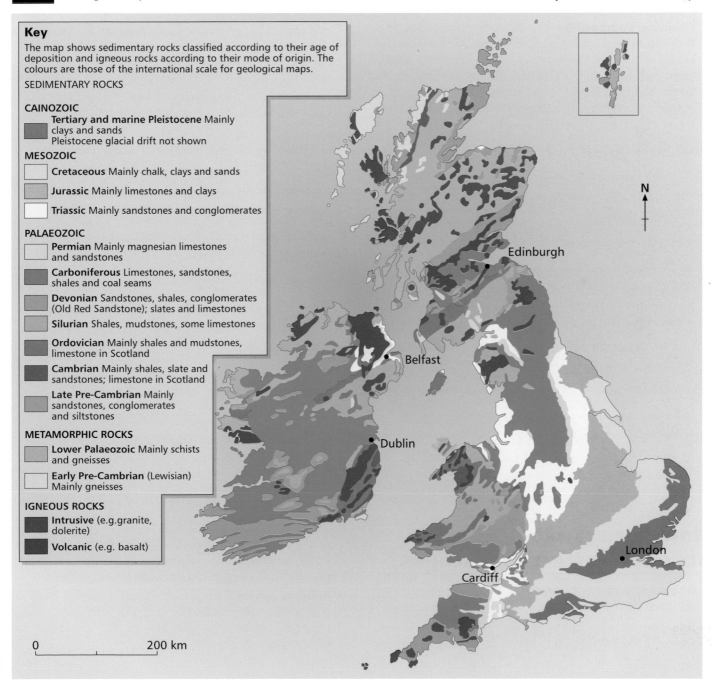

Key

The map shows sedimentary rocks classified according to their age of deposition and igneous rocks according to their mode of origin. The colours are those of the international scale for geological maps.

SEDIMENTARY ROCKS

CAINOZOIC

Tertiary and marine Pleistocene Mainly clays and sands
Pleistocene glacial drift not shown

MESOZOIC

Cretaceous Mainly chalk, clays and sands

Jurassic Mainly limestones and clays

Triassic Mainly sandstones and conglomerates

PALAEOZOIC

Permian Mainly magnesian limestones and sandstones

Carboniferous Limestones, sandstones, shales and coal seams

Devonian Sandstones, shales, conglomerates (Old Red Sandstone); slates and limestones

Silurian Shales, mudstones, some limestones

Ordovician Mainly shales and mudstones, limestone in Scotland

Cambrian Mainly shales, slate and sandstones; limestone in Scotland

Late Pre-Cambrian Mainly sandstones, conglomerates and siltstones

METAMORPHIC ROCKS

Lower Palaeozoic Mainly schists and gneisses

Early Pre-Cambrian (Lewisian) Mainly gneisses

IGNEOUS ROCKS

Intrusive (e.g.granite, dolerite)

Volcanic (e.g. basalt)

Edinburgh

Belfast

Dublin

London

Cardiff

N

0   200 km

How does rock type affect landscapes?

The term **lithology** is often used to describe individual characteristics of a rock. There are several aspects of rock lithology:

- **Rock strength (or hardness)** In igneous and metamorphic rocks the individual minerals tend to be welded together and interlocking (**crystalline**). This generally results in great physical strength. Sedimentary rocks, however, tend to have their individual grains stuck together by a cement which may be a sand, a clay or a mineral that has precipitated out of solution. Some cements are very strong but others are weak. Many sandstones, for example, crumble very easily despite the fact that the individual sand grains may themselves be very strong.

- **Chemical composition** The chemical composition of a rock can be very important. Limestones, for example, are made of calcium carbonate ($CaCO_3$) which is readily dissolved by acidic rainwater (a process known as carbonation, page 52). Most rocks are made up of a number of different constituents, and if one or more is chemically vulnerable, then the whole rock might eventually break apart. Olivine in basalt is very susceptible to chemical weathering and will turn into a clay. Feldspar in granite is also particularly vulnerable to the process of hydrolysis (see page 50).

- **Colour** If a rock contains a mixture of light- and dark-coloured minerals then there will be differences in the way they respond to heat from the sun. The darker minerals will heat up more rapidly, which can lead to stresses within the rock, possibly contributing towards its break-up.

- **Permeability and porosity** The term **permeability** means the ability of a rock to allow water to pass through it. A rock that transmits water freely is permeable, and one that does not is impermeable. Water may pass through a series of holes or pores in a rock, or it may pass along cracks and bedding planes (Figure 2.55). Examples of permeable rocks include limestones (including chalk) and sandstone. Clay is one of the least permeable rocks. **Porosity** is a measure of the proportion of holes or pores within a rock. A porous rock has a high proportion of pores whereas a non-porous rock has a low proportion. If a rock contains pores equal to half its total volume it is described as being 50 per cent porous. Chalk and sandstone are examples of porous rocks.

NOTING ACTIVITIES

1 Study Figures 2.54 and 2.1 (page 24).

  a In what geological period were the rocks beneath London formed?

  b How many years before present did this period start?

  c If you were to travel from London to Cardiff, would you be travelling over progressively younger or older rocks?

  d Where in the British Isles are the oldest rocks found?

  e During which geological period was chalk formed?

  f Describe the distribution of intrusive igneous rocks in the British Isles.

  g How can this map be used to explain the great scenic variety that exists in the British Isles?

2 What is the difference between rock lithology and geological structure?

3 Why are sedimentary rocks often less physically strong than igneous and metamorphic rocks?

4 Study Figure 2.55.

  a What is meant by the term 'permeability'?

  b Describe the two ways in which water can pass through a rock.

  c How does permeability affect the rate of surface erosion by fluvial processes?

  d What is meant by the term 'porosity'?

  e How does porosity influence the effectiveness of weathering processes?

How does geological structure affect landscapes?

Geological structure concerns large-scale features such as joints, bedding planes, folds and folds.

- **Joints** are fractures or cracks that run through rocks (see Figure 2.55). Most joints form as rocks which are 'stretched' during folding (see Figure 2.57) but they can also form when an igneous rock cools and contracts. Joints represent lines of weakness, allowing water to penetrate into a rock and promoting weathering.

- **Bedding planes** are the junctions between beds of sedimentary rock (see Figure 2.55). Like joints, they represent lines of weakness to be exploited by weathering.

- **Folding** When rocks are put under great pressures they may respond by crumpling or folding (see Figure 2.56). There are many different types of fold, although the two main ones are shown in Figure 2.57. Folding can have profound effects on the landscape. **Anticlines** are upfolds, and they tend to form relative upland areas.

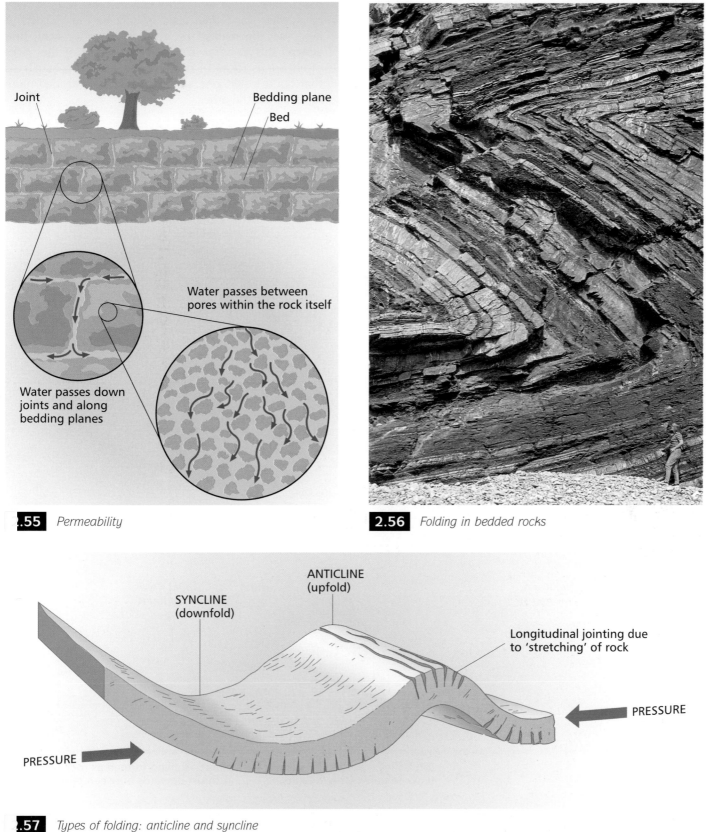

Joint

Bedding plane

Bed

Water passes between pores within the rock itself

Water passes down joints and along bedding planes

.55 *Permeability*

2.56 *Folding in bedded rocks*

SYNCLINE (downfold)

ANTICLINE (upfold)

Longitudinal jointing due to 'stretching' of rock

PRESSURE

PRESSURE

.57 *Types of folding: anticline and syncline*

Synclines are downfolds and tend to form lowland basins. With the passage of time, however, the relationship between fold and landscape becomes far more complicated (see *box* opposite).

- **Faulting** When rocks are put under huge amounts of pressure, for example during tectonic activity, they may crack or fracture causing one section of rock to slip alongside another. The plane along which this movement occurs is called a **fault**. A fault differs from a joint in that some degree of displacement of the rocks has taken place. With a joint there is no such displacement. Look at

Figure 2.58 to discover the main types of fault. Faults have profound effects on the landscape because, like joints but on a much bigger scale, they are lines of weakness that are easily exploited by erosion. At the coast they will be eroded to form caves and eventually bays. Rivers and then glaciers may be fault-guided. A fault may be marked by a steep slope or **fault-line scarp**. This is because the displacement of the rocks often brings rocks of different resistances together so that, as a result, one side of the fault is eroded more rapidly than the other.

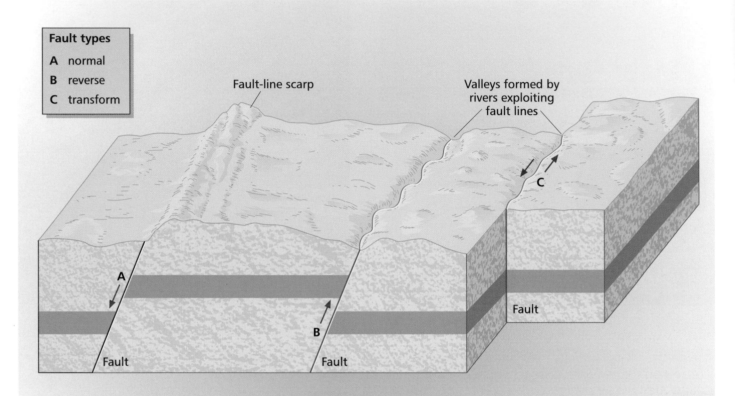

Fault types

A normal

B reverse

C transform

Fault-line scarp

Valleys formed by rivers exploiting fault lines

Fault

A

B

Fault

Fault

Faulting landforms in Namibia

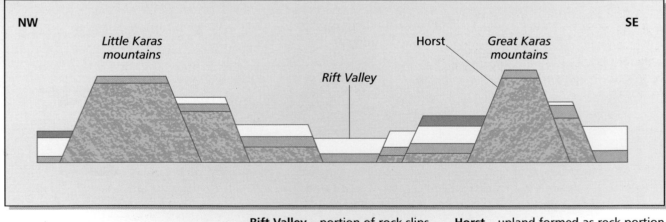

NW

SE

Little Karas mountains

Horst

Great Karas mountains

Rift Valley

Rift Valley – portion of rock slips down in between two faults.

Horst – upland formed as rock portion is thrust upwards between two faults.

2.58 *Faulting*

The development of a folded landscape

When rocks are subjected to pressure, by tectonic activity, they can respond by folding. Tensional joints commonly form at the crest of an anticline where the rocks are being stretched. Weathering and erosion are encouraged, and soon the top of the anticline is removed – this feature then becomes known as a **denuded anticline**.

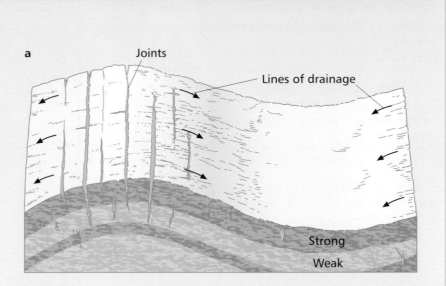

As different rocks become exposed at the surface so their different lithologies begin to play an important role. The tougher bands of rock resist erosion and form **escarpments**, with characteristic **scarp** and **dip** slopes, whilst the weaker bands are eroded to form valleys or **vales**.

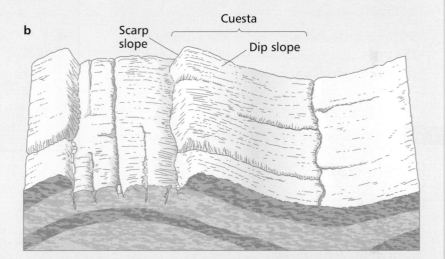

In the syncline, river deposition may lead to the basin becoming infilled. As time passes so the landscape may become even more complex so that the original anticline may form lower land than the syncline. This is described as an **inversion of relief**. Snowdon in North Wales is a good example of an inversion of relief because its rocks are folded in the form of a syncline. Hundreds of millions of years of erosion have left it upstanding, whilst the adjacent anticlinal rocks that would have been very much higher have been removed.

NOTING ACTIVITY

Write a detailed account describing and explaining the sequence of the diagrams in Figure 2.59. Refer to as many features of the landscape as you can.

2.59 Development of a folded landscape

NOTING ACTIVITIES

1 What is the difference between a joint and a fault? Draw a simple diagram to support your answer.

2 Figure 2.56 shows a number of rock structures including beds, bedding planes, joints and folds. Attempt to draw a simple sketch of the photograph and add as many labels as you can identifying the various structural features.

3 Joints, bedding planes and faults represent 'lines of weakness' within a rock. What exactly is meant by this expression?

4 What is the difference between an anticline and a syncline? Draw a simple diagram to clarify your understanding.

5 Study Figure 2.58.

a Use simple diagrams to describe the different types of fault shown.

b Describe the effects that faulting can have on the landscape. Refer to the specific landforms produced by faulting.

STRUCTURED QUESTION

Study Figure 2.60, which is a geological cross-section of the English Channel along the line of the Channel Tunnel. Most of the tunnel runs within the Chalk Marl beds. This is ideal tunnelling material because it is stronger than the Gault Clay underneath and does not fracture like the purer chalk above.

a Name the main geological structure shown between the English and French coasts. *(1)*

b The geological structure was caused by major global forces.

(i) In which direction did these forces operate? *(1)*

(ii) Suggest the global forces that were responsible for the formation of this structure. *(1)*

c Suggest two problems that the geology of the area could have caused for the engineers involved in constructing the tunnel. *(2 x 2)*

d Figure 2.61 is a cross-section through the Vale of Wardour.

(i) Name the geological structure shown between A and B. *(1)*

(ii) Explain how the Vale of Wardour may have been formed. *(3)*

2.60 *Cross-section of the English Channel showing the Channel Tunnel*

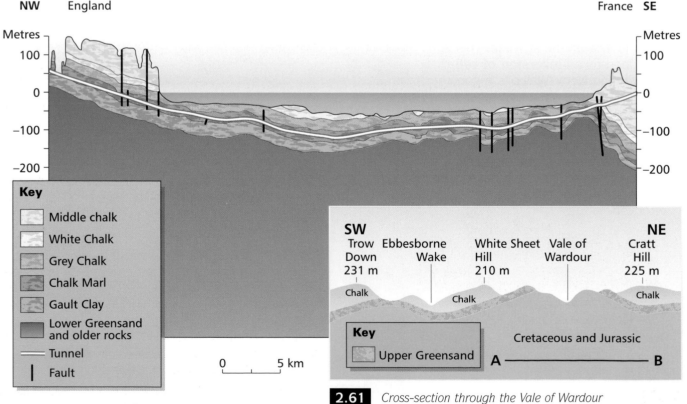

2.61 *Cross-section through the Vale of Wardour*

G Rock types and landscape

Igneous rock landscapes

Study Figure 2.62. It shows some of the common igneous rock forms. Notice that some of the rock forms are extrusive and others intrusive.

Basalt

Basalt is a very fluid lava that can flow for many kilometres before cooling. It often covers vast areas, commonly forming flat, featureless landscapes. If the surrounding rocks are worn away, basalt may form a relatively high plain called a **plateau**, for example the Antrim plateau in Northern Ireland.

When basalt cools it contracts to form joints. If cooling takes place steadily, a regular hexagonal pattern of joints may be formed. When these joints are subsequently exploited by weathering and erosion, columns may be formed (see Figure 2.63). These structures are referred to as **columnar jointing**.

Dolerite

Dolerite is a dense, dark-coloured rock that often forms **sills** and **dykes** (see Figure 2.62). It is a very tough rock and, once the overlying rocks have been eroded away, it often forms relatively high ground.

The Great Whin Sill forms a scarp feature running erratically across north-east England, and covering an area of 2500 km².

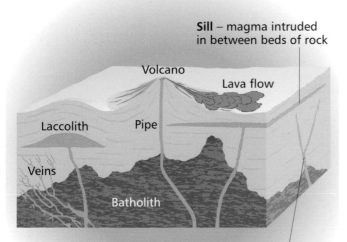

2.62 *Common igneous rock forms*

Sill – magma intruded in between beds of rock

Volcano
Lava flow
Laccolith   Pipe
Veins
Batholith

Dyke – magma intruded across different beds of rocks

2.63 *Columnar jointing in basalt, Antrim, Northern Ireland: the Giant's Causeway*

The Romans made use of its ridge-like qualities by building Hadrian's Wall along part of it (Figure 2.64). The sill cuts across the River Tees in several places, where its relative strength has led to the formation of waterfalls such as High Force and Cauldron Snout.

Granite

Granite forms deep below the ground and is only exposed on the surface after many millions of years of erosion. A physically tough rock, granite is resistant to erosion, and it commonly forms uplands, for example Dartmoor and Bodmin in south-west England. One of the most common features associated with granite is a bare rocky outcrop found on hilltops called a **tor** (Figure 2.65).

Tors are thought to have been formed by weathering deep underground before the granite became exposed on the surface. Despite being physically strong, granite is very vulnerable to chemical weathering. The feldspar readily reacts with acidic water to form a clay, and this chemical change (hydrolysis) weakens the granite causing it to crumble apart. Granite is heavily jointed (see Figure 2.65) and, as Figure 2.66 explains, the density of jointing is believed to have been a critical factor in the formation of tors.

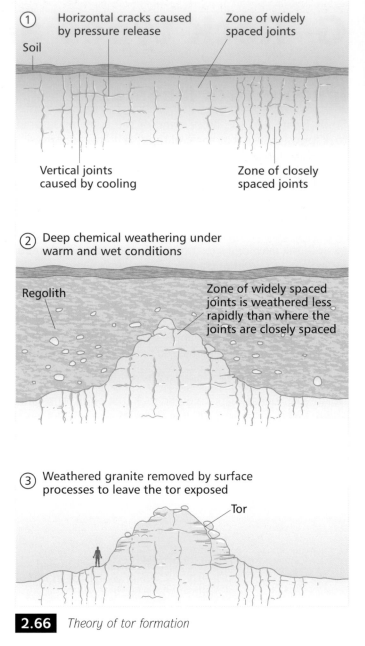

① Horizontal cracks caused by pressure release   Zone of widely spaced joints

Soil

Vertical joints caused by cooling   Zone of closely spaced joints

② Deep chemical weathering under warm and wet conditions

Regolith

Zone of widely spaced joints is weathered less rapidly than where the joints are closely spaced

③ Weathered granite removed by surface processes to leave the tor exposed

Tor

2.66 *Theory of tor formation*

NE   SW

Hadrian's Wall

Dolerite sill

.64 *Hadrian's Wall and the Great Whin Sill*

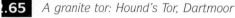**.65** *A granite tor: Hound's Tor, Dartmoor*

NOTING ACTIVITIES

1 Why do basalt lava flows tend to form extensive, flat landscapes?

2 Study Figure 2.63.

  a Draw a sketch to show the columnar jointing in the photograph.

  b Describe briefly how the landscape has formed.

3 Study Figure 2.64. What evidence is there that the dolerite that forms the Great Whin Sill is a relatively resistant rock?

STRUCTURED QUESTION 1

Figure 2.67 shows a variety of igneous features.

a Name the features numbered 1, 2 and 3 on the diagram. *(3)*

b Name a rock type that commonly forms each of the features numbered 1 and 4 on the diagram. *(2)*

c Assume that feature number 3 dips at an angle of a few degrees to the horizontal. Draw a labelled diagram to illustrate the landform which you would expect to develop after the feature had become exposed at the surface, as a result of the erosion of the overlying rocks. *(2)*

Refer to Figure 2.65, which shows a tor on Dartmoor.

d Describe one piece of evidence that the tor has been weathered. *(1)*

e Name and describe two weathering processes that are likely to be acting on the tor at the present time. *(2 x 2)*

f **(i)** What structural feature is S on Figure 2.65? *(1)*

(ii) Explain how these structural features have controlled the shape of the tor above ground. *(2)*

g Outline a suggested explanation for the evolution of tors. *(4)*

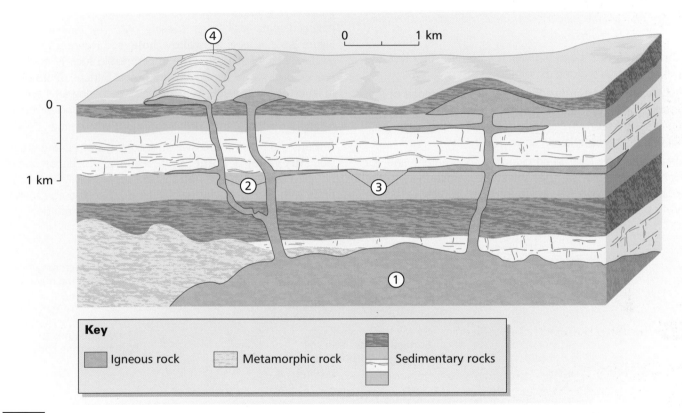

Key

▮ Igneous rock   ▮ Metamorphic rock   ▮ Sedimentary rocks

 2.67   *Features of igneous rock*

Carboniferous limestone landscapes

Carboniferous limestone is a common form of limestone that was formed some 300 million years ago during the Carboniferous geological period (see Figure 2.1 page 24). It outcrops throughout the UK, from the Gower in South Wales to the Pennine Hills in Yorkshire. The limestone in these areas has resulted in a characteristic landscape known as **karst**. Similar landscapes can be found in some limestone areas in other parts of the world.

Several factors combine to create the characteristic limestone scenery shown in Figure 2.68:

- Limestone is physically very strong and will readily form steep slopes without collapsing.

- Limestone has a dense system of joints and bedding planes, and it is highly permeable. Water can travel rapidly through limestone, forming vast cavern systems underground. Limestone is, however, a non-porous rock.

- Composed of calcium carbonate, limestone is highly vulnerable to the chemical weathering process of

2.68 *Typical limestone scenery: near Malham in Yorkshire*

What features are found in limestone landscapes?

Figure 2.69 identifies the main features associated with Carboniferous limestone. Notice that there are both overground and underground features.

Limestone areas frequently exhibit a bare rocky surface criss-crossed by enlarged joints, separating blocks of limestone. This bare surface is called a **limestone pavement**. The enlarged joints are **grykes** and the blocks of limestone in between them are **clints**. Figure 2.70 describes these features and identifies some important controls on their formation.

Limestone surfaces are often pitted with hollows and depressions. Some are simply funnel-like pipes representing enlarged joints, whereas others are large depressions which may or may not have a hole in the bottom. Water flowing off an overlying impermeable rock will tend to flow down an enlarged joint soon after passing onto the limestone (see Figure 2.69). This is a **swallow hole** (see Figure 2.71). Relic swallow holes left dry as the impermeable rock is gradually eroded backwards are called **sink holes**. Surface depressions in limestone can result from surface weathering, or from the collapse of the surface into either an underground cavern or a localised zone of intensive sub-surface weathering.

Perhaps the most impressive limestone feature is the steep-sided **dry valley** or **gorge**. These features, which may or may not have a river in the bottom, are commonly very narrow, often with vertical rocky walls (Figure 2.72). Two contrasting theories have been advanced to explain their formation:

carbonation (see page 52) which gradually dissolves the rock, particularly along its joints.

- Limestone areas tend to have thin soils (as much of it is dissolved during weathering) which are unable to support much vegetation: trees are quite rare and bare rock outcrops commonplace.

- Limestone has long been quarried for cement and for roadstone. Many of the limestone exposures in the Pennine Hills owe as much to human activity as to other processes.

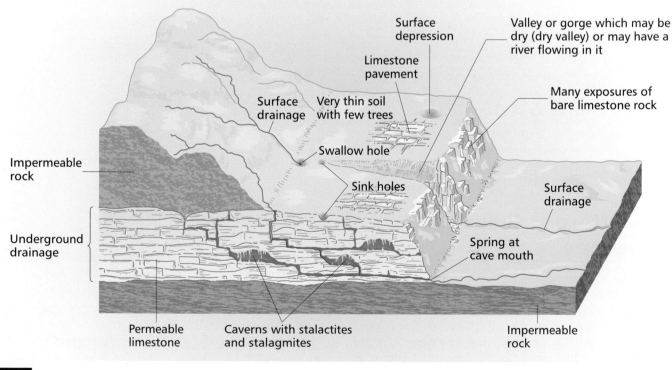

2.69 *Carboniferous limestone scenery*

Glacial erratic left by retreating ice

Smooth surface scoured by ice

Clint – block of limestone surrounded by grykes

Bare, rocky surface as ice stripped off the soil

Vegetation in grykes increases acidity of water in contact with the rock

Grykes – weathered joints in the limestone

2.70 *Limestone pavement: the Burren, Eire*

2.71 *A swallow hole in limestone: Hunt Pot, Yorkshire*

2.72 *A dry valley in limestone: Winnat's Gorge, Derbyshire*

- **Past climatic conditions** During wetter periods, for example at the end of the Ice Age, the water table in limestone may well have been much higher than it is today. This would have caused rivers to flow over the surface, cutting valleys. When the water table began to fall, the rivers would have cut deeper into their valley floors, so forming narrow and steep-sided gorges. Today, if the water table is low and below the level of the valley floor, the valley will be 'dry'.

- **Cavern collapse** This involves the gradual collapse of a series of underground caverns, thereby exposing and revealing a river system underground. Whilst this has certainly happened in some parts of the world, it is not thought to be so widespread in the UK.

As rainwater passes underground through joints and along bedding planes, so weathering and, indeed, erosion by running water, forms pipes, tunnels and **caverns**. Drip-features are common in caverns. The longer, tapered feature extending down from a cavern roof is a **stalactite** whereas the shorter, stubbier feature on the floor of a cavern is a **stalagmite**. They form as water, seeping through the limestone, enters a cavern and then evaporates leaving behind a calcite deposit. A stalagmite will form immediately below a stalactite as water drips off the end and lands on the floor. After many thousands of years, it is possible for both features to join to form a **column** or pillar.

NOTING ACTIVITIES

1 Study Figure 2.69. With the aid of sketches, describe the characteristics of a limestone pavement and comment on the factors that have led to its formation.

2 What are the differences between a swallow hole, a sink hole and a surface depression? Use simple diagrams to illustrate your comparison.

3 Imagine that you have discovered a steep-sided gorge in an isolated outcrop of limestone. What evidence would you look for in trying to establish its possible mode of formation? Aim to write several paragraphs, using diagrams to help.

4 Why are limestone areas often popular tourist destinations?

STRUCTURED QUESTION 2

a Study Figure 2.73. Why are there no streams on the surface of rock C? (1)

b **Shakeholes** are small depressions about 1–3 m deep and 3–5 m in diameter. Figure 2.74 shows a possible sequence of steps in the formation of shakeholes.

(i) What *two* factors determine the precise position of shakeholes? (2)

(ii) Name and describe the process of weathering that is most likely to be responsible for their formation. (2)

c Give *two* possible reasons why the slope at X in Figure 2.73 is so steep. (2)

d There are many dry valleys present in the area. Give *one* possible explanation of the ways in which past conditions have led to the formation of these features. (2)

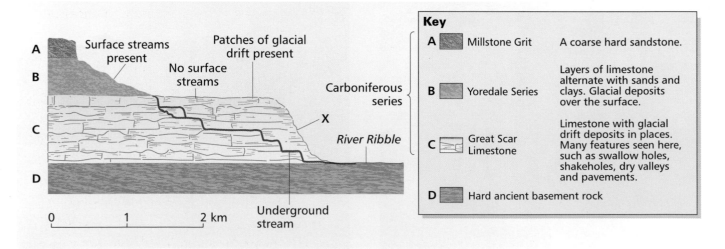

2.73 *Simplified cross-section of the Ingleborough district, in northern England*

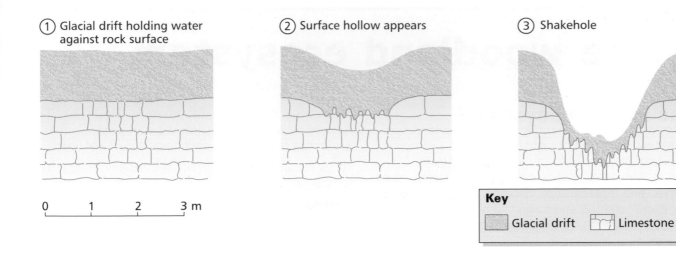

① Glacial drift holding water against rock surface

② Surface hollow appears

③ Shakehole

0   1   2   3 m

Key

Glacial drift   Limestone

2.74 *The formation of a shakehole in heavily jointed limestone*

Chalk landscapes

Chalk is a very pure form of limestone. It is not a particularly strong rock, but its high degree of porosity and permeability means that there are few surface streams, and rates of erosion are slow. This is why chalk tends to form upland areas such as the Wolds in North Yorkshire, and the Chilterns to the north of London (see Figure 2.54 page 63).

Chalk was gently folded during the Alpine orogeny some 20–50 million years ago. This folding has led to the formation of chalk **cuestas**. These features typically have a steep scarp slope and a gentle dip slope (Figure 2.75).

One of the most common features found on the dip slopes of chalk escarpments are **dry valleys** (Figure 2.76). These sometimes steep-sided valleys resemble river valleys in every way except that they contain no river! They were probably formed at some time towards the end of the Ice Age when the chalk was partly frozen, rendering it impermeable. When summer melting occurred, water would have been forced to flow over the surface, carving valleys as it did so. Today, with much lower water tables, these valleys are left 'dry'.

Dry valleys are also found on some scarp slopes. These features, sometimes called **combes**, may well have been formed in much the same way as the dry valleys on the dip slope. However, backwards (headward) erosion by streams (called **spring sapping**) may have been partly responsible.

Chalk has an important economic value as it is used in the manufacture of cement, and quarries are a common sign of human activity in chalk landscapes.

2.76 *A dry valley in a chalk landscape: Incombe Hole, Chilterns*

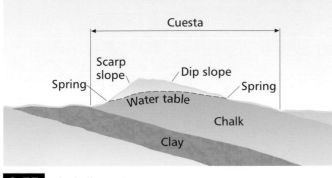

Cuesta

Scarp slope

Dip slope

Spring

Spring

Water table

Chalk

Clay

2.75 *A chalk cuesta*

NOTING ACTIVITIES

1 In what ways have lithology and geological structure combined to cause chalk to form relative upland areas?

2 Study Figure 2.76.

  a Describe the dry valley in the photograph.

  b Outline the possible formation of the feature.

3 In what ways, and for what reasons, does the landscape associated with Carboniferous limestone differ from that of chalk?

A The woodland ecosystem

What is an ecosystem?

An **ecosystem** is a group of organisms living in a particular environment, for example a woodland (Figure 3.1). Ecosystems are often highly complex and involve the interactions and relationships between living organisms (such as plants, animals and fungi) and non-living environmental factors, such as sunshine and precipitation. All the components of an ecosystem are linked together in some way, as is indicated by the arrows in Figure 3.1. These linkages usually refer to particular processes, such as the weathering of rock or the decomposition of leaf litter.

There are a number of important definitions that relate to ecosystems:

- **Biomass** – this is the total weight (usually expressed as kg/m^2) of living organic matter; in other words, the growing plants and trees in Figure 3.1.

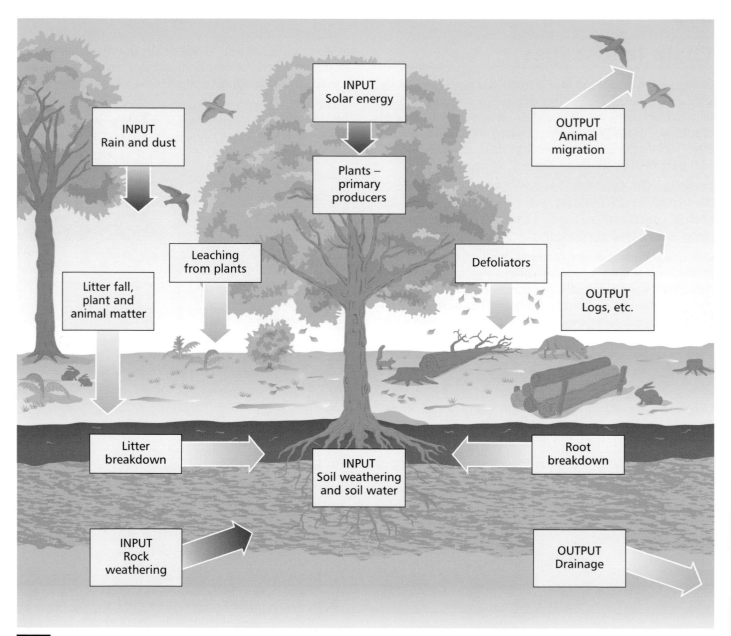

3.1 *A woodland ecosystem*

- **Habitat** – this is a type of home, for example the canopy of a tree where birds and insects make their home.
- **Community** – a group of organisms living in a particular ecosystem are described as a community. In Figure 3.1 this would include the worms, birds, rabbits and foxes.
- **Biotic factors** – these are living components, such as plants and animals.
- **Abiotic factors** – these, in contrast, are non-living environmental factors and include rocks and precipitation.

Living in a woodland ecosystem

Ecosystems have the ability to be self-contained units. Within an ecosystem, such as the woodland in Figure 3.1, there are flows and cycles which enable all the different parts of the ecosystem to survive.

1 Energy flow

The source of all energy is the sun. Plants convert the sun's energy into sugars (carbohydrates) by the process of **photosynthesis**. The total amount of energy absorbed or fixed by green plants is called the **gross primary production** (GPP). Some of the GPP is lost through respiration as plants carry out their normal functions. The rest is used in the production of new leaves. This is known as the **net primary production** (NPP). The productivity of plants is greatest under favourable conditions, i.e. high values of light, warmth, water and nutrients. Tropical rainforests have high rates of productivity whereas Arctic tundra has low rates.

When plants are eaten, the energy that has been converted from the sun is passed on. Study Figure 3.2, which shows a **food chain**. There is a decrease in the amount of energy passing from one link to the next. This is the result of heat loss (e.g. respiration and decomposition) which can be as much as 90 per cent from each level.

Each link or stage of a food chain can be referred to as a **trophic level**. Study Figure 3.3. Notice that the number of individuals, the total biomass and the total productivity fall at each trophic level. Few food chains go beyond four or five levels because there is simply not enough energy to sustain the organisms.

Food chains are a somewhat simplistic way of understanding life in an ecosystem. In reality, there are many more links between individuals, so the concept of a **food web** (Figure 3.4) is more realistic.

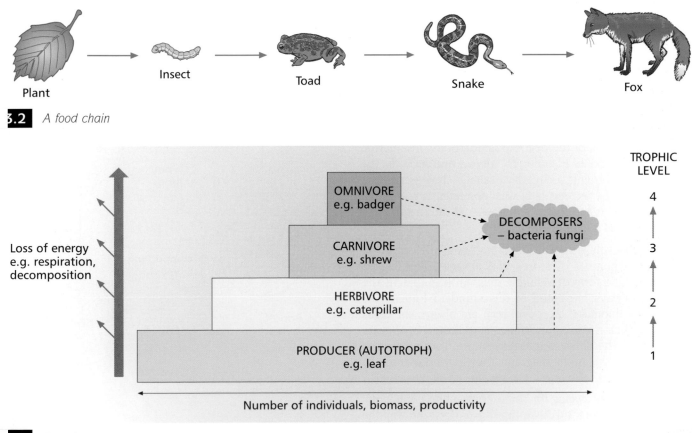

3.2   *A food chain*

Plant   Insect   Toad   Snake   Fox

3.3   *Trophic levels*

TROPHIC LEVEL

OMNIVORE e.g. badger — 4

DECOMPOSERS – bacteria fungi

CARNIVORE e.g. shrew — 3

HERBIVORE e.g. caterpillar — 2

PRODUCER (AUTOTROPH) e.g. leaf — 1

Loss of energy e.g. respiration, decomposition

Number of individuals, biomass, productivity

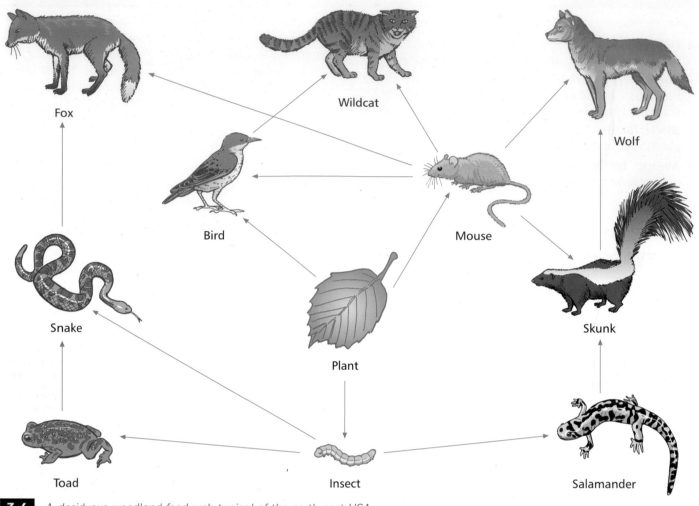

3.4 *A deciduous woodland food web typical of the north-east USA*

2 Nutrient cycles

Nutrients are plant foods. They consist of minerals and chemicals that have been derived from precipitation, the weathering of rock, or the decomposition of organic matter. Plants need nutrients in order to survive and grow. In the natural world, nutrients are readily available; however, house plants have to be fed regularly with commercially available plant foods. Even cut flowers are now sold with a sachet of plant food to be added to the water.

The **nutrient cycle** is vital in sustaining life in an ecosystem. It involves a number of stores and linkages, and is best described in the form of a simple diagram (Figure 3.5). The diagram is always presented in this standard format. The three stores (biomass, litter and soil) are always shown as circles and their relative importance is indicated by the size of the circle. The linkages (processes) are always drawn in the same positions in the diagram. They are represented by arrows. The thicker the arrow, the greater its significance.

Study the nutrient cycles for three contrasting continental ecosystems in Figure 3.6. Take time to understand why the significance of the stores and the linkages vary between the different ecosystems.

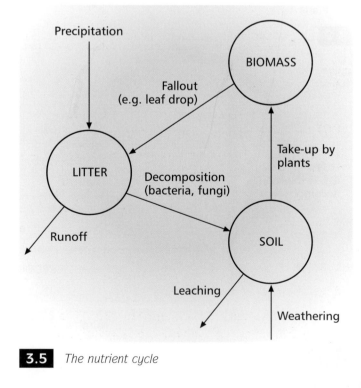

3.5 *The nutrient cycle*

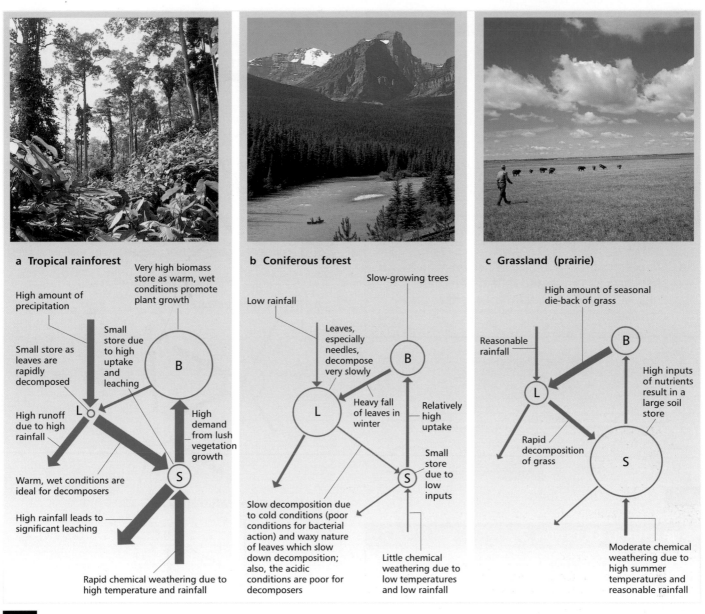

3.6   *Nutrient cycles for three major continental ecosystems*

NOTING ACTIVITIES

1   Define the term 'ecosystem' and give some examples of ecosystems at different scales.

2   **a**   What is meant by GPP?

   b   How does it differ from NPP?

   c   How is energy lost as it passes from one organism to another?

3   **a**   What is a trophic level?

   b   Do you think the pyramid in Figure 3.3 is a good way of illustrating trophic levels? Explain your answer.

4   **a**   What are nutrients?

   b   What are the three main sources of nutrients?

5   Study Figure 3.6. Describe and account for the contrasts in the following between the three ecosystems:

   a   biomass

   b   litter

   c   weathering

   d   decomposition.

STRUCTURED QUESTION 1

Study Figure 3.7, which shows a system of trophic levels (a food chain).

a Identify:

  (i) the input X

  (ii) the flow Y

  (iii) the flow Z. (3)

b Explain the way that energy conversion occurs between X and the autotrophs. (2)

c Explain why the number of individuals and productivity decrease through each higher level in the system. (2)

d Suppose that the numbers of herbivores were suddenly reduced by the spread of disease. Suggest the effects that this might have on the system in Figure 3.1. (4)

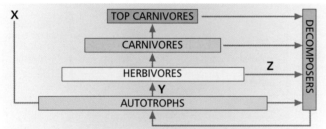

3.7 *System of trophic levels*

STRUCTURED QUESTION 2

Study Figure 3.8, which shows the nutrient cycling system for a northern coniferous forest.

a Identify the nutrient stores X, Y and Z. (3)

b Identify the nutrient flows 1, 2 and 3. (3)

c Explain the differences in the sizes of the nutrient stores. (3)

d Explain the differences in the rates of nutrient flow. (3)

e Explain how the sizes of the nutrient stores and rates of nutrient flow will vary with the seasons. (4)

Figure 3.9 shows changes in the nutrient cycle of a forested area.

f Describe the changes in the stores that occur immediately after clearance. (2)

g Suggest reasons for these changes. (4)

h Why is the cycle in diagram **c** described as being in 'equilibrium'? (2)

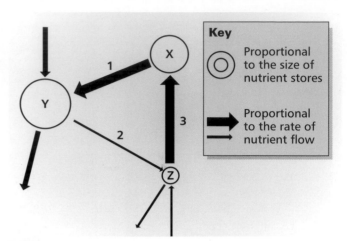

3.8 *Nutrient cycling system in a northern coniferous forest*

a **Before forest clearance**

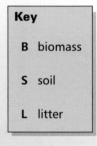

Key

B biomass

S soil

L litter

b **Immediately after clearance**

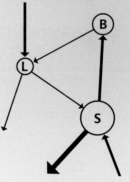

c **Post-clearance equilibrium, pasture**

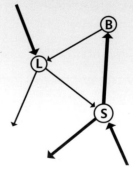

3.9 *Changes in the nutrient cycle of a forested area*

Can human activity affect ecosystems?

Human activity can have very significant effects on ecosystems. For example, people can influence the flow of energy by eliminating one of the trophic levels to increase the amount of energy passing on to a subsequent 'valuable' species. In agriculture or forestry, people are concerned with achieving the maximum potential productivity.

The nutrient cycle can be greatly influenced by the cutting-down of trees, the ploughing-up of pasture or by the addition of fertiliser. For example, deforestation immediately reduces the biomass store. If trees are removed from the ecosystem, there will be a reduction in the litter store. This, in turn, will affect the amount of decomposition that will take place.

People's lives are often dependent on the healthy survival of a natural ecosystem. This is particularly true in many LEDCs. Figure 3.10 illustrates two approaches: one is sustainable, and the other is unbalanced and unsustainable.

NOTING ACTIVITIES

1   Suggest ways in which human activity can influence:

   a   the energy flow

   b   the nutrient cycle.

2   Study Figure 3.10.

   a   Why is the unbalanced cycle unsustainable?

   b   Suggest some possible consequences of the unbalanced cycle.

   c   What may have caused the cycle to become unbalanced? (Suggest human factors and 'natural' factors.)

   d   Suggest measures that could be adopted to make the unbalanced cycle sustainable.

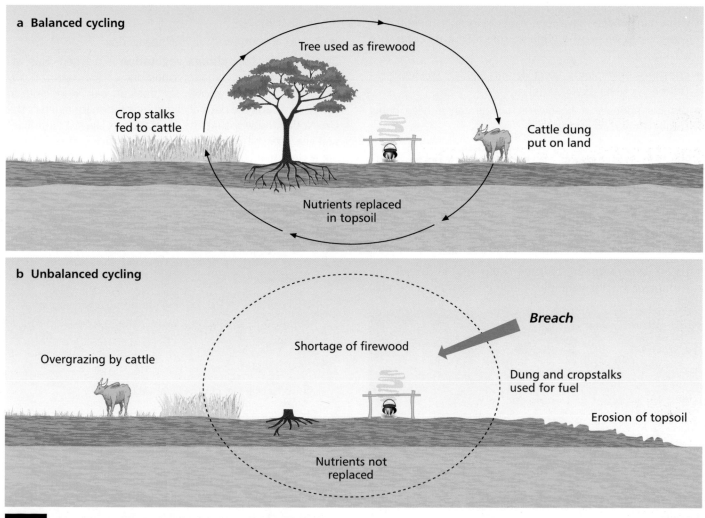

a Balanced cycling

Tree used as firewood

Crop stalks fed to cattle

Cattle dung put on land

Nutrients replaced in topsoil

b Unbalanced cycling

Breach

Shortage of firewood

Overgrazing by cattle

Dung and cropstalks used for fuel

Erosion of topsoil

Nutrients not replaced

3.10   *Nutrient cycling in farming systems in LEDCs*

B Vegetation succession

What is a vegetation succession?

When a plot of cultivated land, for example an allotment, is abandoned, it only takes a matter of a few weeks before weeds begin to take hold (Figure 3.11). Left alone for several years, the weeds will be succeeded by grasses, shrubs and, eventually, trees. This sequence of vegetation is called a **vegetation succession**.

The initial species of vegetation, such as the weeds on an abandoned allotment, are known as **pioneer species**. They are well suited to surviving in relatively harsh environments and often have special adaptations to help them cope:

- bare rock surface plants can survive with negligible amounts of soil by extracting nutrients from precipitation and the atmosphere
- sand dune plants have long roots to enable them to tap water deep below the surface; they can also cope with strong winds
- coastal marsh plants are tolerant of high salt concentrations
- pioneer species in freshwater ponds can survive being totally immersed in water.

What causes a vegetation succession to occur?

The reason why a succession takes place is that the environmental conditions gradually change over time to become more suitable for other plant species to grow.

Initially, the pioneer species help to stabilise the environment by, for example, binding together loose soil.

3.11 *An overgrown allotment*

Roots and leaves trap sediment, causing a slow build-up of a basic soil. Dead leaves and roots decompose to add goodness to the soil. The pioneers also afford protection from strong winds or powerful sunshine, enabling other more delicate species to survive. Essentially, the pioneer species help prepare an area for succession to take place.

As the soil becomes deeper and richer, and the environment becomes less harsh, grasses and then shrubs begin to take a hold. Competition for light results in plants growing ever taller, shading-out smaller plants. Competition for nutrients in the soil often results in different depths of plant root systems.

With each stage of a succession there tends to be an increase in the following:

- biomass
- species diversity (of plants and other organisms)
- productivity
- structural complexity, involving layering.

Eventually, the final **climax vegetation** is reached. This will tend to reflect the climatic conditions of the region. Study Figure 3.12, which shows an ancient oak woodland in Yorkshire. This is the natural climax vegetation for much of the UK. Notice the way that the vegetation is layered (stratified) and see how the roots occupy different layers of soil. There are a great many plant species in this wood-land, illustrating the species diversity that is typical of a climax vegetation.

NOTING ACTIVITIES

1 What is meant by the term 'vegetation succession'?

2 a What are pioneer species and what makes them special?

   b How do pioneer species improve environmental conditions to enable a vegetation succession to occur?

3 Study Figure 3.12.

   a What is the climax vegetation?

   b Make a list of the layers in this woodland.

   c Explain the presence of the different layers of vegetation.

   d Apart from a clear stratification (layering), what are the other characteristics of a climax vegetation compared with earlier stages?

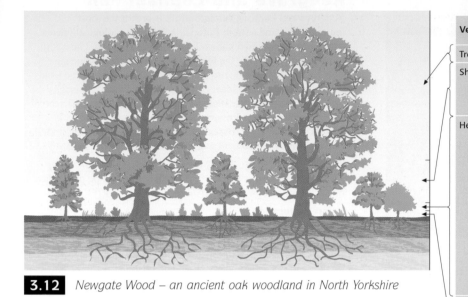

Vegetation		% cover	Average height
Tree layer	Sessile oak	80 (canopy)	14 m
Shrub layer	Bramble	–	1.5 m
	Rose	–	1.5 m
	Rowan	–	2.8 m
Herb layer	Nardus grass	10	25 cm
	Fern bracken	20	60 cm
	Other grasses	10	30 cm
	Wood anemones	6	15 cm
	Wild garlic	5	6 cm
	Dog's mercury	9	20 cm
	Bluebells	4	15 cm
	Primroses	3	10 cm
	Tormentil	3	5 cm
	Bare ground	30	–
Ground layer	Mosses	20	2 cm

3.12 *Newgate Wood – an ancient oak woodland in North Yorkshire*

What are the different types of vegetation succession?

Vegetation succession takes place in a wide range of environments, including rocky surfaces, freshwater lakes and sand dunes. The term **sere** is used to describe the succession associated with a particular environment. Each stage of a succession is called a **seral stage**. The most common seres are:

- lithosere – rock/rock fragments
- hydrosere – fresh water
- halosere – salt water
- psammosere – sand.

A natural succession that proceeds uninterrupted from the pioneer community through to the climax vegetation is called a **primary succession**. However, often a succession is interrupted either naturally, for example by a flood or a fire, or by the actions of people. When the succession resumes, it is a **secondary succession**.

People can interrupt vegetation successions in a number of ways:

- chopping down trees to make way for development (housing, roads, etc.)
- grazing livestock on pastureland
- draining wetland areas for farming
- altering soils and habitats by using artificial chemicals (e.g. fertilisers and pesticides)
- using controlled burning.

Many successions are held in check and are prevented from advancing further by human management. The final stage, which may be grassland or heather moorland, for example, is referred to as a **plagioclimax**. In the UK, most of our vegetation represents a plagioclimax of one sort or another.

NOTING ACTIVITIES

1 Define the following terms:

- primary succession
- secondary succession
- plagioclimax.

2 In what ways can people interrupt natural vegetation successions? Try to suggest some actions not referred to in (**1**).

Primary succession case studies

1 Hydrosere

A hydrosere develops at the edges of a freshwater lake or pond (Figure 3.13). In a fluvial environment, a hydrosere might develop in an oxbow lake.

Figure 3.14 describes the stages in the development of a hydrosere in the UK. The pioneer species are the submerged plants which can survive completely under water. Other plants, such as water lilies, can survive in shallow water with their roots submerged, but their flowering parts on the water surface. These water-based plants collect sediment between their roots and add organic matter to the sediment that collects on the pond floor. As the water becomes shallower, other plant species such as reeds and bullrushes emerge. They trap yet more sediment, eventually causing it to break the surface to form new land. The succession continues with sedges, shrubs and trees (willows and alders) gradually occupying drier land.

3.13 *A hydrosere*

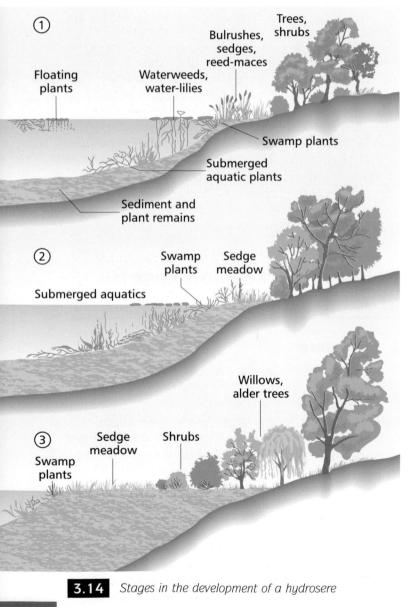

3.14 *Stages in the development of a hydrosere*

Redgrave and Lopham Fen

Redgrave and Lopham Fen in Suffolk is one of the most important wetland sites in Europe, both for its ecological interest and as one of only two known homes of the very rare Great Raft Spider. Over the years, the fen has suffered historical damage. This has been mainly due to the abstraction of water by the local authority to supply the local community, and to land drainage works carried out in the 1960s. There has also been insufficient fen management. The Environment Agency, together with Suffolk Wildlife Trust, Essex and Suffolk Water Company, and English Nature are now all working to restore the fen to its natural state.

The total cost for this will be £3.2 million, and will be partly met by a £1.4 million grant from the European Union LIFE fund.

Before restoration

As the fen was drained, the surface layers of peat rotted and released all their stored nutrients. Many of the rarer plants and animals could not live in the tall, rank fen that developed. Bushes and trees then encroached, and changed the fen from an open species-rich wetland into enclosed scrubland.

After restoration

Essex and Suffolk Water Company, which own the borehole, are looking for another site, and diggers will strip away the old, rotted peat and stack it up to make causeways and footpaths. The scrubby trees will then be taken out to re-create the open fen. Cattle and wetland ponies will be introduced to graze the fen and re-create the habitats that made Redgrave and Lopham internationally famous.

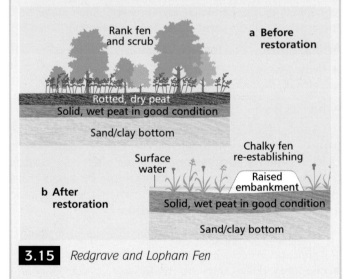

3.15 *Redgrave and Lopham Fen*

Source: *The Guardian, 24 September 1996*

NOTING ACTIVITIES

1 What is so special about Redgrave and Lopham Fen?

2 Describe the causes and the effects of poor management of the area.

3 How has the fen been restored?

The development of a hydrosere in Le Sauget meander, France

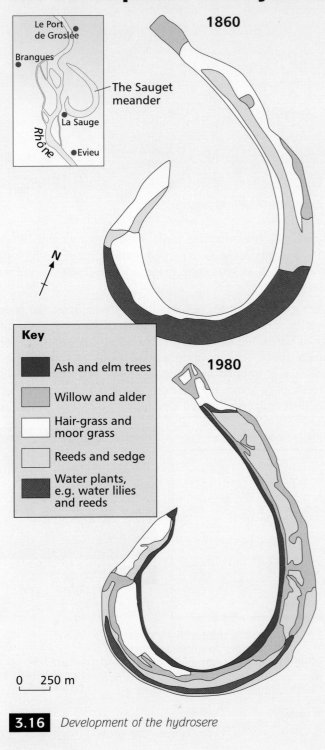

1860

1980

Key

- ■ Ash and elm trees
- ▨ Willow and alder
- □ Hair-grass and moor grass
- ▢ Reeds and sedge
- ■ Water plants, e.g. water lilies and reeds

0   250 m

3.16   *Development of the hydrosere*

Le Sauget meander is in fact an oxbow lake that was cut off in 1690. It is one of a number of oxbow lakes in a highly sinuous stretch of the upper course of the River Rhône in France, near Brangues. As the oxbow lake has gradually silted up over the centuries, a clear vegetation succession has developed, and the oxbow lake forms an excellent example of a hydrosere.

Look carefully at Figure 3.16, and read the following extract about the ecology of the meander.

'In the deepest part of the oxbow lake hydrophytic plants (water plants) such as yellow water lilies are well represented. The terrestrial (land) sequence begins with the development of reeds and sedges. The roots of these plants produce a sort of mattress which develops horizontally, reducing the water surface area. Reeds and sedges can survive in or out of water, and are common in swampy or marshy environments. Over time, silt builds up around their roots, creating ideal conditions for the subsequent species in the succession. Hair-grass and moor-grass occupy the drier land. These grasses are commonly found on wet heaths and moors where the water table fluctuates. In time, the grasses are succeeded by willow and alder, trees that are commonly associated with high water tables.

The margins of the meander have been aggraded by the significant deposition of silt over several centuries. The silt layer is up to 2 m thick overlying alluvial sands. These thick soils, well away from the centre of the oxbow and any last remnants of standing water, have been well colonised by ash and elm trees, which represent the climax vegetation.'

NOTING ACTIVITIES

1   Make a list, in the form of a flow diagram, of the seral stages in this vegetation succession. For each stage, identify the common vegetation species and describe how they are suited to the conditions.

2   Describe the changes in vegetation type and distribution in Le Sauget meander between 1860 and 1980.

3   If the meander is left alone completely (i.e. there is no management), suggest what will happen to the vegetation in the future.

2 Halosere

A halosere develops where there are high concentrations of salt, for example a coastal saltmarsh (Figure 3.17).

Figure 3.18 shows a cross-section through a saltmarsh on the coast of Georgia, USA. The different species represent different stages of the halosere, with the pioneers to the left and the climax vegetation to the right. Saltmarsh plants are adapted to a harsh, semi-aquatic environment and saline soils. Stout stems, small leaves and special adaptations for salt excretion characterise the inhabitants of the tidal creeks and low marshes. The pioneer species include eelgrass

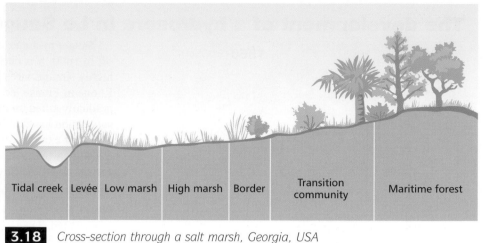

| Tidal creek | Levée | Low marsh | High marsh | Border | Transition community | Maritime forest |

3.17 *Coastal salt marsh vegetation*   **3.18** *Cross-section through a salt marsh, Georgia, USA*

whose aquatic roots help to stabilise the silt and trap debris. Cordgrass grows on the drier levées where conditions are less saline. Inland the high marsh is much drier and has a more sandy, well-drained soil. The conditions are very salty still, and plants tend to be stunted. The border is less salty because the tide rarely reaches it. As a result, there is a greater diversity of plants, which include marsh lavender and rushes. Finally, shrubs and then trees occupy the drier, less salty soils inland.

3 Psammosere

A psammosere is a vegetation succession that develops on sand, most commonly coastal sand dunes.

Sand dunes are an extremely hostile environment for plants. They are very dry (sand drains rapidly) and are exposed to strong salty winds which constantly re-work the sand and cause the dunes to change their shape and position. Sand is naturally infertile and soils here develop very slowly indeed.

Figure 3.19 is a cross-section through sand dunes at Ynyslas, just north of Aberystwyth, Dyfed. As with Figure 3.18, the succession can be seen as progressing with distance inland. The sand adjacent to the sea, where embryo dunes are

forming, represents the most harsh environment. Whilst the majority of this zone comprises bare sand, some plants, such as sea couch grass, are salt-tolerant and have deep roots.

Perhaps the most widespread plant, found on the more mature sand dunes, is **marram grass** (Figure 3.20). Marram grass has curled leaves to prevent water loss and it has very deep roots to tap underground water. As wind-blown material is trapped by the grass, and dead leaves provide nutrients to the newly developing soil, other species begin to take hold. As the soil develops further, low plants are succeeded by shrubs and finally trees.

3.20 *Marram grass on dunes*

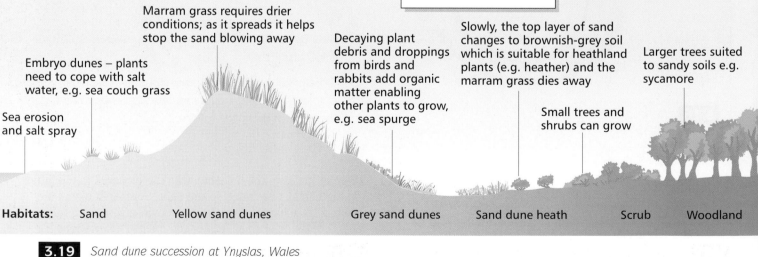

Marram grass requires drier conditions; as it spreads it helps stop the sand blowing away

Embryo dunes – plants need to cope with salt water, e.g. sea couch grass

Sea erosion and salt spray

Decaying plant debris and droppings from birds and rabbits add organic matter enabling other plants to grow, e.g. sea spurge

Slowly, the top layer of sand changes to brownish-grey soil which is suitable for heathland plants (e.g. heather) and the marram grass dies away

Small trees and shrubs can grow

Larger trees suited to sandy soils e.g. sycamore

| Habitats: | Sand | Yellow sand dunes | Grey sand dunes | Sand dune heath | Scrub | Woodland |

3.19 *Sand dune succession at Ynyslas, Wales*

NOTING ACTIVITY

Describe the typical vegetation successions associated with a:

- hydrosere
- halosere
- psammosere.

For each one, you should:

- concentrate particularly on the characteristics and role of the pioneer species
- draw a simple sketch or series of sketches to illustrate the succession
- suggest ways in which human actions may interrupt the succession.

EXTENDED ACTIVITY

Managing Ynyslas sand dunes

This extended activity is all about managing the sand dunes at Ynyslas. Sand dunes are important natural habitats and they are fragile ecosystems which can be easily damaged.

The aim of this activity is for you to make a study of the management issues at Ynyslas.

Read the sections below and then attempt the questions that follow. As an alternative, you could use the questions as a rough guide and write your own neat report, structured as you wish.

Why does Ynyslas need managing?

The sand dunes are a very fragile environment and can be easily damaged by natural processes or by human use. Occasionally strong winds can cause **blowouts** (Figure 3.21).

3.21 *Ynyslas dunes, Wales, showing a blowout*

These lead to erosion of the dunes and can cause damage to both vegetation and wildlife. As the sand dunes are popular with tourists, areas may become trampled, and this can lead to vegetation dying, sand dunes becoming eroded, and wildlife habitats being destroyed.

There are some 300 different flowering plants on the dunes, many of which only occur in such special conditions. There are orchids, marsh hellebore, wild thyme and evening primrose, to name but a few. Mammals such as foxes, rabbits, hedgehogs and stoats are found, as are many birds including ringed plovers, shelducks, meadow pipits, yellowhammers and stonechats. Kestrels, owls and green woodpeckers feed regularly on the dunes, though they have their nests elsewhere.

The wide variety of habitats provides a diversity of conditions. Without careful management, habitats can be destroyed and the ecology of the area will suffer.

What does management involve?

The Ynyslas sand dunes are part of the Dyfi National Nature Reserve. The Reserve is managed by the Countryside Council for Wales, and there is a full-time warden here.

Management of the dunes involves:

- conserving the dunes and the natural habitats
- enabling visitors to enjoy the dunes and to understand the natural environment without causing harm to it.

People are encouraged to explore the area, but there are also a number of clearly marked paths to enable them to walk directly from the car park to the beach, which is what most people want to do. Some of these paths involve sections of boardwalk raised above the dunes so that the dunes are not trampled (see Figure 3.22). Areas that are particularly vulnerable or are being restored, such as where a blowout has occurred (see Figure 3.23), are cordoned-off, with signs

3.22 *Boardwalk over sand dunes at Ynyslas*

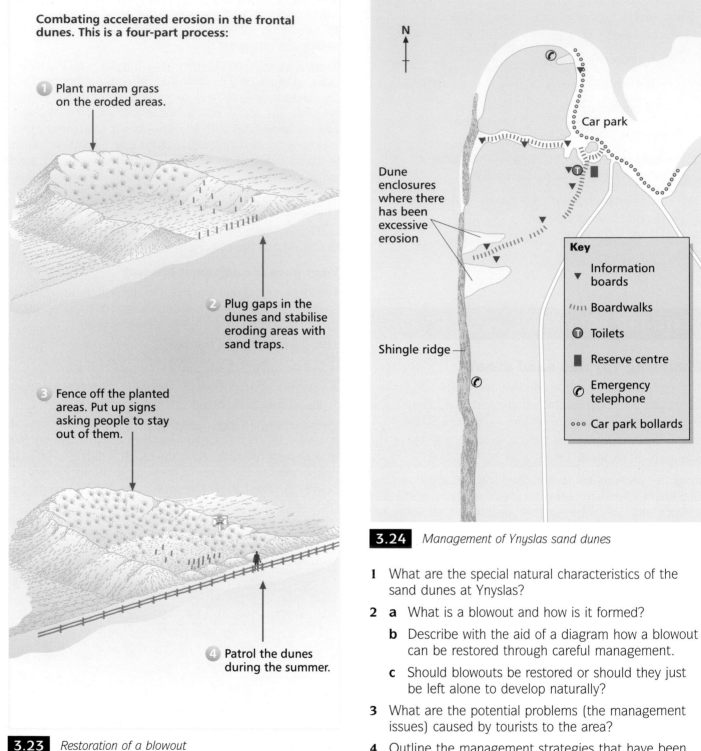

Combating accelerated erosion in the frontal dunes. This is a four-part process:

1 Plant marram grass on the eroded areas.

2 Plug gaps in the dunes and stabilise eroding areas with sand traps.

3 Fence off the planted areas. Put up signs asking people to stay out of them.

4 Patrol the dunes during the summer.

3.23 *Restoration of a blowout*

asking people to keep off. Information boards explain to visitors why it is important to leave these areas alone.

Management at Ynyslas is all about trying to keep a balance between the natural environment, with its diversity of habitats, plants and animals, and the desire of people to use the sand dunes for leisure and recreation. Figure 13.24 shows some of the management measures at Ynyslas.

Car park

Dune enclosures where there has been excessive erosion

Shingle ridge

Key

▼ Information boards

▨ Boardwalks

🅣 Toilets

■ Reserve centre

🕿 Emergency telephone

∘∘∘ Car park bollards

3.24 *Management of Ynyslas sand dunes*

1 What are the special natural characteristics of the sand dunes at Ynyslas?

2 a What is a blowout and how is it formed?

 b Describe with the aid of a diagram how a blowout can be restored through careful management.

 c Should blowouts be restored or should they just be left alone to develop naturally?

3 What are the potential problems (the management issues) caused by tourists to the area?

4 Outline the management strategies that have been employed to minimise the damage done by tourists whilst, at the same time, maximising their accessibility and enjoyment of the sand dunes.

5 Suggest other management strategies that could be adopted for the area and outline their advantages and disadvantages.

STRUCTURED QUESTION 1

Study Figure 3.25.

a Define the term 'sere'. *(1)*

b What type of sere is this? *(1)*

c The graph (Figure 3.26) shows the biomass changes, at point A on Figure 3.25, since the glacier retreated past this point in 1915.

   (i) Define the term 'biomass'. *(1)*

   (ii) Explain the increase in biomass shown in Figure 3.26. *(2)*

d What are the first types of vegetation to colonise the area after deglaciation? *(1)*

e The area around point A on the diagram is now dominated by trees such as silver birch. Give *one* reason why the trees have taken over as the main vegetation type at point A. *(2)*

f The forest developed at the right-hand side of the diagram is often referred to as climax vegetation.

   (i) Define the term 'climax vegetation'. *(1)*

   (ii) Give *one* reason why the climax vegetation at point A might not become permanent. *(2)*

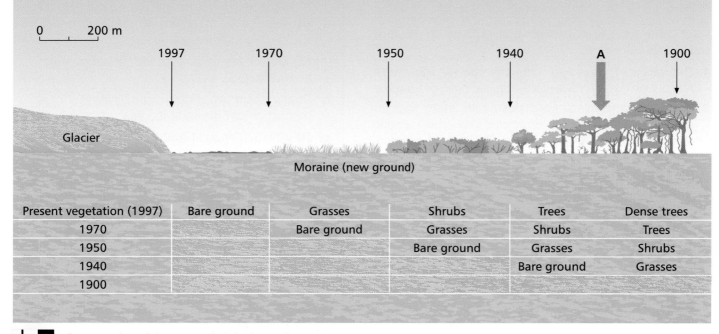

Present vegetation (1997)	Bare ground	Grasses	Shrubs	Trees	Dense trees
1970		Bare ground	Grasses	Shrubs	Trees
1950			Bare ground	Grasses	Shrubs
1940				Bare ground	Grasses
1900					

5   *Cross-section of the snout of a glacier and the sere developed on morainic deposits left behind as the glacier retreated (towards the left of the diagram)*

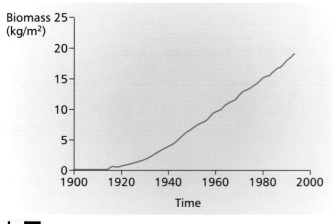

6   *Changes in biomass at point A*

Secondary succession: the heather moorland plagioclimax

The vast purple panoramas that typify the moorlands of Britain (Figure 3.27) create an impression of a completely natural landscape. However, this could not be further from the truth. The characteristic purple flowering heather (*Calluna vulgaris*) is not a natural climax vegetation. It is maintained entirely by human management, which is why it is an excellent example of a plagioclimax.

The natural succession in this environment (see Figure 3.28) results in an oak forest climax vegetation. Notice in Figure 3.28 that heather hardly features at all in any of the seral

89

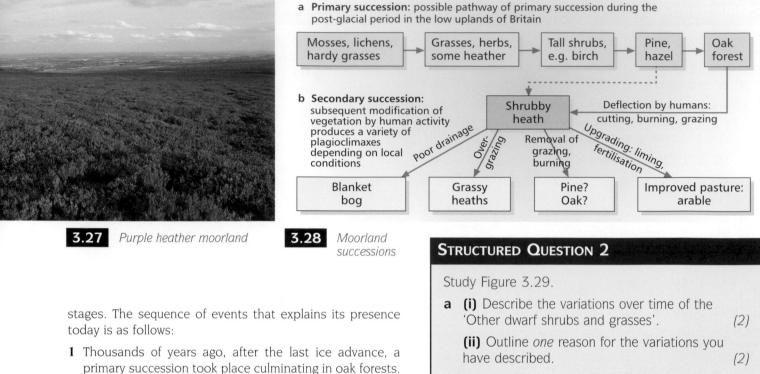

a Primary succession: possible pathway of primary succession during the post-glacial period in the low uplands of Britain

Mosses, lichens, hardy grasses → Grasses, herbs, some heather → Tall shrubs, e.g. birch → Pine, hazel → Oak forest

b Secondary succession: subsequent modification of vegetation by human activity produces a variety of plagioclimaxes depending on local conditions

Shrubby heath

Deflection by humans: cutting, burning, grazing

Poor drainage — Over-grazing — Removal of grazing, burning — Upgrading: liming, fertilisation

| Blanket bog | Grassy heaths | Pine? Oak? | Improved pasture: arable |

3.27 *Purple heather moorland*

3.28 *Moorland successions*

stages. The sequence of events that explains its presence today is as follows:

1 Thousands of years ago, after the last ice advance, a primary succession took place culminating in oak forests.

2 Gradually the forests were cleared to make way for agriculture and other developments. The soils became impoverished and highly infertile. The lack of nutrients and the acidic nature of the soils were, however, ideal for the growth of heather.

3 Young shoots of heather form good grazing for sheep, and the moorlands were widely used by hill farmers. The heather was managed by farmers to maintain its presence, thereby arresting the further development of the secondary succession. Indeed, if the heather were not managed, it would soon be replaced by shrubs and eventually trees (Figure 3.28).

4 Today, through a programme of grazing and periodic burning (to remove the old wood and encourage new shoots), the heather is maintained as a plagioclimax.

3.29 *Some of the main changes during the four-phase* Calluna vulgaris *(heather) development cycle*

STRUCTURED QUESTION 2

Study Figure 3.29.

a (i) Describe the variations over time of the 'Other dwarf shrubs and grasses'. *(2)*

(ii) Outline *one* reason for the variations you have described. *(2)*

b During which two stages is the biomass production of *Calluna* at its greatest? *(1)*

c (i) When, in relation to the four stages, would it be best to burn off the heather? *(1)*

(ii) Give reasons for your answer. *(3)*

Figure 3.30 shows the movement of nutrients in the *Calluna vulgaris* system.

d (i) How does grazing provide an input of nutrients? *(1)*

(ii) How does burning the heather affect nutrient flows? *(2)*

(iii) Outline the importance of water in the nutrient system. *(3)*

e If there was a further decline in sheep farming, and maintenance ceased, what would happen to the heather and why? *(4)*

	Stage of development cycle			
	1 Pioneer	**2** Building	**3** Mature	**4** Degenerate
Mean height (cm)	24.1	52.1	63.2	55.2
Mean age of individuals (years)	5.7	9.0	17.1	24.0
Biomass (g/m^2)				
(a) *Calluna* only	287.2	1507.6	1923.6	1043.2
(b) Other dwarf shrubs and grasses	179.6	41.2	52.0	83.2
(c) Total, all plants	889.2	1702.0	2305.2	1560.8
Net production of young *Calluna* shoots (g/m^2 in one year)	148.8	442.4	363.6	140.8
Light reaching soil surface (% of that in the open)	100.0	2.0	20.0	57.0

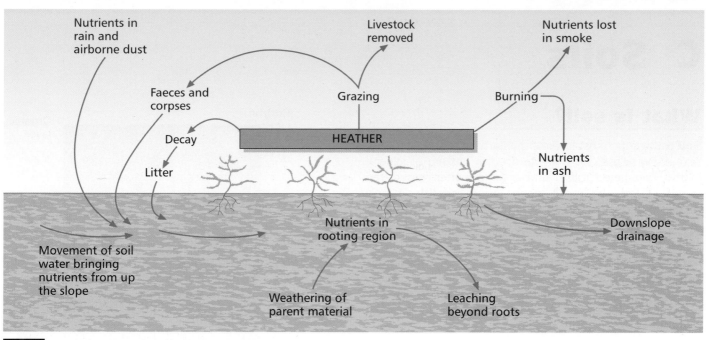

3.30   *Nutrient gains and losses in the* Calluna vulgaris *system*

Human interference in Mediterranean lands

The transitional nature of the climate and the mountainous terrain of much of the Mediterranean has produced a fragile and unstable natural vegetation. The rainfall regime, which fluctuates between droughts and sudden downpours, increases soil erosion. This has helped produce a flora that is four times more diverse than that found in temperate forests.

Little natural vegetation remains, with 75 per cent of the upland vegetation of Mediterranean Europe being created by human actions. There is a long history of human exploitation (10 000 years or more). The natural climax forests of oak and pine, which should cover 80 per cent of the uplands, have been cleared to provide timber for fuel and building materials or to open up new grazing land. The clearance makes the microclimate more extreme, exacerbating the soil erosion problems.

In many places the natural vegetation is undergoing regression (see Figure 3.31) with the rate of decline being dependent on local factors such as topography, climate, soil erosion or human interference. The overall change is from a mesophytic to a xerophytic vegetation; that is, from plants that grow in an averagely moist environment to those that grow in dry surroundings. Much of the Mediterranean land is covered by maquis (scrub). If this is left undisturbed, it will regenerate through natural succession to forest, but if it is overgrazed or otherwise badly managed, it will degenerate into bare soil. In some places, where the critical threshold has been passed, the vegetation will never be able to recover.

One area that has been badly hit by human interference is Corsica. The original forests were used in charcoal and tannin production, and pigs have grazed the young saplings. Other land has been cleared for agriculture, aggravating soil erosion on even the gentlest slopes.

NOTING ACTIVITIES

1   What would be the natural climax vegetation in the upland parts of the Mediterranean?

2   In what ways have people interfered with the succession?

3   What have been some of the consequences of human interference?

4   Comment on the statement that this area has a 'fragile and unstable natural vegetation'.

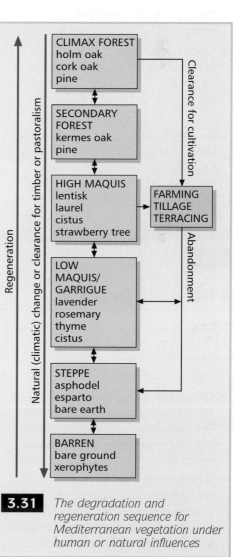

3.31   *The degradation and regeneration sequence for Mediterranean vegetation under human or natural influences*

C Soils

What is soil?

Soil is the uppermost layer of the Earth; it is frequently less than a metre thick. It is made up of a mixture of mineral matter (weathered rocks) and organic matter (e.g. rotted leaves). Weathered rocks alone forms **regolith**: it only becomes a soil when organic matter is added to it.

Soil is clearly essential to life on Earth. It provides the medium for growing food and grazing animals. Soil takes thousands of years to form yet it can be destroyed in a matter of minutes by **soil erosion** (Figure 3.32). Soils need to be managed carefully to ensure sustainability. Overuse of soils can mean they become infertile and can change their structure, making them more vulnerable to wind and water erosion.

3.32 *Soil erosion*

The soil profile

A fully developed soil displays a sequence of layers. Each layer is called a **horizon** and is usually distinguished from its neighbours by being a different colour. Study Figure 3.33. Notice that there are four horizons:

- **O horizon** This is the organic layer. It is made up of leaf litter at various stages of decomposition. Organic matter that has been completely decomposed forms a black, jelly-like substance called **humus**. It contains a lot of valuable nutrients and is an important constituent of soil.

- **A horizon** This is topsoil. It is a combination of weathered rock material from below and organic matter from above. These two constituents are brought together and mixed up by the actions of organisms living within the soil, particularly earthworms. The A horizon is usually the most fertile soil horizon.

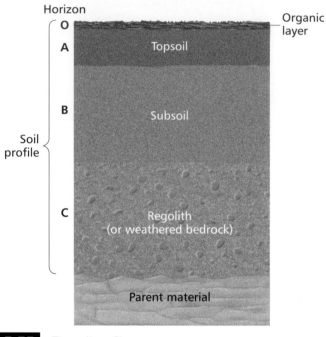

3.33 *The soil profile*

- **B horizon** The subsoil is generally less fertile because it is further from the source of organic matter. It contains a higher proportion of weathered rock material.

- **C horizon** This is the weathered rock material which provides the soil with its mineral matter. Beneath the weathered rock, or regolith, is the unweathered **parent material**. Parent material is commonly solid rock, such as limestone, but it can include deposits such as sands and gravels.

NOTING ACTIVITIES

1 *What is the difference between a regolith and a soil?*

2 *Make a copy of Figure 3.33 and add labels to describe the main characteristics of the four soil horizons.*

What are the properties of soil?

1 Soil composition

Whilst soils are primarily composed of mineral and organic matter, they contain two other important constituents: air and water. Figure 3.34 shows the average composition of an agricultural soil in the UK. Notice that approximately

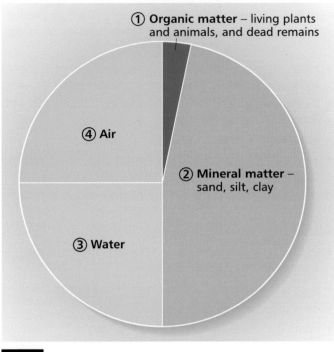

3.34 *Soil composition*

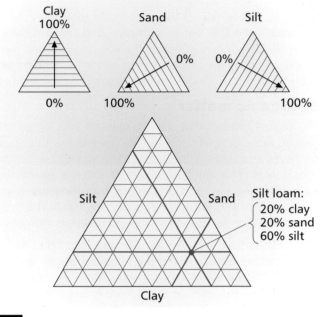

3.35 *Plotting soil texture on a soil texture triangle*

Soil texture	% sand	% silt	% clay
Sand	90	5	5
Loamy sand	85	10	5
Sandy loam	65	25	10
Loam	45	40	15
Silt loam	20	60	20
Clay loam	28	37	35
Clay	25	30	45

3.36 *Common soil textures*

50 per cent is air and water. The higher the proportion of air and water in a soil, the greater the proportion of pore spaces within the soil. Pore spaces are very important because oxygen and water promote plant growth.

2 Soil texture

The texture of a soil describes the arrangement and the sizes of the separate soil particles. Soil texture is determined by the presence of sand, silt and clay-sized particles that have been derived from the parent material. It is the different proportions of these three particle types that determines the texture of a particular soil.

Study Figure 3.35. It is a triangular graph called a **soil texture triangle**. As Figure 3.35 shows, it is possible to locate the texture of any soil sample on this graph.

There are a number of recognised textural types (Figure 3.36):

- Sandy soils are well drained because their large particles create large pore spaces which promote water flow. However, sandy soils are not very fertile.
- Clay soils are much less well drained and can suffer from waterlogging because the small particles fit more closely together. However, they are often fertile because clay particles have the ability to attract chemical elements (nutrients) in the soil.
- Loamy soils, which have a texture roughly midway between a sandy soil and a clay soil, are generally regarded as being of a high quality. They allow some drainage to occur and usually have a high fertility.

Soil texture is largely determined by the nature of the parent material. Rocks like sandstone and granite tend to produce more sand-sized particles than, say, a shale or mudstone which produces higher proportions of clay.

3 Soil structure

Soil structure describes the way in which the individual particles of soil group together to form lumps called **peds**. There are many different types of soil structure (e.g. blocky, prismatic, platy), but the most highly valued by farmers is known as **crumb**. This consists of small, rounded peds that are a good balance of mineral and organic matter. Crumb soils are well drained and form an ideal growing medium for crops.

4 Soil moisture

Water is held in the pore spaces in a soil. After a rainstorm, water infiltrates into the soil and occupies the pore spaces. It then passes down through the soil in response to the force of gravity. When water has stopped flowing through the soil, it is said to be at its **field capacity**. The water that is left behind in the soil is held by tension or a form of 'suction', rather as liquid is held against the inside of a glass bottle.

Plants will make use of the water held loosely in the soil. However, once this water has been used up, and the only water that is left is held fast to the soil particles and cannot be drawn-up by the plants, a state known as **wilting point** is reached. If this condition persists then the plants will die.

5 Organic matter

The organic matter that turns a regolith into a true soil is derived from material deposited on the ground surface, such as leaves and twigs. This material is gradually broken down by soil organisms, such as fungi and bacteria, eventually to form black humus. The process of **humification** is most rapid in warm and wet environments, such as in the tropics, because the decomposers thrive under such conditions.

Study Figure 3.37. It shows the different layers that make up the O horizon. Notice that the A horizon beneath is characterised by being a dark brown colour. This is because a high proportion of this horizon is made up of organic material.

Humus is important for a number of reasons:

- it provides the soil with nutrients, which are released when the organic material is broken down by the soil organisms
- it helps absorb and retain moisture in a soil
- it improves the structure of the soil and helps to bind soil particles together.

The importance of organic matter explains why gardeners apply rotted compost or manure to flowerbeds and vegetable plots in the autumn or the spring.

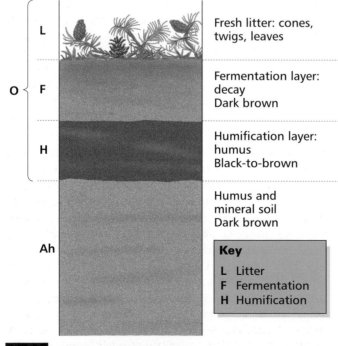

L	Fresh litter: cones, twigs, leaves
F	Fermentation layer: decay Dark brown
H	Humification layer: humus Black-to-brown
Ah	Humus and mineral soil Dark brown

Key
L Litter
F Fermentation
H Humification

3.37 *The O (organic) horizon*

What makes a soil fertile?

A fertile soil is one that is well suited to give a high productivity of agricultural crops. There are a number of factors that combine to result in a fertile soil:

- a high organic content which provides plenty of nutrients and a firm structure to support growing plants
- a balanced mineral content, producing a loamy soil with good drainage and aeration
- a crumb structure
- a neutral or slightly acidic pH value which usually reflects a high nutrient and organic matter content.

Farmers can improve the fertility of their soil in a number of ways:

- using chemicals to add nutrients
- improving soil structure and texture by, for example, adding sand to improve drainage
- altering the acidity by adding manure (acid) or lime (alkaline)
- deep ploughing to mix up the soil
- rotating crops to retain a balance of nutrients; peas, for example, are known as nitrogen-fixing plants because they add nitrogen to the soil.

There is, however, some concern about the excessive use of chemicals and the increasing trend towards intensive agriculture. Chemicals are known to have potentially harmful side-effects to wildlife and can lead to the pollution of water sources (see page 264). There is an increasing desire amongst some consumers to buy 'organic', which involves produce grown or reared without the use of artificial chemicals.

NOTING ACTIVITIES

1 On a copy of the soil texture triangle (Figure 3.35), plot the position of the main soil textures using the information in Figure 3.36.

2 Explain the meaning of the terms 'field capacity' and 'wilting point'.

3 a What is humus?

   b Describe how humus is formed.

   c Why *does* humus form most rapidly in warm, moist environments?

   d Why is humus an important component of soils?

4 How can farmers improve the following soil properties:

   • soil structure

   • soil fertility?

5 Assess the arguments for and against the increased use of chemicals in farming.

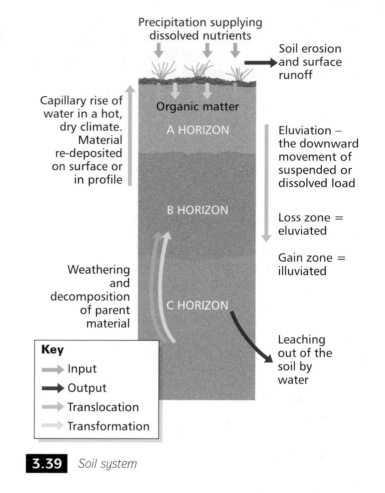

3.39   *Soil system*

STRUCTURED QUESTION 1

Figure 3.38 shows the average values for the composition of soil.

a   Name the constituent that makes up an average of 5 per cent of the soil composition.   *(1)*

b   What is the main source of the mineral matter in the soil?   *(1)*

c   Suggest why the line used in the graph to separate water and air was not drawn as a straight line.   *(2)*

d   **(i)** What is meant by 'soil texture'?   *(1)*

   (ii) Why is soil texture so important in determining the agricultural potential of a soil?   *(4)*

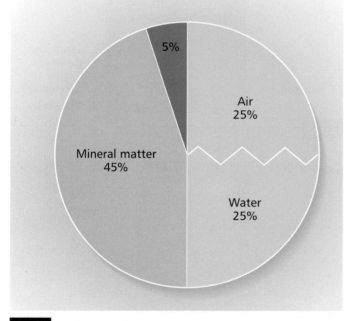

3.38   *Soil composition*

What processes are responsible for forming soils?

Study Figure 3.39, which shows the soil system. Notice that there are four sets of processes in operation:

- inputs   –   these include the weathered parent material and the organic matter
- outputs   –   mainly water and sediment
- transformations –   processes involving change, for example the decomposition of organic matter
- translocations   –   the movement of material through the soil profile.

Notice on Figure 3.39 that a great number of the processes involve the movement of water. Water flow in soils accounts for five very important processes:

1   **Leaching** Rainwater is mildly acidic and, as it passes through the soil, it will dissolve some of the nutrients and transport them downwards. The result of leaching is to leave the A horizon impoverished and lacking in fertility. As most of the dissolved chemicals are bases (alkalis), such as potassium and calcium, the eluviated A horizon can become quite acidic. Leaching is most common when there is heavy rainfall, low rates of evapotranspiration and a well drained soil.

2   **Podsolisation** This is an extreme form of leaching and forms a particular type of soil called a **podsol**. It occurs when the water seeping through the soil is strongly acidic, having picked up organic acids as it passes through the O horizon. It is common beneath coniferous woodland because the needles tend to be very acidic and decompose only slowly. Heavy rainfall and extremely well-drained soils (sands and gravels) promote intense leaching which is capable of dissolving the less soluble elements such as iron and aluminium (collectively called **sesquioxides**). Podsolisation results in very clearly defined horizons (see Figure 3.40), with a pale eluviated A horizon and a reddy-brown, illuviated B horizon, where the sesquioxides have been re-deposited (Figure 3.41). If

a concentrated layer of re-deposited iron forms, it is called an **iron pan**. This will slow down drainage through the soil and may, in fact, lead to waterlogging.

3 **Mechanical downwash** As water drains through the soil, it is capable of carrying with it tiny particles of humus or clay. The transportation of humus will affect the colour of the soil and the clay will affect its texture.

4 **Gleying** If a soil is waterlogged, for example on a floodplain close to a river, it is subjected to anaerobic processes (without oxygen). The process of reduction will cause any iron compounds in the soil to form a grey-coloured clay. Where oxygen remains, for example along plant roots or in animal burrows, oxidation of the iron compounds will result in red 'blotches'. **Gleys** are a common type of soil, and often have a mottled appearance resulting from these isolated pockets of oxygen (see Figure 3.42). Iron pans are also common features of gleys.

5 **Calcification** In drier climates, high rates of evaporation may draw soil water towards the surface. This can result in calcium carbonate being drawn up from parent material to be deposited as nodules in the B horizon. This process can result in good-quality soils which are especially well suited to cereal cultivation.

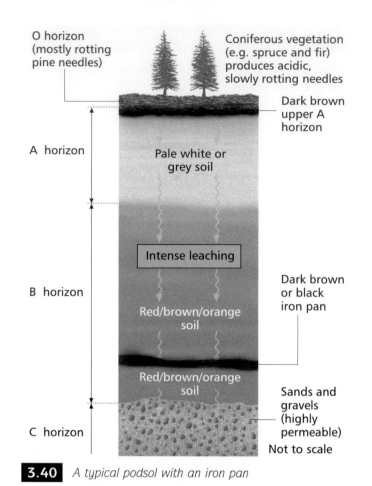

O horizon (mostly rotting pine needles)

Coniferous vegetation (e.g. spruce and fir) produces acidic, slowly rotting needles

Dark brown upper A horizon

A horizon

Pale white or grey soil

Intense leaching

B horizon

Red/brown/orange soil

Dark brown or black iron pan

Red/brown/orange soil

Sands and gravels (highly permeable)

C horizon

Not to scale

3.40 *A typical podsol with an iron pan*

a the surface vegetation: a forest of mixed black and white spruce

b a soil profile

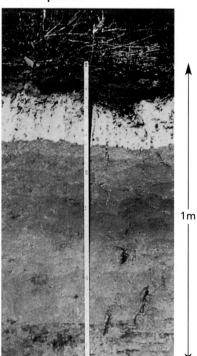

1m

3.41 *A typical podsol, Saskatchewan, Canada*

3.42 *Gley soil, showing mottled features*

NOTING ACTIVITIES

1 What is the difference between:

- eluviation and illuviation

- translocations and transformations?

2 Why does leaching often result in an impoverished A horizon?

3 Study Figure 3.40.

a How does the vegetation type encourage the process of podsolisation?

b What is the parent material and how has it encouraged the formation of the podsol?

c Explain the formation of the iron pan.

d How does a cool, wet climate promote podsolisation?

4 Make a careful sketch of the soil profile in Figure 3.41.

a Draw a box roughly the same shape as the photograph.

b Use a pencil to draw the boundaries of the different horizons. Refer to Figure 3.40 to help you.

c Draw the dark brown blotches concentrated towards the base of the soil profile.

d Now use Figure 3.40 to help you add as many labels as you can to your sketch.

e Complete your sketch by adding a scale and a title.

5 How would you expect mechanical downwash to affect:

- the colour of a soil profile if humus was translocated

- the texture of a soil profile if clay was translocated?

6 Study Figure 3.42. Describe and account for the characteristics of a gley soil.

STRUCTURED QUESTION 2

Figure 3.43 shows the sequence of vegetation and soils on the shore of southern Lake Michigan, USA. This is a freshwater lake.

a What is the term used to describe the sequence of vegetation changes shown along the cross-section?
(1)

b Outline *one* reason why there are no horizons in profile X.
(2)

c Identify *three* factors that have contributed to the development of the podsol, Profile Y.
(3 x 2)

d Explain why the vegetation in the depressions between the dunes (i.e. slacks) changes significantly with distance from the lake shore.
(3)

e Suggest *two* ways in which human activity might modify the development of the sequence of vegetation change shown here.
(4)

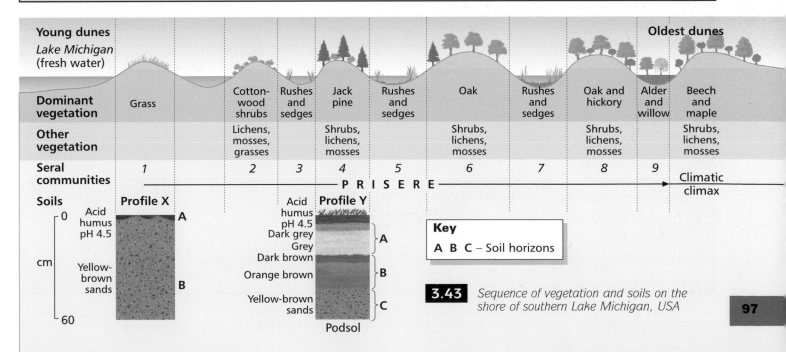

3.43 *Sequence of vegetation and soils on the shore of southern Lake Michigan, USA*

What factors affect soil development?

Several factors affect the development and characteristics of soil:

1 **Climate** Rainfall is a very important influence because a large number of soil processes are water-related, for example leaching and podsolisation. In wet climates water movement tends to be downwards, whereas in a dry climate water is drawn towards the surface (calcification). Warm and wet conditions promote the decomposition of organic matter.

2 **Parent material** The type of parent material affects the acidity, fertility and texture of a soil. For example, a drained but not very nutritious soil. An igneous rock, such as basalt, weathers to produce a clay soil that is generally not so well drained.

3 **Organisms** Soil-living organisms are important in decomposing organic material and in mixing up the components of a soil. Organisms are most abundant in warm and moist environments. They do not like very acidic conditions, such as beneath coniferous woodland.

4 **Relief** The slope of the land affects drainage. Flat areas tend to be poorly drained and prone to waterlogging and the formation of gleys. A steep slope will promote water flow through a soil and thus encourages leaching. The sequence of variation in soil types down a slope is called a **soil catena** (Figure 3.44).

5 **Time** Soils take thousands of years to become fully developed with clear horizons. In the UK, mature soils probably take up to 10 000 years to form – this is the length of time since the last ice advance which scoured the existing soil off many parts of upland Britain. It is the very slow nature of soil formation that makes careful management essential to avoid soil erosion.

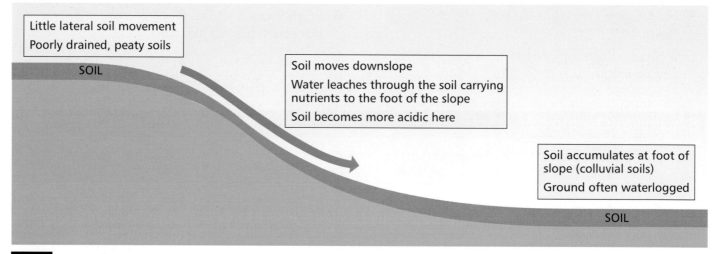

Little lateral soil movement
Poorly drained, peaty soils

SOIL

Soil moves downslope
Water leaches through the soil carrying nutrients to the foot of the slope
Soil becomes more acidic here

Soil accumulates at foot of slope (colluvial soils)
Ground often waterlogged

SOIL

3.44 *A soil catena*

STRUCTURED QUESTION 3

Figure 3.45 shows a soil catena which has developed on the slopes of Bennachie Hill, Aberdeenshire. The parent material is granite.

a What is meant by the term 'parent material'? *(1)*

b What is meant by a 'soil catena'? *(1)*

c Suggest why there is a bleached horizon in the soil at B. *(2)*

d In what way has the vegetation promoted the process of podsolisation at B? *(2)*

e **(i)** Contrast the thicknesses of the two soil profiles at B and C. *(2)*

(ii) Explain the difference that you have observed. *(2)*

f Why has a gley formed at C? *(1)*

g Account for the presence of a mottled grey clay horizon at C. *(2)*

h Suggest two reasons why the process of gleying has occurred at A. *(2)*

i The natural vegetation on Bennachie Hill may be described as a plagioclimax.

(i) What is meant by a 'plagioclimax'? *(1)*

(ii) Suggest how this plagioclimax may have developed. *(3)*

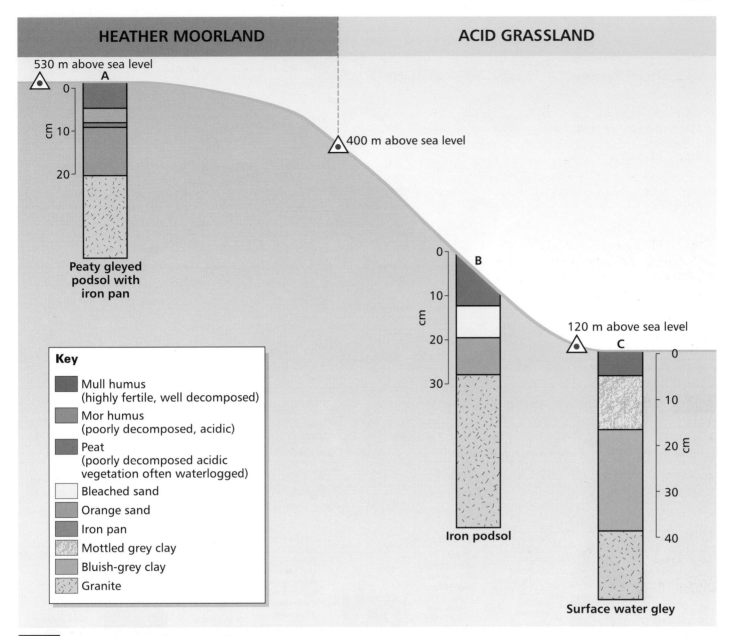

3.45   *Soils on Bennachie Hill, Aberdeenshire, Scotland*

EXTENDED ACTIVITY

Comparing world soils

There are many different types of soil. We have already come across two common soils, podsols and gleys. Podsols form in well-drained soils that are fed by very acidic water. On a global scale, podsols are commonly associated with the zone of coniferous forest that stretches across Canada, northern Europe and into Russia (Figure 3.46). Podsols are an example of a global **zonal soil**.

Gleys do not form a zonal soil. They tend to occur at a much more local scale in response to waterlogging.

In this activity you will compare three widespread zonal soils:

- brown earth (associated with deciduous forest environments)
- chernozem (associated with grasslands, such as the Prairies of North America)
- tropical latosol (associated with tropical rainforest environments).

Soil profiles for each of the soil types are shown in Figure 3.47. Their global distribution is indicated on Figure 3.46.

It is up to you to decide how to structure your comparison of the three soils. However, you should definitely consider the following questions:

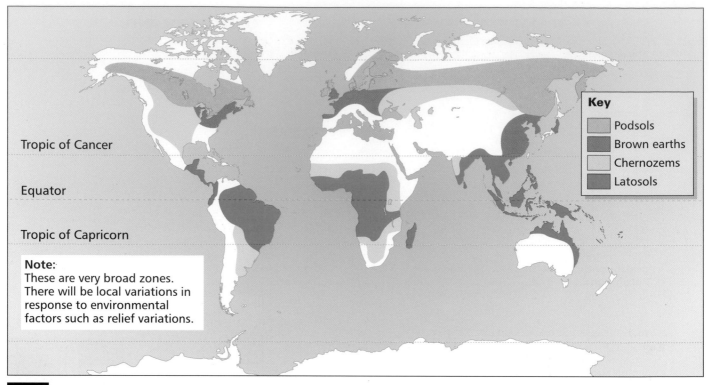

3.46 *Global distribution of selected zonal soils*

Key
- Podsols
- Brown earths
- Chernozems
- Latosols

Tropic of Cancer

Equator

Tropic of Capricorn

Note:
These are very broad zones. There will be local variations in response to environmental factors such as relief variations.

1 How do the soil profiles differ in their characteristics and in the nature of their horizons?

2 What soil-forming processes are in operation, and why?

3 What is the role of climate in the formation of each profile?

4 Each soil type is associated with a particular type of vegetation. Assess the importance of vegetation in the formation of each profile.

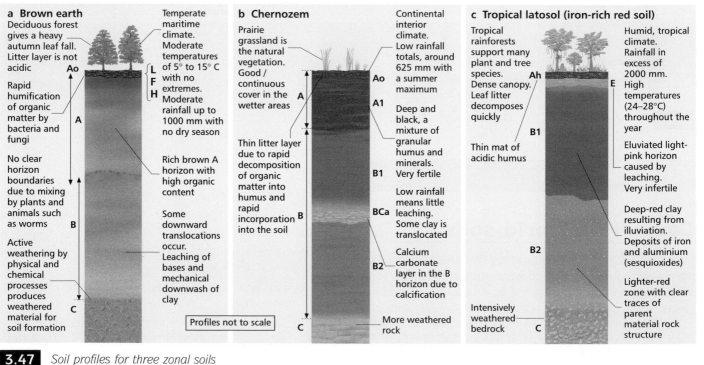

a Brown earth

Deciduous forest gives a heavy autumn leaf fall. Litter layer is not acidic

Rapid humification of organic matter by bacteria and fungi

No clear horizon boundaries due to mixing by plants and animals such as worms

Active weathering by physical and chemical processes produces weathered material for soil formation

Temperate maritime climate. Moderate temperatures of 5° to 15° C with no extremes. Moderate rainfall up to 1000 mm with no dry season

Rich brown A horizon with high organic content

Some downward translocations occur. Leaching of bases and mechanical downwash of clay

b Chernozem

Prairie grassland is the natural vegetation. Good / continuous cover in the wetter areas

Thin litter layer due to rapid decomposition of organic matter into humus and rapid incorporation into the soil

Continental interior climate. Low rainfall totals, around 625 mm with a summer maximum

Deep and black, a mixture of granular humus and minerals. Very fertile

Low rainfall means little leaching. Some clay is translocated

Calcium carbonate layer in the B horizon due to calcification

More weathered rock

Profiles not to scale

c Tropical latosol (iron-rich red soil)

Tropical rainforests support many plant and tree species. Dense canopy. Leaf litter decomposes quickly

Thin mat of acidic humus

Intensively weathered bedrock

Humid, tropical climate. Rainfall in excess of 2000 mm. High temperatures (24–28°C) throughout the year

Eluviated light-pink horizon caused by leaching. Very infertile

Deep-red clay resulting from illuviation. Deposits of iron and aluminium (sesquioxides)

Lighter-red zone with clear traces of parent material rock structure

3.47 *Soil profiles for three zonal soils*

A Population growth

On 12 October 1999, the population of the world reached 6 000 000 000 (6 billion). Just 40 years ago, it was half this number.

Read Figure 4.1. The article makes a number of points about the growth of the world's population:

Overcrowded world faces battle for scarce resources

John Vidal
Environment Editor

At 1.24 am New York time on 12 October the UN Secretary-General will declare there are 6 bn people alive. This should be taken with a pinch of salt because no one knows the real population, but the political point about massive population expansion will be well made.

The 5 billionth human alive today, after all, is not yet a teenager, the 4 billionth is just over 30, and the 2 billionth is still under 70. It took almost all human history to reach 1 billion people.

There will be no great celebration for the 6 billionth babe. The doubling of world population in less than 40 years has been deeply unequally divided. Ninety-seven of every 100 children born today are from the developing world where a combination of factors, mainly to do with extreme and growing poverty and lack of help from the rich, have worked against population control programmes.

Meanwhile, rich countries enjoy stable or declining numbers and some forecasters predict that within 50 years Europe will have a quarter fewer people and Japan 21 million fewer.

But the forecasters have consistently been proved wrong. Twenty years ago it was thought that there would be standing room only eventually on Earth; 10 years ago that, with numbers rising almost 100 million a year, world population would increase to 13 bn within 100 years. Today the increase has slowed to about 78 million a year, and the best estimates from the UN Population Fund are that, great natural and human disasters permitting, we will reach 8.9 bn in 50 years. By then, Africa will have three times as many people as Europe, and the USA will be the only developed country in the world's 20 biggest.

After that, there is growing debate. The UN now thinks population will level off in 120 years' time at about 11 bn, but others argue that because of long-term fertility falls, AIDS and growing wealth, humanity's long-term problem will be too few people.

What might life be like with almost twice as many people? Population increase is not a major problem in itself and, despite much gloomy hype about humanity being unable to feed itself in the future without giant technological breakthroughs, better management of agriculture, water supplies, land reform, fish farming and a move away from a western diet of meat, should be well able to feed the 11 bn. The problems, as today, will be in distribution and access to ever scarcer resources.

Increased population does not necessarily mean ecological crisis but it will not ease already stressed ecosystems and it is bound to add to competition, even conflict, for scarce resources such as fresh water, farming land, minerals and wood.

The rich have been shown to pollute and destabilise the Earth far more than the poor, but land degradation and desertification is a real and growing problem in many countries which can least afford to address them.

The effects of a warming climate over the next 50 years will have an impact mostly on poor populations.

But the 6 billionth baby, like most people, will have a bumpy ride in life. Unless there is a major change in world affairs, he or she will come of age in a poverty-stricken, economically, technologically and demographically polarised world where most people will of necessity live in cities

The UN predicts widespread future food shortages, sanitation and health problems in cities, and researchers link population increases with social tension, and breakdowns in law and order.

Happily, however, the crystal-ball gazers seldom get it right.

4.1 *Extract from* The Guardian, *14 August 1999*

- Population growth is unevenly distributed, with 97 out of every 100 children born today being in the less economically developed world. At the same time, the more economically developed world has seen declining fertility rates and a steadying of population growth rates.

- Estimates about future levels of population are notoriously unreliable. Population growth depends on a large number of factors.

- There is a general agreement that population growth should be controlled. This is because increased numbers of people can put pressure on resources such as land, food supplies, and water.

How has world population grown?

For most of human history, population growth, on average, has remained near zero. This is because the high number of births was cancelled out by widespread diseases, wars and famines (Figure 4.2). The modern growth of the world's population only started in the 18th century with the decline of the death rate in Europe and North America. More people have been added to the world since 1950 than during the entire period of human history before that date. With the world population reaching 6 billion in 1999, the main question is where and how these people will live?

Figure 4.3 shows that population in the developed countries increased by 44 per cent between 1950 and 1990. The figure for developing countries in the same period was 143 per cent. A major concern is that such high rates of population growth could place great pressure on resources, as we will discuss later in this section.

The size of any population over time is clearly related to the balance between the number of people born (**fertility**) and the number of people dying (**mortality**).

| 4.2 | *Famine and disease were a control on population in the past* |

NOTING ACTIVITIES

Re-read Figure 4.1 and answer the following questions.

1 Briefly describe the main features of world population change since 1950.

2 Why should we be suspicious about the accuracy of projections of future world population size?

3 What factors might explain the differences in the levels of fertility between more economically developed countries (MEDCs) and less economically developed countries (LEDCs)?

4 What, according to the newspaper article, are the possible consequences of future increases in the world's population?

What is fertility?

Measurements of fertility indicate the rate at which births are adding new members to a population. The simplest measure is the **crude birth rate**, which is defined as the total number of live births in a year for every 1000 people in the population. The crude birth rate is linked to levels of economic development. In addition, the population

	Population (billions)			% increase	Estimated population (billions)		% increase
	1900	**1950**	**1990**	**1950–90**	**2025**	**2100**	**1990–2100**
Developing countries	1.07	1.68	4.08	143	7.07	10.20	150
Developed countries	0.56	0.84	1.21	44	1.40	1.50	24
World	1.63	2.52	5.30	110	8.47	11.70	121

| 4.3 | *Rates of global population increase* |

structure is important since a youthful population will produce more children than an ageing population. Demographers (people who study population) believe that the crude birth rate is influenced by the following factors:

- the status of women in a country, with high levels of involvement in formal education and employment linked to lower levels of fertility
- religion and social customs
- levels of health care.

Study Figure 4.4. Notice that there are higher levels of fertility in large parts of the less economically developed world and lower levels of fertility in the more economically developed world.

Most developed countries have achieved low levels of fertility (Figure 4.5). There are many reasons for this. These include the decline in the number of people marrying or choosing to delay marriage, increasing education, status and employment of women which means that many women

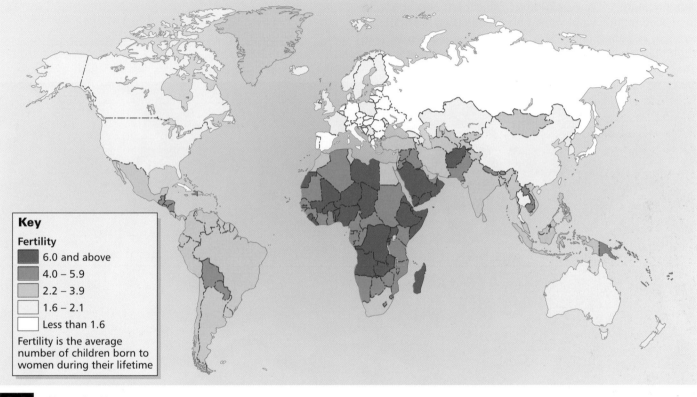

Key

Fertility
- 6.0 and above
- 4.0 – 5.9
- 2.2 – 3.9
- 1.6 – 2.1
- Less than 1.6

Fertility is the average number of children born to women during their lifetime

4.4   *Global fertility levels*

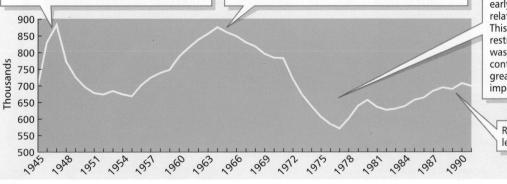

Post-WW2 'baby boom'. Outbreak of war led to a substantial increase in number of marriages. This was compounded by the end of the war. This was a period of recovery and reconstruction with high levels of employment and a feeling of optimism about the future.

Second 'baby boom' linked to earlier marriage, younger child-rearing and increase in births outside marriage. Period of sustained economic growth and full employment, many people experienced rising standard of living. Changing cultural context in which stigma attached to illegitimacy was reduced.

Collapse in fertility levels in 1970s. Fall in number of marriages and trend towards later child-rearing and smaller families. Increased number of abortions following 1967 Abortion Act. Impact of economic recessions in mid-1970s and early 1980s led to unemployment and relative decline in standard of living. This encouraged delay of marriage and restricted family size. Also important was eased availability of oral contraception that allowed women greater control over fertility – increased importance of women in labour force.

Return to 'normal' levels of fertility

4.5   *Births in England and Wales, 1945–91*

choose to have smaller families and start child-rearing later, and the growing material aspirations of the consumer society. There is little evidence of these trends being reversed. The levels of fertility in developing countries are less predictable, due to the diversity in fertility levels, educational attainments, reproductive health and availability of contraception (Figure 4.6).

Demographers agree that the role of women is crucial for any decline in fertility. Studies show that improvements in female education have been of key importance in lowering fertility through delaying marriage and thus the age at which women start to have children. Education also leads to increased female employment and thus raises the status of women, reducing family size and enabling women to plan their fertility. The 1994 International Conference on Population and Development in Cairo recognised that education and social advancement of women are important for reducing fertility.

NOTING ACTIVITIES

1 Write a paragraph to describe the variations in the levels of fertility shown on Figure 4.4. Make sure your paragraph comments on levels of fertility in the following areas: Africa, Western Europe, North America, South America, and South-east Asia.

2 What happened to levels of fertility between 1950 and 2000 in (**a**) LEDCs (**b**) MEDCs?

4.6 *Demographers recognise the importance of raising the status of women if levels of fertility are to be reduced*

STRUCTURED QUESTION 1

Study Figure 4.7, which shows the average fertility of women, and the percentage of women in secondary education, in selected countries.

a Which country has the highest fertility rate? *(1)*

b Which country has the lowest fertility rate? *(1)*

c Do the graphs suggest that there is a relationship between levels of fertility and women's involvement in education? *(2)*

d Why are crude birth rates considered an ineffective way of describing levels of fertility? *(2)*

e What do you think is meant by an 'age-specific fertility rate'? *(2)*

f What factors might account for variations in the fertility rates of different countries? *(4)*

g Improving the status of women in society is generally regarded as crucial in achieving lower levels of fertility. What obstacles might prevent improvements in the status of women in a society? *(6)*

4.7 *Status of women in selected countries*

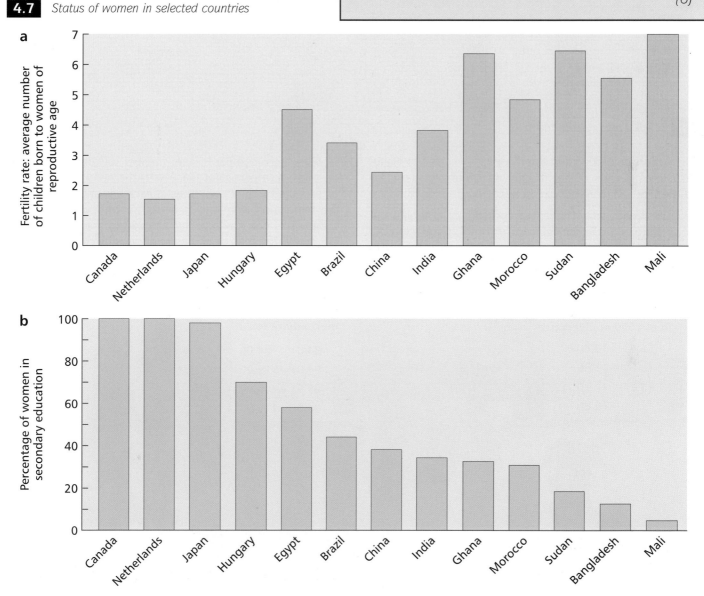

What is mortality?

Measurements of mortality indicate the rate at which people are dying in a population. Again, the simplest measure is the **crude death rate** or the total number of deaths in a year for every 1000 people in the population. As with birth rates, the crude death rate is linked to levels of economic development in a country (Figure 4.8). Countries with low birth rates generally have low death rates. However, the population structure is important since a population with a

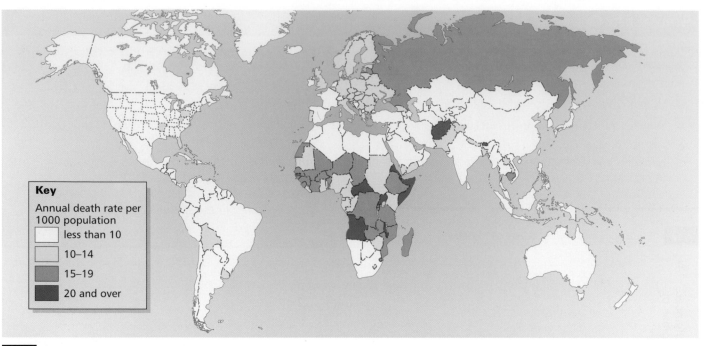

4.8 *Global crude death rates*

high proportion of men and elderly people will lead to higher death rates. Levels of health care availability, social class and the types of work people do will also affect death rates.

The crude death rate is a general measure of mortality, but ignores differences in the age and sex of the population. An alternative measure is the **standardised mortality ratio**

(SMR) which compares the number of deaths actually observed in a particular place with the number that would be expected if the death rates in each age and sex group of the population were the same as for the population as a whole. This can be seen by comparing the two maps in Figure 4.9. The map of crude death rates by county suggests

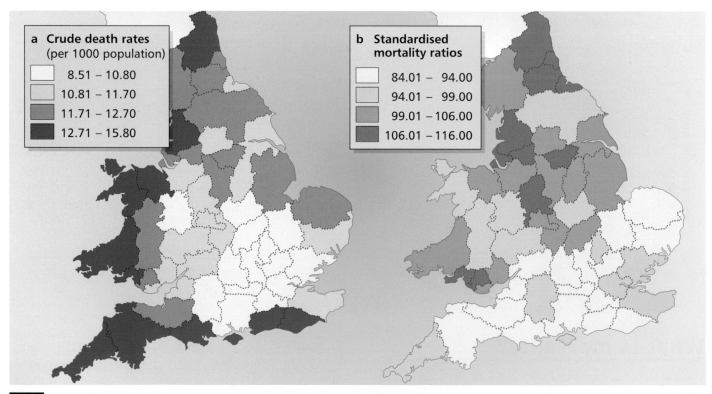

4.9 *Crude death rates and standardised mortality ratios in England and Wales, 1993*

that the highest levels of mortality were found in 1993 along the south coast, in East Anglia, Lancashire and the Scottish Borders. However, these death rates reflect the higher proportions of elderly people in the population of these areas. The map showing standard mortality ratios 'adjusts' for these distortions and allows a direct comparison of mortality rates. See the *box* below for a study of mortality in England and Wales.

Patterns of mortality in England and Wales

Figure 4.9b shows that in England and Wales, there is a clear north–south health divide. The map indicates higher levels of mortality in the industrial regions of northern and western Britain, and lower levels in the south and east.

This pattern is generally thought to be linked to variations in environmental and social conditions. Cities, especially industrial cities, are linked with issues such as atmospheric pollution, inadequate housing and poor social amenities. It is argued that these conditions encourage diseases and unhealthy lifestyles. However, this alone is too simple an explanation. Recent decades have seen improvements in environmental conditions in urban areas, especially with the introduction of the Clean Air Act in 1957. The causes of mortality are linked to social factors including living conditions and life chances. There is a clear social class pattern of mortality, and the geography of mortality can be linked to the distribution of different social groups in Britain.

Some of the factors affecting mortality rates include the following.

- The links between **smoking** and health are well publicised.

Patterns of smoking vary between different social groups. For example, 15 per cent of professionals smoke compared with 40 per cent of unskilled manual workers. This is linked to where people live. The heaviest smoking areas are concentrated in the north and west, including Scotland. The lowest figures are found in the Home Counties, East Anglia and the South West.

- **Exercise and general health** This varies with age, with individuals tending to be most active in their teenage and young adult years. There is a social class element here, with exercise more popular with 'middle class' adults, and the distribution of many sporting facilities is higher in suburban locations as opposed to poorer working-class districts.

- **Obesity and diet** When she was the Minister of State for Health in the mid-1980s, Edwina Currie made a much-publicised comment about the poor diet of northern working-class people being based on fish and chips. Recent research has revealed important variations in the diets of people in different regions. Meat consumption is highest in the North and North West regions and London, and lowest in the South West, Wales and Scotland. Fresh greens, fruit and wholemeal bread are all found more in 'southern' diets. The regional pattern is complicated by social class. An important issue is the types of food available to people. For example, some research on fast food outlets in Liverpool revealed that these were concentrated in poorer parts of the city. The Government's Social Exclusion Unit has noted the problem of 'food deserts' where there is a lack of suitable food outlets.

- **Occupation** In general, the decline of heavy manual labour in Britain's economy has led to a reduction in the link between occupation and mortality. However, stress at work is linked to heart disease, and not working is bad for health, with the unemployed having a significantly worse health record than those in employment. Past occupational patterns continue to influence mortality rates, with the generation in their sixties today having been born in the depression years of the 1930s. Many older people who suffer respiratory problems may have inherited these from their earlier days when cities were more polluted.

How does population change over time?

The difference between the crude birth rate and the crude death rate is the rate of **natural increase**, and this varies widely between countries. Many demographers believe that fertility and mortality rates are closely linked to the level of economic development of a place. They point out that changes linked to industrialisation and urbanisation lead to a **demographic transition** in which high birth and death rates are replaced by low birth and death rates.

The **demographic transition model** describes how the balance between fertility and mortality changes over time. The demographic transition model is shown in Figure 4.10.

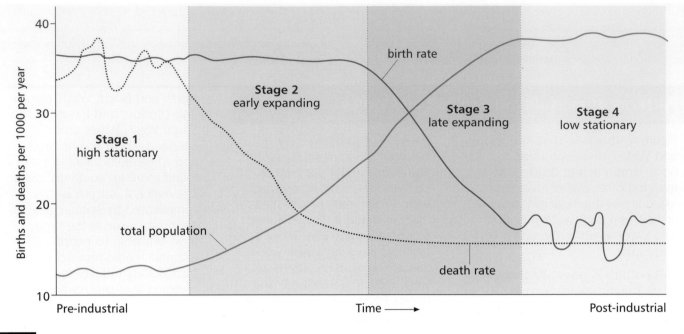

4.10 *Demographic transition model*

The model suggests a series of stages.

- **Stage 1: high stationary or pre-transition stage**

  In this stage birth rates and death rates are high and subject to short-term fluctuations. The death rate is high as a result of checks such as famine and disease. Birth rates are high as people seek to maximise the chance of survival for children. Overall, population growth is static or negligible.

- **Stage 2: early expanding or early transition stage**

  The death rate begins to decline with better living standards, largely due to improvements in nutrition and public health. Famines and epidemics are less frequent. The birth rate remains high, especially as children are a valuable source of family labour and act as security in later life. In this stage population grows at an accelerating rate.

- **Stage 3: late expanding or mid-transition stage**

  In this stage improved technology in agriculture and industry, along with better education systems and legislation controlling child employment, leads to a reduction in the economic and social value of children and an associated decline in the birth rate. Continued improvements in the general standard of living and public health lead to a declining death rate. The rate of population growth begins to fall.

- **Stage 4: low stationary or late transition stage**

  The fourth stage of the demographic transition is characterised by low levels of fertility and mortality. Population growth is minimal, though birth rates may be prone to periodic increases. Some demographers have suggested that a fifth stage needs to be added to the model, where birth rates fall below death rates and lead to negative population growth or population decline.

CASE STUDY

Britain's demographic transition

Britain provides an example of the stages of the demographic transition. Between the middle 18th century and the 1930s Britain underwent a major demographic transformation. Look at Figure 4.11, which shows changing rates of births and deaths in the period since 1700. Prior to the beginning of industrialisation, population was kept in check by high levels of mortality and social and economic restrictions on fertility. Industrialisation removed one of the major constraints to population growth – limited resources – and heralded a new era of rapid population expansion.

The period from the 1740s to the 1880s coincided with the early years of the Industrial Revolution, and during this time the total population increased from 6 to 25 million. There is much debate about the causes of this rapid growth. It has been suggested that the key factor was a fall in the death rate through developments in medicine and surgery or more generally through the contribution of improved public health. On the other hand it could be that there was a rise in the birth rate as people married earlier and improved diets led to increased fertility.

Between 1880 and 1920, death rates fell steeply and this was coupled with a decline in birth rates. As a result, the rate of natural increase was slower. In this period general living standards rose, and there was widespread pressure to limit family size through birth control. The First World War (1914–18) led to a sharp decline in fertility, and this is clearly shown on the graph.

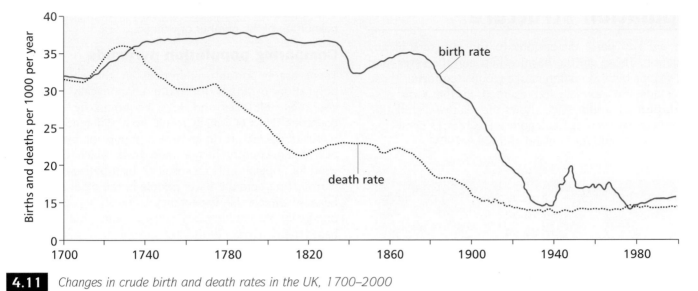

4.11 *Changes in crude birth and death rates in the UK, 1700–2000*

From around 1920 to the present, Britain's population is marked by relatively low birth and death rates. As a result, the rate of natural increase has been slow. Despite this, birth rates have fluctuated in this period, linked to the economic depression of the 1930s, the widespread use of contraception from the 1960s, the legalisation of abortion, the trend towards later marriage, and the 'baby boom' after the Second World War.

NOTING ACTIVITIES

1   Make a copy of Figure 4.10, the demographic transition model. Annotate your diagram to show the different stages of the transition.

2   What factors help to explain the fall in death rates over the period shown on Figure 4.11?

3   Why did the birth rate decline rapidly from 1870 onwards?

4   Figure 4.11 indicates that whilst birth rates are generally low, they still display variations. How might these short-term increases in fertility be explained?

EXTENDED ACTIVITY

Study Figure 4.12 which shows crude birth rates and crude death rates for different continents.

a   On a sketch of Figure 4.12a, plot the figures for Latin America.

b   State the continent which in 1995 had the:

i   highest crude birth rate

ii   highest crude death rate.

a   Crude birth and death rates by continent

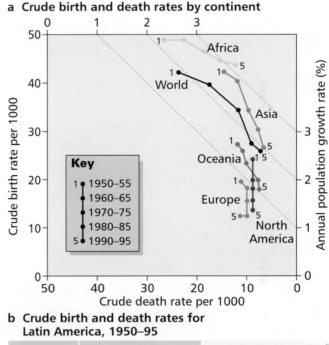

b   Crude birth and death rates for Latin America, 1950–95

Dates	Crude birth rate	Crude death rate
1950–55	37	21
1960–65	35	15
1970–75	32	12
1980–85	28	10
1990–95	25	8

4.12 *Crude birth and death rates in different continents*

c   Give possible reasons for your answers to (**b**).

d   Define the term 'natural increase'.

e   Why might natural increase differ from actual population change?

f   Write a paragraph to summarise the economic implications of differences in population growth rates.

Population structure

There are two basic dimensions to the structure of any population. These are the balance between the sexes, and the divisions between different age groups (**cohorts**). These dimensions are normally represented in the form of a **population pyramid** which shows the relative distribution of numbers between age categories. Figure 4.13 shows the population pyramid for England and Wales 1994.

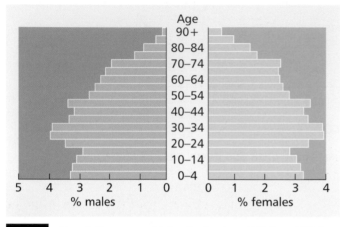

4.13 *Population pyramid for England and Wales, 1994*

Population pyramids, or **age–sex pyramids**, are constructed with the age categories on a central vertical axis. Normally, one-year or five-year categories are used, and the final ages are collapsed into one category (e.g. '90 +'). Figures for males are recorded on the left of the diagram, with females on the right. The horizontal scale may either represent actual numbers, or the percentage of the total population in each age category. The division between the sexes can also be analysed by calculating the male:female ratio, or **sex ratio**.

This expresses the number of men in the population as a proportion of the number of women.

Comparing population pyramids

There are important differences in the shape of the population pyramids for more economically developed countries (MEDCs) and less economically developed countries (LEDCs), which result from differences in their population structure. In general, a population pyramid for a developing country has a wide base which indicates a youthful population (as a result of high birth rates), and a narrow top (there are fewer people in the older age groups because average life expectancy is lower). In contrast, the population pyramid for a developed country has a narrower base (due to a lower birth rate) and a wider top (reflecting a longer average life expectancy).

The age–sex pyramid for Kenya is typical of a country in the less economically developed world experiencing rapid population growth (Figure 4.14a). The wide base indicates that there are large numbers of dependent children aged 0–14 in the total population, which is the result of high levels of fertility. The top of the pyramid is narrow and indicates that a smaller proportion of the population lives to old age. This type of population structure is likely to have a number of important implications:

- Limited resources will be stretched to meet the needs of the large number of dependent children for schooling, nutrition and health care.

- As this group reaches working age, a large number of jobs will need to be created to enable them to support themselves and their families.

- As this group reaches child-bearing age, it is likely that fertility rates will be high, leading to continued high rates of natural population increase.

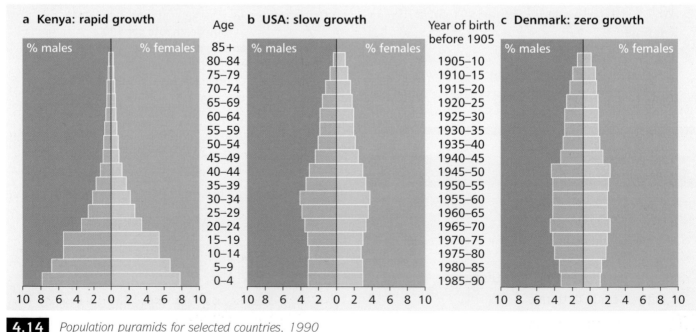

4.14 *Population pyramids for selected countries, 1990*

Dependency ratios

The age–sex pyramid for the USA (Figure 4.14b) is typical of that for a more economically developed country experiencing slow rates of natural increase. The narrow base reflects low birth rates whilst the wider top of the pyramid is the result of people living longer. The 'bulge' in the 30–34 age cohort is a result of the 'baby boom' of the 1960s. The age–sex pyramid for Denmark (Figure 4.14c) is similar to that of the USA. However, it represents a zero-growth pyramid, where birth rates and death rates cancel each other out. In this case, the age–sex pyramid is more like a column than a pyramid, and people are distributed evenly throughout the cohorts. The issues faced by both the USA and Denmark are linked to the prospect of an 'ageing population' where a smaller proportion of the population are of working age and are faced with the task of generating enough wealth to provide high levels of support for the elderly population. The ratio of non-economically active to economically active people in the population is called the **dependency ratio**. It is calculated using the following formula:

$$\text{Dependency ratio} = \frac{\substack{\% \text{ of population} \\ \text{aged } 0–15} + \substack{\% \text{ of population} \\ \text{aged } 65+}}{\% \text{ of population of working age}}$$

EXTENDED ACTIVITY

Study Figure 4.15 which shows population data for Mexico, and Figure 4.16 which shows the age–sex pyramid for the United Kingdom.

1 **a** Use the data in Figure 4.15 to construct a population pyramid for Mexico.

   b Add labels to identify the main features of the pyramid.

Age group	% males	% females
0–4	8.5	8.5
5–9	7.5	7.5
10–14	6.5	6.5
15–19	5.5	5.0
20–24	4.5	4.5
25–29	3.5	3.5
30–34	3.0	2.9
35–39	2.5	2.5
40–44	2.0	2.0
45–49	1.5	2.0
50–54	1.4	1.4
55–59	1.0	1.0
60–64	0.5	0.9
65–69	0.5	0.8
70–74	0.4	0.6
75–79	0.2	0.5
80–84	0.2	0.3
85+	0.1	0.3

4.15   *Age–sex data for Mexico*

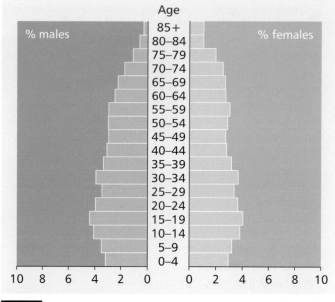

4.16   *Population pyramid for the UK, 1991*

2 Make a copy of the following table and use the Mexican pyramid and data to complete the table.

Age grouping	UK %	Mexico %
0–19	29.3	
20–59	50.9	39.2
60+	19.8	

3 What factors might account for the different proportions of over-60 year olds in Mexico and the United Kingdom?

4 Suggest some of the implications of the differences between the population pyramids of Mexico and the United Kingdom.

NOTING ACTIVITIES

1 What is a population pyramid and how might it be helpful to planning authorities such as central and local government?

2 Draw a sketch of the three population pyramids in Figure 4.14 and add detailed labels (annotations) to describe and explain their shapes.

3 Define the terms 'active population' and 'dependent population'.

4 Why do demographers consider the relationship between these two groups important?

5 What criticisms can be made of the dependency ratio as a measurement?

The implications of Britain's ageing population

Senior citizens enjoying their retirement

In Britain in the 1990s, the gradual ageing of the population has become an increasingly important political issue: the financial problem of finding sufficient resources to provide pension payments. It is linked to the idea of the dependency ratio and the fact that a smaller proportion of the working population will be working to produce the wealth required to support a larger elderly population. The present system of state-provided pensions was introduced in the early years of the 20th century when fertility was significantly higher and life expectancy was lower. The system is based on a 'pay-as-you-go' principle, in which the National Insurance contributions of those in work pay for the pensions of retired people. This was fine so long as the ratio of dependants to those in work was low. In 1950, there was one pensioner for every five people of working age. By 2030 it is estimated that there will be three to every five. A number of solutions have been suggested:

- Reduce the size of individual pensions. This seems unfair to those who have paid National Insurance contributions throughout their working life and expect to receive the benefits of those payments on retirement.

- Raise additional funds through higher taxes. This is currently an unpopular option and governments have tried to avoid any measures that raise taxes, for fear of electoral repercussions.

- Abandon state pensions and replace them with compulsory private pension funds.

These arguments are part of a wider debate about the role of the state in providing support for its citizens. The debate about pensions is just one issue raised by the 'greying' of Britain's population. With increasing age there is an increased demand for medical care and support services. This is especially significant because of increases in life expectancy.

The 1993 National Health Service and Care in the Community Act requires local authorities to assess and provide for the care needs of elderly people. Many authorities have tried to minimise costs by encouraging elderly people to stay in their own homes, supported by visiting care staff. In addition, those elderly people with more than £16 000 assets are required to pay for residential care. This has been particularly controversial, since many have had to sell their homes in order to fund their own care.

The numbers of elderly people in residential care are set to increase by more than 100 000 in the first ten years of the 21st century. This raises important questions about the role of 'informal carers' – usually the children of elderly people – when many are in work and involved in raising their own children. There has been a spectacular increase in private-sector nursing homes, but the cost of residential care is high, ranging between £1000 and £2000 per person each month. The state currently guarantees to pay the cost of care for those who cannot afford it, but this is a politically controversial area, especially when governments are seeking to limit expenditure and keep taxes low.

NOTING ACTIVITIES

1 What factors explain the gradual ageing of Britain's population?

2 Explain what effect this trend will have on the dependency ratio.

3 Why are governments concerned about the ageing of the population?

4 What solutions have been suggested to overcome these problems?

STRUCTURED QUESTION 2

Study Figure 4.17, which shows 1991 population census data for two settlements in Kent: the city of Canterbury, and the coastal town of Herne Bay.

a The data shows the percentage of the population under 20, rather than the usual percentage of people under 15. Can you suggest reasons why more people between the ages of 15 and 20 might be considered part of the dependent population? *(2)*

b In what ways might you expect the crude birth rates and crude death rates in these settlements to differ? *(2)*

c Suggest reasons for the differences in the percentages of the population over 60 years old in Canterbury and Herne Bay. *(3)*

d How might the provision of facilities vary between Canterbury and Herne Bay? *(4)*

e Discuss the social and economic consequences of an ageing population structure for a country such as the UK. *(5)*

	Total population	% of population aged under 20	20–39	40–59	60–79	over 80
Canterbury	37 000	25	30	22	19	4
Herne Bay	32 000	23	23	22	24	8

4.17 *Population data for Canterbury and Herne Bay, 1991*

B People and resources

Are there too many people in the world?

The newspaper article (Figure 4.1 on page 101) suggests that we should be concerned about the rate at which the world's population is growing. The fear is that the number of people will be too great for the amount of food and other resources available. There are, however, other points of view.

The development of human societies depends on the physical resources we find in the world. These resources act as raw materials and energy sources in industrial and agricultural processes. They absorb and transport the by-products of these processes. Resources are also consumed in fulfilling the human needs for shelter and sustaining our lifestyles. The debate about the link between population and resources has been based on the idea of **carrying capacity** which refers to the maximum number of people that can be sustained by an environment without impairing the ability of that environment to sustain itself. The relationship between population size and resources can be expressed in the concept of **optimum population**, which is taken to be where the population of an area is in balance with the resources available. **Overpopulation** refers to a situation where the population of an area is too large to be supported by the resources available.

Thomas Malthus: prophet of doom?

The question about the link between population and resources was raised in the work of an English clergyman named **Thomas Robert Malthus** (Figure 4.18) at the end of the 18th century. His famous essay, published in 1798, was entitled *An Essay on the Principle of Population*. He argued that food is necessary to the existence of human beings, and that 'passion between the sexes' is necessary and constant. From these ideas he noted that as long as people had food to sustain them, there would be a tendency for population to increase. The limits to population growth would be determined by the supply of food. Malthus argued that increases in food production tended to increase in a simple arithmetic fashion (i.e. 1...2...3...4...5... etc.) whilst population tended to increase in a geometric fashion (i.e. 1...2...4...8...16...32... etc.). As Figure 4.19 shows, Malthus was suggesting that population growth would outstrip food production and this would lead to famine and starvation. At a certain point the limit of resources would be reached and the rate of population increase would level off. Up until that point though, Malthus argued, there would be a tendency for population increase. As Malthus put it: 'The power of the

4.18 *Reverend Thomas Malthus*

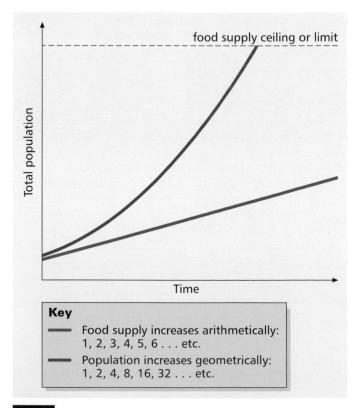

Key

— Food supply increases arithmetically:
1, 2, 3, 4, 5, 6 . . . etc.
— Population increases geometrically:
1, 2, 4, 8, 16, 32 . . . etc.

4.19 *Malthusian ideas on population growth*

population is infinitely greater than the power of the Earth to produce sustenance'.

In order to understand why Malthus adopted such a pessimistic view it is important to understand the context in which he was writing. This was a time of great change in English agriculture and industry. Many people were being displaced from the land by developments in agricultural technology, and seeking to find work in the towns. Many wealthy people were worried that there was a 'surplus' of unnecessary workers. Malthus was seeking to understand this situation.

Although Malthus was writing at the end of the 18th century, more recent writers have shared his pessimism. These are sometimes called **neomalthusians** (or 'new Malthusians') because they share Malthus's perspective that the size of the world's population itself is the cause of problems. Paul Ehrlich, in his book *The Population Bomb* (1968), wrote:

> '*Each year food production in the undeveloped countries falls a bit further behind burgeoning population growth, and people go to bed a little bit hungrier.*'

In the early 1970s the Club of Rome warned that the critical point in world population growth was approaching and that humanity needed to find a state of balance between population and resources. According to this view rapid population growth is the main cause of the problems of the developing world since it leads to poverty, economic stagnation, environmental problems, rapid urbanisation, unemployment and political instability. Since rapid population growth is the cause of the problems, the solution is to persuade people to have fewer children. The main way to achieve this is through family planning.

Was Malthus right?

Malthus's ideas have been disputed. At the time he was writing, Karl Marx and Freidrich Engels argued that the population issue was a false one; they thought that the problem could be solved by new technological developments which would allow increased agricultural production and a more equal distribution of resources. More recently, the agricultural economist Esther Boserup (1965) argued that population growth was an important factor in allowing societies to find innovative ways to increase food supplies. This view suggests that faced with new demands, humans can adapt creatively to solve food supply problems.

Is population growth a cause or a symptom?

An alternative perspective on the population question is that population growth is not the cause of problems, but a symptom. In other words, rather than being poor because they have too many children, people may have many children because they are poor. Children are a valuable

source of labour for families, and may release their mothers for work. In addition, in economies where there is little or no welfare provision, children provide parents with security in old age. In countries where there have been improvements in health care and education, there have been remarkable declines in fertility. Those who favour this view therefore conclude that increasing living standards through fairer social and economic development is the best way to motivate people to have fewer children.

This argument can be taken further to suggest that it is the status of women that provides the clue to controlling population growth. Women who have access to better education and employment opportunities tend to have less need to rely on their children for economic security. Instead of seeking to persuade (or force) women to have fewer children through contraception and family planning, women need to be in control of their reproduction. The best way to achieve this is to improve the social and economic welfare of women. This, of course, requires substantial social change.

How can fertility be reduced?

The size of the world's population and the likelihood of population increase is an important issue at both international and national levels. At an international level most of the effort has been aimed at reducing the number of births worldwide. There are three main reasons for this:

- a concern about the rapidly rising population and the fact that this increase is being experienced more in LEDCs than in MEDCs
- the idea that rapid population growth is linked to social and economic inequality
- the pressure that population growth puts on the availability of resources and the concern that this may lead to environmental degradation and destruction (Figure 4.20).

4.20   *Destruction of the rainforest: the result of overpopulation?*

A global population policy

The United Nations sponsors an international conference on population every ten years. The aim is to develop a global population policy.

1 The Bucharest conference in 1974 was marked by disagreement over the approach to reducing fertility, and whether a population problem even existed. The developed countries argued that a 'population bomb' was about to explode as a result of high birth rates in the developing countries. They advocated family planning policies to reduce levels of fertility. The developing countries argued that the threat came from the developed world which was damaging the environment with its industry and heavy consumption of natural resources. They took the view that the population issue would disappear through economic development and progress.

2 The Mexico City conference in 1984 saw a reversal in these positions as the developing countries pressed for investment in family planning programmes and the developed countries argued that 'development is the best contraceptive'.

3 The most recent conference on population and development took place in Cairo in 1994. The conference recognised that although the global population is still increasing, birth rates in almost every country are falling, and the focus should be on continued reductions in growth rates so that the world population levels off sooner rather than later. The conference reflected the idea that family planning is only one part of the solution, and that reducing population growth is linked to the reduction of poverty and disease, improved educational opportunities (especially for girls and women), and environmentally sustainable development.

The role of women

It is now widely accepted that an important factor in levels of fertility is women's status. In general, countries where women have greater access to education and employment have lower levels of fertility, since women have less need for the economic security and social recognition that children provide. In addition, more equality between men and women is also linked to lower fertility. The 1994 UN conference in Cairo placed an emphasis on improving the rights, opportunities and status of women as the most effective way of reducing global population growth. This is a huge task, since in many countries the status of women is much lower than that of men (Figure 4.21).

4.21 *Women have a vital role of in determining fertility*

National population policies

Some governments have attempted to control population growth. Pro-natalist policies seek to increase natural population increase by encouraging an increase in the birth rate. Anti-natalist policies seek to reduce natural population increase through the reduction of births.

Pro-natalist policies have been pursued by governments seeking to strengthen their political power. For example, the Nazi regime in Germany during the 1930s and 1940s was strongly pro-natalist, generating propaganda about the need to create a master Aryan race. To encourage this, tax allowances were provided for large families, taxes were levied on unmarried adults, and women having abortions were prosecuted. More recently, Israel and Saudi Arabia have encouraged population growth, mainly to strengthen their political power. The upsurge of Islamic fundamentalism has encouraged some governments to adopt pro-natalist policies. In 1984 Malaysia introduced its New Population Policy which encouraged women to 'go for five' in order to catch up with more populous neighbouring states and to prevent ethnic Malays from being outnumbered by the Chinese population.

Anti-natalist policies are more common. India had a family planning programme as early as 1952. The United Nations began to provide services advising population control in 1965. China's one-child policy is probably the best known population control programme. It was enforced through tight community control and a series of rewards and sanctions. Figure 4.22 shows how by the late 1990s it was possible to argue that population policies had led to the defusing of the population time-bomb.

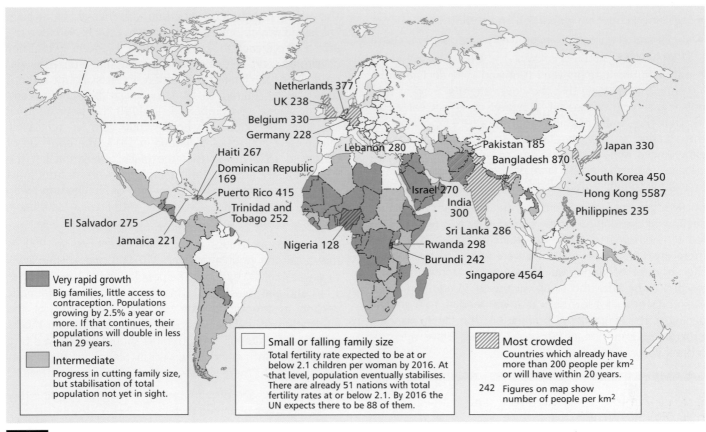

4.22   *Global rates of population increase*

CASE STUDY

China's one-child policy

In the 1950s and 1960s China experienced a decline in its population, due to disasters such as typhoons, flooding and the severe famines that followed. However, from 1963 the country experienced a 'baby boom' and the population grew rapidly. The government was concerned that a rapidly growing population would put pressure on resources, and during the 1970s tried to encourage both family planning and delayed marriage. The popular slogan was 'one is not too few, two will do and three are too many for you'. Communities were encouraged to plan which women should have children and the practice of 'giving birth in turn' became common.

In 1980 the policy of one child per family was officially adopted. The aim was to limit the population to a maximum of 1200 million by the year 2000. There were exceptions to the rule, and it was in urban areas that the policy was most strictly enforced. The policy was enforced by a system of rewards and penalties; for example, parents were offered a 5–10 per cent salary incentive for limiting their families to one child, and a 10 per cent salary reduction for those who produced more than two children.

The policy has slowed down the population growth rate. However, it has also caused tensions. It conflicts with traditional Chinese family values, in which children are seen as a source of happiness and fulfilment and a means of continuing the family line. Concern has been expressed about the long-term effects of the creation of a generation of pampered 'little emperors' with no siblings or cousins. In addition, the Chinese press reported cases of the abandonment or neglect of girl babies, pre-natal testing followed by selective abortion, and instances of female infanticide.

CASE STUDY

Singapore

The National Family Planning Programme was launched in 1966, soon after Singapore became independent. Its role was to plan overall levels of population. In the 1960s the government felt that the rate of natural increase was too high, putting pressure on employment and requiring high levels of investment in housing and services such as health and education. It sought to encourage people to have fewer children through a number of measures:

- the provision of advisory and clinical family planning services
- a publicity programme aimed at educating and motivating people to have fewer children

- legislation concerning abortion and sterilisation
- social and economic disincentives for couples who have large families.

The total fertility rate between 1966 and 1982 declined from 4.5 to 1.7, suggesting that the government's 'Stop at two' policy was very successful. However, it is unclear as to whether the family planning policies were responsible for the decline in fertility. During this period Singapore experienced rapid economic growth and improvements in education, which played a large part in encouraging changes in attitude to family size.

In 1983, government concern that the most highly educated section of the population was producing too few children led to measures to encourage educated women to have more children. These concerns were linked to fears about the shortage of labour which meant that foreign workers needed to be attracted to fill vacancies. In addition, there was concern that a low birth rate was leading to a gradual ageing of the Singaporean population. The family planning programme is no longer 'Stop at two' but 'Have three, or more if you can afford it'.

NOTING ACTIVITIES

1 Write notes to summarise the view of population taken by Thomas Malthus. In what ways have neomalthusians developed his ideas?

2 What criticisms have been made of Malthusian views on population?

3 How have international policies on the question of the world's population developed since 1974?

4 Why is the status of women now regarded as crucial in reducing the rate of global population growth?

5 Give examples of how and why national governments have tried to influence the size of their population.

CASE STUDY

Mauritius

Mauritius provides an example of the delicate relationship between population and resources. Mauritius is an island located in the Indian Ocean about 800 km east of Madagascar (Figure 4.23). It is an extinct volcanic crater. The terrain is mountainous, though in places there is a coastal plain. The tropical climate means that annual rainfall is high, and the volcanic soils are relatively fertile. In the south-west of the island is an area of forest and scrub. Sugar is the main crop, though tea, coffee and vegetables are also grown.

The history of Mauritius is tied up with that of colonialism. The island was colonised by the Dutch, French, and then the British. The French took control of the island in 1715. They developed port activities and roads, and started sugar production. The limited population of the island meant that labour was required to work on sugar plantations, and African slaves were imported. In 1790 the population of Mauritius was 59 000 of whom 49 000 were African slaves. Britain took control of Mauritius in 1815. The abolition of slavery in 1835 meant that the black population left the sugar industry, and the British imported workers from India to work the plantations. These waves of immigration mean that the island is ethnically diverse, with French-Mauritians, Chinese, Muslims, Hindus and Creoles.

Large-scale immigration to the island had ended by 1900. At this time, the population was 370 000 and it then grew slowly to 428 000 by 1939. The population doubled between 1940 and the mid-1970s. This was caused by a dramatic decline in the death rate, largely as a result of the eradication of malaria. The demographic transition model suggests that the fall in death rates should be followed by a decline in birth rates. However, in Mauritius, birth rates remained high. There were a number of reasons for this:

- the average age of marriage remained low, ensuring that fertility rates remained high
- the economy was strong, allowing people to marry early and have large families
- the role of women meant they remained tied to the home.

The rapid growth of the population put great pressure on Mauritius's economy, which was still largely dependent on sugar. In 1953, a government report called for measures to reduce the birth rate. However, this report was out of line with the cultural norms and expectations of the population. The two major religions – Catholicism and Islam – both opposed population control.

In 1963, the Catholic Church withdrew its opposition to government proposals for family planning. From 1965 the Mauritian government funded a range of measures to make family planning widely available to the population. The effect of these measures was to reduce the crude birth rate from its peak of 49.8 in 1950 to around 20 in the mid-1990s. This decline was linked to several other factors:

- an increase in the average age of marriage, especially amongst Hindus in rural Mauritius – this was linked to downturns in the economy from the late 1960s
- a cultural change in the attitude to family size: in the 1970s three children was generally considered to be the optimum number of children; by the 1980s, two was the preferred number
- an increase in the status of women, as female education provision improved and women entered employment beyond the home.

Measures to control population in Mauritius were largely a response to the limited natural resource base of the island.

The Mauritius example points to the limits of the Malthusian view, since it suggests that population size is the result of an inevitable law, but is determined by a complex range of economic, social, political and cultural factors. It also suggests that the size of the resource base is not fixed and unchanging, but depends on the context in which population growth is occurring. For example, in recent decades Mauritius has sought to diversify its economy. It has done this by attracting foreign banks and large transnational corporations, thus making use of the skills and attributes of its population, and by developing a tourist industry.

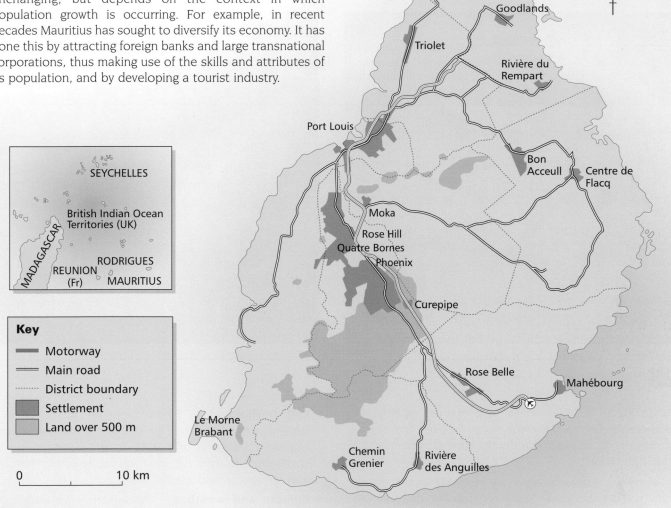

4.23 *Mauritius*

NOTING ACTIVITIES

1   Using the information in this case study and an atlas map to help you, write a paragraph describing the geography of Mauritius. Include the following information:

   – climate

   – relief

   – population size

   – main crops

   – average per capita income.

2   Describe the changes in the population since 1900. How far do these changes reflect the demographic transition?

3   How has Mauritius been able to avoid the problems of overpopulation predicted by Malthus?

C Population distribution

a Environment supporting only a sparse population

b Environment that is densely populated

4.24   *Contrasting environments*

Look at Figure 4.24. It shows two contrasting physical environments. For each environment, consider the factors that make it an attractive or unattractive environment for human habitation.

The world's 6 billion people are not distributed evenly across the Earth's surface. Some general statements can be made about the distribution of the world's population:

- Almost all of the world's inhabitants live on one-tenth of the land area.
- Most live near the edges of land masses, near the oceans and seas, or along easily navigable rivers.
- Approximately 90 per cent of the world's population lives north of the Equator, where the largest proportion of the total land area (63 per cent) is located.
- Most of the world's population lives in temperate, low-lying areas with fertile soils.

Figure 4.25 shows the distribution of the global population. Notice that there are distinct concentrations of people in certain places. There are a number of physical factors that help explain the distribution of the world population. These include the following:

- **Altitude** – the tendency for people to live in low-lying areas is partly explained by human physiology which is intolerant of high altitudes.

- **Soil fertility** – this is an important factor since it influences levels of food production and therefore the carrying capacity of the land. People have tended to concentrate in river valleys which are regularly provided with fertile alluvial deposits.

- **Climate and weather** – the tendency for people to avoid extremes of temperature explains the low levels of population in polar and hot desert regions.

- **Water availability and quality** – the floodplains of large rivers have proved attractive to human settlement, although the hazards of flooding or water-borne diseases may act as a barrier to settlement.

- Type and availability of **natural resources**.

NOTING ACTIVITIES

1   Study Figure 4.25. On a blank outline map of the world, identify areas of the world where human occupancy is dense and where it is sparse.

2   Annotate your map to suggest reasons for these patterns.

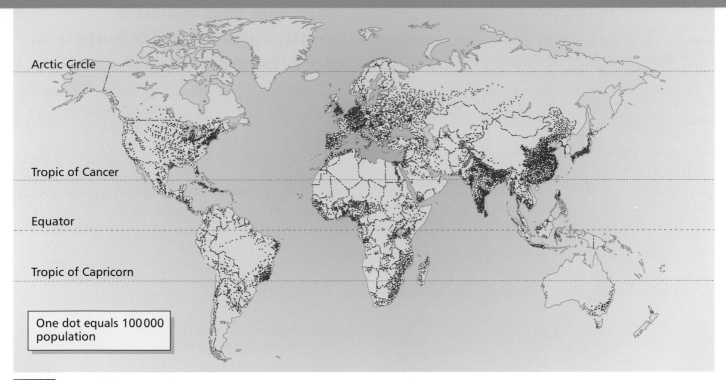

Arctic Circle

Tropic of Cancer

Equator

Tropic of Capricorn

One dot equals 100 000 population

4.25   *Global distribution of population*

Population density

One measure of the distribution of population is **population density**. The crude density is the total number of people divided by the total land area. As Figure 4.26 shows, some parts of the world are very heavily inhabited, while others only sparsely. It is normal to represent population density by drawing a choropleth map. One of the disadvantages of this method is that it hides the variations in population density that exist within an area. For example,

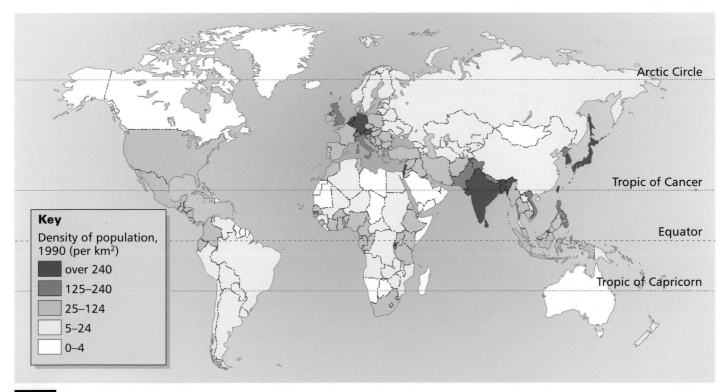

Arctic Circle

Tropic of Cancer

Equator

Tropic of Capricorn

Key
Density of population, 1990 (per km²)

over 240

125–240

25–124

5–24

0–4

4.26   *Global population density*

Australia has an average population density of less than 4 people per km². However, this conceals the fact that Australia's population is largely clustered along the coasts, particularly the east and west coasts (Figure 4.27). This distribution is linked to the pattern of settlement by people of European origin. In the interior, the scattered population is mainly of aboriginal origin. Similarly, Mexico has an average population density of between 125 and 240 people per km², but most of its population is found in the interior of the country away from the mountains and in areas where soil fertility is highest (Figure 4.28).

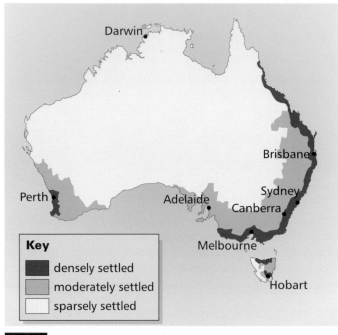

4.27 *Population distribution of Australia*

4.28 *Population distribution of Mexico*

How will the world's population be distributed in the future?

The geographer John Clarke has recently argued that because most future population growth will take place in LEDCs, in the next few decades the world's population will become increasingly concentrated:

- By 2030 the populations of East and South Asia will be over 4.1 billion people, about 1.8 billion more than now. They will comprise 44 per cent of the world's population on just 13 per cent of the world's land area.

- The areas of the Earth that are not presently inhabited will remain uninhabited. These environments are generally too severe for large-scale human occupation.

- Within the inhabited areas of most countries there will be a growing polarisation of population distribution. Economic core areas such as in south-east Brazil and Java (Indonesia) will continue to attract migrants from less developed parts of countries.

- Rapid urban population growth will continue in LEDCs. Between 1950 and 2025 the urban populations of LEDCs will have multiplied more than 14 times. Much of this increase is occurring in mega-cities. This increase has important implications for the provision of housing, infrastructure, employment, education, health and services.

Population distribution: how does it vary at different scales?

It was noted earlier that the map of global population densities conceals important variations within countries. Figure 4.29 shows that much of Britain's population is clustered in quite a small area. In fact, a dominant feature of the population of Britain is the way that it is found largely in urban areas. Some figures will help to illustrate this. In 1801 around 9 million people lived in England and Wales, and one in three lived in towns. By 1851 the population had grown to 18 million, of whom just over half were urban dwellers. By the beginning of the 20th century the population had increased to 32.5 million and 78 per cent lived in towns and cities.

A closer look at the map reveals some distinct features.

- A band of dense population stretches north-westwards from the English Channel across the Thames, through the Midlands and, dividing on each side of the Pennines, continuing into Lancashire and Yorkshire.

- The southern part of this band of high population density is focused around London and the Midlands, and the northern parts focus around Birmingham, Manchester and Leeds.

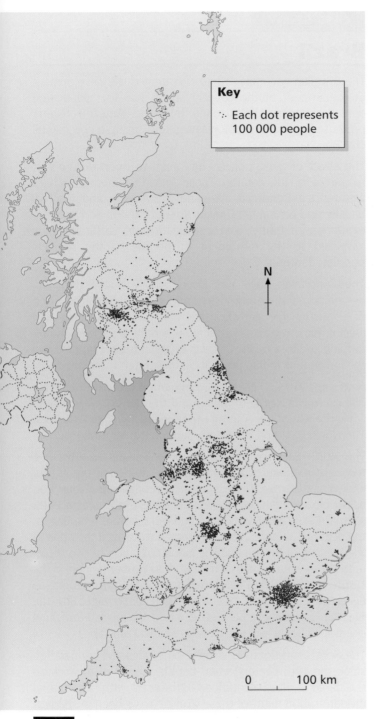

Key

∴ Each dot represents 100 000 people

N

0 ——— 100 km

4.29   *Population distribution of the UK*

- Outside this band there are some other important centres of population, in a line from the South Wales industrial region and Bristol, through the West and East Midlands, to the Humber estuary.
- Further north is the Tyne and Wear region, with Newcastle upon Tyne as its main centre.
- The central belt of Scotland, with Glasgow and Edinburgh as the main cities, dominates the population distribution of Scotland.

NOTING ACTIVITIES

1   What is 'population density'?

2   Why is 'average population density' often an inaccurate description of population density within a country?

3   Suggest reasons for variations in population density within a country.

4   Describe and explain the distribution of population within Britain.

STRUCTURED QUESTION 1

The purpose of this activity is to investigate the population distribution of a region (Wales), and to suggest reasons for that distribution.

a   Study Figure 4.30a. On a copy of Figure 4.30b, draw a choropleth map to represent this data. Use the following categories:

- 149 or under
- 150–299
- 300–599
- 600–999
- 1000 or over.

b   Describe the pattern of population density in Wales as shown by your map.

c   With the help of an atlas and Figure 4.30c, suggest how the following factors might have influenced the distribution of the population:

- relief
- soil fertility
- climate
- natural resources.

Population per km²

Blaenau Gwent	670	Monmouthshire	102
Bridgend	529	Neath Port Talbot	316
Caerphilly	608	Newport	720
Cardiff	2250	Pembrokeshire	71
Ceredigion	39	Powys	24
Conwy	98	Rhondda, Cynon, Taff	566
Denbighshire	109	Swansea	609
Flintshire	331	Torfaen	718
Gwynedd	46	Vale of Glamorgan	356
Isle of Anglesey	94	Wrexham	248
Merthyr Tydfil	523		

4.30a   *Population density in Wales*

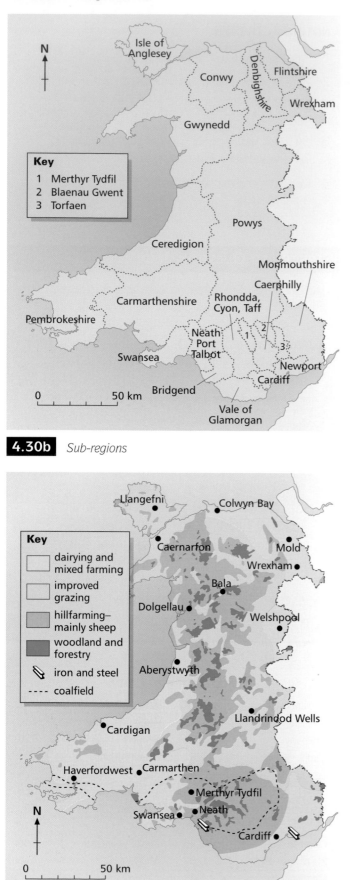

4.30b *Sub-regions*

4.30c *Land use*

CASE STUDY

Brazil

The Brazilian government has encouraged the development of the interior of the country, which has low population densities. It has built roads into the Amazon region, promoted agricultural colonisation and set up a new capital city, Brasilia.

The effect of these policies has been to open up the vast resources of the country. These include minerals such as cassiterite in Rondonia, manganese in Amapa, and iron ore, bauxite and gold in the Carajas area of Para. In addition, the extraction of timber and introduction of cattle ranching have led to the movement of people to the sparsely populated interior.

These policies have not been wholly successful. The destruction of the Amazonian rainforest and the introduction of unsustainable activities such as cattle ranching and logging have replaced sustainable activities such as rubber tapping and tree crops. The way of life of rubber tappers and the indigenous Indian population have been threatened.

Between 1970 and 1990 the population growth rates were highest in the Northern region and the Central West region, while the rest of the country experienced lower growth rates. The North East and the South had a net migration loss.

The North East has a tradition of out-migration. This is a response to frequent droughts in the interior of the region, known as the *sertao*. There is a tradition of seasonal migration as people leave the countryside and travel to the coastal cities such as Recife and Fortaleza to find work, before returning for the next rainy season. These workers have played their part in the building of Brasilia and the industrialisation of São Paulo. In the 19th century many left to work as rubber tappers in the North and in the 1980s many moved to Amazonia as agricultural colonists. This move was encouraged by the Brazilian government because it eased the pressure of rural overpopulation which was causing political tensions.

The South and South East have been areas of in-migration since the 19th century, when people moved in to work on the large coffee plantations. However, with agricultural modernisation many of these workers have been pushed off the land and a large number have moved to Mato Grosso and Rondonia.

These movements have been linked to an increase in the level of urbanisation. In 1940 less than one-third of the population lived in cities, but by 1970 this had increased to around half. Today more than 75 per cent of the Brazilian population live in towns and cities. However, as Figure 4.31 shows, levels of urbanisation vary greatly between regions. Most of the urban growth has been concentrated in the two large cities of Rio de Janeiro and São Paulo which act as the centres for specialised industrialised regions such as Belo Horizonte (metal manufacture) and Salvador (petrochemicals).

a   Regions

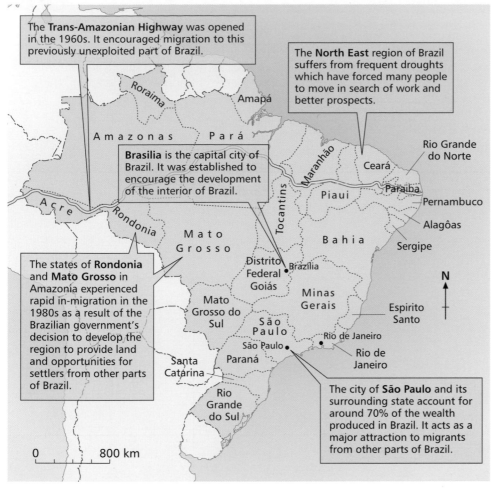

The **Trans-Amazonian Highway** was opened in the 1960s. It encouraged migration to this previously unexploited part of Brazil.

The **North East** region of Brazil suffers from frequent droughts which have forced many people to move in search of work and better prospects.

Brasilia is the capital city of Brazil. It was established to encourage the development of the interior of Brazil.

The states of **Rondonia** and **Mato Grosso** in Amazonia experienced rapid in-migration in the 1980s as a result of the Brazilian government's decision to develop the region to provide land and opportunities for settlers from other parts of Brazil.

The city of **São Paulo** and its surrounding state account for around 70% of the wealth produced in Brazil. It acts as a major attraction to migrants from other parts of Brazil.

Roraima · Amapá · Amazonas · Pará · Acre · Rondonia · Maranhão · Ceará · Rio Grande do Norte · Paraíba · Pernambuco · Alagôas · Sergipe · Tocantins · Piaui · Bahia · Mato Grosso · Distrito Federal · Brasilia · Goiás · Minas Gerais · Mato Grosso do Sul · São Paulo · Espirito Santo · Rio de Janeiro · São Paulo · Paraná · Santa Catarina · Rio Grande do Sul

0   800 km

N

NOTING ACTIVITIES

Study Figure 4.31.

1   On a copy of Figure 4.31a, mark the main flows of population movement in the period 1970–90.

2   Describe the main areas which have experienced increases in population in this period.

3   Suggest reasons for the changes you have described.

4   Use the labels in Figure 4.31a and the case study to annotate your map to explain changes in the distribution of population in Brazil. (See Figures 4.31 b and c.)

b   Population 1970

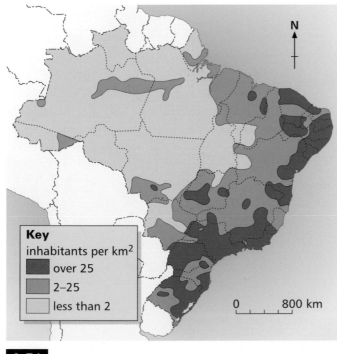

Key
inhabitants per km²
- over 25
- 2–25
- less than 2

0   800 km   N

c   Population 1990

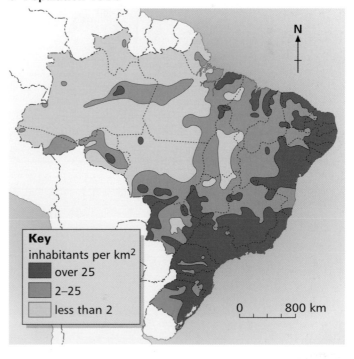

Key
inhabitants per km²
- over 25
- 2–25
- less than 2

0   800 km   N

4.31   *Brazil*

STRUCTURED QUESTION 2

Study Figure 4.32a.

a On a copy of Figure 4.32b, sketch the population densities from the coast at X to the interior at Y. *(4)*

b Suggest reasons for:

(i) the high population densities on the coast *(2)*

(ii) the low population density in the interior. *(2)*

c Why do you think there is a relatively high population at Z? *(2)*

d Suggest reasons why the map might not present an accurate picture of population density in the country. *(2)*

e For a named country, or a region of a named country, describe and explain why the population has changed over time. *(5)*

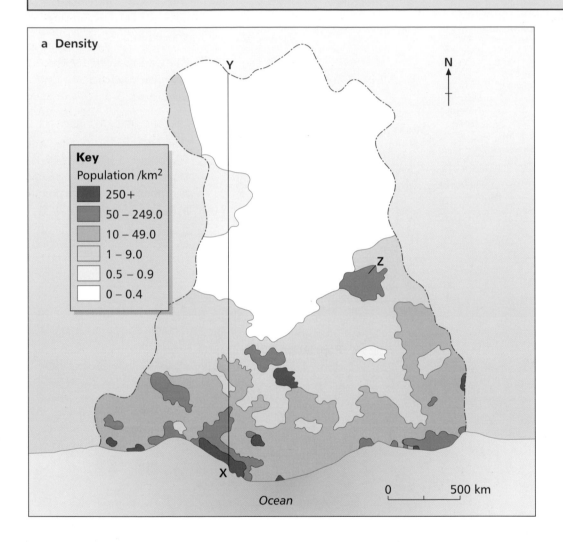

a Density

Key
Population /km²

- 250+
- 50 – 249.0
- 10 – 49.0
- 1 – 9.0
- 0.5 – 0.9
- 0 – 0.4

N

Ocean

0   500 km

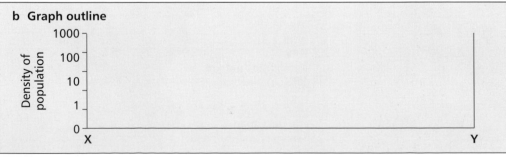

b Graph outline

Density of population: 1000, 100, 10, 1, 0

X   Y

4.32 *Population in an LEDC*

D A mobile world?

This chapter is concerned with issues relating to migration. **Migration** can be defined as:

The movement of a person (a migrant) between two places for a certain period of time.

This apparently simple definition raises some important questions. How far and for how long does someone have to move to be considered a migrant? Is it possible to say that the movement of a person to live and work in a foreign country is the same as a trip to the corner shop? Generally, geographers consider migration to be movement across a boundary of an areal unit (political boundary). Two main types of migration are considered:

- **Internal migration** This is where a boundary within a country is crossed, for instance, a person moving from Greater London to a neighbouring county such as Surrey. People moving into an areal unit are called **in-migrants**, whilst those moving out are called **out-migrants**.

- **International migration** This is the movement of people across national borders, such as between Germany and Poland. The movers into a country are **immigrants**, while the people moving out are **emigrants**.

People do not make decisions to move from one place to another lightly, since migration often involves a radical change in their lives. For example, think how often people talking about their lives will mention moving from one place to another as a significant event. A simple classification of migrations is shown in Figure 4.33. The classification distinguishes between migrations motivated by economic reasons and those driven by social and political factors. Any migration can be located at any point along the continuum depending on the balance of forces. In reality, the motivation to migrate is a combination of the two. In addition, migrations can be either voluntary or involuntary (forced). Thus any point on the diagram represents a unique form of migration.

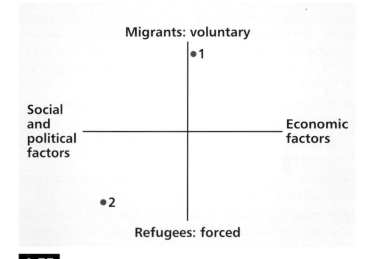

4.33 *A typology of migration*

Refugees fleeing from war

African slaves being sold

Models of migration

1 Ravenstein (1876)

The most influential figure in the study of migration is the cartographer Ravenstein. He published a series of papers outlining a number of laws of migration. Ravenstein's laws, or hypotheses, were published in the *Geographical Magazine* of 1876. They were based on British census data showing place of birth, for 1871 and 1881. The laws were these:

- The majority of migrants go only a short distance.
- Migration proceeds step by step.
- Migrants moving long distances generally go to large centres of population and industry.
- People who live in towns are less likely to migrate than people who live in rural areas.
- Females are more likely to migrate than males within the country of their birth. But males are more likely to migrate to another country.
- Most migrants are adults. Families rarely migrate out of their country of birth.
- Large towns grow more by migration than by natural increase.
- Migration increases in volume as industry and trade develops and transport improves.
- The major direction of migration is from agricultural areas to the centres of industry and trade.
- The major causes of migration are economic.

2 The gravity model

Geographers have attempted to tests Ravenstein's laws. The most influential attempt is the gravity model. It was based on the law of universal gravitation derived by Isaac Newton, which states that the gravitational force (attraction) between two objects is directly proportional to their masses and inversely proportional to the square of the distance between them.

A simpler way of remembering this is:

> *The number of migrants moving between A and B is equal to the population of the origin, multiplied by the population of the destination, divided by the distance between them.*

At first sight the gravity law has a certain logic, since the number of potential migrants is bigger if the sources and destinations are large, and we would expect distance to play a part. However, attempts to apply the law quickly revealed that it was much more than distance and numbers of people that determined migration flows.

3 Zelinsky's model of mobility transition (1971)

Zelinsky developed a general model to show the changes that take place in the rates and scale of migration as a society is transformed over time (Figure 4.34). His model of mobility transition suggests that as societies go through the demographic transition they also experience changes in the nature of personal mobility:

- In pre-modern traditional society there was limited migration and what did occur tended to be local, between rural places.
- In the early transitional society there is a dramatic change marked by movement from the countryside to cities.
- As society develops into a modern state levels of migration are high, but with a largely urban population; the vast majority of migration is between urban areas.
- In late modern industrial societies there is some reversal of patterns of migration until in post-industrial societies

there is a general decline in migration as information is increasingly circulated through communications media.

The model has been seen as quite useful in describing events in the developed world. However, it is flawed because it is over-generalised and relies on a vague idea of 'modernisation' as the driving force behind change.

These three models focus on measuring, quantifying and presenting the patterns of migration flows. However, they tell us little of the mechanisms behind individual acts of migration. They do not tell us *why* people decide to migrate and what this decision means to them.

4   Behavioural models

Behavioural approaches focus on the subjective or personal evaluations made by people in decisions about where to live and whether to migrate. Figure 4.35 is an

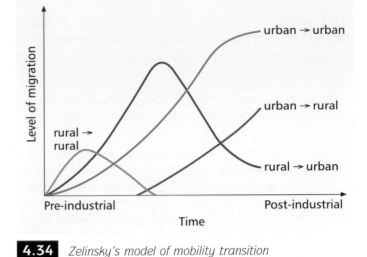

4.34   *Zelinsky's model of mobility transition*

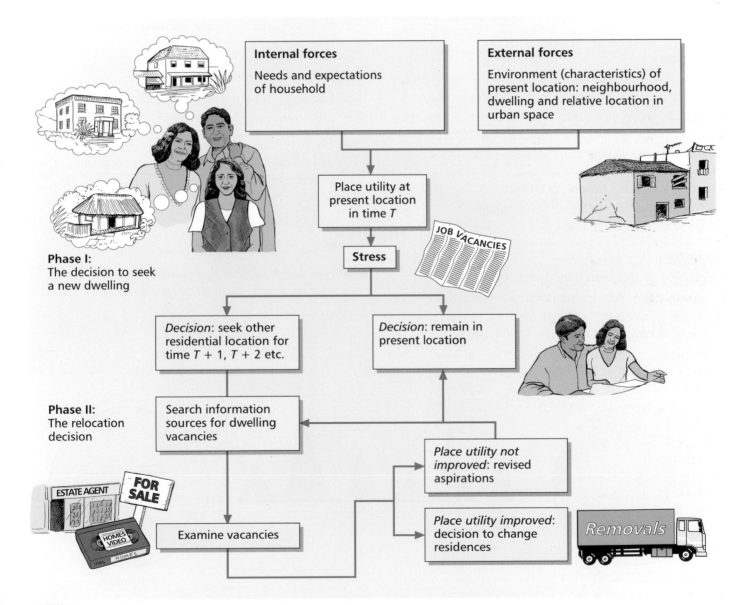

4.35   *Clark's model of migration decision*

example of a behavioural model (after Clark 1986). The **place utility** of a location experienced by an individual is a combination of both internal factors (e.g. the needs and preferences of the household) and external factors (e.g. type of environment, quality of neighbourhood, etc.). These create **stress** which may prompt individuals to evaluate their level of place utility. Where individuals perceive a decline in place utility they may decide to seek to move.

Another example of a behavioural model is Lee's **intervening obstacles** model (Figure 4.36). Both the origin and destination have 'push' and 'pull' factors. However, any simple comparison is complicated by the presence of intervening obstacles such as family responsibilities at the origin, or the high cost of moving, which may prevent migration occurring.

Behavioural approaches to migration suggest the decision about whether to move or stay depends upon the balance of push factors and pull factors. Examples of these include:

Push factors

- Decline in resources (such as coal) or the prices they attract; decrease in demand for a product or service.
- Loss of employment due to incompetence, changing employers' needs, automation or mechanisation.
- Discriminatory treatment on grounds of politics, religion or ethnicity.
- Cultural alienation from a community.
- Poor marriage or employment opportunities.
- Natural or human catastrophe.

Pull factors

- Improved employment opportunities.
- Opportunities for higher income, specialisation or training.
- Preferable environment or living conditions.

- Movement as a result of dependency on someone else who has moved, such as a spouse.
- Novel, rich or varied cultural, intellectual or recreational environment.

Considering push and pull factors is too simplistic to explain observed migrations. In addition, attention needs to be paid to intervening obstacles that can impede migrations. Understanding these obstacles suggests a need to understand the motives and meaning that people attach to migration, and the wider contexts in which migration takes place.

NOTING ACTIVITIES

1 What do you understand by the term 'migration'? Give some examples to illustrate different types of migration.

2 Distinguish between the following terms:

 a in-migration and immigration

 b out-migration and emigration.

3 Why might it be difficult to gather reliable data with which to study patterns of migration?

4 Classical studies of migration attempt to produce general laws about the process of migration. Describe and comment on the usefulness of the following models:

 a the gravity model

 b Zelinsky's model of mobility transition.

5 How do behavioural approaches to the study of migration differ from classical approaches?

6 Define the term 'place utility'.

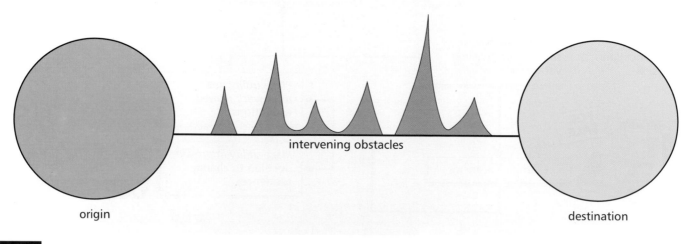

origin   intervening obstacles   destination

4.36 *Lee's intervening obstacles model*

Regional migration in England and Wales

This case study offers an account of the ways in which the population distribution of England and Wales has changed as a result of migration. It suggests that the forces driving this regional migration are linked to changes in the nature of the economy.

1851–1911

This period was one in which rural areas in Britain experienced prolonged and heavy depopulation. Areas such as the south-west of England lost population to the south Midlands, as did other agricultural areas such as East Anglia, rural Wales, Lincolnshire, the Yorkshire Wolds and the northern Pennines. This depopulation was linked to the decline in the demand for farm labour as mechanisation occurred. Farm workers were poorly paid compared with workers in the industrial cities, and decreasing job security led many to move to the rapidly growing and industrialising cities (Figure 4.37). These cities were on the coalfields and coasts (ports) of northern England and South Wales, and London, where there was a growing demand for labour, a variety of work opportunities and opportunities for educational and social advancement. A significant factor of migration at this time was the increased number of Irish-born people moving into the towns and cities of England and Wales.

1911–45

This period saw the continued depopulation of the remotest and most agriculturally dependent rural areas

4.37   *1851–1911: people moved to the cities*

such as Exmoor, central Wales, northern East Anglia, the Fens, large areas of Lincolnshire, and the Pennines. Also significant during this period was the loss of population from large conurbations and coalfield industrial towns. This was in part due to the desire of some people to live in rural residences and to continue to work in towns, and in part the result of the planned resettlement of sections of the population from inner-city areas. The coalfield areas lost population due to the stagnation or decline of the staple industries of these areas, such as textiles, shipbuilding, coalmining and heavy engineering.

Areas with a net gain of population were parts of the Midlands and the south-east of England (Figure 4.38). This

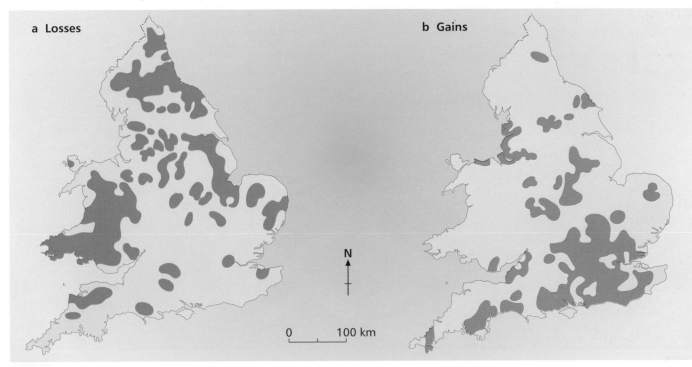

a Losses

b Gains

N

0   100 km

4.38   *Population change in England and Wales, 1921–47*

became known as the 'drift to the south' from the declining industrial regions to the new growth industries in manufacturing and services.

1945–2000

The period since the Second World War has been dominated by a relative shift of population away from the older manufacturing areas in the north and west of Britain towards the relatively prosperous areas of the south and east. At the same time, there has a been a shift away from the south-east towards the south-west and East Anglia. These patterns of migration are socially selective. In-migration tends to be dominated by young adults and contains a disproportionate share of skilled and young professionals. Migration to the south-west and East Anglia tends to be dominated by 'middle-aged professionals and managers' with their families, and by retirement migration.

These trends are linked to changes in Britain's economy over the past 30 years which have led to a decline in employment opportunities in the former industrial regions and an expansion in 'new' jobs in the service and tertiary sectors of the south. In addition, from the 1950s there was a policy of moving large numbers of people and jobs out of London into New Towns around the edge of London (Figure 4.39).

This period is marked by the expansion of housing and employment beyond the green belts of large cities as widespread ownership of cars and improved road systems brought more flexibility in movements to work. Around London, 'expanded towns' were added to the New Towns programme. Finally, an important feature of the post-war period was the immigration of people from the New Commonwealth and Pakistan in response to labour shortages in key industries. Most of these migrants settled in large cities and conurbations, near sources of employment.

NOTING ACTIVITIES

1 For each of the three periods of migration in England and Wales:

   a identify the main origins and destinations

   b suggest the main causes of migration.

2 Suggest some negative effects of out-migration on a region.

3 Suggest some positive and negative effects of in-migration on a region.

4.39 *A New Town: Harlow, in Essex*

STRUCTURED QUESTION 1

Study Figure 4.40, which shows the contribution of natural increase and net migration to population change in British counties from 1981 to 1991.

a Define the following terms:

   (i) natural increase

   (ii) net migration. *(2)*

b Name one county that experienced both natural decrease and net migration loss between 1981 and 1991. *(1)*

c Name the counties that experienced:

   (i) the highest natural increase between 1981 and 1991

   (ii) the highest net migration gain between 1981 and 1991. *(2)*

d In what ways might net migration loss alter the population structure of a county or region? *(3)*

e Suggest reasons for the natural decrease in population and net migrational gain in the counties along the south coast of England between 1981 and 1991. *(5)*

Study Figure 4.41.

f What evidence is there to suggest that the pattern of population change in this period was dominated by:

   (i) a north to south drift

   (ii) an urban to rural shift? *(6)*

g Suggest reasons for the pattern of population change in the period 1981–97. *(5)*

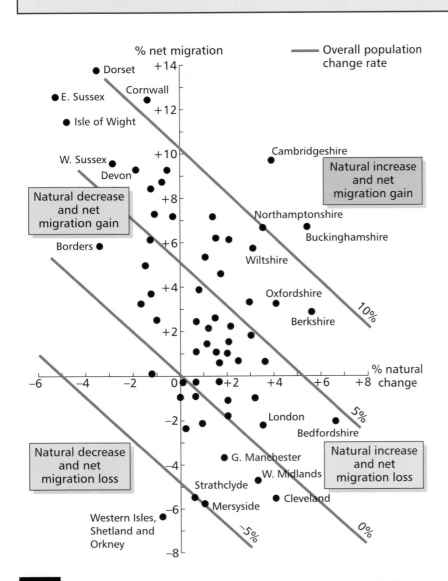

4.40 *Contribution of natural increase and net migration to population change in British counties, 1981–91*

4.41 *Population change in the UK, 1981–97*

Voluntary migration

One of the major explanations for voluntary migration is the existence of greater economic opportunities in certain places, which prompts individuals to seek out a 'better life' for themselves. Examples of this type of economic migration include the immigration of workers to the oil-rich Arab countries of the Middle East at a time of rising oil revenues, and the illegal crossing of the border between Mexico and the USA – the so-called 'Tortilla Curtain'. As the *box* below describes, one of the major migrations in the post-war period has been from the relatively economically less developed countries of southern Europe and North Africa to the more developed economies of western Europe. It has been argued that the growth of a global economy leads to greater distinctions between core and periphery, and workers migrate in search of better opportunities.

Migration to western Europe since 1945

Since 1945, millions of people have migrated from the underdeveloped parts of southern Europe, Africa, Asia, and the Americas to western Europe, in search of employment and better living standards. Nearly all the developed countries of western Europe have experienced large-scale immigration at the same time. Immigrant workers have become a necessity for the economies of the receiving countries.

The causes of these movements are varied and complex. But there are some general features which apply in almost all cases. It is possible to distinguish between the **pull factors** that have attracted migrants to certain west European countries, and the **push factors** that have caused them to leave their home countries.

The pull factors are a combination of economic, demographic and social developments in western Europe during the post-war period. There has been very rapid and almost continuous economic growth in most countries. Post-war reconstruction quickly absorbed the returning soldiers and there was soon a marked shortage of labour. At the same time population growth rates have been slow so that the labour force is not growing fast. In fact, each worker now has to support a growing number of inactive persons. Older people form an increasing proportion of the population, and the average length of time people spend in full-time education has increased.

An important social factor is that as the aspirations of the indigenous population of western European countries have risen, fewer have been willing to undertake work in unpleasant, unskilled manual jobs.

The push factors that cause migrants to leave their countries of origin are unemployment, poverty and underdevelopment. The countries of origin all tend to have higher rates of population growth. They also have low levels of per capita income and slow economic growth. In addition, there are great inequalities between different regions within these countries, between rural and urban areas, and between different social classes. For instance, Italy has a prosperous and fast-growing industrial economy in the north, but in the south there is a stagnant, backward agricultural economy. It is from southern areas such as Calabria and Sicily that the overwhelming majority of Italian emigrants come.

The causes of emigration are linked to underdevelopment. Uneven patterns of development between richer countries and poorer countries result in a surplus of people who cannot find employment in their own country and who are faced with a choice between poverty and near starvation at home, or emigration to western Europe where industry urgently needs labour.

CASE STUDY

French car workers

In the 1950s and 1960s the French automobile industry suffered from a shortage of workers. This labour shortage resulted from the low birth rates of the 1930s and 1940s. Previous labour shortages had been solved by recruiting workers from other European countries such as Belgium, Spain, Poland and Italy. The French were linked to the North African countries of Tunisia, Morocco and Algeria because these were former colonies. In the 1950s and 1960s employers sent agents to remote villages in these countries and sought out men who were willing to leave their homes to work in French factories.

For French firms, the North African workers were a cheap labour source (Figure 4.42). They worked for low wages and

4.42   *African worker in a French car factory*

did not have to be given unemployment or health care benefits. Work in the car industry was hard, involving long hours and repetition. However, these workers, who did not speak much French, were unlikely to get involved in trade union activities and had few employment rights, so were easy to hire and fire.

The attraction of working in France was linked to the lack of job opportunities for migrants in their own countries, and many were attracted to France by the prospect of economic prosperity.

Many workers did not return to their own countries, and once established in France, were joined by their families. This led to a substantial increase in the country's foreign-born population during the period 1967–74. By the mid-1970s the global recession, along with increased mechanisation, meant that employment opportunities were scarce. The French government began to restrict immigration. In 1977 the French government began to attempt to repatriate the workers and their families by offering them financial incentives to return. Some 45 000 people were repatriated, but many still remained. As France has attempted to manage economic changes, the North African population has suffered the brunt of policies to reduce welfare spending, and are more likely to live in poor-quality housing and to be unemployed. To make matters worse, recent years have seen the rebirth of the Front National, which campaigns on a xenophobic platform.

Despite these problems, many second-generation North Africans who were born in France see themselves as French-North Africans (they call themselves 'Beurs') and have little attachment to their 'homeland'.

What are the effects of voluntary migration?

Migration brings with it both advantages and disadvantages. For the receiving country, migrants are a source of labour, often relatively cheap, which allows it to achieve economic growth. For the migrants there is the prospect of greater material prosperity both for themselves and, through the return of earnings (**remittances**) to their home country, for their families. These remittances also help the economy of the source country since they increase incomes and the demand for goods and services. These advantages must be weighed against the personal costs involved in migration. The process of leaving one's home and family for another country is often painful and disorienting, especially where there are strong cultural differences, and many migrants face xenophobia and prejudice. In Britain, the experiences of first- and second-generation migrants from the New Commonwealth and Pakistan have been explored in various forms of popular culture, such as films, novels and music.

135

STRUCTURED QUESTION 2

Study Figure 4.43, which shows world migrant labour flows.

a Define the term 'labour migration'. *(1)*

b Using the map, identify two sources of labour migration and two destinations. *(4)*

c To what extent does the map support the idea that labour migration usually involves the movement of people from LEDCs to MEDCs? *(3)*

d Labour migration is often described as age and gender 'selective'. What do you understand by this term? *(2)*

e For each of the migration flows A, B and C, suggest:

  (i) the likely reasons for the flow

  (ii) the likely characteristics of the migrants involved. *(6)*

f What are the advantages and disadvantages of international labour migration for:

  (i) the host country

  (ii) the source country? *(4)*

Retirement migration

The idea of an 'ageing population' is discussed earlier in this section. Increasing economic prosperity in old age (enabled by the growth of private and state pensions) and increasing longevity (life expectancy) mean that people are spending longer periods in retirement. An important development is the growth of retirement migration. Figure 4.44 shows the migration flows of the elderly in England and Wales. The map shows two major patterns:

- Most retirement migration is over a short distance; for example, those living in the southern and western suburbs of London rarely look to move north and opt instead for nearby seaside resorts on the south coast. Those to the north and east of London are more likely to move to nearby East Anglia. Similarly, those leaving Manchester or Liverpool often move to the Lancashire or North Wales coasts.

- In addition to these short-distance migrations, the map shows evidence of long-distance flows. However, these tend to affect three areas: the south-west, the south, and East Anglia.

The result of these shifts is the development of a series of distinct retirement areas (Figure 4.45). When studying these maps it is important to avoid falling into the trap of thinking that *all* elderly people are involved in this process. Like any process of migration, retirement migration is socially selective, which means that only those who are able to

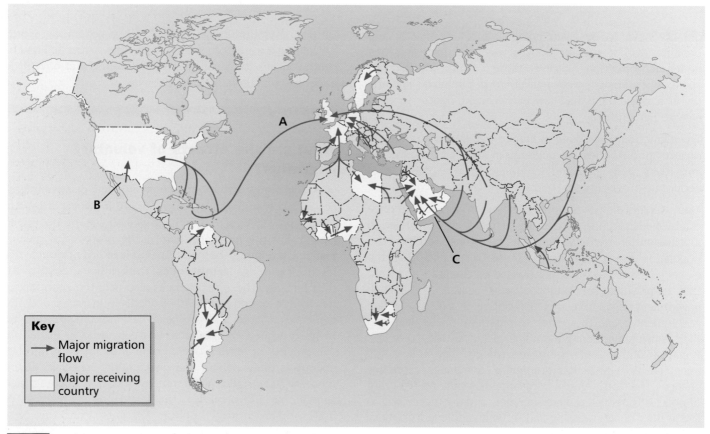

4.43 *Global labour migration flows*

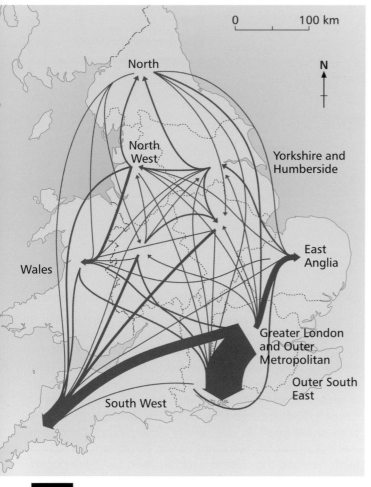

Retirement areas are those with a significantly higher than average proportion of people of retirement age and which are experiencing a growth in the proportion of elderly people.

4.44   *Migration flows of the elderly in England and Wales*

4.45   *Retirement areas in England and Wales*

afford to relocate are able to realise their desire to do so. Those unable to move risk being left behind as a residual population in towns and cities, either in suburbs which take on an increasingly 'old' feel, or in inner areas of cities that are more and more promoted for their youthful character and vibrancy. It is interesting to reflect that the only group at which property developers can legally direct their promotion of 'exclusive development' are the elderly – they could not do so on grounds of race or gender, for example.

EXTENDED ACTIVITY

Study Figure 4.45.

1 Make a copy of the map. Using an atlas to help you, label your map to show the main retirement areas.

2 What do you notice about the location of retirement areas?

3 Suggest reasons for the location of retirement areas in England and Wales.

4 What implications might such retirement migration have for places that experience it?

Forced migration

Forced migration occurs when the decision to relocate is made by people other than the migrants themselves. Examples of forced migrations include the Atlantic slave trade from the late 16th to the early 19th centuries when more than 10 million Africans were transported to work on plantations in North and South America and the Caribbean; and more recently, in 1983, Nigeria in West Africa expelled 2 million foreign workers, and a further 750 000 in 1985, the purpose being to reduce unemployment amongst its own people during the recession that followed the 1970s oil boom.

CASE STUDY

Forced migration in Indonesia

The term **transmigration** is used in Indonesia to describe the relocation of 6 million Indonesians from densely populated parts of the country such as Java, Bali, Madura and Lombok to some of the less densely populated 'outer islands' such as Sumatra, Sulawesi, Kalimantan and Irian Jaya (Figure 4.46).

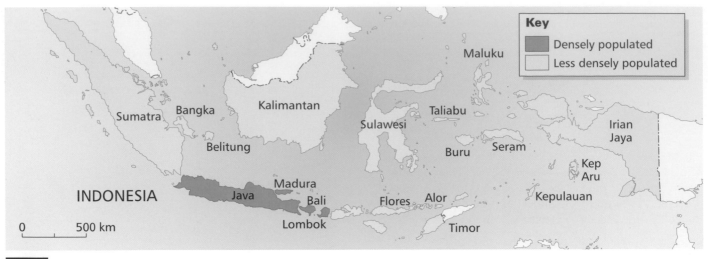

 4.46 *Indonesia: areas of most and least population density*

This resettlement began in 1905 under Dutch colonial rule and gathered pace after Indonesia gained its independence after 1945. In 1969 this policy was taken to a new level with the introduction of the government's 'transmigration programme'. This programme gained support from a number of international development agencies such as the World Bank, and assistance was given by governments of Netherlands, France and Germany. The programme has been controversial. Land on the island of Irian Jaya was taken by force in order to provide space for settlers from Java. The result has been growing unrest between Indonesian armed forces and the nationalist Irianese. Reports suggest that villages have been bombed, and people tortured and killed, with over 20 000 Irianese fleeing their homes and seeking refuge in Papua New Guinea. The Indonesian government wants to settle and assimilate all of Indonesia's tribal peoples, which includes moving Irian Jaya's entire indigenous population of 800 000 into resettlement sites on the island.

Transmigration is also going on elsewhere in Indonesia. In 1975 East Timor was seized by the army to provide areas for further resettlement from Java. In 1999 the island was the centre of violent clashes after the East Timorese voted for independence from Indonesia (Figure 4.47). The resettlement programme has also had damaging effects on the natural environment, including the loss of large areas of tropical rainforest in one of the most biologically diverse areas of the world. Sumatra has lost 2.3 million hectares of rainforest, and the cleared land has been rapidly degraded. It is estimated that 300 000 people are living in dire economic conditions, and lack any basic infrastructure such as clinics, schools and roads. They also suffer from malaria and other diseases. Many people are moving back to the towns and cities.

 4.47 *Conflict in East Timor*

The refugee 'crisis'

It is now common to talk about a 'refugee crisis' (Figure 4.48). **Refugees** are defined by the United Nations High Commission for Refugees (UNHCR) as:

'persons who owing to well-founded fear of persecution for reasons of race, religion, nationality or political opinions, are outside their country of origin and cannot, or owing to such fear, do not wish to avail themselves of the protection of that country'.

Figure 4.49 indicates that there been a dramatic growth in the estimated worldwide number of refugees since the mid-1970s. The graph shows two pronounced increases between 1980 and 1983 and 1988 and 1991. The rise in the number of global refugees is the result of a number of factors:

- changes in international communications have made it easier for potential migrants to gain information about possible destinations

- improvements in transport have made it easier and cheaper for potential migrants to escape their situation

- the increased number of states in the world means that movements that were once within a state are now classified as international flows

- the development of welfare states in many developed countries and the establishment of supra-national humanitarian bodies such as UNHCR and UNICEF allow for larger numbers of refugees to be accommodated.

Figure 4.50 describes the plight of such refugees.

4.48   *Refugees on the move*

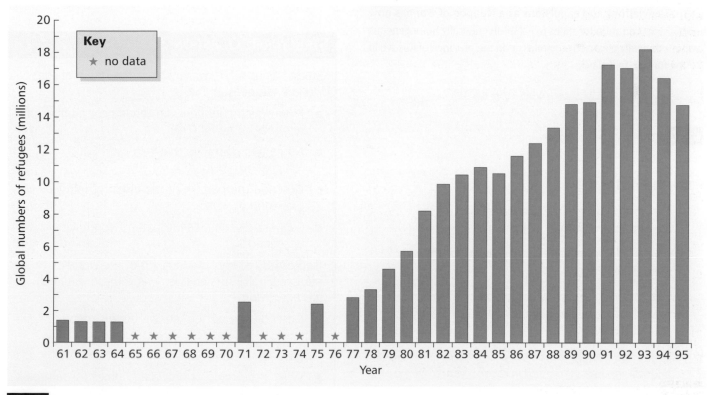

4.49   *The growing number of refugees, 1961–95*

The plight of refugees

THE HISTORY OF THE PAST 50 YEARS, and the 1990s in particular, has been the forced exodus of populations from their homes and great waves of people fleeing state-inspired ethnic-terrorism, ethnic cleansing and human rights abuses. It is estimated that there are 25 million people who have been forced to leave their countries, and another 25 million who are now internally displaced in their own countries but unable to return to their lands or villages.

But official figures are only the tip of the iceberg. Many more millions have been displaced by the partitions of countries, ecological disasters and massive development programmes. Mostly, these people do not figure in the statistics.

Few of these refugees return to their homes for good. Only 250 000 people of the millions displaced in the conflicts in the Balkans in the 1990s have returned, and very few of these to their original homes.

Major examples of the displacement of people in this century include the Jews and Armenians fleeing genocide, massive forced migrations in the Soviet Union under Stalin, millions of people fleeing communism, and more than 20 countries torn apart by populations fleeing oppression.

The increasing problem of refugees is the result of an awareness of governments that it is easier than ever to make people flee. New forms of warfare, the spread of light weapons and cheap landmines can easily panic populations. The use of mass evictions and expulsions as a weapon of war has now become common for states to establish culturally homogeneous (or ethnically cleansed) societies, as in the province of Kosovo in the former Yugoslavia.

The speed at which these flights and expulsions are taking place is quickening. Examples of these 'flash migrations' in recent years include the exodus of more than a million Iraqi Kurds, the Kosovan Albanians, over a million Rwandan citizens, and 2 million people within and from Liberia. The people most affected are the most vulnerable and victimised. These include minority groups, stateless people, and people with little or no political representation.

Added to these problems is the fact that refugees face a cold welcome in a world that feels overburdened with displaced people. Globally, there is mounting rejection of refugees, and states have been quick to erect physical and administrative barriers. People seeking asylum may spend years in fear of expulsion, be denied the right to settle permanently, be unable to work or denied welfare benefits.

For those who are able to return, life may be harder than ever. The trauma of flight and exile can be matched by the return which, although often imagined as joyful, can be as harrowing as exile itself. Humanitarian agencies report that people returning can feel as insecure as when they left, may have over-optimistic hopes of what they will find, and often lack land and resources with which to rebuild their lives. In most cases the social and physical infrastructure has been destroyed, work is scarce and there is little support from government agencies.

Source: based on 'The endless diaspora', The Guardian, 2 April 1999

4.50 *The plight of refugees*

NOTING ACTIVITIES

1 Explain the difference between 'voluntary migration' and 'forced migration'.

2 Give examples to illustrate both voluntary and forced migration.

3 Since the 1970s, many countries have been reluctant to accept economic migrants. Why do you think this is?

4 It has been suggested that the number of forced migrations has increased in recent years. Suggest possible reasons for this increase.

STRUCTURED QUESTION 3

Study Figure 4.51, which shows the origin of the world's refugees in 1995.

a From which continent did the largest number of refugees come in 1995? *(1)*

b Name two countries that had over 3000 refugees in 1995. *(2)*

c Describe the pattern of the origin of refugees in the world in 1995. *(2)*

d Suggest reasons for the pattern you have described. *(4)*

Figure 4.52 shows the location of the world's refugees in 1995 – that is, where these refugees are now living.

e Describe the pattern shown by this map. *(2)*

f Suggest possible reasons for the pattern you have described. *(4)*

g Why is the issue of providing homes for refugees considered a politically sensitive issue? *(4)*

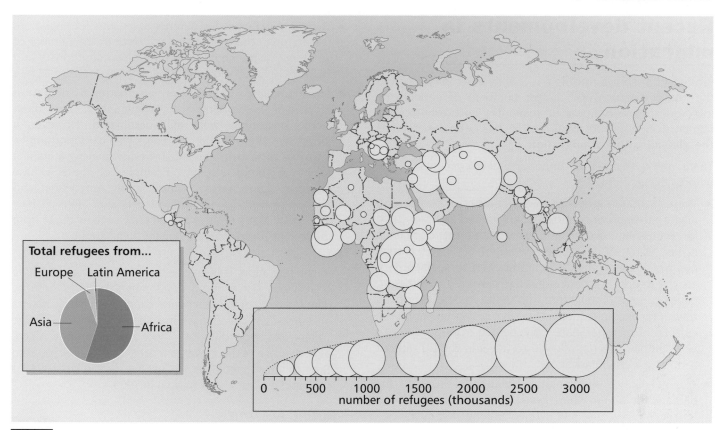

4.51 *Origin of the world's refugees, 1995*

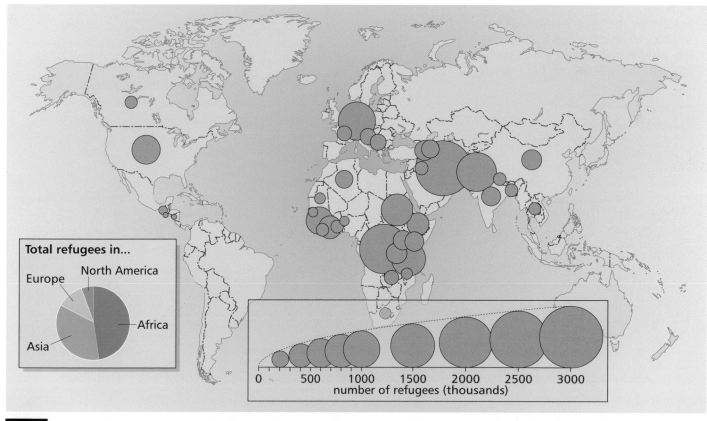

4.52 *Location of the world's refugees, 1995*

Recent developments in migration

Figure 4.53 shows the main channels of migration in the contemporary world. Larger numbers of nations are involved in these flows of migrants, and ever-growing numbers of people are involved. This has been described as **the globalisation of migration**.

There are a number of important developments in the nature of migration:

- There has been a decline in the number of people who migrate to another country and are immediately accepted for permanent settlement and granted full citizenship rights. Many MEDCs have attempted to limit the total number of immigrants who are allowed to seek entry or are granted it. A recent trend has been for countries to develop joint immigration policies. In the European Union, for example, there is freedom of movement for the citizens of member states, but people from outside this political bloc are not allowed to travel in the EU without acquiring a visa in advance.

- A large part of international migration is linked to the demand for labour. Where low-skill labour is required, this is strictly monitored and defined as 'contract migration'. Examples of this include *gastarbeiters* (guest workers) in Germany, oilworkers in the Gulf and, increasingly, labour in the 'Tiger' economies of South-east Asia.

- Flows of migrants are linked to disparities in economic wealth, as unskilled labour seeks work. Due to tighter restrictions, much of this migration is 'hidden' and illegal. It is estimated that there are as many as 5.5 million illegal aliens in the USA, mostly from Mexico.

- There are an increasing number of reasons for migration. For example, there has been a growth in the number of temporary movements by business-people, tourists and students.

- An increasing number of international migration movements are undertaken by women. In the past, most labour migrations were male-dominated. However, a growth in the number of employment opportunities available to women has led to women playing a leading role in migration streams. Examples of these include Filipinos, Cape Verdians and South Americans who migrate to work as domestic helpers and carers in Italy and Spain. Many refugee movements, such as those from the former Yugoslavia, are female-dominated.

- International migration is more and more dominated by forced migration. This may be the result of political, ethnic or religious persecution; development projects such as dam-building; or environmental change. Such movements are sometimes called 'flash migrations' – they are of large volume but short duration. These are common in Africa, as in the case of the movement of one million people from Rwanda into Congo (Zaire) in just five days in July 1994, or the rapid evacuation of Kosovo in 1999.

NOTING ACTIVITIES

1 What is meant by the term 'the globalisation of migration'?

2 Look at Figure 4.53. On a copy of the map, annotate some of the arrows to explain the reasons for some of the migrations shown (use examples from this chapter).

3 Over the course of your studies, use newspapers, journals, CD-ROMs and internet sites to build up a resource file on examples of migration.

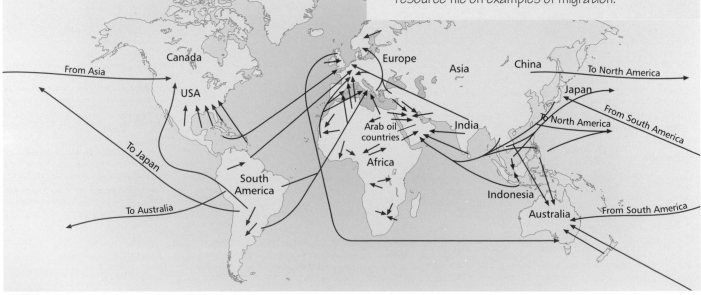

4.53 *Major global migration flows since 1973*

A An introduction

Look at Figure 5.1. The advertisement appeared in the magazine *Country Living* in the mid-1990s. The magazine was one of the fastest-growing in terms of sales throughout the 1980s and 1990s, despite the fact that Britain is a predominantly urban country – four out of five Britons live in towns or cities. Given this fact, it is tempting to suggest that the magazine is popular because it promotes an image of rural living, away from the worries and anxieties associated with urban life. This is also the way the advertisement for Milton Keynes works. Look carefully at the text: it is full of comments about the 'idyllic English pub', 'community', 'peaceful rolling countryside', 'village greens', 'thatched pubs', and so on.

As geographers, though, we might want to ask some questions about the images and words in this advert. For instance, if all these managing directors have moved in to the villages around Milton Keynes and are now the 'locals', what has happened to all the original 'locals'? There are other questions we could ask about the types of work these people do, and where they buy goods and services. In fact, a simple picture such as this raises a whole set of geographical questions about the nature of rural settlement.

This section of the book is about the ways in which geography can contribute to an understanding of rural areas.

Mac Makino, MBD of New Wave Logistics (UK) Ltd, enjoys a quiet drink with his wife, Nobuko, at The Thatched Inn, Adstock.

Gloria James (centre), General Manager of Click Systems Ltd, drops in to meet friends at Ye Olde Swan Tavern, Wroughton-on-the-Green. Legendary haunt of Dick Turpin.

Malcolm Brighton (left), MD of DRS Plc, chats with a neighbour outside The Swan, a 16th century inn at Milton Keynes village.

MD of Datanews Ltd, Stella Hitner (right), with her husband Bob and a friend, relaxing over drinks outside The Three Locks, Stoke Hammond.

Friedhelm Stellet (right), Chief Executive GB of SCHÜCO International KG, unwinds in the garden of The Bell at Beachampton.

The Black Horse at Great Linford. Favourite watering hole of Dick Jones, General Manager of Dana Distribution, Milton Keynes, and his wife Pamela.

The idyllic English village pub.

The focus of the community; a "drop in, see a friend" sort of place.

A dying breed, you may say, but not if you live within 10 miles of Milton Keynes.

Whether you're in one of the 13 original villages that nestle within the city itself or in one of the typical English villages that lie in the peaceful rolling countryside nearby, you'll find the kind of village pubs that you thought were fast disappearing.

Locals.

Pubs on village greens. Pubs next to quiet picturesque canals. And a 16th century thatched pub right in the very heart of Milton Keynes, with ancient beams and peaceful garden haunted only by birdsong and convivial atmosphere.

No wonder the lucky locals who have chosen to relocate their business to Milton Keynes seem pleased with themselves.

They are able to enjoy an ideal lifestyle. English village life at its best combined with a business environment that is world famous for its modernity and efficiency.

Yet the two are only a few miles apart. And a world away from everywhere else.

To find out more about relocating your business to Milton Keynes, call CNT on 0800 721 721.

The City of **Milton Keynes** ◆

5.1 *'Come to Milton Keynes'*

What is 'rural'?

Look at Figure 5.2. It shows a series of scenes that might be described as 'rural'. However, it is difficult to reach agreement on what is meant by 'rural'. Geographers have spent a lot of time attempting to define what they mean by the term. Various definitions have been suggested.

1 Definitions based on population

One common way to define 'rural' is based on a measure of population size or density. For example, Denmark, France, Italy, Spain and Sweden all classify rural areas as those administrative units that fall below a defined threshold (minimum) population. A problem with this is that the **threshold** population used varies greatly. Italy and Spain, for example, classify rural areas as those that have a population of less than 10 000 inhabitants. But in France the figure is 2000 and in Ireland it is 100. With reference to England, the Countryside Agency defines a settlement as 'rural' if it has a population of 10 000 or less. Figure 5.3 shows the size and number of rural settlements in England. Notice that the majority of rural settlements in England have a population of less than 500. As we will see, the small size of many rural settlements means that they have some unique features.

Settlement size	Number of settlements	Total population within settlement size band (millions)
<500	13 142	1.7
500–999	1 581	1.1
1000–2999	1 330	2.3
3000–9999	682	3.7
Total >10 000	16 735	8.8

5.3 *Rural settlements in England*

2 Definitions based on function

Geographers have assumed that rural areas share the following characteristics:

- they are dominated by extensive land uses (such as agriculture and forestry), or large open spaces of undeveloped land
- they contain smaller, lower-order settlements (such as hamlets and villages)
- they are based on a way of life characterised by a close-knit community.

Defining the 'rural' according to function has been useful in allowing geographers to produce objective measures. For example, the geographer Paul Cloke devised an **index of rurality** which involved looking at a series of statistical measures of rurality (Figure 5.4).

From his analysis Cloke classified places into four categories, ranging from 'extreme rural' to 'extreme non-rural'. Figure 5.5 shows the results of such an analysis using information from the 1991 Census.

5.2 *What is 'rural Britain'?*

•	**Occupancy rates**	The proportion of households or dwellings actually in occupation.
•	**Commuting**	The percentage of residents who work outside the local area.
•	**Female population**	The proportion of the total population who are women aged 15–44.
•	**Amenities**	The percentage of households with exclusive use of fixed bath and inside WC.
•	**Population density**	The number of people per km².
•	**Agricultural employment**	The proportion of the workforce employed in agriculture.
•	**Elderly population**	The percentage of the population aged 65 or over.
•	**Remoteness**	The distance from the nearest settlement of more than 50 000 people.

5.4 *Definitions of some indicators of rurality*

3 Definitions that focus on the forces affecting 'rural' areas

One of the problems of the functional approach is that it focuses on the characteristics of the rural area itself, whilst much of what happens to rural areas depends on events outside of them. For instance, the decision of a transnational corporation to close one of its branch plants in a rural area often rests with a company's headquarters in a large urban area or even in a different country.

4 Definitions based on perception

Defining 'rural' by the functions that go on there has been challenged by geographers, who view the rural as a social construct. By this they mean that 'rural' means different things to different people. It all depends on an individual's **perception**. They are interested in the meanings that people attach to rural areas and how they influence their behaviour. As we will see later, the belief that rural areas are places where the pace of life is slower, where the quality of life is better, has led to some people seeking to leave towns and cities and live in rural areas.

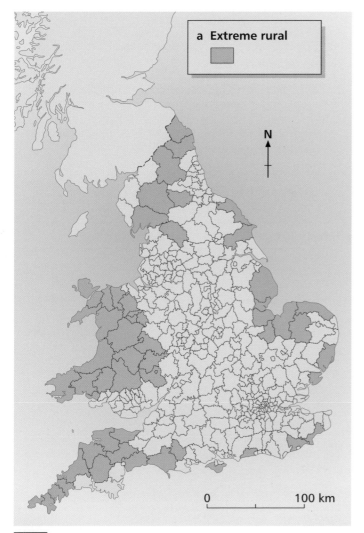

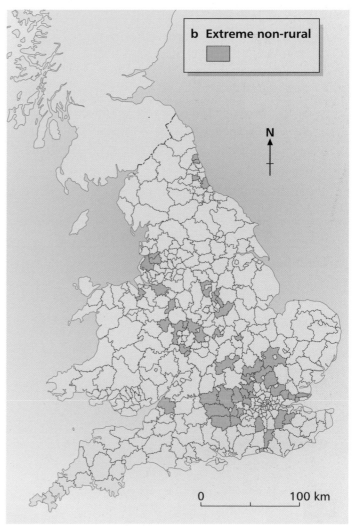

5.5 *Cloke's 'extreme rural' and 'extreme non-rural' areas (1991)*

145

NOTING ACTIVITIES

1 **a** Make a copy of Figure 5.6. Look at the list of words under the following headings. Place them into what you think are the correct boxes on your copy of Figure 5.6.

Land use	Landscape	Population	Quality of life
farms	green	densely populated	friendly
factories	dereliction	sparsely populated	crime
industry	open	immigrants	lonely
agriculture	natural	commuters	traditional
forestry	chimneys	wealthy	modern
offices	sky	retired	noisy
crops	clean	sparse	safe
cattle	trees	homeless	healthy
villages	built-up	farmers	peaceful
hedgerows	fresh air	middle class	boring
traffic	congested	villagers	poor
towns			old-fashioned

b Compare your results with those of other people in your class.

c Comment on the ease with which you were able to allocate each term into the boxes.

2 In groups of two or three, discuss the meaning of the term 'rural'. Attempt to write a short definition of the term.

3 Summarise the ways in which geographers have defined the term 'rural'. Comment on the advantages and disadvantages of each.

4 For a rural settlement with which you are familiar, outline a series of measurements you could undertake to construct an index of rurality.

5 Study Figure 5.5.

a Using an atlas to help you, identify the parts of England and Wales described as being 'extreme rural'.

b Why are many 'extreme rural' areas found at the periphery of England and Wales?

c With the aid of an atlas, comment on the pattern of 'extreme non-rural areas'.

d How useful do you think it is (for planning purposes) to distinguish between areas with different degrees of rurality?

6 Some geographers have suggested that 'rural' is a social construct.

a Ask a series of people to say what they understand by the term.

b Were their answers similar to the definitions used by geographers?

c Where do you think such 'everyday' understandings of the term 'rural' come from?

	Definitely rural	Definitely urban	Both/Neither
Land use			
Landscape			
Population			
Quality of life			

5.6 *Rurality grid*

B Rural settlement patterns

Studies of rural settlement have looked at the origins of settlement, attempting to answer the question of why villages grew up in particular places. In prehistoric times the density of lowland vegetation meant that settlements occupied the higher ground. These **sites** were dry, accessible and easy to defend. Over time, as agricultural technology improved, vegetation was cleared and land in valleys was cultivated. Settlement developed on the lower ground which offered the advantages of shelter, water and deeper soils. Often, proximity to a spring or a wood was an advantage. Later, as trade developed between communities, sites were adopted for commercial reasons, for example because they were at a crossroads.

Two main types of settlement pattern emerged. The first is a **nucleated** pattern (Figure 5.7), where individual dwellings are grouped or packed closely together. The second is a **dispersed** pattern (Figure 5.8) where individual dwellings are spread out. Figure 5.9 discusses some of the factors that promote the nucleation of settlements, along with examples of where such factors operate. The ideas of site and the

5.7 *Nucleated settlement*

5.8 *Dispersed settlement*

Defence

This was an important factor in the past, when outlaw bands roamed freely and farmers defended themselves by grouping together in villages. More recent examples are the Kenyan Mau Mau emergency which led to over 1 million people gathering together to find safety, or in the conflict between Israel and Palestine, ongoing after 1948, where a large number of farmers live in concentrated settlements and commute to the fields as a response to guerrilla attacks by Palestinian forces. Where peace and stability exist, this reason for nucleation is unimportant.

Family and clan ties

Where the settlement of an area was made by people who were blood relatives or had strong ties, a nucleated pattern of settlement was likely. These settlements are easily identified by place-name evidence. For example, many places in western Europe have the suffixes *–ingen*, *–inge*, *–ing* and *–ange*. These are Germanic and mean 'the people of'. Where settlement was by individual pioneer families where blood ties and group belonging were weaker, a more dispersed pattern of settlement developed.

Abundance of water

Where water was scarce, nucleated patterns of settlement developed. For example, areas of permeable rock such as limestone encouraged clusters of settlements where water was available from deep wells or springs, as on the South Downs in Wiltshire. In marshy areas, settlement clustered on available 'dry points'.

Patterns of inheritance

Areas of nucleated settlement may also be related to practices of inheritance. Where land is divided equally between sons and daughters of landowners, a nucleated pattern will develop as successive generations build houses on the same site. Where inheritance is decided on the law of primogeniture (land is passed to the eldest son or daughter) a dispersed pattern will emerge as other family members move away to build farmsteads.

Political or religious reasons

In the former Soviet Union the government had a policy of concentrating people into villages of 1000 or more. This was partly to improve service provision but also a means of maintaining ideological control of the peasantry. The same pattern was true of communist China, which established large communes in order to increase food production. An example of a religious reason for nucleation is the Mormon colonisation of the USA.

5.9 *Factors favouring the nucleation of settlement*

degree of nucleation are useful in understanding the way in which settlement developed.

In addition, geographers have classified rural settlements according to the arrangement of houses within them (**morphology**, or form). Figure 5.10 is one attempt to do this.

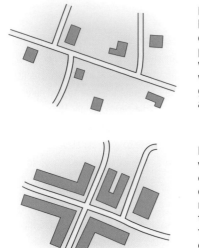

Fragmented or loose-knit villages have no original nucleus. They probably developed where areas of woodland were gradually cleared for agriculture.

Nucleated or clustered villages form at route centres. They may originally reflect the need for defence, or the place where farming was carried out communally.

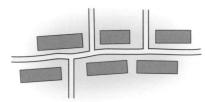

Linear villages are elongated along roads, rivers, ridges or valleys.

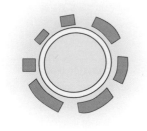

Open villages are nucleated villages around a central open space, such as a pond or a green. In Britain these were probably linked with the need for defence.

0   500 m

5.10 *A classification of settlement form*

NOTING ACTIVITIES

1 **a** What is the 'site' of a settlement?

   b With reference to your own home settlement, discuss the advantages of its site for early settlers.

2 Study Figures 5.7 and 5.8. Compare the settlement patterns of the two villages, using appropriate terminology.

Central places

A rural community may be regarded as the population living within the **sphere of influence**, or **catchment area**, of a central place upon which the people are largely dependent. In these central places are found shops, supermarkets, banks, the offices of commercial businesses, and services such as secondary schools, health centres and hospitals. These services are only found in larger centres because they require a certain size of population to make their provision economically viable.

During the 20th century, and especially since the widespread availability of motor transport, the catchment areas, or sphere of influence, of larger central places have expanded, and functions such as shops and services have become concentrated in larger settlements. This has meant that many villages have lost many of the functions which they possessed in the early part of this century. The result of these processes has been the development of a complex **hierarchy** of centres, from hamlets and small villages at the lower end, to large towns at the other. Apart from a few houses or farmsteads, the hamlet or small village usually contains little more than a church, a pub, a post-box and telephone kiosk. The larger urban centres provide more specialised, less frequently required functions.

EXTENDED ACTIVITY

The purpose of this activity is to allow you to consider some of the factors that affect the level of service provision in a rural area.

Figure 5.11 is a map showing the county of Northamptonshire, its major towns, roads and rural settlements, and Figure 5.12 shows the population sizes for 11 villages in Northamptonshire and the services available in these villages.

1 Draw a graph to show the relationship between population size and the number of services (functions) available. Plot the number of services available on the *y* axis and the population size on the *x* axis.

2 Write a paragraph summarising the results of your graph. What is the relationship between population size and number of functions?

3 On Figure 5.12, look at the provision of bus services in the villages. Why do you think most villages have a shopping bus service but only one has an evening bus service?

Now study Figure 5.13, which shows the percentage of villages in Northamptonshire with named functions.

4 Using the following categories, place each function in a copy of the classification grid (Figure 5.14). The first function has been done for you.

> 80%	high presence
50%–79%	medium presence
20%–49%	low presence
< 20%	poor presence

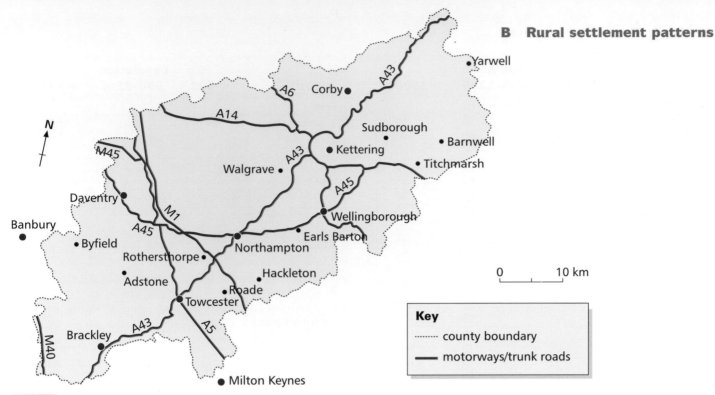

5.11   *Location of selected villages in Northamptonshire*

	Adstone	Sudborough	Yarwell	Barnwell	Rothersthorpe	Titchmarsh	Walgrave	Hackleton	Byfield	Roade	Earls Barton
Population (1991)	**93**	**131**	**246**	**370**	**415**	**546**	**747**	**785**	**1205**	**2176**	**4810**
General store			*	*		*	*	*	*	*	*
Non-food shop								*	*	*	*
Post office			*	*	*	*	*	*	*	*	*
Bank										*	*
Public house		*	*	*	*	*	*	*	*	*	*
Petrol station							*	*	*	*	*
Mobile food shop	*	*	*	*		*	*		*	*	
Mobile library	*	*	*	*	*	*	*	*	*	*	
Shopping bus service		*	*	*	*	*	*	*	*	*	*
Journey-to-work bus			*		*		*	*	*	*	*
Evening bus service											*
Sunday bus service											*
Post box	*	*	*	*	*	*	*	*	*	*	*
Village noticeboard	*	*	*	*	*	*	*	*	*	*	*
Village hall	*	*	*		*	*	*	*	*	*	*
Public library											*
Doctors' surgery							*	*	*	*	*
Police station											*
Primary school				*	*	*	*	*	*	*	*
Secondary school										*	
Play area			*				*	*	*	*	
Sports ground			*	*	*		*	*	*	*	*

5.12   *Availability of functions in selected villages in Northamptonshire*

Facility	Population < 600	600–1200	> 1200
General store	32	88	100
Non-food shop	6	33	88
Post office	44	95	100
Bank	0	0	15
Public house	67	98	98
Petrol station	13	43	83
Mobile food shop	75	80	80
Mobile library	94	100	93
Shopping bus service	93	100	100
Journey-to-work bus	44	93	85
Evening bus service	1	13	29
Sunday bus service	7	25	27
Post box	100	100	100
Village noticeboard	95	98	100
Village hall	74	98	95
Public library	0	0	29
Doctors' surgery	8	38	68
Police station	0	8	22
Primary school	25	95	100
Secondary school	0	3	15
Play area	42	85	100
Sports ground	30	90	95

5.13 *Percentage of villages in Northamptonshire with named facility, by size of settlement, 1992*

5 Identify functions where the provision is **high** in all-sized villages.

6 Identify functions where the provision is **low** or **poor** in the smallest villages, but **high** in the largest villages.

7 What factors might explain the level of service provision in rural settlements in Northamptonshire?

Central place theory

The Northamptonshire example reveals that there is a clear **settlement hierarchy** in which, at the top, large centres are able to provide a greater number of functions, whilst at the bottom, small centres have very few functions. This tendency for the development of settlement hierarchies was explored by the German geographer Walter Christaller in the 1930s. Christaller developed a **central place theory** to explain the size and spacing of settlements as a result of people's shopping behaviour. In his studies of southern Germany, Christaller noticed that there were large numbers of smaller places which offered a limited number of shops and services (or functions). He noted that these were located at quite short distances from one another. Large towns, however, were fewer and farther between, but offered a much greater number of functions, attracting shoppers from much greater distances.

In order to explain this observed pattern, Christaller used the ideas of the range and threshold of a product or service (Figure 5.15). The **range** is the maximum distance that

Group	Population < 600	600–1200	> 1200
Retail			
1 General store	low	high	high
2			
3			
4			
5			
Medical			
1			
Educational			
1			
2			
3			
4			
Public transport			
1			
2			
3			
4			
Community			
1			
2			
3			
4			
5			
6			
7			
8			

5.14 *Classification grid*

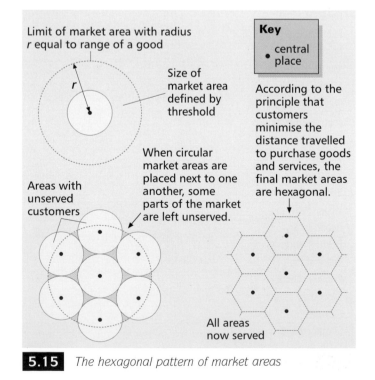

5.15 *The hexagonal pattern of market areas*

customers will normally travel to obtain a good or service. High-order goods and services are those that are relatively expensive and purchased infrequently (for example, specialised medical treatment or furniture); they have the greatest range. Low-order goods and services are less expensive and are required at frequent intervals (for example, a newspaper or hairdresser); they have a very short range. The **threshold** of a good or service is the minimum market area with enough potential customers to make the provision of that good or service profitable. High-order services such as a university have a large threshold, whilst a grocery store has a low threshold.

From these ideas, Christaller worked out the theoretical pattern of settlements. He based his theory on a number of assumptions:

- There is an unbounded uniform plain on which there is equal ease of transport in all directions. Transport costs increase in direct proportion to the distance travelled, and there is only one type of transport.
- Population is evenly distributed over the plain.
- Central places are located on the plain to provide goods and services to their hinterlands (surrounding areas).
- Consumers visit the nearest central place in order to obtain the goods and services they require.
- The suppliers of goods and services act as 'economic men' by attempting to maximise their profits by locating on the plain to obtain the largest possible market.
- Some central places offer a greater range of functions (goods and services) and higher-order functions. Others provide a smaller range of functions and lower-order functions.
- All consumers have the same income and the same demand for goods and services.

How does central place theory work?

Study Figure 5.15. It shows that each central place should, in theory, have a circular trade area. However, if three or more circles are put together, unserved spaces will exist. In order to eliminate these spaces the circular market areas must overlap. Since people in these overlapping zones will visit their nearest centre (remember the assumption of travelling the minimum distance), the final market areas will be hexagonal. The hexagonal pattern is the most efficient way of packing market areas onto the plain so that every resident is served.

Christaller identified three patterns of settlement based on different principles. These were described by **k values** which indicate the number of lower-order centres dominated by a higher-order settlement (Figure 5.16). The most commonly used k values are:

- $k = 3$ – the **marketing principle** is that which most efficiently serves the needs of consumers
- $k = 4$ – the **traffic principle** maximises the number of settlements along straight lines - this increases accessibility
- $k = 7$ – the **administrative principle** is the most efficient pattern for providing government over an area.

---

k = 3

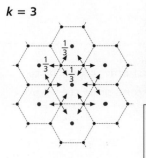

In the $k = 3$ network, each higher-order centre serves the equivalent of three lower-order centres. This is made up of the whole of the market for its own centre, plus $\frac{1}{3}$ of the market shares of the six surrounding lower-order centres.

i.e. $1 + \frac{1}{3} + \frac{1}{3} + \frac{1}{3} + \frac{1}{3} + \frac{1}{3} + \frac{1}{3} = 3$

Key
- higher-order centre
- lower-order centre
- ← direction and proportion of custom from lower-order centres to higher-order ones

k = 7

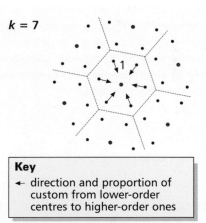

Key
- ← direction and proportion of custom from lower-order centres to higher-order ones

In the $k = 7$ network is based on the administrative principle in which one higher-order centre serves the population of six lower-order centres. In this case the higher-order centre serves its own population plus that of six lower-order centres.

i.e. $1 + 1 + 1 + 1 + 1 + 1 + 1 = 7$

In the $k = 4$ network, transport between settlements is the important principle. The higher-order centre serves its own market plus one-half of the population of the six lower-order centres.

i.e. $1 + \frac{1}{2} + \frac{1}{2} + \frac{1}{2} + \frac{1}{2} + \frac{1}{2} + \frac{1}{2} = 4$

k = 4

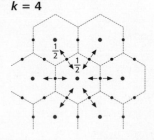

5.16   *k-values*

Does central place theory apply to the 'real world'?

Christaller's model has been very influential in the study of settlement patterns since the 1960s. It provides a model against which actual settlement patterns can be compared. In addition, it was used by planners in the post-war period to make decisions about which settlements to expand and which to allow to contract. The model has attracted much criticism, since it is based on a set of conditions that rarely exist in reality.

- The model assumes that services are the only reason for settlement. It does not recognise that other functions (such as manufacturing industry) also create employment and population.

- It offers a one-dimensional picture of human behaviour. In practice, consumers do not always visit their nearest store.

- Christaller's theory was set in the 1930s and does not acknowledge the importance of recent social and economic changes such as population mobility, urban growth, the emergence of hypermarkets and the importance of planning regulations in the location of service centres.

- The model assumes that economic factors are the most important in affecting settlement patterns. In doing so it neglects important historical influences.

- The evidence that Christaller's model describes actual settlement patterns is patchy.

Like all models, Christaller's central place theory can be criticised as being a simplification of reality. Perhaps the most important thing to recognise is that Christaller's model represents a rural landscape of the past, and is of limited use in understanding the processes at work in rural Britain in the present. As a result, geographers have looked elsewhere in their attempts to understand the nature of rural areas (see *box* opposite).

NOTING ACTIVITIES

1   Define the following terms:

   a   threshold   **c**   high-order good

   b   range   **d**   low-order good.

2   Summarise the assumptions on which Christaller's central place theory is based. Comment on how realistic you think they are.

3   With the aid of a simple diagram, explain how the *k* = 3 pattern of settlement can be derived from Christaller's theory.

4   Briefly outline what is meant by the 'traffic principle' and 'administrative principle'.

STRUCTURED QUESTION

Study Figure 5.17, which shows market areas based on Christaller's central place model.

a   In what ways might a second-order central place differ from a first-order central place?   *(2)*

b   Figure 5.17a shows the position of first-order centres according to the marketing principle (*k* = 3). Explain what is meant by the 'marketing principle'.   *(2)*

c   On a copy of Figure 5.17b, mark the position of first-order centres according to the traffic principle (*k* = 4).   *(3)*

d   Explain what is meant by the 'traffic principle'.   *(3)*

e   Comment on the strengths and weaknesses of Christaller's central place theory.   *(5)*

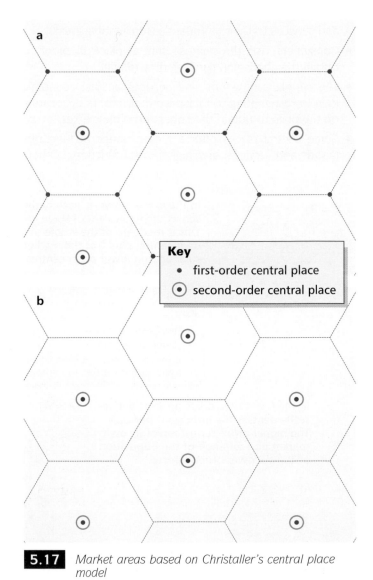

Key
- first-order central place
- ⊙ second-order central place

5.17   *Market areas based on Christaller's central place model*

The search for order

One example of the search for order in settlement patterns is the use of a technique known as **nearest neighbour analysis**. This attempts to measure the distribution of settlements according to whether they are clustered, random or regular. An example of nearest neighbour analysis is given in Figure 5.18. Nearest neighbour analysis is a technique used by geographers to search for order and regularity in the landscape. It is commonly used to compare patterns of settlement distribution in different places. However, it does not offer any explanation for the patterns – this is best achieved by careful study of maps, along with an understanding of the historical reasons for settlement.

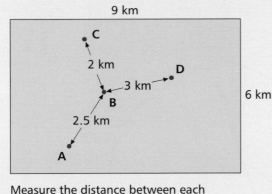

9 km

6 km

clustered   random   regular

The *Rn* statistic will produce a result between 0 and 2.15.

1   Measure the distance between each settlement and its nearest neighbour:

A → B = 2.5 km
B → C = 2 km
C → B = 2 km
D → B = 3 km

2   Calculate the average distance between settlements:

$$2.5 + 2 + 2 + 3 = \frac{9.5}{4} = 2.4$$

3   Calculate the size of the area:

9 km × 6 km = 54 km²

4   Apply the following formula:

$$Rn = 2\bar{d}\sqrt{\frac{n}{A}}$$ where   *Rn* = the nearest neighbour index
A = the size of the area
$\bar{d}$ = the average (mean) distance between settlements
n = the number of settlements

i.e. $$Rn = 2 \times 2.4\sqrt{\frac{4}{54}}$$

$$= 4.8 \times 0.27$$

$$Rn = 1.29$$

This figure suggests a fairly random distribution.

5.18   *Nearest neighbour analysis*

Rural settlement patterns in Norfolk

The aim of this case study is to provide an example of how some of the ideas in this chapter apply to actual settlement patterns in rural areas.

Figure 5.19 is a map showing the location of some of the main settlements in Norfolk. Rural settlement patterns vary throughout Norfolk, and there are five distinct areas. The following paragraphs describe the main features of settlement patterns in these areas. As you read them, locate the places mentioned on the map.

North West Norfolk is an area of arable and grassland landscapes. It is dominated by large estates, which means that there are relatively few settlements. The area is quite dry, which in the past limited the amount of water available. As a result, settlement developed in the valleys of the rivers Burn, Stiffkey and upper Nar. The villages of the Creakes, Walsingham and West Acre are examples of river valley villages which developed where pastureland was available for grazing. Until the 18th century, the area was used mainly for grazing sheep, so settlement is sparse. Villages are widely spaced (about 4–5 miles apart) and connected by straight roads. They are often nucleated settlements, clustered around village greens or ponds, such as the 'pond' villages of Stanhoe, Docking and Great Massingham. The transport network in the area is undeveloped and the villages have escaped pressure for development. There is visitor pressure on country houses such as Sandringham, which puts pressure on rural roads. In addition, the villages of Heacham, Snettisham and Dersingham have attracted large numbers of caravans, beach huts and bungalows.

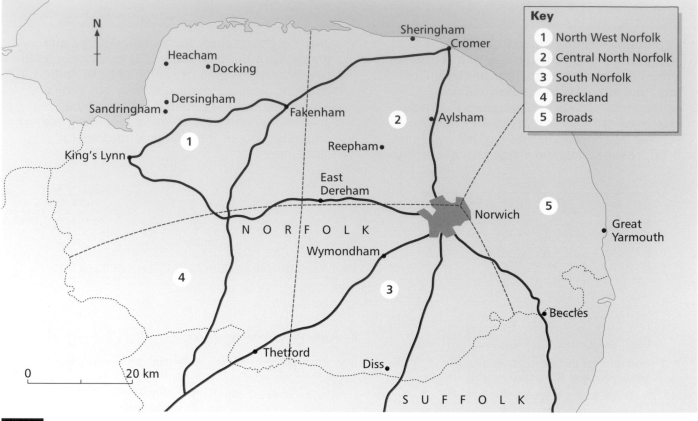

5.19 *Sketch map of Norfolk*

Central North Norfolk is the area to the north-west of Norwich. It is a rural area largely dependant on by arable farming. In terms of settlement, it is dominated by Norwich, and there is no other large centre of population. There are relatively few towns, but those such as Aylsham and East Dereham are active centres, each with its own identity. Larger villages, such as Hingham and Reepham, have more functions than usual because they are relatively distant from other centres. To the north of this region, the settlement pattern is dominated by small villages scattered among the dense network of narrow, twisting lanes. On the coast, Cromer and Sheringham are the major settlements,

5.20 *A rural settlement in Norfolk: Aylsham*

established as holiday resorts in the early 19th century. In recent years, the settlements close to Norwich along the A47 and A1067 have experienced 'creeping suburbanisation' and the agglomeration (merging) of once distinct villages.

South Norfolk occupies a large area of central East Anglia to the south of Norwich. It is a flat boulder clay area and is used for arable farming, apart from pasture in the river valleys. The settlement pattern is closely related to the river valleys, where slopes were easier to drain and cultivate than the poorly drained clays. South Norfolk has a high density of dispersed villages. Many of its towns, such as Wymondham, Diss and Long Stratton, are historic centres which have experienced in-filling since the Second World War. These areas, to the south of Norwich, have developed into an extensive commuter belt, and the area faces development pressure along its main transport routes (the A14 corridor). To the south-east of Bungay the landscape is dominated by large areas of land which in the past were used for common grazing. As a result, few settlements developed.

Breckland has a long history of settlement. However, it is now sparsely populated. It is composed of poor, free-draining sands which limited the supply of water. As a result, the settlement pattern evolved around the availability of water, and the area is crossed by a number of rivers whose valleys are marked by a string of nucleated villages. On the uplands there are only occasional farmhouses and hamlets. Thetford is the only town of significant size in this area. It was one of the earliest London overspill towns and now supports a wide range of industries. The air bases at Lakenheath, Honington and Mildenhall have boosted the local population since 1945.

The Norfolk **Broads** is a low-lying area of land on the eastern edge of East Anglia, between Norwich and the North Sea coast. The Broads are areas of open water around which there are extensive areas of marshland. The main agricultural land use is cattle grazing, although there has been an increase in arable farming in recent years. The area is sparsely populated, with large stretches of marshland lacking any settlement. There are few roads, and these cling to the valley sides, linking the villages that cluster on the edge, where the land is drier. Settlements are rare in the marshes, where isolated farmhouses stand remote and alone. The main settlements are Great Yarmouth which has a long history as a seaport and is now a popular holiday resort, and Beccles, on the southern edge of the Broads.

For this activity you will need to use an atlas map of Norfolk.

1 a Make a large outline sketch map of Norfolk (see Figure 5.19).

 b On your map, locate and label all the settlements mentioned in the text. Add other details, including:
  • rivers
  • upland areas
  • road numbers
  • forests.

 c Using the notes in this case study, annotate your map to illustrate the pattern of settlement in Norfolk.

 d Write a summary to accompany your map to explain the patterns you have described. Use the ideas about site, form and nucleation to help you.

Study Figure 5.21, then answer the following questions.

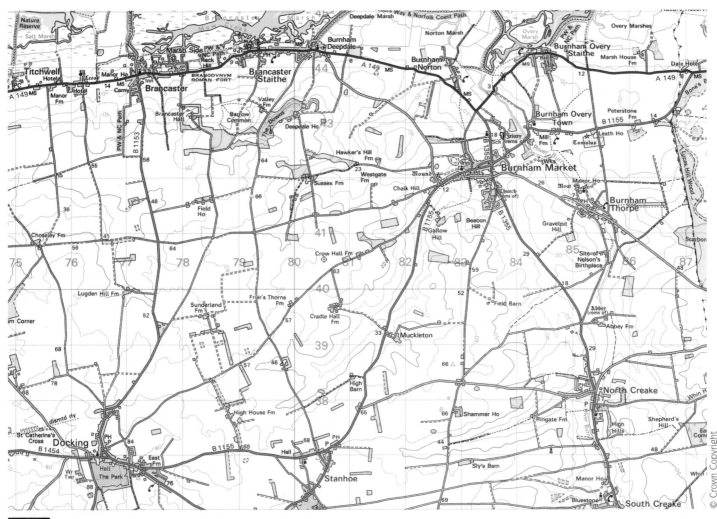

5.21 *OS map extract of part of North West Norfolk*

2 a Describe the main features of the settlement pattern of the area shown on the map.

b Suggest reasons for the sites occupied by:

(i) South Creake (8635)

(ii) Brancaster (7744)

(iii) Burnham Market (8342).

c Describe the settlement form (morphology) of Stanhoe (8036).

d Suggest ways in which the functions of the settlements of Brancaster and Burnham Market may have changed in the last 25 years.

3 Figure 5.22 is an outline of the area covered by the Ordnance Survey map. Follow the procedure for the nearest neighbour analysis (see *box* on page 153). Is the pattern dispersed, random or clustered?

4 Look at Figure 5.23. It shows the changing market areas of some settlements in Norfolk. Some have expanded and others have contracted.

a (i) Describe the changes in the market areas of Swaffham and North Walsham.

(ii) Give possible reasons why changes in these market areas might have occurred.

b What changes would you expect to find since 1971 in:

(i) the functions

(ii) the market areas of the settlements identified in Figure 5.23?

c How might these changes affect some rural dwellers more than others?

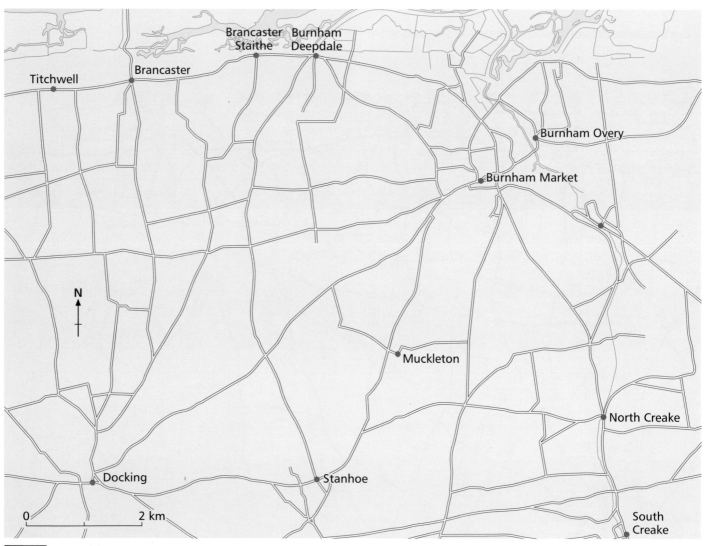

5.22 *Settlement pattern in north-west Norfolk*

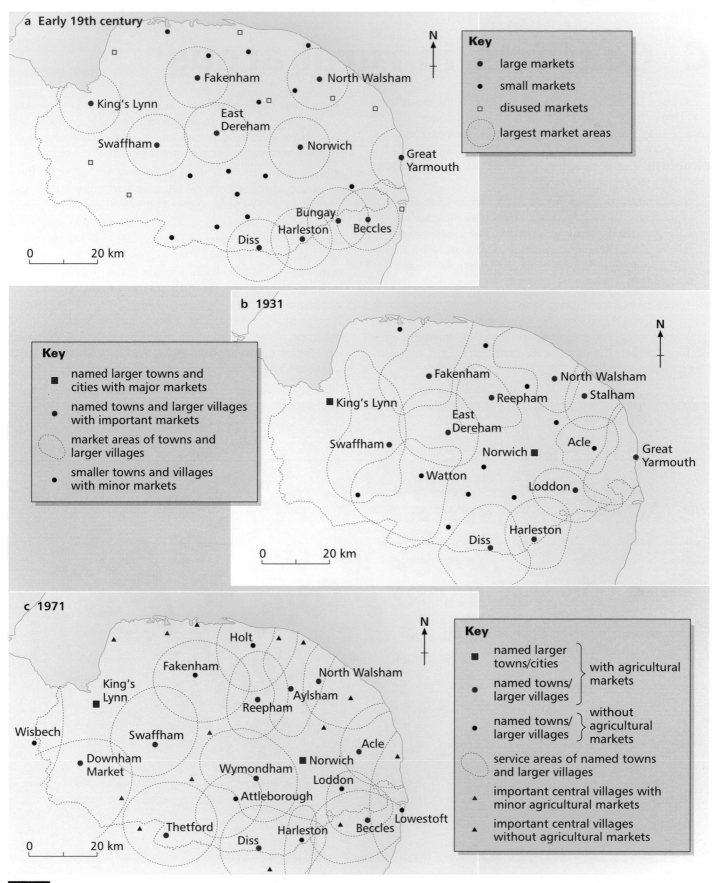

a Early 19th century

Key
- large markets
- small markets
- □ disused markets
- ⬭ largest market areas

King's Lynn
Fakenham
North Walsham
East Dereham
Swaffham
Norwich
Great Yarmouth
Bungay
Harleston
Beccles
Diss

0   20 km

N

b 1931

Key
- ■ named larger towns and cities with major markets
- ● named towns and larger villages with important markets
- ⬭ market areas of towns and larger villages
- ● smaller towns and villages with minor markets

King's Lynn
Fakenham
North Walsham
Reepham
Stalham
East Dereham
Swaffham
Acle
Norwich
Great Yarmouth
Watton
Loddon
Harleston
Diss

0   20 km

N

c 1971

Key
- ■ named larger towns/cities } with agricultural markets
- ● named towns/larger villages }
- ● named towns/larger villages } without agricultural markets
- ⬭ service areas of named towns and larger villages
- ▲ important central villages with minor agricultural markets
- ▲ important central villages without agricultural markets

Holt
Fakenham
King's Lynn
North Walsham
Reepham
Aylsham
Wisbech
Swaffham
Acle
Downham Market
Norwich
Wymondham
Loddon
Attleborough
Lowestoft
Thetford
Harleston
Beccles
Diss

N

0   20 km

5.23 *Changing market areas in Norfolk: early 19th century to 1971*

C Conflicts in rural areas

Look back to the advertisement for Milton Keynes (Figure 5.1 page 143). It attempts to offer an image of the countryside as a place with a community spirit and a retreat from urban tensions – in other words, it is an image of a **rural idyll**. The geographer John Short (1991) talks about the 'myth' of the countryside:

> *'The countryside is pictured as a less-hurried lifestyle where people follow the seasons rather than the stock market, where they have more time for one another and exist in a more organic community where people have a place and an authentic role. The countryside has become the refuge from modernity.'*

This picture-postcard image of rural life is common in popular culture. However, Short uses the idea of 'myth' and calls upon geographers to see through this 'romantic' view of rural areas. There are two main problems with this 'romantic' view:

- To see rural areas as harmonious and close-knit communities tends to ignore the very real conflicts and tensions that exist in rural areas, especially those to do with conservation, employment growth, and housing.

- It tends to see rural areas as distinct and separate from other parts of society. In fact, what happens in rural areas is very closely linked to events in other places.

The view of rural Britain as a place of harmony and social tranquillity was shattered by the Countryside March, which took place on 1 March 1998 (Figure 5.24). Although the march was stimulated by opposition to the proposed ban on hunting, it took on a wider range of issues, such as opposition to new house-building, and the economic problems of farmers. The march was in many ways a reflection of the important changes that have taken place in rural Britain in recent years. Figure 5.25 summarises some of the issues facing parts of rural Britain.

5.24 *Countryside March, 1 March 1998*

'No quick fixes' on path to green and pleasant land

By Brian Groom

A portrait of a nation in danger of loving its countryside to death was unveiled yesterday by the Countryside Agency, the government's new advisory body for rural England. It comes into existence on Thursday.

Ewen Cameron, chairman, said in launching the Agency there were 'no quick fixes' to countryside problems, but he believed the Agency could make a difference. Its top priorities were: showing how to tackle rural disadvantage; improving rural transport while taming the impact of traffic growth; demonstrating a more sustainable approach to agriculture; and increasing access to the countryside.

In *The State of the Countryside 1999*, a statistical profile, the Agency demonstrates that in many ways rural England is a great economic success.

Its population grew by 6.9 per cent between 1981 and 1991, compared with a 3 per cent national increase. The number of jobs in rural districts increased by 8.1 per cent between 1991 and 1996, compared with 3.7 per cent nationally.

Unemployment rates are lower – 4.4 per cent in rural districts last August compared with a 6.4 per cent English average.

And the countryside has adapted so well to the decline of agriculture that its mix of industries and services now mirrors the national economy: 39 per cent of employment in rural districts is in the service sector, compared with 40 per cent nationally.

The countryside contains 9.3 million people, a fifth of the total population. It is so attractive that most people want to live there: 89 per cent of rural dwellers say they are content, but fewer than half of town and suburb dwellers, and one in five city dwellers, are happy with their place of residence.

That is where the problems start. The number of households is forecast to grow by a quarter in the two decades to 2011, but high prices are creating a shortage of affordable homes. Two-thirds of households cannot afford to buy a big enough home, and in the early 1990s, 16 000 families in rural districts were accepted as priority homeless each year.

In 1997, 75 per cent of rural parishes had no daily bus service, 42 per cent had no shop, 49 per cent had no school, 83 per cent had no GP based in the parish, and 29 per cent had no pub.

Traffic grew three times as fast on rural A roads between 1981 and 1997 as on urban roads. Country roads accounted for 55 per cent of all road deaths, compared with 41 per cent in towns and 4 per cent on motorways.

Peripheral rural counties have the lowest wage levels in England. Cornwall has an average weekly wage £88 below the national average, and is among other rural counties in the bottom 10 for gross domestic product per head.

Mr Cameron said: 'Many people believe the countryside is threatened as never before and that it has changed for the worse in the last 20 years.'

With a staff of 380 and a £50 million-a-year budget, the Agency must rely on partnerships with other bodies, from Brussels to parish councils, to have an impact.

Mr Cameron hopes to create a 'best practice' forum with rural representatives from the eight regional development agencies that also come into being this week.

The Agency has been formed by merging the

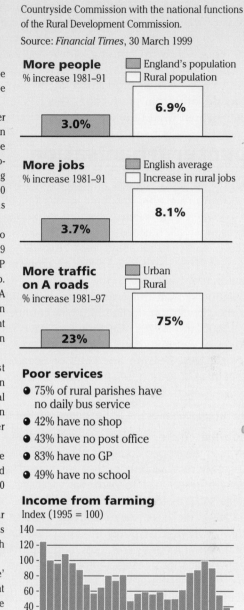

Countryside Commission with the national functions of the Rural Development Commission.

Source: *Financial Times*, 30 March 1999

More people
% increase 1981–91

- England's population
- Rural population

3.0% 6.9%

More jobs
% increase 1981–91

- English average
- Increase in rural jobs

3.7% 8.1%

More traffic on A roads
% increase 1981–97

- Urban
- Rural

23% 75%

Poor services

- 75% of rural parishes have no daily bus service
- 42% have no shop
- 43% have no post office
- 83% have no GP
- 49% have no school

Income from farming
Index (1995 = 100)

5.25 *Pressures on rural Britain*

Changes in agriculture

The State of the Countryside 1999 stated that 88 per cent of England is 'countryside', meaning that it is free from residential, industrial and transport use. In terms of land use, most of rural Britain is agricultural. However, farming as an economic activity has undergone important changes since 1945. These changes are linked to processes of **specialisation** (where farms focus on producing one or maybe two products) and **intensification** (where more production is squeezed from the same area of land). The result has been fewer people engaged in farming, and larger farm sizes. In addition, changes in agricultural policies at both national and European Union levels have encouraged many farmers to find other ways to increase their income.

The central belief of post-war land use policy – that rural land should be preserved for agriculture – has started to erode, leading some geographers to talk of a **'post-productivist' countryside**. The result of this change has been to open up a whole series of arguments and debates about what is appropriate use of rural areas. The main conflict has been about the role of farmers. In the period after the Second World War, the main goal of agricultural policy was to increase food production, and farmers were given financial support to modernise their activities. As a result, rural areas were equated with farming. However, now that food production in Europe has expanded, farmers have been asked to reduce their production. Many have been paid to **set aside** land from production. This has led to important debates about what rural areas should be used for.

Counterurbanisation

The reduced importance of farming in rural areas has opened the door to new rural residents as old agricultural buildings and cottages have been converted into 'executive' houses and even small businesses. After decades of population decline, some rural areas have experienced population increases. This pattern is known as **counterurbanisation**. Counterurbanisation was the term used to describe the population growth that was occurring in rural US counties, and the population decline in urban counties there, during the 1960s and 1970s. Geographers have reported similar patterns of counterurbanisation in north-west Europe, Japan, Australia, New Zealand and parts of Scandinavia. Migrants usually have a secure, relatively well-paid job (which means they are likely to be middle-class and middle-aged).

There are two major sets of explanations of this trend:

- **'Job-led'** There has been a significant change in the type of work in developed countries. Whereas firms used to cluster together, more and more they are dispersing, because they are no longer tied to particular locations. The shift from large-scale mass production to small firms allows them to locate in more remote rural areas. In these areas they can take advantage of pools of 'green' labour (largely female and non-unionised), low rents, and improving services and communications. These firms would find it difficult to recruit managers in rural areas, though, and, as a result, these people are likely to move with the firm to the rural locations. The process in which manufacturing and service activities are more and more found in rural areas is known as the **'industrialisation of the countryside'**.

- **'People-led'** This suggests that counterurbanisation is the result of the preference of people for rural life. A recent survey of British migrants to rural areas revealed several reasons for their move:

  - the rural area was more open and less crowded; people said they no longer felt hemmed in by houses – there was a more 'human' scale to things

  - it was quieter and more tranquil, with less traffic noise and less hustle and bustle

  - the area was cleaner, with fresh air and less traffic pollution

  - the area allowed people to escape from the 'rat race' and society in general

  - there was a slower pace of life in the area, with more time for people

  - the area had more community and identity, a sense of togetherness, and was less impersonal

  - it was seen as an area of less crime, fewer social problems and less vandalism – people said they felt safer at night

  - the environment was better for children's upbringing

  - there were fewer non-white people in the area

  - there was less 'nightlife' and fewer 'sporty' types.

One of the results of this preference for rural living has been an increased pressure for house building. For example, England's population is expected to grow from 49.1 million to 52.5 million between 1999 and 2021. In addition, higher rates of divorce and later marriage will increase the demand for housing. This means that a large number of new homes will need to be built. Estimates suggest that 3.8 million new homes will be required between 1996 and 2025. Where will these new houses be built? One proposal is to build on **brownfield** sites (land previously used for industry – Figure 5.26a). However, the desire for people to live away from large urban areas is unlikely to disappear, which means that there is pressure to build on greenfield sites in the countryside (Figure 5.26b).

a Brownfield site

b Greenfield site

5.26 *Building sites*

NOTING ACTIVITIES

1 What is meant by the term 'counterurbanisation'?

2 What is the difference between 'job-led' and 'people-led' counterurbanisation?

3 Study the list of reasons given by British migrants for moving to rural areas.

   a To what extent are these reasons perceptions rather than reality?

   b Which reasons do you think are likely to be most influential in encouraging migration? Why?

   c From your own knowledge of your local region, suggest reasons why people have moved into rural areas.

4 Increased house building is just one issue associated with counterurbanisation. Can you suggest other impacts, both positive and negative?

Leisure and tourism

Many people now have the money and leisure time to spend in the countryside. The expansion of leisure parks, conservation holidays and so on have focused on rural areas. The effect has been to make rural areas the 'playground' for Britain's urban population. There are a number of trends that help to explain the increased pressure on rural areas for leisure and tourism.

- Real incomes for the majority of people have grown steadily over the last two or three decades, with more available for leisure spending.
- Car ownership is still increasing. More than 70 per cent of households in the UK possess a car. What is more, people are travelling further.
- The popularity of the countryside for visitors has grown. 1.3 billion day trips were made to the English countryside in 1996. Seventy-six per cent of the population of England visited the countryside at least once in 1990.
- A higher proportion of the UK population is in intellectually skilled employment and many more people have further/higher education qualifications. Surveys suggest that such people are twice as likely to visit the countryside as are semi- or unskilled workers.
- The countryside is associated with peace and tranquillity. Many people desire the open spaces of the countryside, in contrast to the 'stress and pollution' of urban areas.

These factors have led to increased pressure on the countryside and introduce a whole series of conflicts. Some of the most important conflicts in recent years are summarised in Figure 5.27.

One of the key factors leading to increased pressure on rural areas from leisure and tourism is increased personal mobility. With high rates of car ownership, and an increased demand for leisure, rural areas are coming under increased pressure from visitors. A report in the early 1990s forecast that traffic on rural roads is likely to increase by between 127 per cent and 267 per cent by 2025. What is more, whilst traffic in urban areas is likely to be restricted through measures such as road pricing and traffic calming, this may encourage more car owners to move out of the cities, with the effect of making the problem worse in rural areas.

Since 1945, a number of measures have been taken to make rural areas more accessible to people for leisure and recreation. These are shown on Figure 5.28. They include:

- seven designated National Parks in the more remote areas, and two specially designated areas in the Norfolk Broads and the New Forest
- 34 Areas of Outstanding Natural Beauty (AONB)
- 29 Heritage Coasts
- around 192 000 km of public rights of way, including ten National Trails
- a new National Forest and 12 Community Forests.

These measures are the result of social changes which have given rise to new patterns of leisure supply and demand as society becomes increasingly mobile and people look to gain satisfaction from a wide range of leisure pursuits. The range of designated areas shown on the map dates from policy decisions made after the Second World War (1939–45). It was assumed that these were 'special' places that needed protection. Other rural areas were assumed to be safe in the hands of farmers. However, in recent decades it is recognised that the constant drive to produce more food is no longer necessary, and farmers have been encouraged to diversify their activities. In addition, people increasingly look to rural areas as sources of leisure.

NOTING ACTIVITIES

1 Why are increasing numbers of people visiting the countryside for leisure and recreation?

2 Suggest some problems and conflicts that might result from the increased use of rural areas for leisure and recreation.

3 With reference to a local rural area popular with visitors, identify some of the potential conflicts. In what ways has the management authority acted to help reduce these conflicts?

Coastal marinas

Since the late 1980s there has been an increased demand for new marinas and linked developments at coastal estuary locations in England and Wales. Groups such as the RSPB and WWF raised concerns about the impact of such schemes on wildlife habitats and breeding grounds.
Example: Cardiff Bay redevelopment scheme was approved by Parliament in 1993. This involved a major transformation of the estuary, and the wholesale destruction of an SSSI, by a barrage across the bay.

Access to private land

Most of rural Britain is privately owned. The issue of general access to the countryside – 'the right to roam' – is a source of friction between landowners and the general public.

Holiday cabins

The economic pressure on farmers to diversify has led many to create on-farm tourist facilities, such as holiday cabins.
Example: a plan for a 10-cabin development in the Cotswolds AONB near Great Witcombe (Gloucestershire) – permission for the development was granted despite opposition from Tewkesbury District Council. Such applications often cause controversy because they involve changes in land use conflicting with previously established planning policies.

Golf courses

The 1980s saw a rapid expansion in new golf course development, mainly on lower-grade farmland of decreasing productive value. These tend to be local conflicts, but raise questions because they are often in areas of Green Belt. The most notable conflict occurred over a proposal for a golf course at Catholes in the Yorkshire Dales National Park, which attracted strong opposition from the Ramblers Association and other conservation groups.

Mountain bikes

Since 1992 the English Lake District has been the site of conflict between mountain bikers and other recreational users (walkers and climbers) over what is 'appropriate' use of the area. There have been similar conflicts in Snowdonia – particularly on Snowdon itself – leading to voluntary bans on biking on certain slopes.

War games/paintball

The growth of war games is a result of farm diversification. It usually takes place in woodlands.
Example: Bonny Wood, near Stowmarket (Suffolk), a 15 ha woodland SSSI, which has been used for war/paintball games. The Nature Conservancy Council (now English Nature) found that damage had been done to ancient oak and ash woodland, and particularly to the abundant ground vegetation and shrubs.

Holiday villages

Center Parcs Holiday Village at Longleat, Wiltshire is a 160 ha development which gained planning permission in 1992. The site lies within the Cranbourne Chase AONB. There are similar ventures at Sherwood and Elvedon, and proposals for others at Market Weighton (Yorkshire), Eden Valley (Cumbria) and West Wood (Kent).

 5.27 *Land use and leisure conflict in the 1990s*

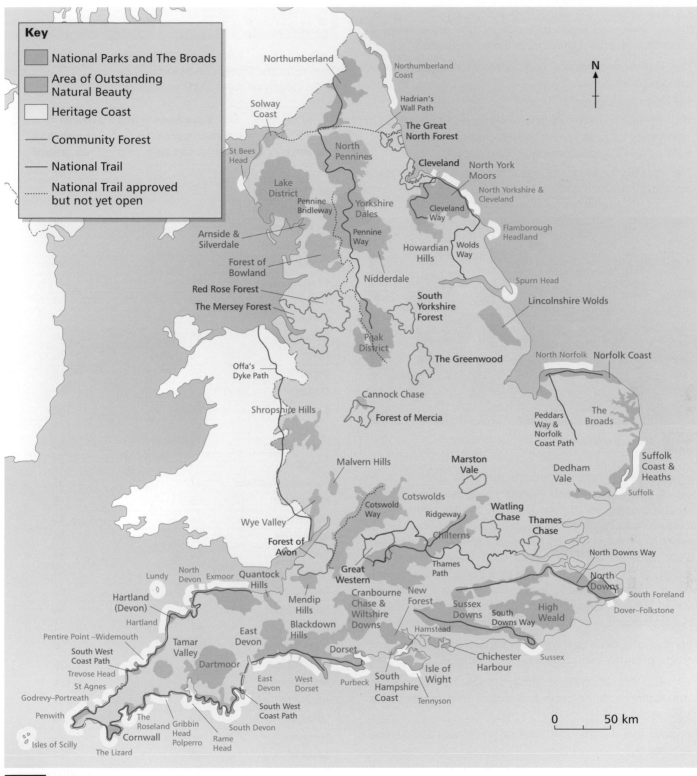

Key

▨	National Parks and The Broads
▨	Area of Outstanding Natural Beauty
☐	Heritage Coast
—	Community Forest
—	National Trail
⋯	National Trail approved but not yet open

5.28   *Protected areas in England*

Transport

For many people, the car is a symbol of freedom and mobility. Increased levels of personal income and leisure time have led to increased car use. Over the last 20 years the distance travelled by car has increased by 55 per cent, while walking, cycling and bus travel have all declined (by 20 per cent, 25 per cent and 38 per cent respectively). The government forecasts that traffic could increase by a further 36–84 per cent by the year 2031.

The growth of rural traffic is greater than in urban areas, a fact explained by declining levels of public transport in rural areas, high levels of car ownership and the growth of leisure activities. Increased levels of traffic in rural areas create a number of problems (Figure 5.29). These include:

- the loss of landscapes and habitats through road/rail construction

- pollution from motor vehicles

- the gradual erosion of the distinctive character and tranquillity of rural areas through traffic increases, night lighting and standardisation of road design

- the growing isolation of the 22 per cent of the rural population without access to a car.

5.29 *Rural traffic congestion*

The question of how to reduce traffic in rural areas is complex. Whilst in general people recognise the need to protect the character of rural areas, moves to reduce traffic through restrictions such as road-pricing or quotas on the number of vehicles allowed in an area at any one time are fiercely opposed. An example of such resistance was the response to the suggestion by the Lake District National Park Authority that peak traffic in the National Park should be rationed. No progress with this suggestion was made, despite growing problems in this environmentally sensitive area.

A number of measures have been suggested to reduce the problem of traffic in rural areas. These include:

- prevent the spread of urban areas by halting the development of housing and out-of-town shopping in rural areas

- impose speed limits of 40 mph on rural roads and 20 mph in villages

- set national targets to reduce traffic levels by the year 2025

- increase transport choice by improving rural bus and rail services.

As noted above, the car is a symbol of freedom and progress for many people. It is also vital in rural areas for gaining access to health care, education, retailing and other services. This suggests that attempts to manage and reduce traffic in rural areas in the future are likely to face resistance.

EXTENDED ACTIVITY

Look at Figure 5.30. It presents data published by the Council for the Protection of Rural England (CPRE) for future traffic growth on rural roads in England.

| County | Forecast traffic growth % | |
	with measures to reduce car use	business as usual
Avon	48	107
Bedfordshire	121	265
Berkshire	49	99
Buckinghamshire	53	107
Cambridgeshire	53	140
Cheshire	45	90
Cleveland	104	221
Cornwall	48	94
Cumbria	42	82
Derbyshire	45	96
Devon	46	90
Dorset	57	115
Durham	41	81
East Sussex	65	145
Essex	52	112
Gloucestershire	46	90
Hampshire	53	113
Hereford & Worcester	47	92
Hertfordshire	54	118
Humberside	44	85
Isle of Wight	43	84
Kent	44	91
Lancashire	41	93
Leicestershire	47	102
Lincolnshire	48	94
Norfolk	47	91
North Yorkshire	46	90
Northamptonshire	49	96
Nottinghamshire	56	134
Northumberland	43	84
Oxfordshire	49	110
Shropshire	48	94
Somerset	48	94
Staffordshire	45	104
Suffolk	46	105
Surrey	49	107
West Sussex	60	127
Warwickshire	44	93
Wiltshire	49	100

5.30 *Traffic forecasts for rural England, 1996–2031*

1 Using an outline map of England and Wales or a copy of Figure 5.31, choose a method to show the figures for those counties that face the greatest pressure from traffic increase assuming 'business as usual'.

2 Suggest reasons for the pattern on your map.

3 On a second outline map of England and Wales, shade the counties where measures to reduce traffic would be most effective.

4 Why might the data on future traffic levels be unreliable?

5 Suggest reasons why governments might be reluctant to implement policies to reduce levels of traffic in rural areas.

Economic changes

In addition to population changes and new patterns of leisure activity, rural areas in Britain have experienced significant changes in economic activity. The long-term decline of agricultural employment has meant that less than 2 per cent of employment in Britain is currently in agriculture. Thus, although rural areas are dominated by agricultural landscapes, their economies depend less and less on agricultural employment.

Geographers noted the increases in rural manufacturing jobs throughout the 1980s. This was known as the **'urban–rural shift'**, and the growth in rural manufacturing employment was in marked contrast to the situation in many large urban areas which were experiencing job losses.

Some limited rural areas have been attractive to high-tech or other industrial activities, especially those close to motorway access or where greenfield sites are available (Figure 5.32). This can be explained by the availability of cheaper land and lower rents, and also the desire to live and work in an attractive environment. In addition, government agencies have attempted to attract investment by providing subsidies for new firms.

Rural areas have also experienced service-sector growth in recent decades (Figure 5.33). In the 1960s and 1970s there was an increase in public sector employment such as health, education and local government. More recently, new information technology has led to new forms of employment. On a small scale, there is the growth of homeworking or 'telecottaging', where people work from home using modern technology such as e-mail to communicate with others. On a larger scale, new service sector jobs such as call centres

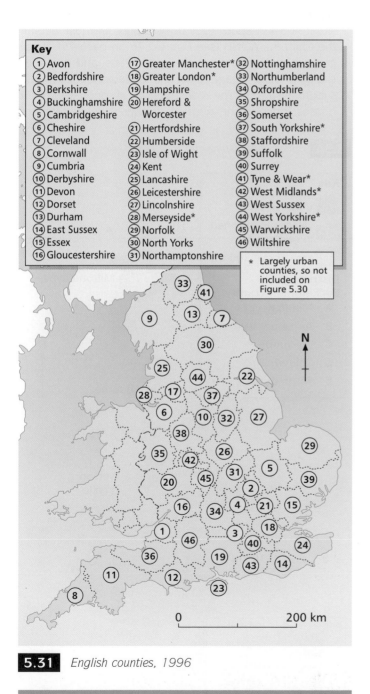

Key

① Avon	⑰ Greater Manchester*	㉜ Nottinghamshire
② Bedfordshire	⑱ Greater London*	㉝ Northumberland
③ Berkshire	⑲ Hampshire	㉞ Oxfordshire
④ Buckinghamshire	⑳ Hereford &	㉟ Shropshire
⑤ Cambridgeshire	Worcester	㊱ Somerset
⑥ Cheshire	㉑ Hertfordshire	㊲ South Yorkshire*
⑦ Cleveland	㉒ Humberside	㊳ Staffordshire
⑧ Cornwall	㉓ Isle of Wight	㊴ Suffolk
⑨ Cumbria	㉔ Kent	㊵ Surrey
⑩ Derbyshire	㉕ Lancashire	㊶ Tyne & Wear*
⑪ Devon	㉖ Leicestershire	㊷ West Midlands*
⑫ Dorset	㉗ Lincolnshire	㊸ West Sussex
⑬ Durham	㉘ Merseyside*	㊹ West Yorkshire*
⑭ East Sussex	㉙ Norfolk	㊺ Warwickshire
⑮ Essex	㉚ North Yorks	㊻ Wiltshire
⑯ Gloucestershire	㉛ Northamptonshire	

* Largely urban counties, so not included on Figure 5.30

N

0   200 km

5.31 *English counties, 1996*

5.32 *Modern high-tech business in a rural setting: the workers in this building design and make speech synthesisers*

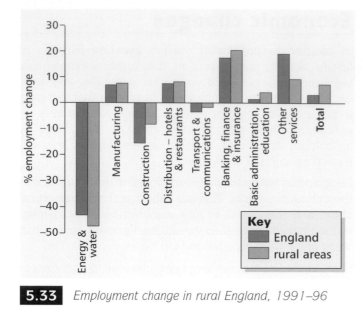

5.33 *Employment change in rural England, 1991–96*

dealing with customer transactions such as insurance claims and air ticket administration are attracted to rural areas in search of inexpensive labour forces and low-rent offices.

Development pressures in Aylesbury Vale, Buckinghamshire

Buckinghamshire is a county in south-east England (Figure 5.34). It covers an area 80 km from north to south and 40 km from east to west. In the north lies the new town of Milton Keynes, while in the centre of the county is Aylesbury Vale district and the main town, Aylesbury. To the south of the Chilterns escarpment lie the districts of South Bucks, Chiltern and Wycombe. Aylesbury Vale is a good example of a rural area that is experiencing pressure from development.

The county has experienced rapid population growth. During the 1970s it was the fastest-growing county in the UK, with a population increase of nearly 20 per cent between 1971 and 1981. Planning controls in the south of the county have meant that pressure for development has been concentrated in the north of the county. Aylesbury Vale has been subject to almost continual growth pressure since the 1950s. Initially this was the result of the relocation of people from London after the Second World War, but the momentum has been maintained by the fact that the area is a favoured location for the new service sector and high-tech industries.

By the late 1970s the policy was to restrain development in South Bucks and to channel new urban growth to selected locations in the remainder of the county, namely Milton Keynes, Aylesbury, Newport Pagnell, Olney and Buckingham. The plan was to maintain the division between town and countryside. Agricultural land was to be safeguarded and

development in the villages was to be restricted to within their boundaries. The continued economic prosperity and population growth has meant that areas such as Aylesbury Vale have faced pressure to accommodate new housing. The way these developments are dealt with can be seen by looking at two examples of specific settlements.

- **Swanbourne** is located in the most 'rural' part of Aylesbury Vale (see Figure 5.34). It has a population of 400, which has remained virtually static for 20 years. Many of the village buildings have remained unchanged since the 17th century. This can be explained by the dominance of the landlord of the estate who has resisted large-scale development in and around the village.

- **Weston Turville** (Figures 5.34 and 5.35) is a large village situated 3 km south-east of Aylesbury and just outside the Chilterns AONB (see Figure 2.76 on page 75). It has grown significantly since the early 1980s, and is a good example of a village that has undergone substantial suburbanisation. The construction of one site of 58 houses 'opened the door' to further development and

5.34 *Buckinghamshire and Aylesbury Vale*

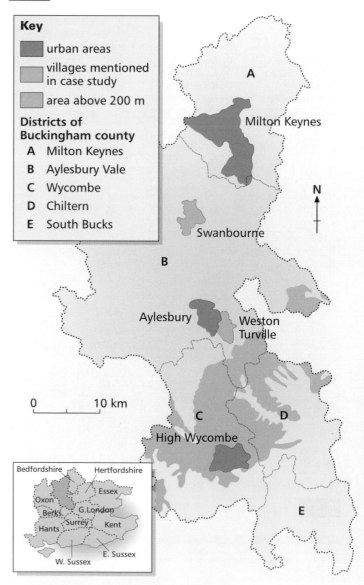

5.35 *Weston Turville, on the edge of the Chilterns AONB*

means that any 'unwelcome' developments will be vigorously opposed through the planning system. This leads to some interesting issues for geographers.

Aylesbury is attractive to more wealthy 'incomers' because it fits their image of rural life, which means life in a 'community'. However, this 'community' seeks to exclude all those elements that these residents wanted to move away from. As a result, the planning system is being used to protect the interests of people who have spent a lot of money in 'escaping' to the countryside. These concerns seem a world away from the experiences of people living in less affluent and more remote rural areas.

altered the character of the village. Many of the farms that were part of the settlement have been converted to housing, and there is now only one working farm left.

The pressure for further development of green space has caused conflict in Weston Turville, and parts of the village have now been designated as conservation areas. However, the continued growth of Aylesbury is likely to create pressure for more development. Recently a public inquiry rejected proposals for a planned superstore on the edge of the area, and there is talk of a southern bypass around Aylesbury which would cut into Green Belt land.

There are other pressures for development in Aylesbury Vale. For example, calls for farmers to diversify their activities and seek alternative uses for land have coincided with population growth and the upsurge in the popularity of leisure pursuits in rural areas. One of the most attractive of these is golf. Between 1987 and 1992, Aylesbury Vale District Council received 24 applications for new golfing facilities. Turning their land over to golf courses is attractive to farmers because it takes up extensive amounts of land and makes use of existing land management skills.

In their study of Aylesbury Vale, the geographers Jonathan Murdoch and Terry Marsden (1994) suggest that the struggle to preserve the character of Aylesbury Vale (and other places like it) will not be easy. The 'picture postcard' appeal of these places makes them prone to pressures for new development. However, the social make-up of these areas

NOTING ACTIVITIES

1   Suggest reasons for the rapid population growth experienced in Buckinghamshire since the 1970s.

2   What social, economic and environmental problems might arise as a result of rapid population growth in the rural parts of the county?

3   An important issue in Aylesbury Vale is the increased demand for housing. Three possible solutions are:

-   to accommodate the demand for new housing by 'in-filling' areas of the town of Aylesbury where space is available

-   to accommodate new development by distributing it around villages in the district

-   to build a completely new settlement.

a Copy and complete Figure 5.36. On your table, mark ticks in appropriate places to evaluate how far each of the three options above satisfies the criteria shown.

b From your evaluation, suggest which of the three options *you* would favour. Give reasons for your answer.

	Infill and develop areas within and around Aylesbury	Scatter development in villages in the district	Build a completely new settlement
Protect Green Belt land			
Conserve the character of rural settlements			
Reduce the area of derelict land in towns			
Reduce levels of air pollution in towns			
Reduce the need for people to travel			

5.36 *How to deal with rural population growth*

Issues in remote rural Wales

Remote rural areas, such as large parts of Scotland, mid-Wales, the northern Pennines, the south-west peninsula and much of the east coast of England, have traditionally been dominated by agriculture. They experienced population decline through much of the 20th century, as people left the land. In recent decades some areas have attracted in-migrants (often retired people or second-home-owners), whilst the younger age groups continue to leave. In these areas, a number of issues arise. These include the future of the rural economy, given that these places tend to be physically remote and located a long way from consumer markets. The lack of employment opportunities is exacerbated by the fact that average wage levels tend to be lower, leading to lower household incomes. Linked to this is the issue of service provision. In some areas, there are tensions between incomers and the indigenous (local) population over housing, since many younger people are unable to afford housing. This issue has been made worse by a reduction in the amount of new housing built by local authorities.

People living in rural Wales have experienced several particular problems in recent years.

1 Population changes

For most of the 20th century, rural Wales experienced population decline, as people left to find work in towns and cities both in Wales and in the rest of Britain. However, by the 1970s and early 1980s census data indicated population

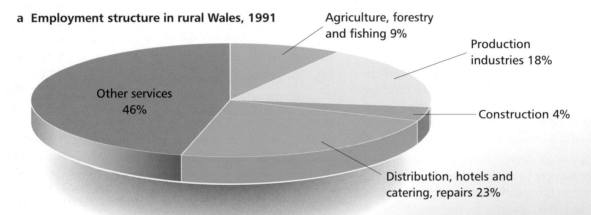

a Employment structure in rural Wales, 1991

Agriculture, forestry and fishing 9%

Production industries 18%

Construction 4%

Distribution, hotels and catering, repairs 23%

Other services 46%

b Average weekly earnings of adult full-time workers (male) 1991, expressed as a percentage of the national average

Clwyd West	83.5
Dyfed	80.8
Gwynedd	83.6
Powys	77.3
Wales	87.7
Great Britain	100.0

c Main problems facing young people seeking accommodation (percentage)

No problems	9.9
Shortage of affordable housing for purchase	38.9
Shortage of affordable housing for rent	16.1
Shortage of council housing	11.5
Competition from affluent migrants	9.0
Difficulty in obtaining mortgages	6.3
Other	8.4

5.37 *Information on rural Wales*

increases in many parts of rural Wales. For example, the population of rural Wales increased by 37 000 persons between 1981 and 1991, which represents a 6 per cent growth in the overall rural population. This population growth is the result of in-migration, particularly of people from England. Indeed, by 1991, 30 per cent of the population of rural Wales had been born in England, an increase of almost one-third since 1981. It is important to note the selective nature of these population movements, since the overall increase in population hides a net out-migration of younger people from the Welsh countryside. In fact, the changing social make-up of rural Wales has led to fears that the cultural character of rural Wales is under threat, especially through the decline of native Welsh speakers.

2 Low economic status

Figure 5.37 shows the employment structure of rural Wales. Services are clearly the dominant sector of employment, accounting for 69 per cent of all jobs. This hides the true picture, though, since the growth of service sector employment has been geographically uneven, and strongest in the more accessible parts of the countryside. In addition, many of these jobs are low-quality work, aimed at and taken up by women. This is reflected in income statistics. Overall, the average gross domestic product (GDP) per head for Wales is only 85 per cent of the average for the UK as a whole. Counties such as Dyfed, Powys and Gwynedd fare even worse, with GDP per head being 19 per cent below the European Union average.

3 Loss of services

Transport provision has undergone considerable change in rural Wales since the 1960s. Rail services have been reduced so that many parts of Wales are no longer served. The de-regulation of bus services in the early 1980s has also led to reduced services in rural areas. These changes must be set alongside high rates of car ownership, but it is important to pay attention to who has access to the use of cars. Research findings suggest that it is generally the people with lowest incomes, women, the young and elderly people who are 'transport-poor' and therefore more likely to suffer from isolation.

The *Lifestyles in Rural Wales* survey (1994), which looked at four rural areas in different parts of Wales, found that many people reported a general decline in the level of service provision. Of 185 villages with a population of less than 3000, 66 per cent lacked a permanent shop, 21 per cent had no post office, 31 per cent did not have a primary school, and 15 per cent were without a bus service.

4 Poverty and deprivation

It is important to avoid the impression that life in rural Wales is a never-ending tale of doom and gloom. However, the *Lifestyles in Rural Wales* survey did provide evidence of poverty and deprivation. This can be seen through the case of Corris, a village in Gwynedd. It has experienced economic change in the post-war period. The slate quarries that provided the economic base of the community have closed and employment in agriculture and forestry has also declined. In addition, there have been important cultural and social changes as the result of the in-migration of English-speakers both as full-time residents and second-home owners. The problem of low income was particularly felt by Welsh-speaking families, and reflected the lack of well-paid jobs within reasonable travelling distance. There is a lack of rented housing and affordable housing, which makes it difficult for young people to stay in the area. Problems of poor transport have meant that some people have had to rely on expensive village shops or mobile shops.

NOTING ACTIVITIES

1 Describe the likely effects of population changes in rural Wales.

2 Summarise the data on the economic character-istics of rural Wales shown on Figure 5.37.

3 In what way would the loss of services in rural Wales affect the following groups more than others:

   a elderly people

   b people with low incomes?

4 Suggest measures that the government might take to reduce the impacts of poverty in parts of rural Wales.

5 Copy and complete the grid below to show how the problems faced by pressured rural areas (such as Aylesbury Vale) and remote rural areas (such as rural Wales) differ.

	Pressured rural areas	**Remote rural areas**
Economic		
Social		
Environmental		

STRUCTURED QUESTION

Study Figures 5.38a, b and c, which show changing land use in and on the outskirts of Bournemouth in southern England.

1 Use the maps to identify *four* changes that have occurred in the land use of the Littledown area.

(4)

2 For each of the changes you have identified, suggest reasons for the changes. *(4)*

3 Suggest ways in which these changes might have caused conflicts between different groups of people in the area. *(6)*

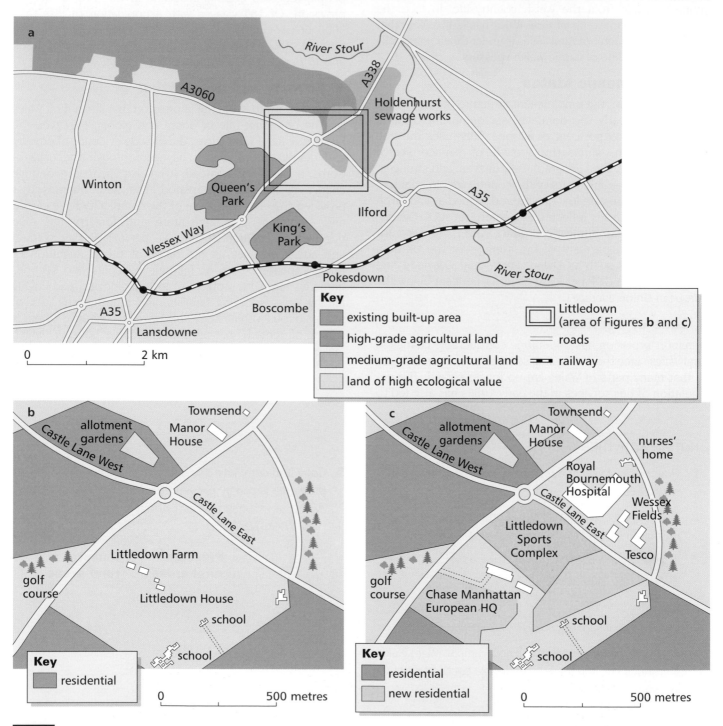

5.38 *Changing land use: Bournemouth*

D The future of rural Britain

In this chapter we consider some of the current developments in rural policy in Britain. The Countryside March in 1998 (see Figure 5.24) highlighted the political significance of rural issues, and showed that governments cannot afford to ignore the concerns of people living in rural areas.

The new Labour Government (elected in 1997) outlined its vision for rural Britain:

- a living countryside with thriving rural communities where all people have access to services such as health care, schools and shops
- a working countryside which contributes to national prosperity, with a mix of businesses, jobs and homes, reducing the need to commute long distances
- strong relationships between town and country
- a properly protected environment
- the countryside should be for all – this means that access should be available so that the character of the countryside can be enjoyed widely.

As a vision for the future of the countryside this is an attractive list. However, there are some important questions to ask about it, especially in the light of what we have studied in this section. For example, we have seen that many remote rural areas lack basic services. As a result many of these areas are suffering out-migration or depopulation. The same is true of employment: some parts of rural Britain are likely to be attractive to new businesses, while others are less attractive. These are important issues and they are not easily resolved (Figure 5.39).

Who cares for the countryside?

A number of organisations and interest groups (groups that campaign on specific issues) are concerned with the future development of rural areas.

1 The Countryside Agency

Governments seek to achieve their visions through a range of policies designed for rural areas. The **Countryside Agency** was formed in April 1999. It combined the Rural Development Commission and the Countryside Commission. The Countryside Commission was the government's countryside and landscape adviser. It aimed to make sure the countryside was protected. It was responsible for designating Natural Parks and Areas of Outstanding Natural Beauty (AONBs), defining parts of the coast that were of special value (Heritage Coasts) and managing England's National Trails. The Rural Development Commission was the government's agency for economic and social development in rural England. Its main role was to stimulate job creation and the provision of essential services in the countryside. In April 1999 the two commissions were joined together into a new agency – the Countryside Agency – with responsibility for advising the government and taking action on issues relating to the environmental, economic and social well-being of the English countryside.

5.39 *The countryside: a living, working community*

The aims of the Countryside Agency are:

- to conserve and enhance the countryside
- to promote social equity and economic opportunity for the people who live there
- to help everyone, wherever they live, to enjoy this national asset.

2 Regional Development Agencies

In April 1999 a number of Regional Development Agencies were introduced. They have taken on many of the rural regeneration programmes of the Rural Development Commission. The Regional Development Agencies are responsible for a number of schemes aimed to improve the quality of life in rural areas.

Rural Development Areas

Rural Development Areas (RDAs) cover parts of 29 counties of England. They cover about 35 per cent of the land area of England and include about 6 per cent of the population (2.75 million people). They were introduced in 1984. Rural Development Areas include those parts of rural Britain in greatest economic and social need (Figure 5.40). Most of these areas are located in remote rural areas.

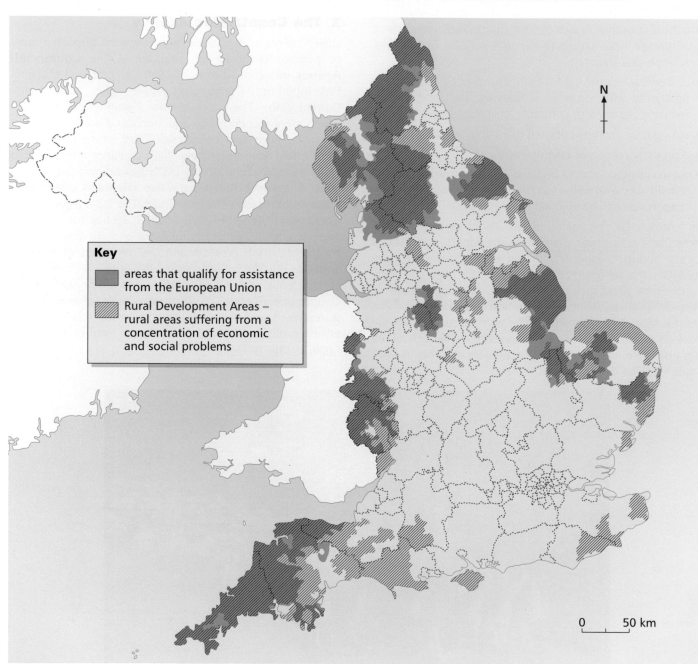

Key

- areas that qualify for assistance from the European Union
- Rural Development Areas – rural areas suffering from a concentration of economic and social problems

0 50 km

5.40 *Rural Development Areas in England, and areas that qualify for assistance from the EU*

Figure 5.41 shows the Rural Development Area for Devon in south-west England. This is a priority area because of its relative economic and social disadvantage. The Rural Development Commission funded projects in these areas. These projects are designed to increase the range of economic activities in the local economy, and improve the availability of training opportunities. These economic objectives are linked to a range of social goals such as assisting in the regeneration of rural settlements, increasing the supply of affordable housing, improving access to community facilities and services, and addressing the problems experienced by disadvantaged groups.

Rural challenge

The Rural Development Commission organises the Rural Challenge, which is designed to stimulate economic and social regeneration in rural communities in England. Rural Challenge was introduced in 1995 and takes the form of an annual competition providing prizes of up to £1 million. It encourages partnerships of local public, private, voluntary and community groups in Rural Development Areas to put forward innovative ways of tackling economic and social problems. Six prizes are awarded each year and projects are selected in a series of staged competitions. There is a requirement that private, public and community sectors are represented in the bid and that the schemes have the support of the local community.

3 The Countryside Council for Wales

This is the body responsible for offering the government advice on issues as they affect rural Wales. It seeks to maintain the natural beauty of the land and coast of Wales by providing help and advice to those responsible for its management. It seeks to promote the idea of a sustainable countryside by balancing the demands for employment and economic prosperity with the desire to maintain landscape character and wildlife.

The role of the CCW is likely to become more important in the future with the establishment of a National Assembly for Wales. In 1999 the CCW presented a report to the National Assembly which pointed to the environmental degradation that has occurred in recent years. The report gave examples of environmental problems. These included:

- the loss of ecologically diverse meadows in low-lying areas as a result of increased arable farming
- the loss of moorland and hill grassland due to overgrazing by livestock, and conifer planting
- conifer planting, which has led to the acidification of the soil which has then affected rivers and lakes
- the development of wind farms, which has affected the character of the Welsh countryside.

4 The European dimension

Many decisions that directly affect the lives of people in rural Britain, such as the level of financial support for farming, are decided at a European scale. Indeed, one of the key questions facing European decision-makers is how to reduce the level of support for agriculture without destroying the nature of rural communities throughout

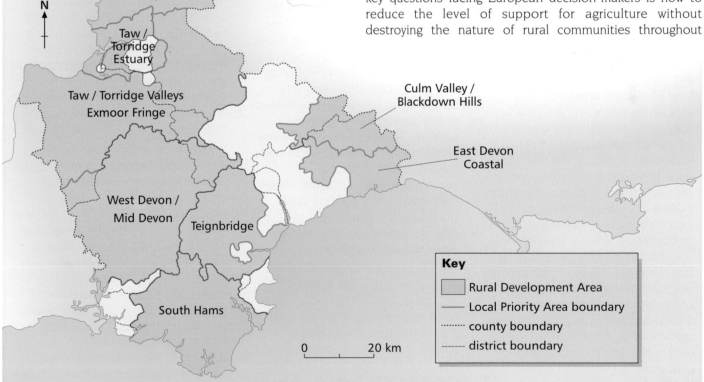

5.41 *Rural Development Areas in Devon*

Europe. The aim is to develop an integrated rural strategy that allows farmers to diversify their activities but still make a living in rural settlements.

The European Union has a programme entitled *The Future of Rural Society*, which identifies three types of rural area and the issues associated with them:

1 Rural areas near to large urban centres that are experiencing the pressures of modern development. In these areas, the priority is to strengthen measures to protect the rural environment.

2 Rural areas in decline which would benefit from economic diversification.

3 Marginal and remote rural areas which are experiencing depopulation and will continue to do so without high levels of financial assistance.

Figure 5.40 shows areas in Britain that are eligible for EU assistance ('Objective 5b funding'). This is money available specifically for the regeneration of rural areas.

NOTING ACTIVITY

Make notes on the aims and activities of the following organisations:

* the Countryside Agency
* Regional Development Agencies
* the European Union
* Countryside Council for Wales.

Getting Over the Food Mountain

Professor Philip Lowe discusses the challenges facing the Countryside Agency as rural Britain enters the post-productivist era.

'The new Countryside Agency builds on the legacy of its predecessors but presents new opportunities. The Rural Development Commission (RDC) and the Countryside Commission (CC) have done much to promote rural development and countryside management, but were formed for reasons that no longer exist.

They were established in the belief that food production was the key function of rural areas and that agricultural policy was the priority. They were established at a time when agriculture was the major force in rural areas. Ironically, the work of the Rural Development Commission and the Countryside Commission actually worked to undermine the idea that the countryside = farming = food production. The Rural Development Commission did this by reminding us that there is a diverse rural economy beyond farming. The Countryside Commission revealed the environmental damage wrought by modern agriculture. In different ways, they both challenged the dominance of agriculture in the countryside.

Rural areas are no longer defined by their economic function but by their geography. They are affected by the same pressures that affect urban areas. These include pressures for housing, transport, services and employment. The test for the Countryside Agency will be to create a contemporary vision of rurality that recognises the diversity of rural circumstances and is widely shared by rural and urban people. This vision must also acknowledge the needs of future generations. The main role of the Countryside Agency will be to celebrate and conserve what is valuable about rural areas. This will involve identifying and championing the distinctive features of the countryside in different regions. These include the physical landscape, but also the social and cultural characteristics that contribute to the distinctive values of rural areas (local building styles, traditions, arts, crafts and skills, dialects, country sports, festivals and regional food and drinks).

With the economic decline of agriculture and other primary industries, new economic activities need to be attracted to the rural areas. This rural diversification should build on local distinctiveness and include tourism, crafts, environmental services, regional foods, cultural industries and rural services and infrastructures. Localities and regions need to formulate their own positive vision for the future of the rural economy.

An important task for the new Countryside Agency is to help bridge the rural–urban divide. The government's commitment to opening up access to the countryside is vital. Urban people must have the opportunity to enjoy the countryside, but it is important that this benefits rural areas too, and that rural people do not come to feel like props in a rustic theme park. Rural people must also have the opportunity to fully participate in the modern world. This requires improved mobility, and investment in advanced telecommunications for rural areas.

Where does all this leave farming? Increasingly, farmers will need to put delivering a clean, healthy and attractive environment over producing food. This is what modern society wants from rural areas, and agriculture is the primary force in managing the physical environment. A move towards a more resource-conserving future will give more emphasis to the role of rural areas as sites for the supply, use and replenishment of renewable resources. This may involve new types of production such as biomass, energy crops and wind farms.'

A sustainable countryside?

Most discussions on the future of rural areas now refer to the need to create **sustainable rural environments**. Sustainability is an important idea. One influential definition of sustainable development is:

'*development that meets the needs of the present without compromising the ability of future generations to meet their own needs*'.

The Rio Earth Summit in 1992 led the majority of the world's governments to adopt the principle of sustainable development. The British Government published a *Strategy for Sustainable Development* in 1994, which set out ideas about improving the planning system to reflect environmental concerns. Whilst there is general agreement that sustainability is a desirable goal, reaching agreement about how this can be achieved is more difficult. One checklist is offered below for what sustainable development would look like in rural areas.

To achieve sustainable development there needs to be:

- an expansion of the total biomass through increasing forest areas, tree planting along field boundaries, and protection of areas of natural vegetation
- increased composting of organic wastes and reduced use of artificial fertilisers
- increased production of energy from renewable sources such as wind, wave, tide and geothermal
- reduced consumption of fossil fuels
- improved access and public transport in terms of frequency and convenience
- reductions in long-distance travelling
- growing self-sufficiency of the local economy – for example, a greater variety of local job opportunities and supply of daily goods from local sources
- publication of regular audits on waste, pollution, energy and water.

NOTING ACTIVITIES

1   What do you understand by the term 'sustainability'?

2   Many uses of rural land are unsustainable. This may be a result of excessive use (e.g. footpath erosion around a lake) or poor management (e.g. pollution of a river by a local factory). Make a copy of Figure 5.42. With other members of your group, think of ways in which rural land uses might be unsustainable, and suggest ways in which they might be made more sustainable.

Land use	Possible unsustainable use	Possible sustainable use
Agriculture		
Transport	Congestion on rural roads	Limit need to travel by encouraging home work, locate services in villages
	Noise pollution and safety problems on rural roads	Speed restrictions through villages
Forestry and woodland		
Mineral extraction		
Industry		
Housing		
Recreation		

5.42   *Rural land use: sustainable, or not?*

Land use planning

The land use planning system determines what development can take place in rural areas and where development is located. The planning system is made up of several levels or tiers:

- Central government issues national planning guidance to local authorities. This guidance provides the broad aims and objectives the government wishes to see adopted.
- The central government may also provide regional planning guidance which sets out the objectives at a regional level.
- Upper-tier local authorities or county councils draw up Structure Plans which indicate where development of housing, industry, retailing and leisure should be focused.
- Lower-tier local authorities or district councils produce Local Plans with more detailed and site-specific information.

Structure Plans

Structure Plans are important because they set out policies for future development within a county. They are based on a geographical survey of the area which covers issues such as land use changes, population trends, economic changes and transport networks. The aim is to build up an impression of the key problems or issues facing the area in the coming years and which require a solution. Structure Plans cover a wide range of issues. These include:

- population
- settlement patterns
- employment
- industry
- housing
- retailing
- transport
- conservation
- leisure and recreation
- minerals.

Having identified the main issues facing the county, planners decide upon policies to solve them. Planners are required by law to consult individuals and interested organisations in the community at various stages. Once the Structure Plan is completed it is submitted to the Secretary of State for the Environment who will usually arrange for an 'Examination in Public' to be held into any major controversial issues.

CASE STUDY

The Bedfordshire Structure Plan 2011

Bedfordshire is a small, densely populated county. In 1991 its population was 533 000, the majority living in the larger towns. Between 1951 and 1991 the population of the county increased by 64 per cent.

The Structure Plan 2011 identifies a need for more housing. This is due to an expected 9 per cent increase in the population and the trend towards smaller household size as the number of retired and single people increases. The demand for new housing will put pressure on the existing Green Belt. Three-quarters of Bedfordshire's area is agricultural land, much of which is good quality, and it contains parts of the Chilterns AONB.

Although parts of Bedfordshire are affluent, there are pockets of high unemployment and the economy performed less well in the 1990s. The Structure Plan identifies the need to provide employment opportunities through attracting new industries.

There was an increase in car use for travel to work from 58 per cent to 71 per cent between 1981 and 1991. Car ownership increased while the number of people using buses, cycling and walking declined. This has led to problems of congestion.

In response to these issues, the Bedfordshire Structure Plan 2011 identifies a number of key objectives. These include the needs:

- to contribute to sustainable development whilst strengthening the local economy
- to concentrate development in the main urban areas and within two strategic corridors
- to identify targets and indicators for assessing the progress towards sustainable development.

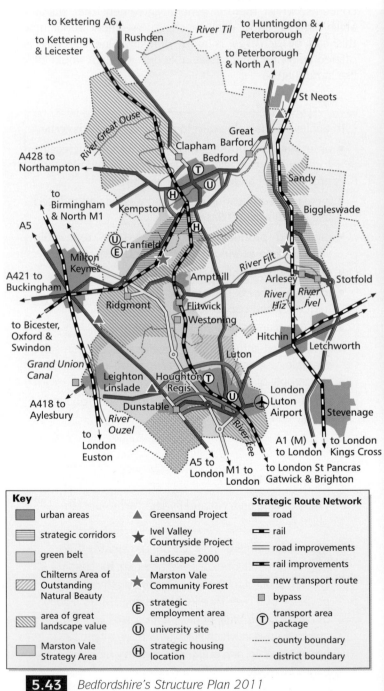

Key

urban areas	
strategic corridors	
green belt	
Chilterns Area of Outstanding Natural Beauty	
area of great landscape value	
Marston Vale Strategy Area	

▲ Greensand Project
★ Ivel Valley Countryside Project
▲ Landscape 2000
★ Marston Vale Community Forest
Ⓔ strategic employment area
Ⓤ university site
Ⓗ strategic housing location

Strategic Route Network
- road
- rail
- road improvements
- rail improvements
- new transport route
- bypass
- Ⓣ transport area package
- county boundary
- district boundary

5.43 *Bedfordshire's Structure Plan 2011*

Figure 5.43 is a map showing part of Bedfordshire's Structure Plan. The main features of the Plan are these:

- The Southern Bedfordshire Green Belt is to be maintained.
- 43 000 new dwellings are to be provided between 1991 and 2011, mainly in two Strategic Growth Corridors: one south of Bedford in the centre of the county, the other in the east of Bedfordshire.

- Town centres are to be maintained, whilst edge-of-town developments are to be discouraged.
- Development of housing is to be closely linked with the expansion of employment and integrated with transport routes.
- Mineral extraction is to be minimised, and efficiency, re-use and re-cycling of waste will be encouraged.

CASE STUDY

Devon County Structure Plan 2011

The Devon County Structure Plan identifies three main types of area. These are:

- main areas of economic activity
- rural areas where development will be constrained
- rural areas where the aim is to encourage economic diversification.

These areas are shown on Figure 5.44.

The Structure Plan sets out the need for 74 500 new dwellings and 775 ha of employment land in the county between 1995 and 2011. The major part of this development is to be concentrated in the five main areas of economic activity. The main urban areas of Plymouth and Exmouth are identified as particularly pressured, and two new communities are planned to meet the housing and employment needs of these areas. Focusing development in this way will relieve pressure on other rural areas. A large part of the county, which is predominantly rural, is identified as being in need of new investment in economic activity.

In addition, the Structure Plan seeks to reduce the need to travel and promote the use of public transport, and to preserve the quality of existing landscapes within the county.

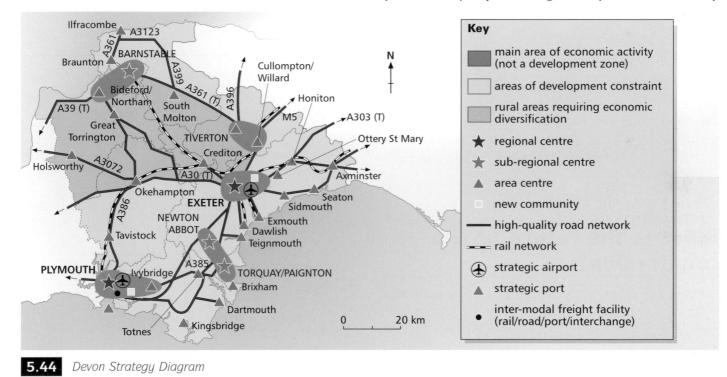

5.44 *Devon Strategy Diagram*

Do Structure Plans work?

As we have seen, Structure Plans indicate where local authorities would like to see particular kinds of development taking place. However, in reality plans can be overruled. Developers can apply to develop sites that are not shown on

a plan and still hope to get approval. They may be successful for a variety of reasons:

- Planning applications are decided with regard to 'all material considerations', of which the Structure Plan is just one. Others include highway and traffic requirements, the design and density of the proposed development, and the impact on the landscape.

- Economic pressures may be difficult to resist even if the plan seeks to limit urban development. This is often the case with land at the edge of towns which is sought after for commuter housing but is not located within a Green Belt.

- Although Structure Plans may seek to limit urban development in an area, the decision may be overruled by the national government. This has been the case in recent years. Perhaps the most notable example is Stevenage Borough, where the Structure Plan limited new housing development on agricultural land but the government ruled that it would have to incorporate additional housing into its plan.

NOTING ACTIVITIES

1   What do you understand by the term 'sustainable development'?

2   Suggest a set of indicators that could be used to measure how sustainable a rural area is.

3   What is a 'Structure Plan'?

4   In what ways do the Structure Plans for Bedfordshire and Devon reflect the problems of pressured rural areas and remote rural areas?

5   Why might there be a gap between the aims of a Structure Plan and the reality of development in an area?

6   Obtain a copy of the Structure Plan for your own area. Use the checklist on page 175 to evaluate the extent to which it is moving towards sustainability.

What is the future of the countryside?

The new millennium is marked by uncertainty as to the future of rural areas. The 'rediscovery' of poverty and deprivation in rural areas has challenged the idea that society is simply on a continual path to progress, leading to a better life for all. Instead, we are more aware of the inequalities in quality of life experienced by people in rural areas. For the majority of people, who live in urban areas, it is comforting to know that there is a quieter, more relaxed, more 'natural' environment beyond the city.

It is difficult to make generalisations about the nature of rural areas; they are marked more by difference than similarity. In addition, the consensus that was once held, that rural areas should mainly be concerned with agricultural production, has broken down, so that there is no longer any agreement as to what rural areas should be used for (this is best seen in the conflicts over leisure).

Finally, geographers are coming to realise that different people experience rural areas in very different ways (some as a place of fear, others as a place of spirituality), so that talking about what rural areas are like (or should be like) is very difficult.

NOTING ACTIVITIES

1   Read the whole of this section again. When you have done so, make brief notes as to what you think are its main themes. Compare your notes with those of a fellow student. Did you note the same things? In what ways were your understandings different?

2   a   Prepare a short presentation with the title 'The Future of Rural Britain'. Break down your presentation into a series of sections. Possible subheadings include:

- the state of rural Britain

- issues facing rural Britain

- government policies towards rural Britain

- the next 20 years.

b   Present your presentation to the other members of your group. Use it as the basis for a group discussion.

A Global energy: the atmospheric motor

The **atmosphere** is a layer of gases (nitrogen 78 per cent, oxygen 20.95 per cent and carbon dioxide 0.03 per cent), liquids (e.g. water) and solids (e.g. dust) that forms an envelope around the Earth. It is held in place by the force of gravity. The active or 'weather-producing' layer of the atmosphere is limited to the lowest layer or **troposphere** which extends to 16 km at the Equator and 8 km at the poles.

The weather has a major influence on physical processes (including weathering, mass movement, ecosystems and soil formation), as well as on human activities such as agriculture and tourism. The atmosphere is highly dynamic and behaves as a fluid. It is extraordinarily complex, and no computer has yet been able to model accurately the intricacies of its workings.

The atmosphere can be studied at two scales:

- **Weather** – this describes the state of the atmosphere at a particular place and at a particular time. It includes descriptions of temperature, wind speed, wind direction and precipitation. The weather is a precise record of actual conditions and varies constantly.

- **Climate** – this is the average weather over a significant period of time, usually 30–35 years. It is a statistical average and does not relate to any *particular* time.

The energy budget

At the heart of all atmospheric processes is the heat energy derived from the sun. It is this energy that drives the atmosphere and is responsible for life on Earth.

The source of energy is the sun, which emits **short-wave** radiation. The amount of radiation that reaches the outer atmosphere is termed the **solar constant**. As the radiation passes through the atmosphere, it is depleted in a number of ways (see Figure 6.1). It may be **reflected** off the upper surface of clouds, or **scattered** by the air (it is the scattering of the blue wavelengths that gives us blue sky). It may be **absorbed** by liquids and gases; for example, ozone in the atmosphere absorbs ultraviolet rays which would be harmful to people. Eventually, about half of the solar constant reaches the Earth's surface.

On reaching the Earth's surface, some of the radiation is immediately reflected back into the atmosphere. This is called the **albedo effect**. The albedo of a surface is its reflectivity – a white surface (such as ice) has a high albedo and a dark surface (such as trees or water) has a low albedo. Some of the radiation warms the Earth and, in common with all warm bodies, heat is re-radiated. This is called **terrestrial**

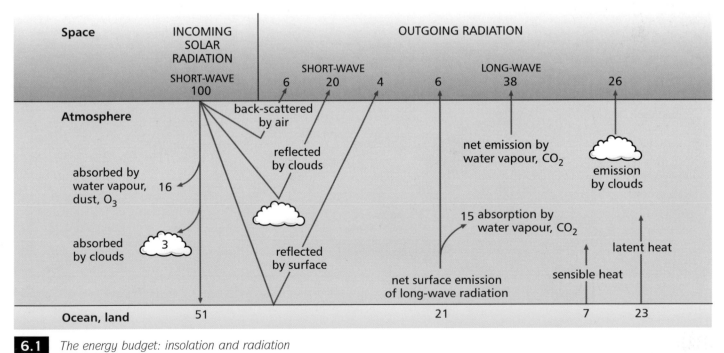

6.1 *The energy budget: insolation and radiation*

radiation and, in contrast with solar radiation, is **long-wave** energy.

Energy is also lost to the atmosphere by the transfer of **latent heat**. When water evaporates, heat is used up and stored as latent heat. It is released when the water vapour condenses to form cloud droplets. **Sensible heat transfer** is the transfer of heat by the processes of **convection** (rising hot air) and **conduction** (the transfer of heat by being in contact with a warm surface).

Is solar radiation distributed evenly across the Earth's surface?

Whilst the overall energy budget forms a balance, whereby the inputs equal the outputs, the amount of solar radiation reaching the surface (called **insolation**) is not spread evenly across the world.

As Figure 6.2 shows, much more insolation is received at the Equator than at the poles. The main cause of this is the effect of the curvature of the Earth. The amount of insolation reaching the surface is greatest when the sun's rays approach at right-angles (i.e. the sun is directly overhead). As the angle of the rays becomes increasingly oblique, the energy decreases as it is dispersed over a greater area (Figure 6.3). This effect is similar to that of varying the angle of a torch beam on a flat surface. Therefore the curvature of the Earth results in less energy being received at the poles than at the Equator. This inequality is increased by the fact that the rays pass through a greater thickness of atmosphere (see Figure 6.3)

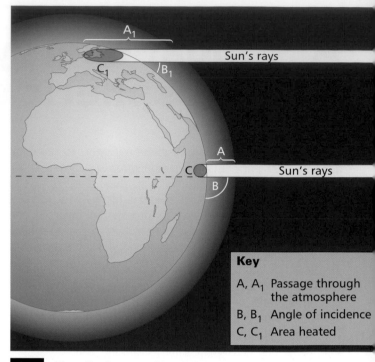

Key

A, A_1 Passage through the atmosphere

B, B_1 Angle of incidence

C, C_1 Area heated

6.3   *The effectiveness of insolation*

as they approach the poles, therefore losing more energy through scattering, reflection and absorption. The high albedo at the poles and the massive seasonal variations, involving 24 hours of winter darkness, also reduce the amount of insolation available for absorption. At the Equator, seasonal differences are minimal, and the sun is

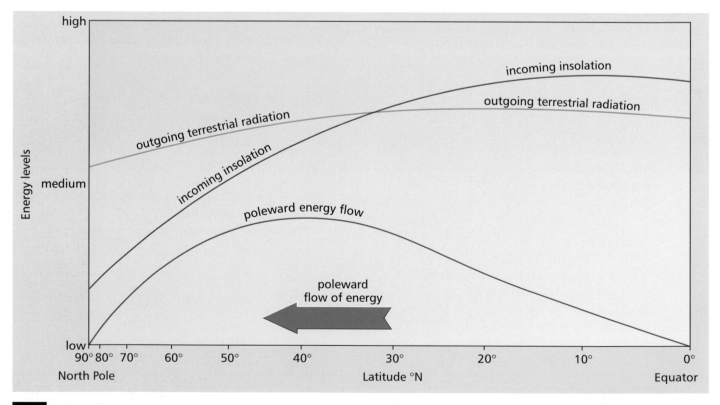

6.2   *Incoming solar radiation, outgoing terrestrial radiation and poleward energy flows, by latitude, for the northern hemisphere*

always more or less 'straight up', producing concentrated energy on the Earth's surface.

Why are there seasons?

The Earth's axis is not in alignment with that of the sun. This means that the level of solar radiation varies over the year at all latitudes and creates a seasonal pattern, including winter/summer and wet/dry seasons. This effect is least marked at the Equator as the sun is overhead there twice a year (21 March and 21 September – these are the **solstices**), so these areas have eight 'weak' seasons a year. At the poles, however, there is an extreme contrast, with 24 hours of darkness in the winter and 24 hours of sunlight in the summer. The **equinoxes** are when the sun is overhead at the Tropics – in the UK the longest day is on 21 June (the sun is overhead at the Tropic of Cancer, $23\frac{1}{2}°$ North) and the

shortest day is on 21 December (the sun is overhead at the Tropic of Capricorn, $23\frac{1}{2}°$ South).

How is balance achieved?

On the face of it, the information in Figure 6.2 suggests that the Equator is warming up and the poles are getting colder. This is, of course, not the case because there is – more or less – a balance. It is achieved by the horizontal transfer of energy, which can take a number of forms:

- warm ocean currents move heat towards the poles and cold ocean currents transfer cooler water Equatorwards
- the major trade winds also result in a net transfer of heat from the Equator to the poles
- major storms, such as hurricanes, transfer heat polewards.

NOTING ACTIVITIES

1   What, essentially, is the difference between weather and climate?

2   Study Figure 6.1.

   a How much solar radiation reaches the surface?

   b What has happened to the rest of the incoming radiation?

   c What is the technical term for radiation reflected from a surface?

   d How does the amount of reflected radiation vary according to the surface type?

   e What would happen to the energy budget if the amount of CO_2 in the atmosphere increased?

   f What is meant by the term 'sensible heat transfer', and how does it operate?

   g Use actual figures to help you explain how the energy budget forms a balance.

3   Explain, using a diagram to help, how the curvature of the Earth is largely responsible for the uneven distribution of insolation at the Earth's surface.

4   How is a latitudinal heat balance maintained despite the uneven distribution of incoming radiation?

STRUCTURED QUESTION 1

Figure 6.4 shows a simplified version of the energy exchanges occurring in the atmosphere–land system on a cloudless sunny day.

a Draw a similar diagram to show the corresponding energy exchanges, simplified for a cloudless night.   *(3)*

b Suggest three changes in the relative magnitude of the energy exchanges showing in Figure 6.4 that would occur if the cloudless sunny day budget were for the atmosphere–ocean system.   *(3)*

c Suggest two ways in which human activity may modify such energy exchanges, and explain why the modification occurs.   *(2)*

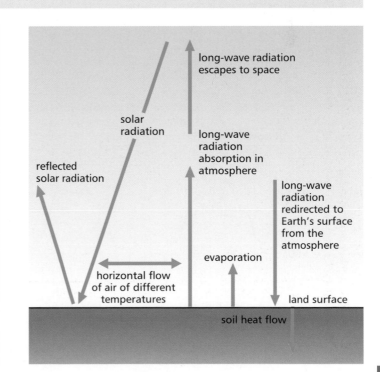

 Simplified diagram to show the energy exchanges in the atmosphere–land system on a cloudless, sunny day

STRUCTURED QUESTION 2

Figure 6.5 shows variations of net radiation throughout the year at a series of weather stations. The latitude and longitude of each station is given. Net radiation is defined as 'incoming radiation minus outgoing radiation'.

a What is the highest value of net radiation at:

- Novolazarevskaya
- Hamburg? *(1)*

b Why is the highest mean monthly value of net radiation found in the Antarctic at Novolazarevskaya in December? *(2)*

c Describe the annual trend of net radiation values at Yangambi. *(2)*

d Why do monthly values of net radiation vary less throughout the year at Yangambi and Khormaksar, which are both tropical stations? *(2)*

e Why do some stations experience negative values of net radiation at certain times of the year? *(2)*

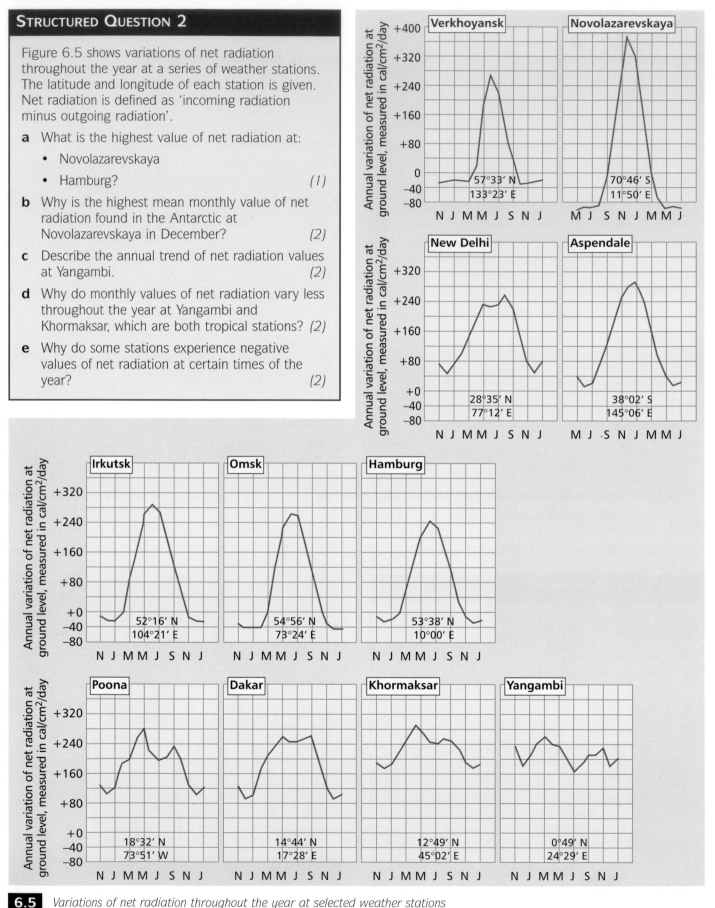

6.5 *Variations of net radiation throughout the year at selected weather stations*

Why are the temperature zones not regular?

Whilst it is possible to identify broad latitudinal temperature patterns, there are many irregularities. This is due to several different factors.

Land and sea

The surface of the Earth is not uniform, and this influences its response to solar radiation (its **thermal efficiency**). Land masses absorb short-wave energy and radiate long-wave energy more rapidly than water (rivers, lakes and the oceans), causing more extreme temperatures than are found at the same latitude over the oceans. For this reason land areas have hotter summers and colder winters than areas near the sea. This effect is most marked in the northern hemisphere, which contains 62 per cent of the global land areas. The differences in thermal efficiency give rise to two main climatic types: **continental** and **maritime**.

Albedo

The colour of the surface also influences the proportion of energy absorbed, and this effect is termed the **albedo** (the proportion of incoming radiation reflected by the surface). On average 32 per cent of solar radiation is reflected from the surface as short-wave energy but this varies, from as much as 90 per cent over fresh snow or white ice, to as little as 10 per cent over dark-green coniferous forests (see Figure 6.6).

Ocean currents

As a fluid with a low thermal efficiency, water forms an effective mechanism for the transfer of energy across latitudes. Major, long-term, flows of water are termed **ocean currents** and these have a strong influence on atmospheric temperature, as heat is either released (a warm ocean current) or absorbed (a cold ocean current). The major oceans are characterised by cells of currents that transfer energy from the equatorial regions towards the poles and return cold water from the poles to the low latitudes (see Figure 6.7). Winter temperatures in the UK and north-west Europe are significantly affected by the warm **North Atlantic Drift**. The disruption of normal currents in the Pacific in 1998/99, called El Niño, significantly affected weather patterns in South and Central America.

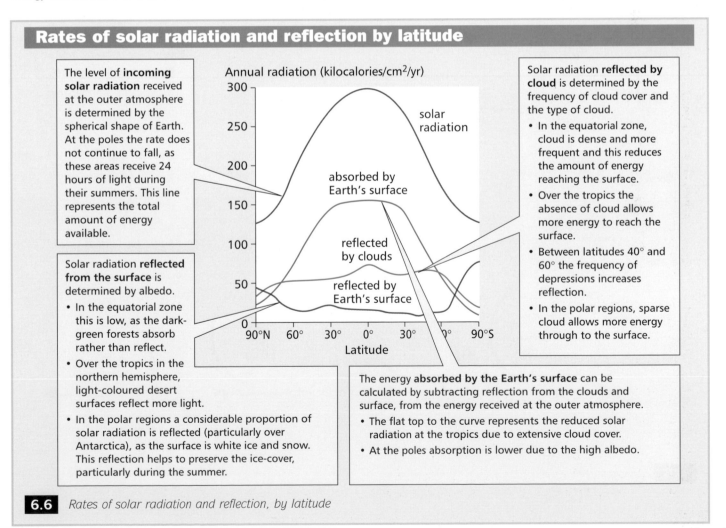

Rates of solar radiation and reflection by latitude

The level of **incoming solar radiation** received at the outer atmosphere is determined by the spherical shape of Earth. At the poles the rate does not continue to fall, as these areas receive 24 hours of light during their summers. This line represents the total amount of energy available.

Solar radiation **reflected from the surface** is determined by albedo.

- In the equatorial zone this is low, as the dark-green forests absorb rather than reflect.
- Over the tropics in the northern hemisphere, light-coloured desert surfaces reflect more light.
- In the polar regions a considerable proportion of solar radiation is reflected (particularly over Antarctica), as the surface is white ice and snow. This reflection helps to preserve the ice-cover, particularly during the summer.

Solar radiation **reflected by cloud** is determined by the frequency of cloud cover and the type of cloud.

- In the equatorial zone, cloud is dense and more frequent and this reduces the amount of energy reaching the surface.
- Over the tropics the absence of cloud allows more energy to reach the surface.
- Between latitudes 40° and 60° the frequency of depressions increases reflection.
- In the polar regions, sparse cloud allows more energy through to the surface.

The energy **absorbed by the Earth's surface** can be calculated by subtracting reflection from the clouds and surface, from the energy received at the outer atmosphere.

- The flat top to the curve represents the reduced solar radiation at the tropics due to extensive cloud cover.
- At the poles absorption is lower due to the high albedo.

Annual radiation (kilocalories/cm^2/yr)

solar radiation

absorbed by Earth's surface

reflected by clouds

reflected by Earth's surface

6.6   *Rates of solar radiation and reflection, by latitude*

Ocean currents and sea surface temperatures

Cold ocean currents can be seen on Figure 6.7 as extensions of blue towards the Equator. In the northern hemisphere these include the Labrador Current (between Greenland and Canada), the Californian Current, and the cold Canaries Current off the west coast of North Africa.

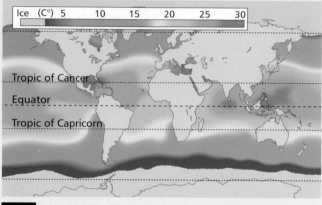

 6.7 *Sea surface temperatures and ocean currents. 15 August 1999*

The North Atlantic Drift is not particularly evident in August, as the North Atlantic is heated by solar radiation. In January it is more prominent.

The latitudinal pattern of temperature is evident in Figure 6.7, with the zone of maximum heat to the north of the Equator (this image was taken in August). In January these zones migrate southwards and warm ocean currents in the north become more prominent.

Shallow seas isolated from the major oceans are considerably hotter. This is because the shallow water heats more rapidly and there is less circulation to diffuse the energy.

Cold water extends towards the Equator along the west coasts of South America (the Peruvian cold current), Africa (the Benguela ocean current) and Australia. The cold water is driven by the prevailing westerly wind and is particularly strong off Chile and Peru where it occupies a deep ocean trench (Figure 6.8).

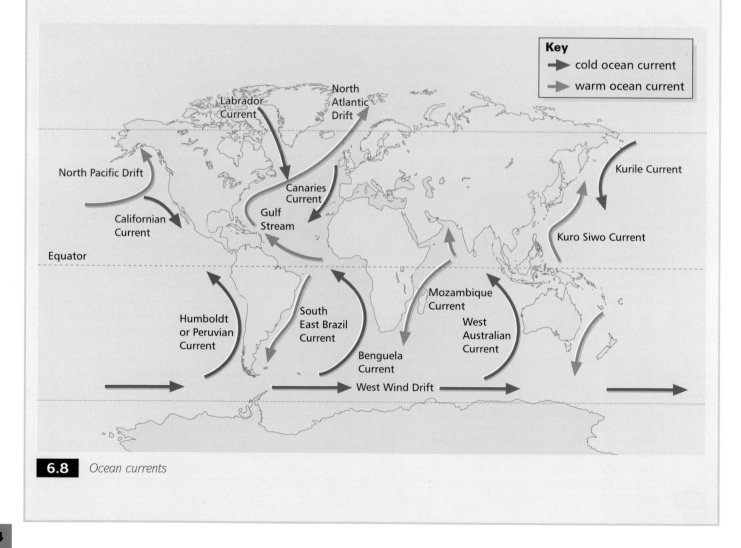

6.8 *Ocean currents*

An El Niño event

In an El Niño event, hot tropical water extends as a surface flow to the coast of South and Central America (Figure 6.9, lower image). This brings rain to the coasts of Peru and Chile, areas typically among the driest places on Earth. In 1986 such an event caused widespread disruption of fishing and agriculture and caused the Atacama Desert to bloom. This illustrates the importance of ocean currents in regulating temperatures and rainfall.

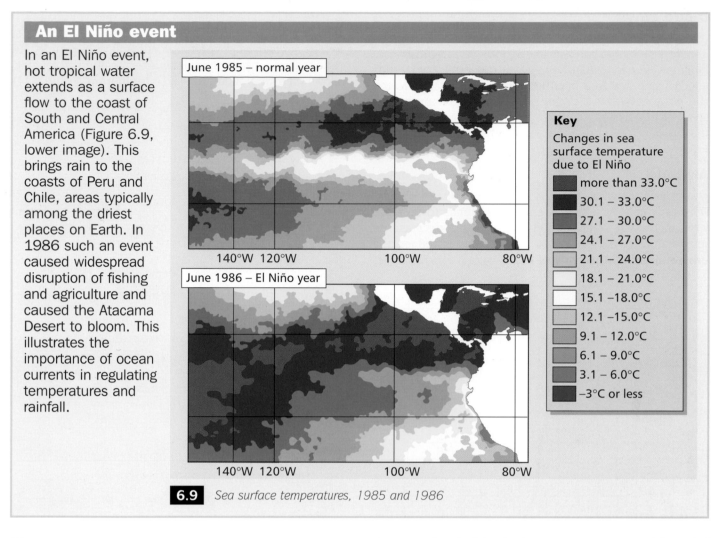

6.9 *Sea surface temperatures, 1985 and 1986*

Air masses

An **air mass** is a large volume of air (often continental in scale) that is relatively uniform in terms of temperature and humidity. Air is heated or cooled by the surface, and when air masses move away from their source region, they transfer energy. Air has a high thermal efficiency (it loses and gains heat rapidly), so the energy transfer is less efficient than that of ocean currents. Air masses are closely associated with the temperature zones and the distribution of land and sea, and they are identified by both temperature (*Equatorial*, *Tropical* and *Polar*) and water content (*Continental* and *Maritime*). One of the main reasons for the frequent changes in British weather is the location of the British Isles at the junction of five major air masses (Figure 6.10). During the passage of a depression it is quite possible that three of these air masses will influence Britain during the course of a day, with a significant effect on temperature.

Air mass	Direction of source	Source region	Temperature	Humidity	Purity
Arctic	northerly	Arctic	very cold	dry	clean
Polar maritime	north-west	North Atlantic	cold	wet	clean
Polar continental	north-east and east	Scandinavia and Siberia	cold in winter, warm in summer	dry	clean if north-east but polluted from east
Tropical maritime	south-west	south-west Atlantic	warm in winter but cool in summer	wet	clean
Tropical continental	south-east and south	southern Europe	warm/hot	dry	polluted

6.10 *Air masses affecting the UK*

Cloud cover

Cloud consists of either ice crystals or small water droplets, and forms a physical barrier to both solar (short-wave) radiation and terrestrial (long-wave) radiation. Locally this effect may be noticed by a decrease in temperature as the sun is obscured by a cloud, but globally it is apparent in areas of frequent cloud cover or frequent clear skies, as in the humid tropics and the subtropical high deserts (see Figure 6.6).

Altitude

The heat in the atmosphere is gained from terrestrial long-wave radiation. As a result, temperature tends to decrease with increasing altitude. On average the decrease in temperature is 0.6°C per 100 m, and this is termed the **environmental lapse rate** or **ELR** (Figure 6.11). This effect

is noticeable when climbing mountains and is apparent in snow settling and accumulating above the 'snow line', as well as in changes in vegetation and soil with altitude. The relationship between altitude and temperature is generally shown on a type of graph called a **tephigram** (Figure 6.11).

NOTING ACTIVITIES

1 Why *do land areas have hotter summers and colder winters than coastal areas?*

2 Study Figure 6.6.

   a *Make a copy of the graph, showing the two lines that relate to 'reflection'.*

   b *Add labels to explain the trends of these lines.*

3 *What effect do ocean currents have on the climate of:*

   a *the west coast of South America*

   b *the west coast of the USA?*

4 *What is an 'El Niño event', and what effect does it have on global weather?*

5 *Study Figure 6.10.*

   a *Draw a sketch map to plot the courses of the five main air masses affecting the UK.*

   b *Add labels to describe the characteristics of each air mass.*

6 *What is a temperature inversion, and how is it formed?*

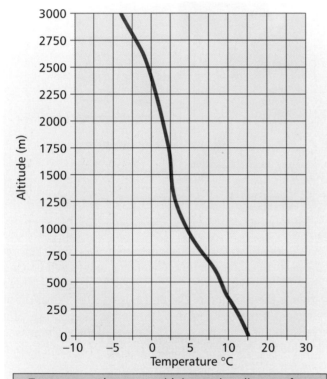

- Temperature decreases with increasing distance from the surface.
- The average rate of decrease is 0.6°C per 100 m.
- The line showing the rate of cooling is generally not uniform, representing different layers of air in the lower atmosphere.
- Occasionally, the temperature of the air may increase rather than decrease. This is called a **temperature inversion** and is most common close to the ground. When the ground surface cools during a calm, clear night, air in contact with it is cooled by conduction. This lowest layer of air may become cooler than the air immediately above it, forming an inversion. Inversions are commonly associated with pockets of frost and fog.

6.11 *The environmental lapse rate*

EXTENDED ACTIVITY

Study Figure 6.13.

1 Using a world map outline, make a copy of the temperature anomalies for July.

2 On your map, add the ocean currents from Figure 6.8, using a blue pencil for the cold currents and a red one for the warm currents. Label the currents.

3 Either by using annotations or a key, identify and account for some of the anomalies shown on Figure 6.13. Try to find ones that are caused by different factors. Use an atlas to help you describe the locations accurately.

4 Write a paragraph suggesting which factors seem to be the most important in causing temperature anomalies in July.

Global temperature patterns

One of the direct effects of the energy budget is to determine global temperature patterns. The pattern of temperature is extremely complex – take a look at maps in an atlas – and one of the most useful types of map plots temperature anomalies. A temperature anomaly is the difference between recorded temperature (°C) and the mean for that latitude (Figures 6.12 and 6.13).

January

The extension of the Gulf Stream across the North Atlantic is called the North Atlantic Drift. This carries warm water (approximately 13°C) towards the coast of Norway and lifts temperatures to 24°C above the latitudinal mean. The coastal areas of north-west Europe have an equable climate with mild winters.

During winter the continental interiors radiate heat, causing temperatures to fall well below the average for the latitude. In eastern Siberia, January temperatures are 24° below the average for that latitude, and the absolute minimum may reach −70°C. This gives interiors a severe climate with very cold winters.

In January, maximum solar radiation occurs over the Tropic of Capricorn. The temperatures off the west coasts of South America, Africa and Australia are below (4°C) the average of the latitude due to cold ocean currents.

The continental interiors of the southern continents heat up in January and become hotter than their latitudinal means. This effect is less marked than in Asia in July, as the land masses of the south are smaller. In general, temperature anomalies are greater in the northern hemisphere.

6 2 *January temperature anomalies*

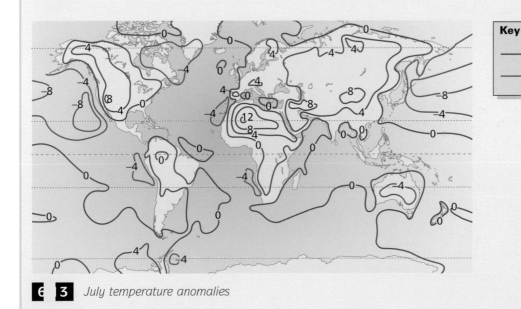

Key

—— positive anomaly

— negative anomaly

6 3 *July temperature anomalies*

STRUCTURED QUESTION 3

Study Figure 6.12, which shows world temperature anomalies for January.

a (i) What is the maximum temperature anomaly over the north-east Atlantic (area A)? *(1)*

(ii) Outline one reason for the temperature anomaly at area A. *(2)*

b Describe and explain the temperature anomalies at the following locations:

(i) off western South America (area B) *(3)*

(ii) in central Russia (area C). *(3)*

c Study Figure 6.14, which shows global surface-received solar radiation.

(i) Describe the location of the areas of maximum surface-received solar radiation. *(2)*

(ii) Why are the areas of maximum surface-received radiation not on the Equator? *(2)*

(iii) Explain why not all the surface-received solar radiation is available for heating the ground surface. *(2)*

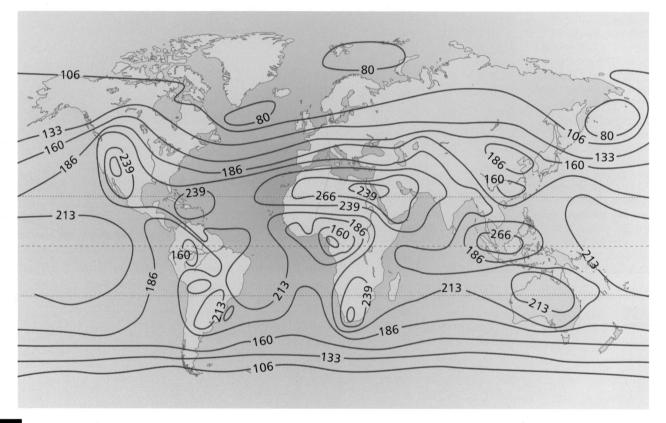

6.14 *Worldwide distribution of annually averaged global surface-received solar radiation (Watts/m²)*

B Pressure and winds

What is atmospheric pressure?

The atmosphere is retained around the Earth by the force of gravity. Generally, air pressure decreases with increasing height, but it also varies horizontally due to variations in surface temperature.

Pressure is measured by a **barometer** and is expressed in **millibars** (mb). The average sea-level pressure is 1013 mb but pressures can rise to over 1040 mb when cold air is descending, or fall as low as 950 mb when local hot air is rising. Pressure is plotted on a synoptic chart in the form of **isobars** (iso = a line joining points of equal value, bar = barometric pressure in millibars).

The differential heating of the Earth's surface – which we looked at in chapter 6A – is sufficient to create a pattern of pressure cells, often simplified as the **tricellular model** (Figure 6.15). This is an essential foundation to the understanding of both weather and climate. The three cells are relatively separate circulatory systems and form two major **high pressure zones** and two major **low pressure zones** at the surface. Low pressure zones occur where surface air is rising, and high pressure systems occur where air is descending (see Figure 6.15).

1 The Hadley cell

Surface air is heated in the low latitudes and rises, forming a major belt of low pressure called the **equatorial low**. Surface air rises to over 16 km and at the **tropopause** (the upper limit of active weather) it diverges and descends over the tropics. This cold descending air forms belts of high pressure called the **sub-tropical highs**. The descending air diverges at the surface with air flowing back into the Hadley cell (the trade winds) and moving towards the poles to form the Ferrel cell.

2 The Ferrel cell

In the weakest of the three cells, air rises in the colder regions around 60° latitude and moves through the upper atmosphere, descending at around 30° latitude.

3 The Polar cell

Air descends over the colder poles and flows outwards towards the Ferrel cell. The zone of convergence between the Polar cell and the Ferrel cell is called the **Polar front**, and this forms a zone of low pressure.

The three-cell model shows the relationship between temperature, pressure and major wind systems. The location of the cells varies in a seasonal cycle as the temperature zones move with the position of the sun.

What is wind?

Wind is the movement of air from an area of high pressure to one of low pressure. It results from a difference in pressure between two points – this is termed the **pressure gradient**. The greater the difference in pressure, the stronger the wind. On a weather map, a steep pressure gradient (and strong winds) is shown by isobars that are very close together.

Wind direction is always identified by its source: for example, air in a westerly wind moves from the west to the

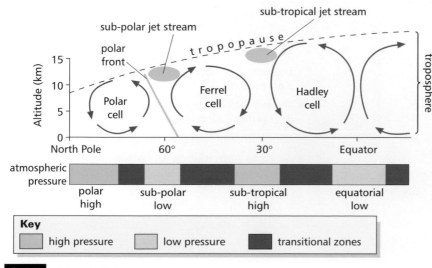

> **High pressure belts** Air in the upper atmosphere is converging (moving together) and is sinking. This causes lower atmospheric divergence (air moves apart). The air descends because it is cold and dense, and few clouds are formed.
>
> **Low pressure belts** Air is rising with lower atmosphere convergence and upper atmosphere divergence.
>
> **Transitional zones** between the high and low pressure systems: these regions will be seasonally affected by the highs and lows as the pressure belts migrate with the seasons (overhead sun).
>
> **Jet streams** In the upper atmosphere there are two important wind systems called **jets**. These winds are not directly experienced at the surface but do influence weather. The high-altitude winds (10 to 15 km from the surface) have a velocity of between 200 and 400 km/h, and the two main flows are located between the three circulatory cells.

6.15 *The tricellular model*

Section 6 The Atmosphere

east. This convention is used because it tells us much about the character of the air that is currently influencing our weather. The source region determines the air temperature, the water content (relative humidity) and the purity, in terms of pollution.

Why does wind direction vary?

The direction of the wind on the Earth's surface is determined by three main factors.

1 Pressure gradient

Air flows from high to low pressure across a **gradient** (the difference in pressure divided by the distance) and this

movement is referred to as a **pressure gradient wind**. This explains the general direction of the major surface winds (including the trade winds). At a more local scale, pressure gradient winds are common in coastal areas as land and sea breezes (Figure 6.16).

2 The Coriolis effect

At a global scale this simple pattern of wind direction is modified by the rotation of the Earth, an effect known as the **Coriolis effect**. This causes a deflection of the air flow in the northern hemisphere to the right and in the southern hemisphere to the left (Figure 6.17). The deflection results in a prevailing westerly airstream in both hemispheres.

Sea breeze

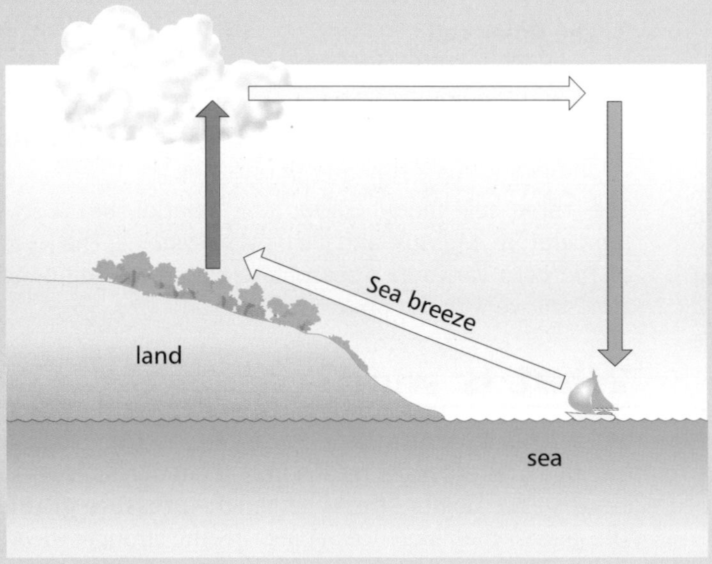

- The land heats rapidly, reaching peak temperatures in the late afternoon, especially in summer.
- The air above is heated and begins to rise.
- A local area of low pressure forms over land.
- The sea heats less rapidly and the cooler surface maintains a low air temperature.
- Air flows from high to low pressure (from sea to land) as a cool, moist sea breeze.

Land breeze

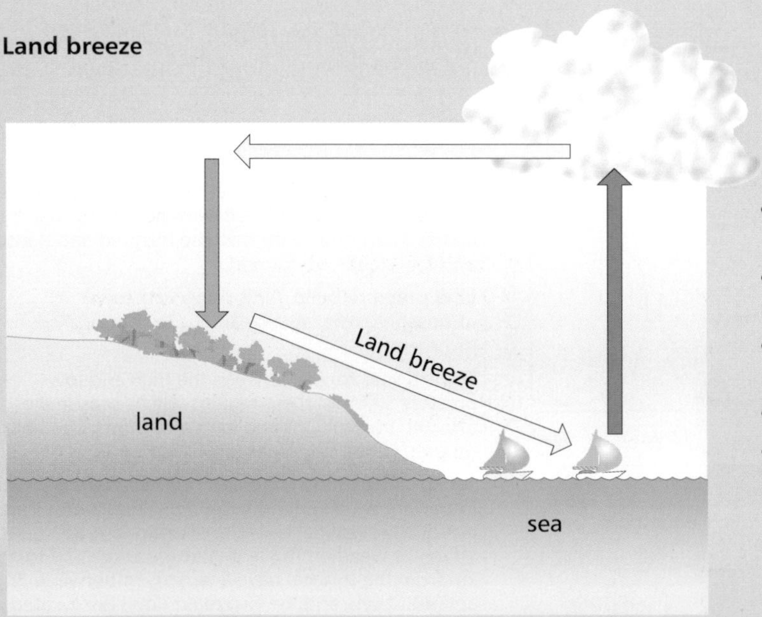

- During the night the land radiates heat rapidly and cools the air above.
- The colder air sinks to form a local area of high pressure.
- The sea retains heat gained slowly during the day, and the air above remains relatively warm.
- A local area of low pressure forms over the sea.
- Air flows from high to low pressure (from land to sea) as a cool land breeze, most apparent in the early morning.

6.16 *Land and sea breezes*

Cyclonic systems

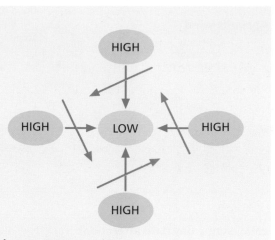

Low pressure systems

- Air flows from area of high pressure to area of low pressure across the pressure gradient.
- In the northern hemisphere this is deflected to the right by the Coriolis effect.
- The result is an anticlockwise circulation of air around the low.

Key
→ pressure gradient winds
→ geostrophic winds

Anticyclonic systems

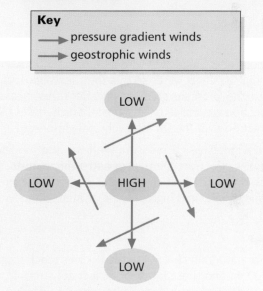

High pressure systems

- Air flows from area of high pressure to area of low pressure across the pressure gradient.
- In the northern hemisphere the flow is deflected to the right by the Coriolis effect.
- The result is a clockwise circulation of air around the high.

6.17   *The Coriolis effect (in the northern hemisphere)*

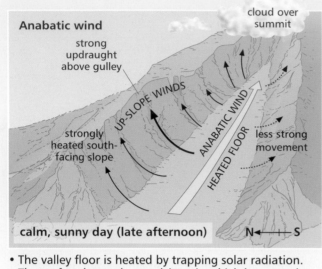

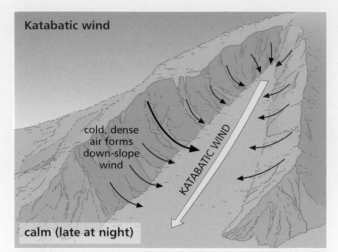

Anabatic wind

strong updraught above gulley

cloud over summit

UP-SLOPE WINDS

ANABATIC WIND

HEATED FLOOR

strongly heated south-facing slope

less strong movement

calm, sunny day (late afternoon)   N ← | | → S

- The valley floor is heated by trapping solar radiation.
- The surface heats the overlying air, which becomes less dense.
- Air rises from the valley floor and flows up the valley sides as an anabatic wind.
- This process can be strong enough to trigger storms in valley regions.

Katabatic wind

cold, dense air forms down-slope wind

KATABATIC WIND

calm (late at night)

- Radiation from the upper slopes rapidly cools the surface.
- The air is cooled by the surface and becomes more dense.
- The cold air flows down the valley side as a shallow surface layer or katabatic wind.
- The dense air collects on the valley floor, possibly forming valley fog.

6.18   *Valley winds*

The Coriolis effect increases with latitude, and by 50–55°N it is sufficient to deflect winds so that they flow almost at right-angles to the pressure gradient. In effect this means that wind direction is parallel to the isobars, which makes plotting wind direction over the UK relatively straightforward.

Winds moving across a pressure gradient and modified by the Coriolis effect are termed **geostrophic winds**. These form the spirals or vortices evident in both low pressure systems (cyclonic) and high pressure systems (anticyclones, Figure 6.17).

3 Relief

The major mountain chains like the Rockies reach heights of over 6000 m, and these block or deflect geostrophic winds. Even the relatively low Alps often separate wind systems between northern Europe and the Mediterranean.

In mountain areas a pressure gradient may arise, from differential heating and cooling, between the floor of a valley and the upper slopes. This forms a valley wind system of **anabatic** and **katabatic flows** (Figure 6.18), with direction determined by the slopes.

Why does wind speed vary?

Wind speed is measured in knots (1 knot = 1.8 km/h), and is determined by four main factors.

1 Pressure gradient

Wind speed is generally determined by the steepness of the pressure gradient as evident from the spacing of the isobars: open spacing indicates a low wind speed whilst close spacing generates a greater pressure gradient and a greater wind speed. Wind speeds of over 50 knots are unusual in the UK as there is generally not enough energy to create a sufficient pressure gradient. In the tropics, where there is much greater solar energy, wind speeds can exceed 100 knots (see page 216).

2 Friction

Wind, like water, is slowed by friction with the surface. The effect of friction is greater over land than over sea.

3 Turbulence

Internal friction produced by turbulence also reduces wind speed. In coastal areas, strong winds are common as air passing over the sea retains its laminar flow. Inland, wind speed rapidly falls as laminar flow is disrupted by the turbulence created as air passes over relief features, vegetation and buildings. Trees are often planted to provide shelter for houses and settlements, and hedgerows once provided a useful natural defence to aeolian soil erosion.

4 Local factors

Local obstructions may force wind into a narrower space and accelerate velocity. This can occur in 'canyon' streets in the CBD that are lined with high buildings or, as in Figure 6.19, between the cooling towers of a power station.

6.19 *One effect of local funnelling. Shortly after the cooling towers of the Ferrybridge power station in south Yorkshire were constructed, they were exposed to a gusty wind (November 1965). These 114 m structures were built in a line, which allowed wind to be forced between the towers. The resulting pressure differences caused three of the towers to 'explode', and the remainder had to be demolished because they were considered to be structurally unsafe.*

NOTING ACTIVITIES

1 Study Figure 6.15.

   a Describe the operation of the Hadley cell, and explain how it is responsible for creating pressure belts on the Earth's surface.

   b What are 'jet streams', and where are they found?

2 **a** Use an atlas map showing the physical geography of the world to make a list of the major deserts. Write down the approximate latitude of each desert.

   b In which of the pressure belts shown on Figure 6.15 are most of the deserts found?

   c Suggest reasons why this is the case.

3 Describe, with the aid of diagrams, the operation of the following local winds:

   a land and sea breezes

   b anabatic and katabatic winds.

Patterns of pressure and wind

Study Figure 6.20, which shows global pressure patterns and prevailing winds in January. Although the patterns are rather complex, it is not too difficult to make sense of them (see the annotations). There are a number of major features which you should notice:

* The pressure belts described in Figure 6.15 are quite clearly identifiable, especially the sub-tropical high in the northern hemisphere.

* Pressure zones become intensified where there are strong 'local' influences, e.g. ocean currents or large continental areas.

* Winds move out of areas of high pressure (anticyclones) and into areas of low pressure (cyclones).

* Winds circulate in a curved fashion as determined by the Coriolis effect.

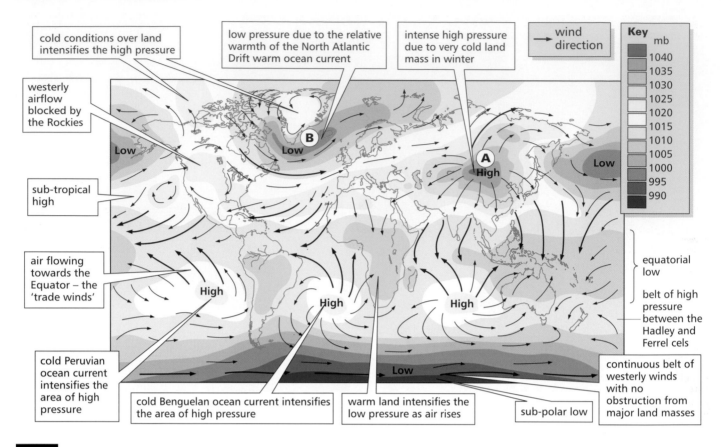

cold conditions over land intensifies the high pressure

low pressure due to the relative warmth of the North Atlantic Drift warm ocean current

intense high pressure due to very cold land mass in winter

wind direction

Key mb

1040
1035
1030
1025
1020
1015
1010
1005
1000
995
990

westerly airflow blocked by the Rockies

sub-tropical high

air flowing towards the Equator – the 'trade winds'

cold Peruvian ocean current intensifies the area of high pressure

cold Benguelan ocean current intensifies the area of high pressure

warm land intensifies the low pressure as air rises

sub-polar low

equatorial low

belt of high pressure between the Hadley and Ferrel cels

continuous belt of westerly winds with no obstruction from major land masses

6.20 *Pattern of pressure and winds in January*

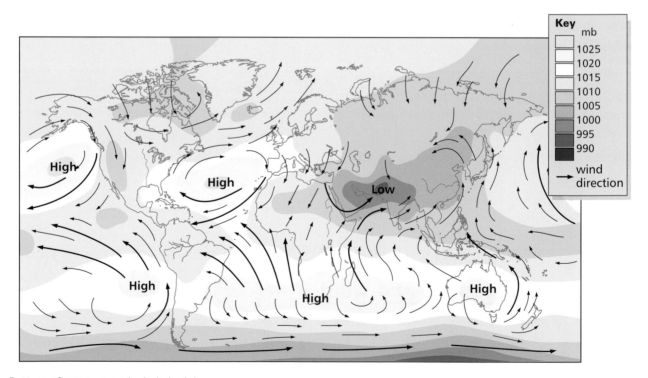

Key mb

1025
1020
1015
1010
1005
1000
995
990

wind direction

6.21 *Pattern of pressure and winds in July*

EXTENDED ACTIVITY

Study Figure 6.21, which shows pressure patterns and prevailing winds in July.

a On a blank world map, plot the following details:

- major areas of high and low pressure (use two different colours)

- major winds.

b Add annotations, similar to those in Figure 6.20, to describe and account for the major pressure zones and wind directions. You will need to refer to earlier maps of temperature and ocean currents to help you.

STRUCTURED QUESTION

Figure 6.20 shows the pattern of pressure and surface winds in January. Pressures are given in millibars reduced to sea level.

a (i) What name is given to the pressure pattern labelled A? *(1)*

(ii) What name is given to the pressure pattern labelled B? *(1)*

b (i) Describe the direction of the winds circulating in the pressure pattern A. *(2)*

(ii) Outline *two* reasons for the pattern of winds you have described. *(4)*

c Why are the highest atmospheric pressures to be found over Asia at this time of year? *(2)*

d Both wind and pressure patterns are more complex in the northern hemisphere than in the southern hemisphere. Suggest *two* reasons why this is so. *(4)*

e (i) Describe the pressure and wind patterns over South-east Asia. *(2)*

(ii) Explain how your description of these patterns would be different for July (Figure 6.21). *(4)*

C Water in the air

What is the hydrological cycle?

Clouds are the visible component of water in the atmosphere. They are formed of very small particles of water (< 0.04 mm diameter) and ice. Clouds are part of the **hydrological cycle** (Figure 6.22) as water is condensed, having been evaporated from the surface. Eventually, the water in clouds returns to the surface as precipitation. The majority of precipitation falls into the oceans, as they make up most of the Earth's surface. Of that which falls on the land, a high proportion is eventually transferred to the oceans by rivers. Some is stored as ice or soaks into the underlying bedrock.

The hydrological cycle is an example of a closed system in that there are no inputs from, or outputs into, external systems: it is completely self-contained.

What is humidity?

All air contains some water in the form of vapour. This is gained through the processes of evaporation and transpiration from water at the surface. **Evaporation** involves the transfer of water from liquid to vapour with a loss of energy (latent heat). The rate of evaporation is controlled by three main factors:

- **temperature** – increases with increasing heat
- **wind speed** – wet air is moved away from the surface so that evaporation increases with higher wind speeds
- **humidity** – when the air near the surface is already wet it can absorb less new water vapour.

The water vapour added to the air from the surface is pure, as bases and other solutes are left in the oceans, lakes and rivers; this process is termed **distillation**.

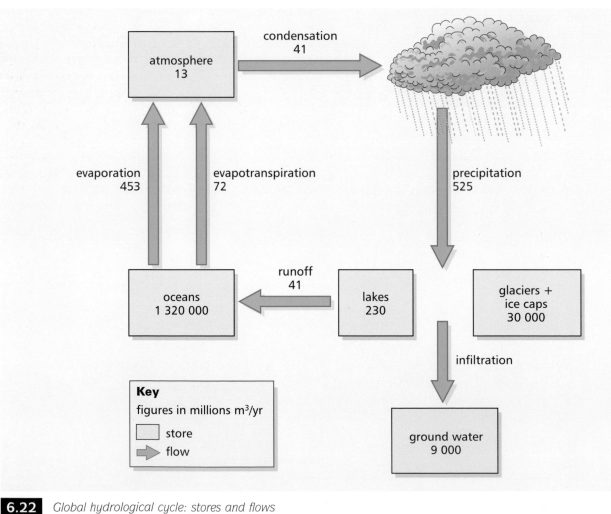

6.22 *Global hydrological cycle: stores and flows*

The amount of water vapour in the air is measured in two ways:

- **Absolute humidity (AH)** is a measure of water vapour in the air as a percentage of the air volume. It is largely determined by the nature of the surface under the air mass. Continental air masses tend to be significantly drier than maritime air masses because there is less water available.

- **Relative humidity (RH)** is the more important measure and represents the water vapour in the air as a percentage of the total water vapour that the air can hold at a particular temperature before it is saturated. Quite simply, the RH indicates how near a volume of air is to water-bearing capacity.

How does relative humidity change?

The amount of water vapour that air can hold is controlled by temperature. Warm air has a greater capacity than cold air and, therefore, as air is heated its relative humidity decreases even though no water vapour is being lost. Conversely, as air is cooled its relative humidity increases, even though no water vapour is being added (Figure 6.23).

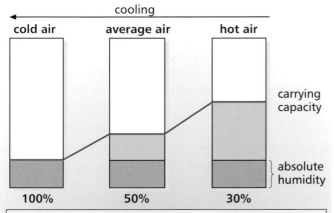

cooling

| cold air | average air | hot air |
| 100% | 50% | 30% |

carrying capacity

absolute humidity

Absolute humidity: the actual amount of water vapour in the air. In this example the AH does not change as no water vapour is added or lost.

Carrying capacity: the total amount of water vapour that the air can carry at that temperature.

100% relative humidity
Cooling has reduced the carrying capacity to the same volume as the absolute humidity. The air has reached dew point (i.e. the temperature at which it condenses) and will condense to form cloud droplets.

50% relative humidity
The air is halfway to its carrying capacity. The AH has not changed but as the air is warmer, it can potentially hold more water.

30% relative humidity
The carrying capacity has increased further as the air has been heated. The air can now potentially absorb even more water vapour.

6.23 *A volume of air at different temperatures, showing the relationship between absolute humidity, carrying capacity and relative humidity*

NOTING ACTIVITIES

1 Study Figure 6.22.

   a Calculate the percentage of total water:

   (i) stored in the clouds

   (ii) stored in the oceans.

   b Describe the pattern of circulation shown in the diagram.

   c The amount of water evaporated from the oceans exceeds the amount of water that falls into the oceans. How do you account for this fact?

   d Why is the hydrological cycle described as a 'closed system'?

2 What is the difference between 'absolute' and 'relative' humidity?

What is condensation?

When air has cooled sufficiently for the RH to reach 100% it is said to be **saturated** or to have reached its carrying capacity. This critical point can be referred to by the temperature the air must fall to before it condenses (its **dew point**) or the altitude at which this temperature is reached (its **condensation level**). At this point the water vapour molecules fuse around a condensation nucleus (a speck of dust, for example) to produce a cloud droplet (0.01–0.04 mm diameter) with the release of latent heat. In the atmosphere, the altitude at which this occurs is evident as the base of the cloud; this is the condensation level (Figure 6.24).

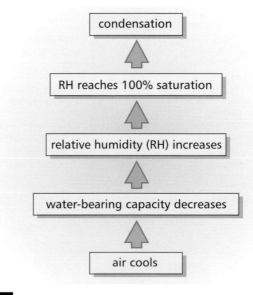

condensation

↑

RH reaches 100% saturation

↑

relative humidity (RH) increases

↑

water-bearing capacity decreases

↑

air cools

 6.24 *The condensation process*

How does air cool?

Water vapour condenses to form cloud droplets because the air is cooled. This occurs through two main processes.

1 Adiabatic temperature change

The atmosphere decreases in density and pressure with increasing height (see page 189). As a parcel of air rises through the atmosphere it expands because there is less external pressure. It is this expansion that causes **adiabatic** cooling. As air descends through the atmosphere the surrounding air pressure increases and the parcel is compressed; this causes the air to heat adiabatically. The adiabatic process is a change in temperature due to a change in pressure without the external transfer of energy. This is a very important process in the atmosphere (and is also the basis of refrigeration and air-conditioning systems) as it is the temperature of the air that determines its relative humidity (Figure 6.25).

The rate or speed at which adiabatic temperature change occurs is not constant: it varies according to the water content of the air:

- **dry adiabatic lapse rate** or **DALR** – this is the rapid rate of temperature change that occurs in unsaturated air at 1.0° per 100 m of vertical displacement

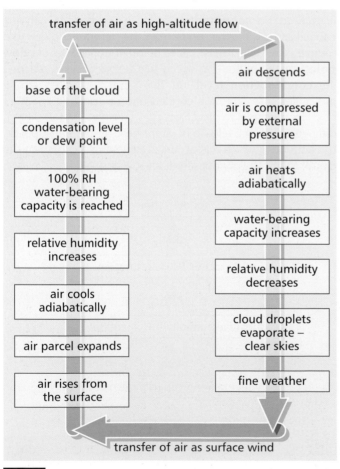

transfer of air as high-altitude flow

base of the cloud		air descends
condensation level or dew point		air is compressed by external pressure
100% RH water-bearing capacity is reached		air heats adiabatically
relative humidity increases		water-bearing capacity increases
air cools adiabatically		relative humidity decreases
air parcel expands		cloud droplets evaporate – clear skies
air rises from the surface		fine weather

transfer of air as surface wind

6.25 *Changes in relative humidity during ascent and descent*

- **saturated adiabatic lapse rate** or **SALR** – the slower rate of temperature change in air that has reached 100% relative humidity; this change is at approximately 0.5°C per 100 m.

These rates are typically plotted on to graphs called **tephigrams** (see Figure 6.11 page 186) to show the change of temperature in parcels of air moving vertically through the atmosphere.

2 Surface cooling

Air may also be cooled by contact with a colder surface. This includes other air masses, a colder sea or a colder land surface. The horizontal movement of air over a surface is termed **advection** and this causes condensation and cloud near the surface, including advection fog and mist. Unlike adiabatic cooling it does not involve the vertical movement of air nor the resulting changes in pressure. This is the effect that occurs on a cold mirror when running a hot tap in the bathroom during the winter.

What is cloud?

Clouds consist of water droplets that are sufficiently small to remain suspended in the air through external friction (< 0.04 mm diameter). Clouds store water, and the movement of cloud by wind transfers water between regions, in particular from over the oceans to the land masses. Clouds may develop vertically to as much as 16 km depth and, as temperatures decrease with altitude, middle-level cloud often contains snowflakes, and upper clouds ice particles.

Clouds are very important to weather forecasters (Figure 6.26).

- As the visible element of weather, cloud supplies substantial information about what is happening in the atmosphere.
- The form or shape of clouds tells us about humidity, stability and the presence of air masses and fronts.
- The speed at which cloud moves shows the wind direction and speed.
- Cloud density and height is used to predict precipitation.
- Clouds are directly visible on satellite photographs, and cloud radar is now used in some countries, including the UK, to provide extremely accurate short-term weather forecasts.

Why do some regions have few clouds?

Extensive areas in the tropics and at the poles rarely have cloud cover, because air in these regions is descending. This movement causes compression, adiabatic heating, an increase in water-bearing capacity and a decrease in relative humidity, so condensation does not occur and clouds do not develop (see Figure 6.25). As a result, cloud seldom forms, as the air cannot cool to dew point, and the typical weather is clear skies, allowing both uninterrupted solar radiation and uninterrupted terrestrial radiation.

Air descends because in the upper atmosphere it is colder and denser than the surrounding air. Globally, this effect is produced by the adiabatic cooling of air rising in the Hadley cell and by the low solar radiation experienced at the poles. The high pressure zones migrate in a seasonal cycle but they are relatively constant, and feature on satellite photographs as large areas devoid of cloud.

The global distribution of cloud types and the cloud-free high pressure zones are particularly clear on water vapour satellite images, which give a three-dimensional effect, showing concentrations of water vapour and cloud in the atmosphere (Figure 6.27).

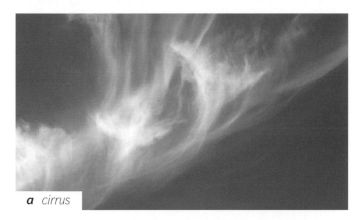

a cirrus

b cumulus

c stratus

6.26 *Clouds are classified by shape into three main categories: cirriform, cumiform and stratiform On satellite images the shape of cloud formations and the absence of cloud give an instant impression. The shape and location of cloud can be used to identify the active processes and to predict weather. Clouds represent a balance between convection forces which move cloud droplets upwards, and advection forces which move cloud droplets horizontally.*

Water vapour is frequently used to create satellite images. It shows the concentration of water in the atmosphere. This image was taken from a satellite in late September.

The central white area marks a dense area of water vapour which is being evaporated from the Earth's surface by powerful solar radiation at the Equator. The dark areas in the sub-tropics of the northern hemisphere return this water to the Earth's surface as rain.

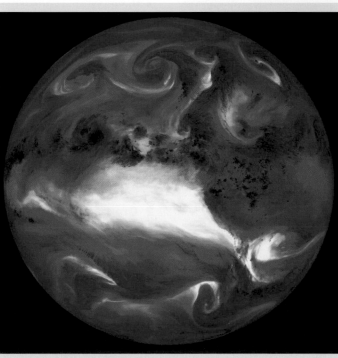

6.27 *Global water vapour*

The curved and convoluted vortices or spirals are typical of cyclonic systems or depressions that develop along the Polar front. Well-developed weather systems are visible as white swirls in the temperate latitudes of both hemispheres.

There is an area of low water vapour and clear skies at the junction of the Hadley and Ferrel cells, which is an area of high pressure. Here air is descending, being compressed, and heating adiabatically so that the water-bearing capacity increases. Under these conditions relative humidity is low, causing an absence of cloud and rain.

NOTING ACTIVITIES

1 Describe the changes in a parcel of air as it ascends vertically through the atmosphere.

2 What is the difference between the DALR and the SALR?

3 What is 'advection', and how can it lead to the cooling of air and the formation of fog and mist?

4 Study Figure 6.27.

   a Comment on the nature of the cloud cover that corresponds to the equatorward limb of the Hadley cell.

   b Describe the nature of the cloud cover that corresponds to the Polar front in the southern hemisphere.

   c Explain why there is so little cloud across much of southern Africa and Australia.

What is precipitation?

Precipitation is a general term for all water moving through the atmosphere in liquid or solid form towards the Earth's surface. Rain is just one type of precipitation; others include drizzle, snow, sleet, hail and dew.

The process of rain droplet formation is still not completely understood, but three mechanisms are thought to operate:

1 Coalescence

Cloud droplets are constantly moving within a cloud due to turbulence, thermals and down-draughts. During this movement cloud droplets collide, fuse and increase in mass. As the larger droplets move downwards through the cloud, further collision occurs and the droplets become larger and accelerate. In tall clouds this will lead to an ever-increasing proportion of large water droplets in the lower cloud, producing rain.

2 Bergeron–Findeisen process

Through the adiabatic cooling of air during ascent and with increasing distance from the surface, the upper parts of clouds are generally well below freezing point. Even at the Equator the upper cloud may be as cold at −65°C. Ice particles in clouds develop around freezing nuclei (salt and fine soil particles) and these attract super-cooled water vapour and increase in mass. Initially the ice particles maintain altitude through turbulence, but at a critical mass they fall towards the surface, either melting to form rain droplets or remaining frozen to fall as snow or hail.

3 Seeder-feeder

This is a composite theory that was developed to explain rain formation in clouds lacking the low temperatures and vertical development required for the Bergeron–Findeisen and coalescence processes. An upper layer of cloud produces a light rain through the Bergeron–Findeisen process, and this falls into a lower band of cloud, triggering rapid coalescence.

Why does precipitation occur?

Air is pulled towards the surface by gravity, and energy is required for it to rise. There are three main mechanisms causing the upward movement of air: convection, orographic and cyclonic. Although these mechanisms share many features in common, they have different causes and, in terms of cloud and precipitation, different consequences.

1 Convection

Convection occurs when a local surface of land or sea is heated to a higher temperature than the surrounding areas. The heat reduces the density of the air and it starts to rise vertically. As the air rises, cooling occurs and eventually condensation forms clouds. When a pocket of air becomes warmer and less dense than the surrounding **environmental air**, it is said to be unstable, and the general atmospheric condition is referred to as **instability** (Figure 6.28).

Convection is a common process, as variations in surface, albedo and relief cause some local areas to heat more rapidly than others. The resulting thermal currents of rising parcels of air may not be sufficient to always cause clouds and rain, but frequently there is sufficient energy to give rise to towering clouds and possibly to the development of thunderstorms, hurricanes and tornadoes.

Convection is generally associated with:

- the equatorial zone where energy from insolation is high
- continental areas where land surfaces heat rapidly
- valley floors where solar radiation is trapped
- dark surfaces where albedo is low and long-wave terrestrial radiation is high
- clear skies (anticyclonic) when solar radiation is enhanced
- the late afternoon, when the surface has fully heated and is radiating maximum energy
- urban areas where energy is released from and created in the heat island.

2 Orographic or relief rainfall

The energy to force air to rise can also be provided by wind, and this leads to the formation of cloud and precipitation on the windward flanks of mountains and hills. The incoming air is forced to rise by the physical obstruction, even though it has the same temperature and density as the surrounding air. This is termed **conditional instability**, as the rising air behaves as though it were 'unstable' but the cause is the 'conditional' presence of the relief (Figure 6.29).

Orographic cloud and precipitation are associated with:

- relief of sufficient altitude to cause cooling of the incoming air to dew point
- a prevailing wind, generally off the sea (maritime air has a high relative humidity).

3 Frontal systems

A front is a boundary between two different air masses. Look back at Figure 6.27 (page 199) to see the location of the Polar front which forms the boundary between cold air from the poles and warm air from the tropics. Fronts tend to be very active areas of cloud and precipitation formation (as experienced frequently in the UK). At a front the warmer air becomes highly unstable in relation to the colder air mass, and it rises. This causes adiabatic cooling, condensation and rain droplet formation. In addition, the cold denser air undercuts the warmer air mass and forces it to rise in a similar manner to the physical obstruction in orographic rainfall. Fronts extend for considerable distances and the cloud and rain generally shows a characteristic linear shape, allowing the approximate position of the front to be identified. Frontal cloud and rain tends to be persistent and of medium intensity, dispersing as the front passes over. The weather patterns associated with frontal systems are discussed in greater detail in chapter 6D.

Where does it rain?

Cooling, condensation and water droplet formation require energy, and this is not uniformly distributed. Globally, average annual precipitation varies from over 30 000 mm (a staggering 30 m) on the south-facing slopes of the Himalayas in northern India, to rainless years in parts of the Atacama Desert of Chile and Peru. The global pattern reflects the influence of latitude, land/sea differences, relief, ocean currents and jet streams (Figure 6.30).

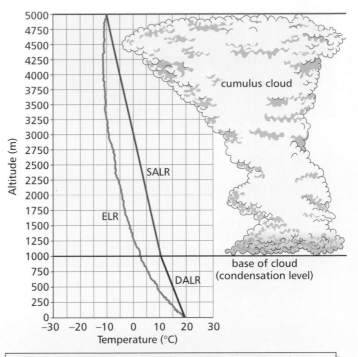

- The local pocket of air is warmer and less dense than the environmental air. It is unstable and will rise.
- As it rises it cools adiabatically at the DALR. All air with a relative humidity less than 100% is regarded as being 'dry'.
- Condensation level: the altitude at which water vapour condenses to form cloud droplets. This is visible as the base of the cloud. This is the altitude at which temperatures fall to dew point.
- Air cools at the SALR above condensation level. Latent heat is released during condensation, so reducing the rate of cooling.
- The top of the clouds is where rising air cools to the same temperature and pressure as the environmental air mass. Stability is restored. This is called the tropopause.

6.28 *Convection: instability*

Moist air from the South Atlantic moves onshore with the prevailing wind. This air is stable as it has the same temperature and pressure as the surrounding air.

The moist maritime air is forced to rise by the major relief feature of Table Mountain (the escarpment rises 1000 m). As the air rises it expands, cools adiabatically and condenses to form cloud on the upper slopes of the mountain. The cloud is layered and does not have the vertical development of convection systems.

Coalescence and Bergeron–Findeisen processes are limited by the cloud height, and precipitation is generally a low-intensity drizzle. However, this may continue for some time, as its duration is determined by the wind.

On the leeward side of the mountain the air sinks. It is now highly stable as it has cooled below the temperature of the surrounding air. As the air descends it adiabatically heats and the cloud disperses. This produces a dry, warm wind and forms a rainshadow area in the lee of the mountains.

6.29 *Orographic rainfall: the 'Tablecloth' – rainclouds over Table Mountain, South Africa*

NOTING ACTIVITIES

1 Describe, with the aid of a diagram, how convection is caused by atmospheric instability.

2 Draw a sketch of a mountain range, similar to the Tablecloth in Figure 6.29, and add annotations to describe the process of orographic uplift. Include the following terms in your annotations:

- adiabatic
- prevailing wind
- cloud
- rainshadow.

3 Why is heavy rain often associated with frontal uplift?

4 Study Figure 6.30.

a (i) With the help of an atlas, identify the parts of the world that have an average rainfall in excess of 2000 mm.

(ii) With reference to earlier maps and diagrams, try to account for the location of these areas.

b Complete the same exercise as in a above, but for those areas with a rainfall of less than 250 mm.

c What seems to be the most important factor or factors determining the amount of rainfall at a particular place?

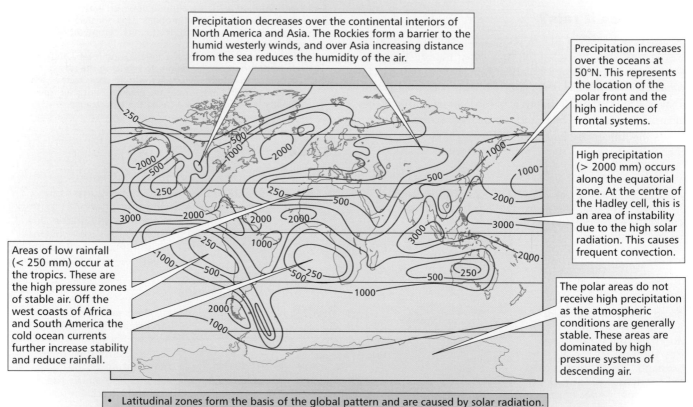

Precipitation decreases over the continental interiors of North America and Asia. The Rockies form a barrier to the humid westerly winds, and over Asia increasing distance from the sea reduces the humidity of the air.

Precipitation increases over the oceans at 50°N. This represents the location of the polar front and the high incidence of frontal systems.

High precipitation (> 2000 mm) occurs along the equatorial zone. At the centre of the Hadley cell, this is an area of instability due to the high solar radiation. This causes frequent convection.

Areas of low rainfall (< 250 mm) occur at the tropics. These are the high pressure zones of stable air. Off the west coasts of Africa and South America the cold ocean currents further increase stability and reduce rainfall.

The polar areas do not receive high precipitation as the atmospheric conditions are generally stable. These areas are dominated by high pressure systems of descending air.

- Latitudinal zones form the basis of the global pattern and are caused by solar radiation.
- The latitudinal pattern is distorted by land/sea differences, and by the effects of ocean currents and major relief features.

6.30 *Global pattern of average precipitation (mm)*

STRUCTURED QUESTION

Examine Figures 6.31 and 6.32. Use both maps to answer the following.

a Locate Area A on Figure 6.31. Explain why there is an area of highest one-hour rainfall total here. *(2)*

b Locate Area B on Figure 6.31. Explain why there are areas of highest one-hour rainfall total here. *(2)*

c Locate Area C on Figure 6.31.

(i) Describe the pattern of maximum one-hour rainfall likely to occur here in a five-year period. *(2)*

(ii) Explain why the maximum one-hour rainfall total is low. *(3)*

d Does Figure 6.31 give an accurate representation of the pattern of rainfall in the British Isles? *(3)*

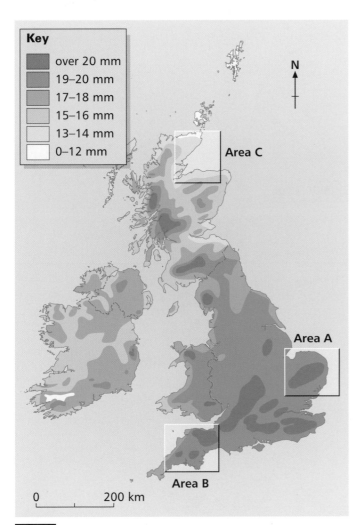

6.31 *Maximum one-hour rainfall likely to occur once in any five-year period in the British Isles*

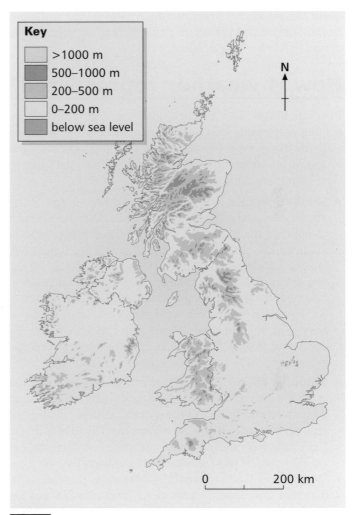

6.32 *Relief map of the British Isles*

D Understanding weather: the UK

The UK is classified as having a temperate maritime climate with mild winters and cool summers. However, day-to-day weather is very changeable, and the prediction of weather over more than four days is something of a mystical art. In many regions of the world, weather patterns are extremely predictable because pressure patterns tend to remain more or less stable for long periods of time.

How is weather represented?

To understand and predict weather, the atmospheric conditions must be described. This is achieved in two main ways: by using images (usually from satellites in space), and by studying synoptic charts (constructed from data gained from ground stations and satellites).

- **Satellite images** These provide a graphic image of the atmosphere at a given point of time, and they have helped to improve the accuracy of weather forecasting. Satellite images show cloud (visible images) and water vapour (WV images – see Figure 6.27 page 199) at night and during the day (infrared images). However, such images, at present, cannot show surface temperatures (unless there is a clear sky), nor can they show wind direction, wind speed or atmospheric pressure. Winds can be calculated by comparing images over time but this has only limited use in prediction without a knowledge of pressure systems.

- **Synoptic charts** These are weather maps constructed using data from satellites, cloud radar, aircraft, ships and ground stations, and they include full details of atmospheric conditions as experienced on the surface. The charts have evolved over time to give an accurate and rapid picture of the atmospheric conditions over an area. The symbols used to describe temperature, pressure, wind direction and speed, cloud cover, precipitation and fog combine choropleth, isoline and numerical techniques in

a very efficient manner (see Figure 6.33). Synoptic charts are both descriptive of weather conditions at a given point of time and predictive in that they allow the forecast of weather over the next few hours and days.

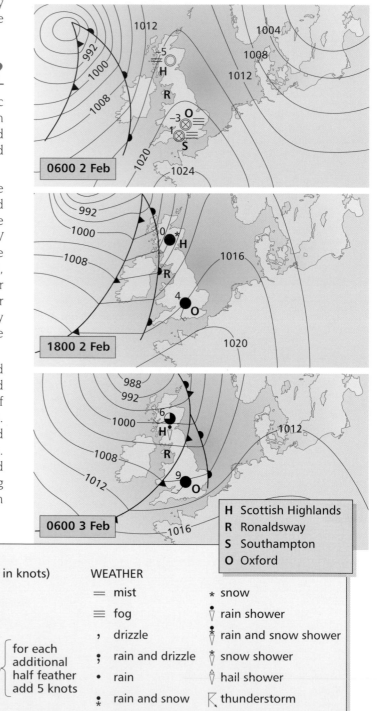

6.33 *Weather conditions over the British Isles in February*

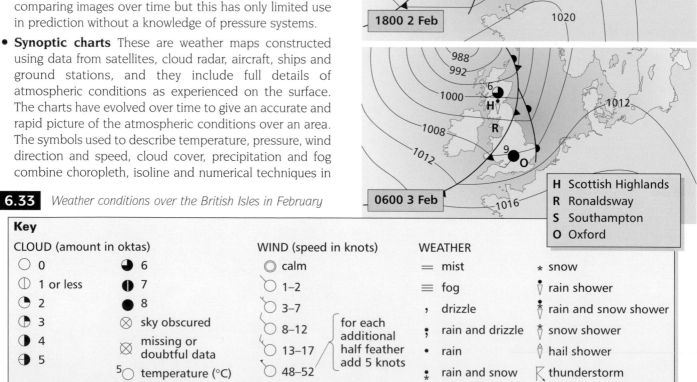

H	Scottish Highlands
R	Ronaldsway
S	Southampton
O	Oxford

Key

CLOUD (amount in oktas)

○ 0
◐ 6
◑ 1 or less
◕ 7
◐ 2
● 8
◕ 3
⊗ sky obscured
◑ 4
⊠ missing or doubtful data
◐ 5
⁵○ temperature (°C)

WIND (speed in knots)

◎ calm
1–2
3–7
8–12
13–17 — for each additional half feather add 5 knots
48–52

WEATHER

= mist
≡ fog
, drizzle
; rain and drizzle
• rain
⁎ rain and snow
∗ snow
▽ rain shower
⁂ rain and snow shower
✳ snow shower
△ hail shower
⟨ thunderstorm

How to interpret a synoptic chart

The circle at the centre of the data marks the location of a weather station. Those over the sea show data collected from ships or aircraft.

Temperature The figure to the left or upper left of the station shows temperature in degrees Celsius (°C). This is recorded at a set time on a thermometer located in a Stevenson screen.

Pressure Pressure is shown by isobars – continuous lines that link points of equal atmospheric pressure. Isobars tend to be circular in shape and are at 4 mb intervals. These can be regarded as contours showing the areas of low and high pressure. Fronts are shown using the symbols:

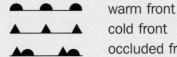

warm front

cold front

occluded front.

Wind direction Wind direction is shown by the line or 'arrow' extending from the station. The line points in the direction the wind has come from. The stations in the key under 'Wind' have a north-westerly wind. Wind directions are helpful in determining which air mass is affecting an area. For example, if the general wind direction is from the north-west, the air mass is probably polar maritime (see Figure 6.10 on page 185).

Wind speed Wind speed is shown by 'feathers' added to the wind arrow. Calm conditions are shown as a double circle around the station.

Cloud cover Cloud cover is shown by progressive shading of the station symbol in eighths or oktas. This shows, at a glance, the distribution of cloud. When the sky is obscured by fog or mist, the symbol X is used.

Precipitation Precipitation is shown through the use of symbols. These are generally placed to the left of the station.

Example: The weather at Oxford (O) at 0600 on 3 February:

- temperature 9°C
- cloud cover 8 oktas
- wind direction SW
- wind speed 13–17 knots
- no precipitation.

What are the types of weather in the UK?

The UK is located at 51–55° North of the Equator, and lies on the western margins of Europe. At this location it lies under the convergence zone of five major air masses (see Figure 6.10), each with very different characteristics. It is the movement of these air masses and the interaction between them that produces a typically chaotic and changeable pattern of weather. In general terms the UK experiences three main types of weather and these have an average seasonal pattern:

- spring and autumn frontal systems
- winter high pressure systems
- summer high pressure systems.

Frontal systems

When you hear during a weather forecast that 'the day will start clear and bright but rain will move in from the west...' this generally means that a frontal system is approaching (Figure 6.35). Frontal systems or **depressions** are a major source of cloud and rain over the UK. Look at the satellite image 6.34, which shows the extensive heavy cloud associated with a passing frontal system and depression. These front systems are associated with the Polar front, the contact zone between polar air masses moving in from the north, and tropical air from the south (see Figure 6.15 page 189). The contact between polar and tropical air is modified by the effect of **Rossby waves** to create an undulating, or meandering, pattern. Rossby waves are upper atmosphere winds (part of the polar jet stream) that exert a frictional

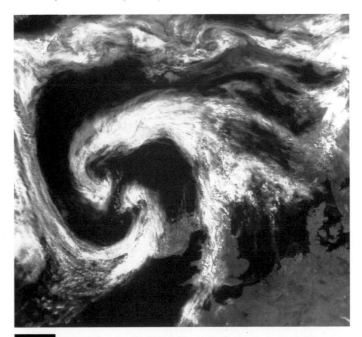

6.34 *Satellite image showing cloud associated with a depression and frontal systems*

drag on the lower atmosphere, and are unpredictable in their course and meander.

A depression over the UK occurs when tropical air is pushed northwards and intrudes into the polar air mass. The tropical air (warm sector) is surrounded by polar air and becomes extremely unstable. Where tropical air is pushing into the polar air a **warm front** is formed and where polar air pushes into the warm sector a **cold front** forms. Along both these fronts warm air rises, cools adiabatically and condenses to form cloud and precipitation.

Depressions have a distinctive pattern of weather, which is described in Figure 6.35.

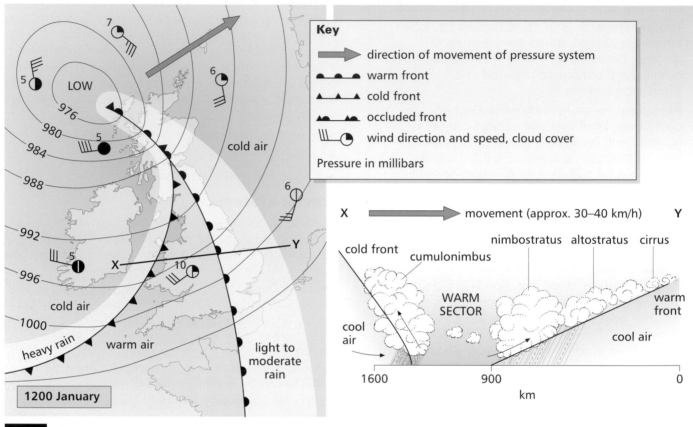

6.35 *The weather associated with a frontal system over the UK*

The sequence of weather associated with a depression

The warm front extends along the eastern side of England and into western France. The cold front extends westwards over the Atlantic. The depression has started to occlude (the cold front catches up with the warm front) to the north-west. Look carefully at Figure 6.35 to see the differences in weather characteristics on either side of the fronts.

- Air temperature shows a distinctive pattern. Ahead of the warm front, the air is cool. Behind the warm front, temperatures increase significantly, only to fall back again behind the cold front.

- Air pressure is determined by temperature and in the warm sector, pressure falls. It is this low pressure of tropical air contrasting with the higher pressure of the polar air that causes air in the warm sector to become unstable and to rise, producing cloud and rain.

- Wind speed varies as the fronts pass. Speed is highest in the turbulent air caused by the vertical movement of tropical maritime air at the warm and cold fronts. Winds at the cold front in particular are often very strong and gusty. Notice how wind speeds are reflected in the closeness of the isobars.

- Precipitation is typically in two bursts. The warm front brings a prolonged period of light to moderate rainfall whereas the cold front brings a shorter period of heavy rainfall, sometimes with thunder. Where the fronts have become occluded (over western Scotland) there is a single period of moderate rainfall. Rainfall intensity is highest at the cold front due to the greater vertical development of the cloud.

What causes fine weather?

In the summer and winter the polar front migrates to the north and south of the UK, bringing cloud and rain to these latitudes. The UK typically experiences weather associated with the tropical and polar high pressure systems, which bring relatively stable conditions. In high pressure systems cold upper atmospheric air is converging and descending. As the air descends it is compressed, it heats adiabatically and the relative humidity decreases (see page 197). The decreasing relative humidity reduces the incidence of condensation, cloud formation and precipitation, giving clear skies and fine weather.

Summer anticyclones

As the Polar front moves north, the Azores high pressure system (an anticyclone) extends northwards from the south-west to cover much of the UK. The sinking air stifles cloud formation so there are clear skies, and the long day-length (17 hours in June) results in high surface temperatures.

The wind direction within an anticyclone in the northern hemisphere is clockwise, and winds are generally light due to the low pressure gradient. In combination with a stable atmosphere this tends to cause the build-up of pollutants in the air near the surface – ozone, pollen, and nitric oxide from car exhausts. Typically the wind is southerly (as the Azores high is centred to the south-west of the UK) and this brings in polluted air from industry in northern France and north Italy.

In inland areas of the UK the land surface may heat rapidly during the day, and in late afternoon local pockets of air become unstable and rise. These may be sufficiently unstable to cool to the dew point temperature, condense and form towering cumulonimbus clouds with associated thunder and lightning, and intense precipitation. This is most common in East Anglia and the South East, where summer maximums of rainfall are recorded.

Winter anticyclones

In the winter the Polar front moves southwards and the polar high pressure systems from the north dominate over the UK. These systems are formed from cold polar air descending through the atmosphere causing atmospheric stability and clear skies. Whilst the descending air is adiabatically heated it remains cold, though the skies are clear of cloud. The sun is low in the sky and day-length is short so that the surface does not heat and, during the long night, significant terrestrial radiation reduces temperature further to a minimum just before dawn. This progressive cooling of the surface tends to produce **inversions**, with frosts and fogs.

As the surface radiates long-wave energy into the clear sky, the surface temperature falls rapidly. The air above the surface is cooled and reaches dew point temperature, causing condensation and the formation of cloud droplets. In autumn and spring the cooling may be sufficient to only cool the surface to freezing point (termed a **ground frost**) whilst in winter the freezing temperatures extend to the overlying air, producing an **air frost** as well, possibly with **freezing fog**. Fogs formed in this manner are termed **radiation fogs** and are most common in inland areas where the land surface radiates heat most rapidly. On floodplains this effect is often enhanced by cold, dense air flowing down the adjacent slopes and collecting on the valley floor (see Figure 6.36) For a layer of cold air to develop at the surface, low wind speeds or calm conditions

6.36 *Radiation fog*

are required (typical of anticyclonic systems). This cold layer generally dissipates during the early morning as solar radiation heats the surface.

In urban and industrial areas, smoke and pollution can be added to the stagnant surface layer increasing its density and creating a **smog**. Fortunately this is now rare in the UK as emissions are strictly controlled under the Clean Air Act 1956, but in December 1952 the Great Smog of London is estimated to have indirectly killed 4000 people. Fog and frost remain major hazards in the UK as they disrupt communications and are a frequent cause of road deaths and accidents.

NOTING ACTIVITIES

1 Identify the similarities and differences between anticyclonic weather in summer and winter. Consider the following aspects:

   a temperature

   b precipitation

   c winds

   d other phenomena.

2 Study Figure 6.36.

   a What is radiation fog?

   b How is radiation fog formed?

   c Why is it common in valley bottoms?

   d How can fog be a hazard to human activity?

Why does weather vary within the UK?

The UK is a relatively small country, with an area of 242 000 km^2, but there is a significant variation in weather between regions. These differences are sufficient to be expressed in averages for the different regions. The variations are caused by a combination of factors, including:

- a north–south orientation extending across 10 degrees of latitude
- a location on the Polar front
- a warm ocean current to the west (the North Atlantic Drift) and a cold North Sea to the east with winter average temperatures of 13°C and 4°C respectively
- mountainous terrain to the north and west and lowlands to the east and south
- a contrast between inland central areas and coastal areas
- depressions often affecting only a limited area of the country.

Temperature patterns in the UK

January

The general pattern is of increasing temperature from east to west. The west coast has a **maritime** climate with mild winters whilst the east is more **continental** and has colder winters. With low solar radiation the main source of heat in the west is from the North Atlantic Drift (average temperature in winter is 13°C) with heat transferred over the land by the prevailing westerly winds.

Relief also reduces temperature, as in northern Scotland and the Wicklow Mountains in Ireland.

July

The general pattern is of decreasing temperatures from south to north. There is higher solar radiation due to the greater frequency of anticyclones associated with air masses moving up from the south.

The effect of relief is again evident, reducing temperatures over Wales and northern England.

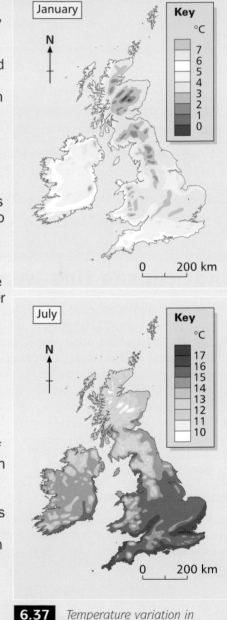

6.37 *Temperature variation in the UK*

Precipitation patterns in England and Wales

The west coast has a higher than average rainfall as the prevailing wind is westerly, introducing moist maritime air. Orographic rainfall gives high totals in the Lake District, North Wales and also in the mountains of Scotland (some areas in North Wales receive rain on more than 300 days a year).

Rainfall increases over the Pennines as the upland area is sufficient to force air to rise, condense and form cloud and rain (conditional instability). To the east, a rainshadow area forms where air descends on the leeward side and becomes stable.

Rainfall over central England is mainly cyclonic, from frontal systems that move over the entire country. This area has spring and autumn maximums and an average total of 600–800 mm.

Rainfall in the south-east is mainly convectional. The region heats during summer anticyclones, causing instability and storms. Unlike the rest of the UK, East Anglia has a summer maximum of precipitation.

The South Downs, rising to 400 m, provide a sufficient obstacle to cause orographic rainfall. This gives the hills 50 per cent more rain than the relatively flat area of East Anglia directly to the north.

Rainfall in the west is both cyclonic and orographic. Rain occurs at all times of the year, with a maximum in winter.

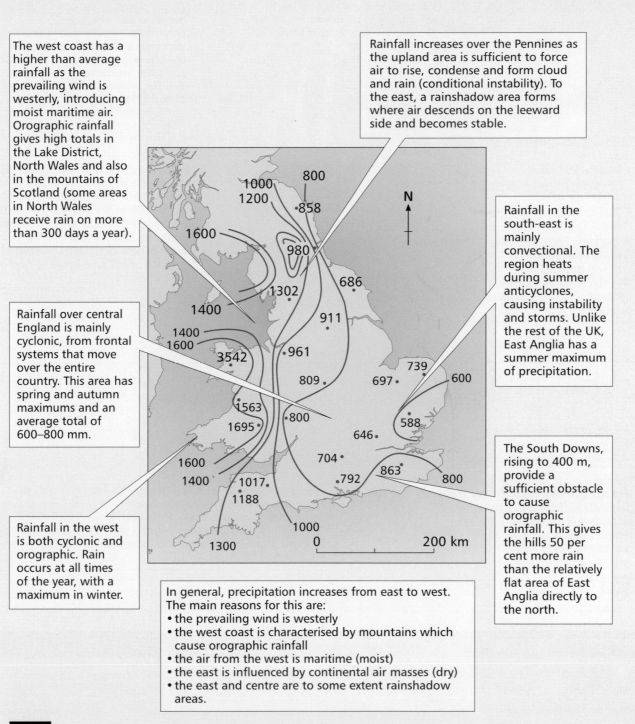

In general, precipitation increases from east to west. The main reasons for this are:
- the prevailing wind is westerly
- the west coast is characterised by mountains which cause orographic rainfall
- the air from the west is maritime (moist)
- the east is influenced by continental air masses (dry)
- the east and centre are to some extent rainshadow areas.

 Average annual precipitation (mm) in England and Wales

Fog and mist in the UK

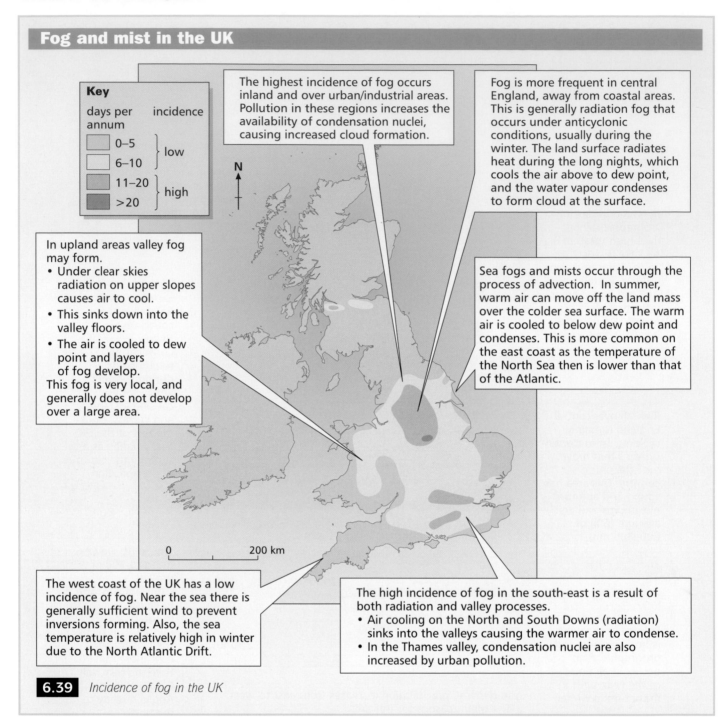

Key

days per annum — incidence

0–5	low
6–10	
11–20	high
>20	

N

The highest incidence of fog occurs inland and over urban/industrial areas. Pollution in these regions increases the availability of condensation nuclei, causing increased cloud formation.

Fog is more frequent in central England, away from coastal areas. This is generally radiation fog that occurs under anticyclonic conditions, usually during the winter. The land surface radiates heat during the long nights, which cools the air above to dew point, and the water vapour condenses to form cloud at the surface.

In upland areas valley fog may form.
• Under clear skies radiation on upper slopes causes air to cool.
• This sinks down into the valley floors.
• The air is cooled to dew point and layers of fog develop.
This fog is very local, and generally does not develop over a large area.

Sea fogs and mists occur through the process of advection. In summer, warm air can move off the land mass over the colder sea surface. The warm air is cooled to below dew point and condenses. This is more common on the east coast as the temperature of the North Sea then is lower than that of the Atlantic.

0 ——— 200 km

The west coast of the UK has a low incidence of fog. Near the sea there is generally sufficient wind to prevent inversions forming. Also, the sea temperature is relatively high in winter due to the North Atlantic Drift.

The high incidence of fog in the south-east is a result of both radiation and valley processes.
• Air cooling on the North and South Downs (radiation) sinks into the valleys causing the warmer air to condense.
• In the Thames valley, condensation nuclei are also increased by urban pollution.

6.39 *Incidence of fog in the UK*

STRUCTURED QUESTION 1

Study Figure 6.39, which shows the incidence of fog in the UK.

a Define the term 'fog'. *(1)*

b Describe the pattern shown on the map. *(3)*

c Suggest reasons for the 'high' incidence of fog in the areas indicated. *(3)*

d Why is fog common in anticyclonic conditions? *(2)*

e Why does fog commonly form in autumn and spring? *(2)*

f In what ways can human activity lead to an increase in the incidence of fog? *(2)*

g What problems does fog pose for human activities? *(2)*

Urban climates

The UK is a highly urbanised country. In 1991, 91 per cent of the population lived in towns and cities, and urban areas now cover 11 per cent of the land area. This has created its own pattern of climate, as urban areas have a distinctive albedo and also generate heat from economic activity and domestic heating. The artificial, built surfaces of urban areas have a low albedo and absorb short-wave energy, thus releasing greater long-wave energy to heat the air above. Further energy is released from urban activities, leading to an increase in air temperature above cities in contrast to the surrounding rural areas; this effect is termed a **heat island** (Figure 6.40). Heat islands generally develop under anticyclonic conditions, when there are clear skies and low wind speeds. The clear skies allow increased insolation and radiation. Calm conditions prevent the air from mixing or being dispersed. Heat islands increase in strength with increasing size of the city.

In addition to modifying temperature, urban areas can also cause increased precipitation as the pollutants released from the concentrated economic activity provide more hygroscopic nuclei and the heat increases atmospheric instability. Cloud and rain may intensify as they pass over a city and the subsequent precipitation may become more acidic. Fog over cities may also become denser as the surfaces radiate heat rapidly and the pollutants increase condensation nuclei. These effects are most apparent under anticyclonic conditions when wind speeds are low and mixing of air is minimal. A depression passing over a city has sufficient wind speed to mix rural and urban air, dispersing any differences in temperature or air quality. In cities, winds tend to be gustier, with the buildings interfering with the general wind direction, causing gusts and lulls. You will probably have experienced this yourself within a city, particularly if there are tall buildings and narrow streets.

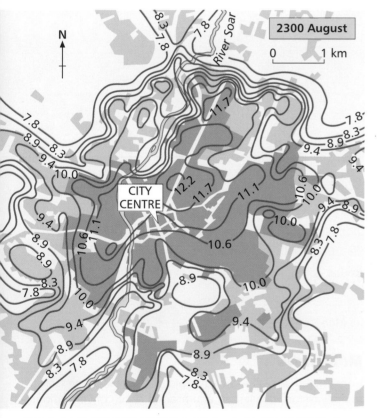

2300 August

0   1 km

Key

–8.3– isotherm (°C)

high building density of the central districts

low building density, mainly suburbs

non built-up land

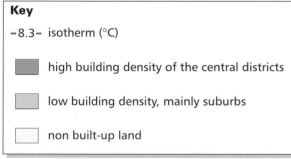

Urban climate: a heat island in Leicester

NOTING ACTIVITIES

1   Study Figure 6.40.

   a What do you understand by the term 'heat island'?

   b What is the temperature difference between the centre of the city and the outskirts?

   c Describe the rate of temperature change with distance out of the city centre.

   d What is the evidence that the built-up area is directly responsible for the heat island effect?

   e Why do cities tend to be warmer than rural areas?

2   Apart from the heat island effect, how else do city climates differ from the surrounding countryside?

EXTENDED ACTIVITY

Carry out a study of the climate of the UK. To do this you will need to consider the information about temperature and rainfall included in this chapter.

1   Complete simplified maps of the patterns of temperature (January and July) and rainfall. Show the areas of highest and lowest values.

2   Add labels to your maps describing and explaining the patterns.

STRUCTURED QUESTION 2

Study the maps in Figure 6.33 on page 204, which show weather conditions over the British Isles in February.

a Name the pressure pattern feature to the north of the British Isles at 0600 on 3 February. *(1)*

b Name the front that runs through western Ireland at 0600 on 2 February. *(1)*

c What is the name of the air mass that is affecting most of England at 0600 on 3 February? *(1)*

d (i) Describe the weather conditions at Southampton (S) at 0600 on 2 February. *(2)*

(ii) Suggest reasons for the weather you have described. *(4)*

e Describe the changes in precipitation type across the Scottish Highlands (H) between 1800 on 2 February and 0600 on 3 February. *(2)*

f Figure 6.41 shows how precipitation intensity changed at Ronaldsway (R on Figure 6.33) between 0600 on 2 February and 0600 on 3 February. Describe and explain the changes in precipitation intensity shown on the graph. *(4)*

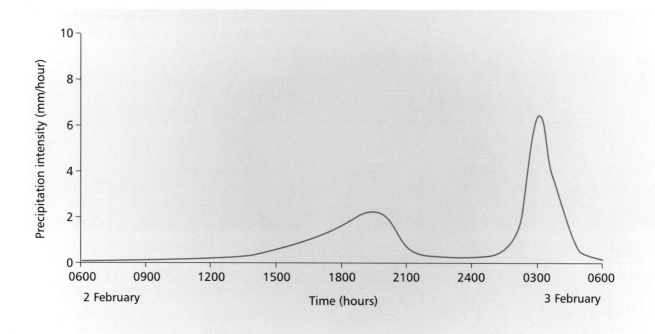

6.41 *Changing precipitation intensity at Ronaldsway airport, Isle of Man*

The economic costs of weather

In the pre-industrial and pre-technological UK, economic systems were in a working relationship with weather. Farmers used the winter frosts to break up the soil, reducing labour costs, and work was seasonal. Many people lived near to their place of work, reducing the need for transport. Goods were often produced locally, reducing transport distance. In the 20th century, the increasing complexity and application of technology has changed this relationship and 'freed man from nature', but this has made our systems more vulnerable to extreme conditions, potentially causing havoc. The economy is now so dependent on communication systems that any disruption has an immediate and major economic impact.

Motorways

Constructed for all-weather rapid movement of vehicles, motorways are densest in and around large concentrations of industry and population. In these areas increased atmospheric pollution provides more hygroscopic nuclei, which increases water droplet and fog formation under winter anticyclonic conditions. In the UK the main motorway network is inland, increasing radiation, and motorways also follow low-lying land which creates advection fog as cold air flows down-slope causing condensation on the floor of the valley. With higher speeds of traffic movement and a false perception of safety, all too frequently the result is major accidents involving several or many vehicles. The response has been to implement warning systems and lighting on vulnerable stretches as, for example, on the M6 across the Cheshire Plain.

Road icing

This occurs when water vapour condenses directly onto a cold surface and is then frozen either as crystals (visible white ice) or as a film (the harder-to-spot black ice). Icing of roads is fairly predictable as it is caused by clear skies, low air temperatures and dew point temperatures; these can be predicted with some accuracy. Road surfaces are often dark in colour and this gives a low albedo, increasing heating during the day but also increasing the rate of cooling through radiation at night. Main routes, including motorways, bus routes and radial urban roads, are generally gritted (a mixture of sand and salt) before the frost occurs. On minor untreated roads the likelihood of accidents is reduced by lower speeds and the greater local knowledge of conditions. However, because iced or snow-covered roads are unusual in the UK, few drivers know how to respond when these conditions are encountered.

Railways

The rail network is complex, and also increasingly dependent on technology rather than manual labour to operate signalling and points systems. This has made the network vulnerable to the 'wrong kind' of snow. Very cold, fine particles are blown into junction boxes, and short-circuit electrical systems. Conversely, large wet snowflakes adhere to rails and overhead cables, causing loss of traction and loss of power pick-up, bringing the network to a standstill. Such types of snow are relatively rare in the UK and it is not considered economically justifiable to protect against them. High winds can also disrupt the supply of power by overhead cables, as occurred in south-east England after the storm of 1987.

Buildings

During the winter, heating costs are high and considerable heat is radiated from buildings as long-wave energy. This contributes to the formation of urban heat islands but can be reduced by insulation. Air is an inefficient conductor of heat and can be used to separate the hot interior surfaces of buildings from the cold external air, reducing heat loss by up to 75 per cent. The use of fibre-glass in attics, double glazing of windows and wall-cavity foam all use trapped air to reduce heat loss. Such measures are expensive and become more popular when energy prices rise (i.e. the cost of heating buildings).

E The average as meaningless

Atmospheric processes are relatively constant, and much of global weather fits a predictable pattern; this justifies the use of averages to identify climates. For much of the time the atmosphere behaves as expected, and this allows people to adopt methods of agriculture, architecture, living patterns and even dress to both cope with and exploit the conditions. Constancy also allows soils and vegetation systems to develop and reach an equilibrium with climate to form zonal soils and biomes.

However, at periodic and often unpredictable intervals the average is swept aside and is replaced by extreme weather patterns that shatter the equilibrium and often trigger rapid and dramatic physical and human change. These are termed **natural events** and include hurricanes, tornadoes, floods, drought and smogs, all representing major deviations from expected weather conditions. The frequency of such events – the **recurrence interval** – may be as great as once in 250 years but their effect on human and physical geography is extremely significant.

In this chapter we examine some examples of recent extreme weather events. Check out the Stanley Thornes website to find more examples.

The Great Storm of 1987 in England

The Meteorological Office had been predicting severe weather from five days before 15 October. During the 15th, warnings of heavy rain were issued for the South and South East on television and radio, but there was no mention of high winds. Storms at this time of the year are typical, as the Polar front lies over southern UK, but this depression was to be different.

On the night of 15/16 October a fast-moving depression caused the most severe storm damage to south-east England for 200 years. Very cold air from the North Atlantic (6°C to the west of Ireland) was pushing as far southwards as Portugal and warm air from the Canaries (22°C in Portugal) was pushing northwards over Iberia. The result was a strong cold front and a deep low pressure cell. Atmospheric pressure fell as low as 956 mb (Figure 6.42). The resulting pressure gradient caused gusts of wind at over 100 knots, though the sustained surface wind speed did not exceed the critical value of 64 knots necessary for 'Hurricane' status.

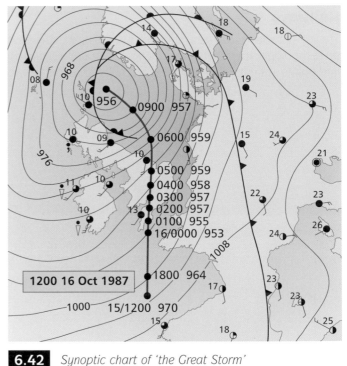

6.42 *Synoptic chart of 'the Great Storm'*

The physical impact

- 15 million trees are thought to have been blown down or severely damaged.
- Broadleaf trees were particularly badly hit, with the equivalent of two years of UK production destroyed.
- Non-native trees were hard-hit, and important scientific specimens were lost, including several at Kew and Ventnor.
- Orchards in Kent and Essex were devastated, including some with rare traditional English species.

- Tree damage was frequently caused by snapping in coniferous species but many broadleaf trees were uprooted (the fate of six of the seven oak trees at Sevenoaks). The frequency of uprooting was caused by the heavy rainfall during the weeks before the storm which had saturated the ground giving support roots less grip and greater lubrication.

The social impact

- At least 18 people were killed as a direct result of the storm. Peak wind velocities were in the early hours of the morning, and this probably reduced the death toll.
- Houses and cars were damaged throughout the South and South East, and the majority of schools were shut. In Suffolk all 360 schools were closed, with 30 damaged by falling trees.
- The London Fire Brigade dealt with 6000 calls in 24 hours.
- In the South East, millions were terrified by the high wind speeds and the extent of the structural damage.

The economic impact

- Roads and rail services throughout the South, South East and East Anglia were paralysed by fallen trees.
- Power cables and pylons were damaged by the wind and by falling trees which cut electricity to widespread areas. The Underground was brought to a halt.
- For two days the employed population struggled to get to work; many factories and offices were forced to close temporarily.
- Shipping in the Channel came to a standstill and the port of Dover closed for the first time in the 20th century. Cross-Channel services were suspended (Figure 6.43).
- The 'forced' release of large volumes of timber onto the market stimulated carpentry and reduced the price of temperate hardwoods.

NOTING ACTIVITIES

1 a Why was the 'storm of 1987' so severe?

   b The media referred to the storm as a 'hurricane' (see Figure 6.43). Was this accurate?

2 Make a list of the *immediate* (short-term) effects of the storm, and another of the *longer-term* effects.

3 Do you think the storm would have attracted so much attention if it had occurred away from the south-east of England?

6.43 *A Channel ferry beached by the storm*

Hurricane Mitch in Central America, 1998

In late October 1998 the Central American republics of Honduras, Nicaragua and Guatemala were devastated by the most severe hurricane of the century, and possibly for the last 200 years. The event had a major physical, social and economic impact on these developing countries, and it illustrates both the power of natural systems and their unpredictability.

How did Hurricane Mitch develop?

Hurricane Mitch developed as a tropical storm over the warm waters of the central Atlantic Ocean (Figure 6.44). When sea surface temperatures rise above 26°C the surface air becomes unstable and a convection cell is formed. The air starts to rotate in an anticlockwise vortex (northern hemisphere) and the system gains further energy from the latent heat released through condensation and water droplet formation.

Mitch started as an average tropical cyclone but soon grew to a rotating mass over 1000 km in diameter and with wind speeds in excess of 280 km/h. The extreme instability caused clouds to build to over 16 km, producing very heavy wind-driven rain.

Typically these systems move north-westwards through the Caribbean and then swing north-eastwards before passing along the east coast of America. As they move over colder water their energy diminishes and they fade to form storms. Mitch did not follow the anticipated route but veered westwards into Central America before turning eastwards towards Florida.

With a recurrence interval as high as once in 250 years, this event radically changed the physical and human geography of the entire region.

This water vapour image clearly shows the 'circular saw' effect of the vortex. The cloud is rotating anticlockwise around the central low pressure.

At the centre of the hurricane there is a clear 'eye' where cold air is descending and the surface is calm. Much damage is caused by the reversal of wind direction as the eye passes over.

Major cloud systems develop around the centre, where warm unstable air is rising rapidly, forming cumulonimbus clouds. It is these that produce the torrential rain.

6.44 *Satellite image of Hurricane Mitch, October 1998*

The physical impact

- Wind speeds exceeded 280 km/h and the system extended for over 1000 km.
- In mountain areas, over 6000 mm of rain fell in 24 hours (year average in UK 850 mm). 250 mm of rainfall adds over 250 000 tonnes of weight to the surface per km^2.
- In low-lying areas rivers flooded to 15 m above normal levels (the second floor of one major hospital was flooded).
- Previously stable slopes failed under the weight of the water, and its lubricating effect caused widespread mass movements. One mudslide on the flanks of the Casitas volcano in Nicaragua engulfed a town killing over 3000 residents. The local mayor described the scene as 'a desert littered with bodies'. Mass movements were responsible for the majority of deaths.

The economic impact

- Damage in Nicaragua was estimated at $1 billion and in Honduras as $4 billion.
- 70 per cent of the important cash crops were totally destroyed, including bananas and coffee.
- In Honduras agriculture employs 66 per cent of the workforce and contributes 25 per cent of GNP.
- In Honduras 90 per cent of roads, bridges and other transport infrastructure was destroyed.
- Nicaragua had a debt of $6 billion and Honduras of $4.2 billion; the French government immediately forgave $134 million of the debt.
- It is estimated that it will take up to 30 years for these countries to fully recover.

The human impact

Over 10 000 people were killed, with many bodies left buried in mud and debris. Thousands more were missing.

- Water supply systems were contaminated and sewerage systems destroyed, increasing the risk and incidence of disease.
- In Honduras 280 000 people were displaced.
- The education system was suspended until March 1999 as the buildings were used as shelters.
- In December 1999 it was estimated that 1.2 million people had been displaced, 70 000 formal homes destroyed and also a further 30 000 informal dwellings.

NOTING ACTIVITIES

1   Describe the typical features of a hurricane (tropical cyclone).

2   Describe the development and course of Hurricane Mitch.

3   **a** Assess the short-term and longer-term impacts of Hurricane Mitch. Present your information in the form of a table.

   b Suggest the *most serious* long-term impacts of the hurricane. Justify your suggestions.

   c Most of the countries that were severely affected were LEDCs. To what extent would this fact have increased the impact of the hurricane?

6.45 *Damage caused by Hurricane Mitch in Honduras, Central America*

The Great UK Drought, 1976

The summer of 1976 came as a shock. Water, previously regarded as a resource in surplus, suddenly became a valuable commodity. The drought started in 1975 with a fairly dry summer, and the winter of 1975–76 was much drier than usual. During the summer of 1976 the synoptic situation was dominated by high pressure systems anchored over the UK (Figure 6.46). These are typical of an English summer but generally alternate with frontal systems, with the latter bringing cloud and rain. A strong and static north-moving Rossby wave maintained the position of the high pressure for over four months. The anticyclones deflected the frontal systems northwards, bringing wet weather to Scandinavia, but over the UK the descending air produced clear skies with temperatures rising above 30°C on a daily basis.

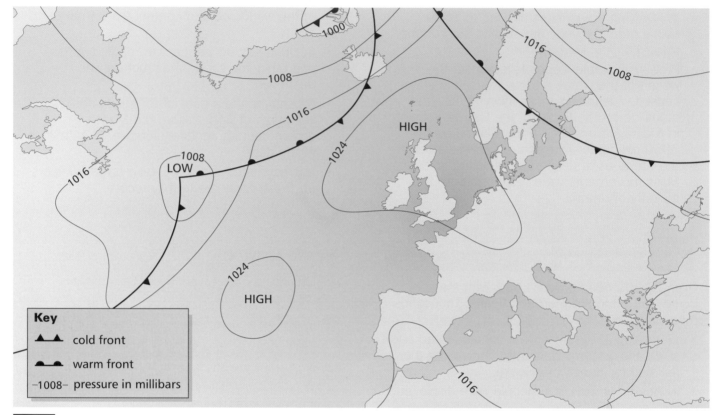

Key

▲▲▲ cold front

●●● warm front

−1008− pressure in millibars

6.46 *Synoptic chart for western Europe, August 1976*

6.47 *The Thames at Wallingford, August 1976*

The physical impact

- May 1975 to September 1976 represented the driest 16 months in known weather history.

- Much of England experienced less than 40 per cent of average rainfall.

- Temperatures and clear skies increased rates of evaporation and transpiration.

- The groundwater level (the water table) fell and reservoirs depleted rapidly.

- Rivers and lakes dried up – the Thames at Wallingford was much reduced (Figure 6.47), and whole ecosystems were lost.

- Fires became a major risk as the vegetation withered and dried. Access to woodland areas was restricted, and major forest fires occurred in South and mid-Wales, Sussex and Hampshire.

The social impact

- Water supply systems are adjusted to *average* conditions, and in much of England water is stored in reservoirs; regular rain replenishes supplies. These reserves began to run dry.
- In the south of England, wells used to abstract groundwater ran dry.
- By August many towns in West Yorkshire, South Wales and south-west England had a total water restriction for most of the day.
- In these areas access to water was limited to a few standpipes; queues were common.
- The Drought Act was passed by Parliament forbidding all non-essential uses, including swimming pools.

The economic impact

- Water supplies for industry and agriculture began to dwindle.
- Priority lists were drawn up, with essential food-processing industries at the top. Steel plants and power stations – large consumers of water – were threatened.
- Alternative sources of water were either too expensive (imports from Ireland) or contaminated (water from coal and tin mines).
- In agriculture, zero growth of grass was reported in the South and South East.
- Crop yields fell, and thousands of cattle were slaughtered to save on scarce fodder.
- Cereal crops and vegetables in central and eastern England largely failed.
- Some food prices doubled, whilst farm costs increased by 20 per cent, because of expensive imports of feed-stock; revenue fell by 10 per cent.
- In the longer term, subsidence of buildings became a serious problem. Clay, previously regarded as a safe foundation, dried out and shrank causing buildings to move and sink. Many residents cut down trees near the house to restrict water loss, but after the water table recharged the following winter, this caused even more damage by swelling clay under the houses. Since 1976 houses on clay foundations have had a higher insurance premium.

NOTING ACTIVITIES

1   Describe the meteorological cause of the drought.

2   Suggest some human factors that may have contributed to the shortages of water.

3   Rank the economic impacts of the drought and give reasons for your ranking.

4   Suggest ways in which authorities could reduce the likelihood of similar impacts associated with drought in the future.

Global warming: myth or reality?

The 1976 drought and a sequence of hotter than average summers in the 1980s helped to fuel the debate that the global climate was changing; it was getting warmer. The concept of 'global warming' has received wide publicity but it remains controversial and the implications remain unclear. The whole issue highlights how little we really know about our atmosphere.

What is global warming?

Carbon dioxide and other gases, e.g. ozone and methane, in the atmosphere stop some of the radiation from the Earth's surface escaping into space. They absorb and emit energy back towards the surface; this is termed the **greenhouse effect**. This mechanism was first noted by Tyndall in 1863, and between 1880 and 1940 global temperatures did increase by 0.25°C. However, between 1940 and 1970 global temperatures fell by 0.2°C and interest in the concept declined.

What are the 'facts'?

Scientists are generally agreed on the following 'facts':

- By analysing air samples trapped in deep Antarctica ice, it has been calculated that carbon dioxide concentrations of 270 parts per million have persisted over the last 10 000 years. This level is termed the **baseline**.
- In 1957 the average level had risen to 315 ppm.
- By 1994 it had reached 350 ppm.

- The main source of carbon dioxide in the atmosphere – burning fossil fuels – has increased dramatically. Between 1850 and 1950, 50 000 million tonnes were burnt; now the same amount is burnt in just 12 years.
- The greenhouse gas, methane, is increasing by 1.2 per cent a year, mainly from natural gas exploitation and the expansion of wet padi fields.

Does it matter?

People are concerned about global warming because of the potentially serious effects that it could have on the physical and natural world:

- The main effect of an increase in greenhouse gases is a global increase in temperature of 2°C by 2025.
- This increase will not be uniform but will probably increase with increasing latitude; the equatorial regions will have a marginal increase whilst the high latitudes will warm more.
- For countries like Canada and Russia this will be beneficial, and land suitable for agriculture will expand; the northern 'wastes' will become more habitable.
- Sea level has been a major focus of attention and, indeed, levels increased by 15 cm in the 20th century, closely following the temperature change. However, this is probably due to the expansion of sea water as it heats rather than to the addition of water to the oceans from melting ice.

- With increased temperature there is increased evaporation over the oceans and this will lead to greater global precipitation. The ice in Greenland and Antarctica is getting thicker as these areas have increased snowfall.
- Increased heat in the atmosphere will also increase wind velocity, so major storm events should increase in frequency. If the Mediterranean heats to over 26°C, hurricanes could develop and devastate the coastal areas.
- The main problem is that meteorologists just do not know and cannot agree on what to expect. The atmosphere is a very complex, balanced system and prediction is, at best, crude. Changes in ocean currents are one problem area – there is evidence that the North Atlantic Drift will be weakened. If this occurred, one result of global warming for the UK would be colder and more severe winters.

What can be done?

The release of greenhouse gases into the atmosphere could be reduced. The Rio Summit in 1992 took steps in this direction.

Controlling greenhouse gases would restrict the way we create wealth, how we live and where we live. It would have major economic, social and political implications. More specifically, industrialisation in less economically developed countries would have to be modified, colonisation of tropical forests would have to cease, and industrial and energy processes in the developed world would have to change.

At the moment the threat of atmospheric change is not sufficient to persuade governments (especially the USA) that economic and social sacrifices must be made; it is a fine balance.

EXTENDED ACTIVITY

Global warming is a huge issue and one that is being constantly updated. Whilst this section provides a brief summary, you should use the Stanley Thornes website to find out more about the issue. Complete a study examining the following:

1 The evidence for global warming.
2 The causes of global warming:
   a the natural mechanisms (the greenhouse effect)
   b the human factors (pollutants).
3 The possible impacts of global warming.
4 Actions being taken by the international community in an attempt to reduce the problems.

A The drainage basin hydrological cycle

A **drainage basin** is the area that is drained by a river and its tributaries. The movement of water in the drainage basin is an important part of the global hydrological cycle. Precipitation transfers water to the ground surface from where the majority of it either percolates into the soil and bedrock or is evaporated. Under extreme conditions water may flow over the surface before reaching a river and then flowing to the ocean or into lakes.

The precise pathways taken and rates of water transfer vary enormously from basin to basin and from time to time, as does the amount of water held in 'store' (in the soil or in the bedrock). It is important to understand the hydrological characteristics of individual drainage basins in order to manage them successfully.

How does water move through the drainage basin system?

Study Figure 7.1, which shows the main components and linkages in the drainage basin system. Notice that this is an example of an 'open' system because it has inputs (precipitation) and outputs (water and sediment). It contrasts with the global hydrological system which is a 'closed' system.

Let's track the route that water takes in a drainage basin.

Interception

Most drainage basins are clothed by one or more forms of vegetation. It may be scrub and bushes in semi-arid regions, grassland in temperate latitudes, or coniferous forest and tundra in the high latitudes. Whatever the type and extent of the vegetation, it will, to some extent, intercept precipitation (Figure 7.2). Water may be held on leaves to then be

Type of vegetation/biome	Loss of water by interception (average per year)
Temperate pine forest	94% if low-intensity rain; 15% if high-intensity
Brazilian evergreen rainforest	66%
Grass	30–60%
Pasture (clover)	40% in growing season
Coniferous forest	30–35%
Temperate deciduous forest	20% with leaves; 17% without leaves
Cereal crops	7–15% in growing season

7.2 *Interception losses for different types of vegetation/biomes*

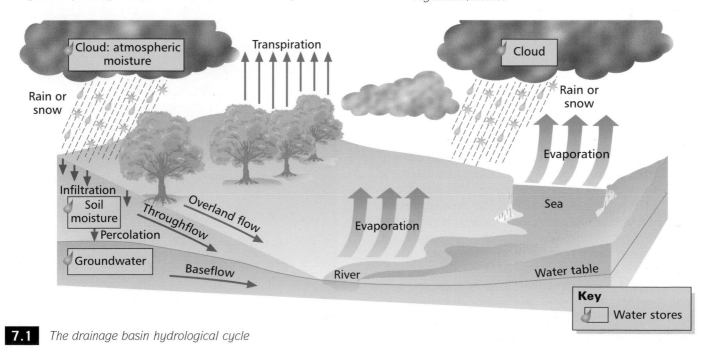

7.1 *The drainage basin hydrological cycle*

evaporated, or it may drip down from leaf to leaf as **throughfall**. Alternatively it may actually flow down tree trunks or plant stems as **stemflow**.

Vegetation has several important roles to play:

- It reduces the impact of raindrops by preventing them landing directly on the ground. This reduces rainsplash, soil erosion and soil compaction (which might encourage overland flow).
- Plants make use of water as they grow, so reducing the amount of water available in the system.
- Plant roots encourage water to pass into the soil and rock, therefore slowing down its rate of transfer.

Essentially, vegetation reduces the amount of water available to pass through the system, and it slows down its transfer.

A study of two similar-sized upland catchments of the River Severn and the River Wye in the 1970s provided evidence on the role of forests in reducing the amount of water that eventually reaches rivers. The Severn catchment contained 67.5 per cent forest whereas the Wye catchment only contained 1.2 per cent. Having received the same amount of rainfall during the study period, there were startling differences in runoff. In the Severn catchment some 38 per cent of the total precipitation was 'lost' (e.g. evaporated) compared with only 17 per cent in the Wye catchment. These variations could only be explained by the fact that the Severn catchment was mostly forest whereas the Wye was mostly grassland.

Interception is not limited to vegetation. Vast areas of the world are urbanised so that precipitation is intercepted by roofs, buildings, roads and other urbanised surfaces. Unlike trees, however, these surfaces with their gutters and drainpipes actually speed up water transfer.

Depression storage

Once on the ground surface, water may collect in hollows and depressions to form small ponds and puddles – this is known as **depression storage**. Water will either evaporate, seep slowly into the soil or flow overland.

Evapotranspiration

This is the combined loss of water by evaporation and transpiration. **Evaporation** is where liquid water turns into a vapour as it is absorbed by the air. It occurs when water is lying on the surface of the ground or on plant leaves, and is most rapid when the air is warm and dry. **Transpiration** is the process by which plants release water through tiny holes called stomata on the underside of their leaves. Once released, the water is available to be evaporated.

Infiltration

A great deal of the water that reaches the ground surface will soak or infiltrate into the soil. The **infiltration capacity** – the rate at which water soaks into the soil – is an extremely

important factor. If it is exceeded then water will be unable to soak away and overland flow might result.

Infiltration capacity may be exceeded during a period of heavy or prolonged rainfall when the soil is simply unable to absorb water at the rate at which it is falling (or melting, if it is snow). A low infiltration capacity will often exist if the soil is particularly thin, frozen or already saturated. Vegetation, particularly in the form of trees, will often promote a high infiltration capacity as the roots will form pathways for water to percolate underground.

Water actually soaks into the soil by a combination of **capillary action** (the attraction of water molecules to soil particles) and gravity, with the latter usually dominating.

Overland flow

Overland flow involves water moving over the ground either in small channels called **rills** which run down into streams (Figure 7.3), or across the whole surface as **sheetflow**. Overland flow is not very common in the UK. It is, however, quite common in some other parts of the world, particularly in semi-arid environments where the soil is often baked hard in between the rare, but torrential, rainstorm events. It most often occurs when the soil has become saturated or frozen, thereby reducing its ability to soak up additional water. Unable to soak away, water simply flows on the surface. Occasionally, torrential rain can lead to overland flow when the soil is unable to soak up the rainfall fast enough.

7.3 *A mountain stream*

Overland flow is an extremely rapid form of water transfer and it is often a significant cause of flooding. Authorities charged with preventing damaging floods do all they can to prevent overland flow occurring.

Soil moisture

Soils vary tremendously in their texture and structure. When a soil has been saturated with water and has then been left to drain, it is said to be holding its **field capacity**. Field capacity is measured as a depth of water, usually in millimetres, in the same way as rainfall. Very porous sandy soils retain little water so are said to have a low field capacity, whereas clay soils, which are often saturated, have a much higher field capacity. Field capacity is quickly reached in a sandy soil but takes much longer to be reached in a clay soil.

During the summer, high rates of evaporation and transpiration may reduce the soil moisture to below its field capacity (Figure 7.4). This leads to a condition known as a **soil moisture deficit**. Under such conditions plants begin to wilt and eventually die, and the ground surface becomes hard and dusty. When rain falls again, the soil is **recharged** as water infiltrates into the soil. In regions where a soil moisture deficit exists more or less permanently, farming is only possible with the use of irrigation.

Throughflow

Water moving through the soil is described as throughflow. Water will often move downhill roughly parallel to the ground surface. It is nothing like as fast as overland flow (water usually travels at between 0.005 and 0.3 m per hour) but it usually accounts for the majority of water transfer to rivers, particularly in temperate latitudes.

Water may move slowly through the soil from pore to pore or it may move more quickly along distinct underground routeways (called **pipes**) formed by plant roots, animal burrows or small cracks (common in clay soils, for example). Occasionally these pipes may be several centimetres in diameter.

Water will continue to pass vertically through the soil until it reaches the **water table** (the upper level of saturated rock and soil) or hits an impermeable layer where it tends to adopt a more horizontal line of flow until eventually it emerges at the ground surface.

Groundwater flow

If the underlying bedrock is permeable, water will slowly soak into it from the soil above. Water transfer through rock (groundwater flow) tends to be extremely slow, often taking tens or hundreds of years. This is because solid rocks usually have fewer pore spaces than unconsolidated rocks or soil. For example, unconsolidated gravel can transmit water at rates of up to 20 000 cm/hr whereas the equivalent figure for consolidated sandstone is only 200 cm/hr.

However, some rocks do transmit water quickly. You have probably heard news reports describing the plight of potholers trapped underground in limestone caverns by rapidly rising water following a sudden storm. Limestones, which are often heavily jointed, are capable of generating very rapid groundwater flow. A study at Cheddar, Avon calculated rates of 583 cm/hr, which is considerably faster than most rates of throughflow. Some igneous rocks, such as granite, can also have quite high rates of groundwater flow due to the presence of joints.

Groundwater flow feeds rivers through their banks and beds. Being a generally slow method of transfer, it carries on

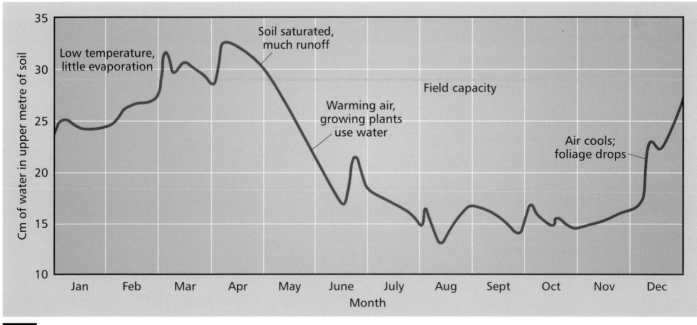

7.4   *Annual cycle of soil moisture*

supplying rivers well after an individual rainfall event. It is primarily responsible for maintaining flows during periods of drought. This steady flow of water is referred to as **baseflow**. It differs from **stormflow** which is far more temporary and 'flashy' in nature, comprising mainly overland flow, throughflow, and direct channel precipitation.

Channel store

The channel store is the river itself. Whilst a small amount of precipitation will fall directly into it, the vast majority of the water arrives through its bed and banks. Rivers form an 'exit' from the drainage basin hydrological system in much the same way as evaporation and transpiration.

NOTING ACTIVITIES

1 Why is the drainage basin hydrological system an example of an 'open' system?

2 Study Figure 7.2, which shows variations in amounts of interception. Comment on the importance of the following in determining the amount of interception:

  a the type of vegetation

  b the time of the year (growing season)

  c the intensity of the rainfall.

3 How might each of the following encourage overland flow:

  a deforestation

  b presence of steep slopes

  c frozen or saturated soil

  d bare rocky surfaces, common in mountain areas

  e urbanised areas, with the associated tarmac surfaces

  f overgrazing, or the use of heavy farm machinery?

4 Study Figure 7.4.

  a What is meant by the term 'field capacity'?

  b Why does it remain the same throughout the year?

  c Make a copy of the graph and add the following labels in their correct places:

  • early spring soil moisture surplus

  • summer soil moisture deficit

  • period of maximum evaporation and transpiration

  • recharge by infiltration of rain and melting snow

  • summer rainstorm.

  d Imagine that the graph has been drawn for land owned by a farmer. Contrast the likely problems faced by the farmer in the winter and in the summer.

STRUCTURED QUESTION

Figure 7.5 describes the hydrology of a hillslope vegetated with mature coniferous trees and grass. Each arrow shows a component of the drainage basin hydrological cycle that might operate when precipitation is in the form of rain and air temperatures have been above freezing for some time.

a Define the following terms:

  (i) stemflow   *(1)*

  (ii) infiltration   *(1)*

  (iii) throughflow.   *(1)*

b **(i)** What is the water table?   *(1)*

  (ii) Suggest how one of the components in Figure 7.5 might lead to a rise in the water table.   *(1)*

  (iii) What will happen to the stream if the water table rises?   *(1)*

c Assume that temperatures had been below 0°C for several weeks and precipitation had been in the form of snow. Describe the effects of these conditions on the operation of:

  (i) interception   *(2)*

  (ii) infiltration.   *(2)*

d Suggest two ways in which the change to sub-zero conditions might at some time increase the flood risk from the stream at the foot of the slope.   *(2)*

e Assume that the land owner is considering chopping down the trees and using the land for arable crops. Suggest what effects this change in land use might have on:

  (i) the hydrology of the slope   *(2)*

  (ii) the risk of flooding.   *(2)*

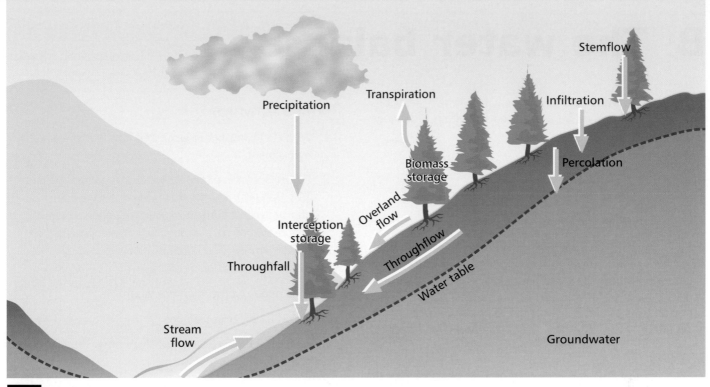

Stemflow

Transpiration

Precipitation

Infiltration

Biomass
storage

Percolation

Overland
flow

Interception
storage

Throughfall

Throughflow

Water table

Stream
flow

Groundwater

7.5   *Hydrology of a vegetated hillslope*

B The water balance

The **water balance** is a simple equation which describes how precipitation may be accounted for within a drainage basin:

$$P = Q + E \pm S$$

where P = precipitation

Q = total streamflow (runoff)

E = evapotranspiration

S = storage (in soil and bedrock).

The water balance is used by hydrologists to plan and manage water supply within a drainage basin. It can be used to suggest possible water shortages for which special measures like hosepipe bans can be implemented to preserve stocks. It has implications too for irrigation, pollution control and flooding.

CASE STUDY

The Rivers Teviot and Colne

Study the graph in Figure 7.6. Notice that the water balance year runs from October to September. This enables the year to start and end when there is minimum natural storage of water, i.e. at the end of the summer and before the winter rains set in. Notice that the graph can be split into three parts representing total precipitation, streamflow and that which is 'lost' (i.e. evapotranspiration and storage).

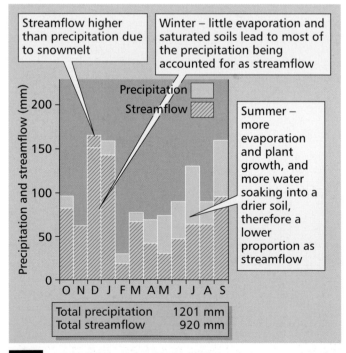

Streamflow higher than precipitation due to snowmelt

Winter – little evaporation and saturated soils lead to most of the precipitation being accounted for as streamflow

Precipitation
Streamflow

Summer – more evaporation and plant growth, and more water soaking into a drier soil, therefore a lower proportion as streamflow

Total precipitation	1201 mm
Total streamflow	920 mm

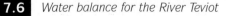

7.6 *Water balance for the River Teviot*

NOTING ACTIVITIES

1 What is the water balance and how is it of value to hydrologists?

2 Study the map of the Teviot drainage basin (Figure 7.7). Suggest how the following drainage basin characteristics might explain why such a high proportion of total precipitation is accounted for by streamflow and relatively little is 'lost' to evapotranspiration or storage:

   a a dense network of river channels

   b relatively steep slopes

   c bedrock consisting mostly of impermeable shales.

3 Study the information for the River Colne basin in southern England (Figures 7.8 and 7.9).

   a Compare the water balance for the River Colne with that of the River Teviot (Figure 7.6).

   b Using evidence from the map (Figure 7.9), suggest why the amount of streamflow barely fluctuates throughout the year despite the variations in precipitation.

   c Suggest reasons, other than those shown in Figure 7.9, that might also help to account for the steady pattern of streamflow in the River Colne.

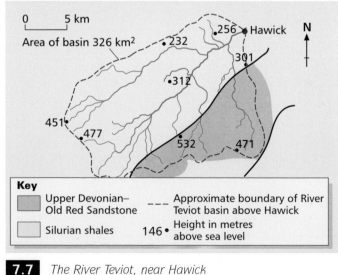

7.7 *The River Teviot, near Hawick*

226

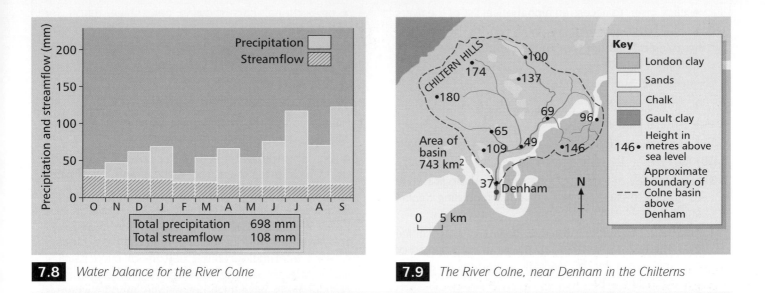

7.8 *Water balance for the River Colne*

Total precipitation 698 mm
Total streamflow 108 mm

7.9 *The River Colne, near Denham in the Chilterns*

IT investigation: patterns of water balance for selected UK river basins

The following investigation should ideally be done 'on-line' by individuals or pairs. However, it is possible to download the information from the Internet as hard copy.

The Institute of Hydrology maintains an archive record of hydrological data for a selection of river basins in the UK. This includes daily values of discharge (which could be used to plot hydrographs) and information regarding peak flows, runoff percentages and precipitation. Whilst it would be possible to conduct a number of IT-linked investigations, this one is concerned with extracting information about the water balance.

Aim of the investigation

- To make a study of water balance patterns across the UK using information obtained from the Internet.

Procedure

1 Study Figure 7.10 which locates 14 river basin gauging stations in the UK.

2 Access the Institute of Hydrology's website at www.nwl.ac.uk/ih/
   - click the 'National Water Archive' icon
   - click the most recent 'Hydrological Data UK Series' Yearbook
   - click 'River Flow and Groundwater Level Data'
   - click 'Riverflows'
   - click 'Map'.

You will notice that there are many gauging stations located on the map, including the 14 listed in Figure 7.11.

To access the data for your 14 sites:
   - return to 'Riverflows'

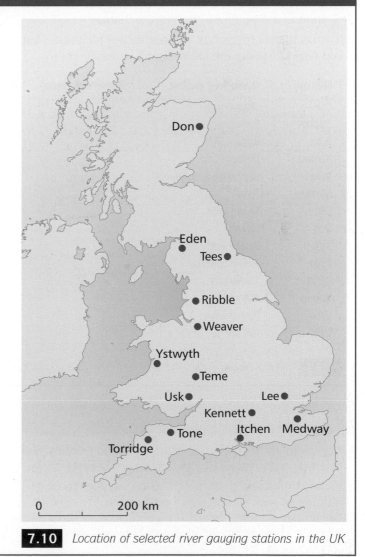

7.10 *Location of selected river gauging stations in the UK*

- click 'river' to access an alphabetical list of the rivers
- click the right-hand column (Y) to access the hydrological data for the required station. Take time to read what is there.

Locate in the 'Summary statistics':

- Annual runoff
- Annual rainfall.

It is this information that you will use in your investigation.

3 Make a copy of Figure 7.11. Use the hydrological data to complete the first two columns, recording rainfall and runoff.

4 Add six more stations for river basins of your choice from the rest of the UK, to give you an even spread of stations across the UK. Mark these stations on a larger copy of Figure 7.10.

5 Now complete the other two columns in your table.

6 Complete one or more maps of your choice to show patterns of water balance across the UK. The most important map is one showing percentage runoff.

It is up to you to choose an appropriate technique, but you might consider the following:

- proportional bars showing percentage runoff for each station
- choropleth colouring
- plotting of isolines to show percentage runoff (e.g. at 20%, 40%, 60%, etc.).

Be prepared to experiment in rough first.

7 Having completed your map(s), write a short written summary addressing the following:

a Describe the patterns of water balance across the UK. Which areas have the highest and the lowest percentage runoff values?

b Use an atlas, and a geological map of the UK (see Figure 2.54 page 63), to try to account for the patterns you have observed. In addition, read the basin characteristics information at the bottom of each gauging station's entry on the webpage. Try to explain why certain areas have high-percentage runoffs whereas other areas have low values. Consider what may be accounting for the 'losses'.

c In what ways do you think this information might be helpful to water resource managers, who are concerned with ensuring water supply and monitoring floods and droughts?

River	Gauging point	Rainfall (mm)	Runoff (mm)	Loss (mm)	Runoff as % rainfall
Usk	Chain Bridge				
Torridge	Torrington				
Tone	Bishops Hull				
Kennett	Theale				
Itchen	Highbridge and Allbrook				
Medway	Teston				
Lee	Feildes Weir				
Don	Parkhill				
Tees	Broken Scar				
Eden	Temple Sowerby				
Ribble	Samlesbury				
Weaver	Ashbrook				
Ystwyth	Pont Llolwyn				
Teme	Tenbury				

7.11 *Water balance data for selected river gauging stations*

EXTENDED ACTIVITIES

1 Complete hydrographs for two contrasting basins and attempt explanations for the differences between them.

2 Obtain information from the same source about the altitude of each station. This could be used to conduct a Spearman's rank correlation coefficient test to see whether there is any relationship between altitude and percentage runoff. A study of altitude and rainfall amount could also be an interesting way of examining the significance of orographic rainfall.

3 Use the archive data to find out about recent flood events. For example, several rivers in the South Midlands (between Warwick, Oxford and Northampton) flooded following torrential rain during Easter 1998. Use the data to discover how the flows during this period differed from the average values.

C Hydrographs

A hydrograph is a graph that shows changes in the discharge of a river over a period of time. A **storm hydrograph** shows the response of a river to a particular rainfall event. These are of great value to authorities charged with monitoring and preventing flooding. Hydrographs can also be drawn to show the pattern of flow over a year. Such graphs are usually referred to as **river regimes**.

The storm hydrograph

During and after a storm event, most of the precipitation will make its way downhill to a river channel. Some of it may flow rapidly as overland flow, whilst the rest flows more slowly as throughflow or, slower still, as groundwater flow. As it reaches the river, it will cause an increase in the volume of streamflow – this is known as the river's **discharge** (measured in 'cumecs' – cubic metres per second).

Look at Figure 7.12, which shows the main characteristics of a storm hydrograph. Notice that the vertical axis shows discharge in cumecs whereas the horizontal axis shows time. Notice also that the hydrograph has been split into two:

- **baseflow** (mostly groundwater flow through rock) remains fairly constant before, during and after a storm: it maintains the flow of a river in between rainfall events.
- **stormflow** (overland flow, throughflow and direct channel precipitation) accounts for the bulk of the hydrograph, resulting in its characteristic shape. Overland flow is usually responsible for most of the rapidly rising limb and the peak. This is because it is the fastest form of water transfer to a river.

Why do storm hydrographs differ?

The shape of an individual storm hydrograph depends on many factors. Dramatic hydrographs, with steep limbs and high peaks, tend to occur when there is significant overland flow, for example if there is a period of torrential rainfall, if the soil is frozen or saturated, if trees have been cut down, or if the basin slopes are steep. Less dramatic, flatter hydrographs may occur if the storm is light and in the summer, when evaporation rates are high, and leaf canopy is at its maximum.

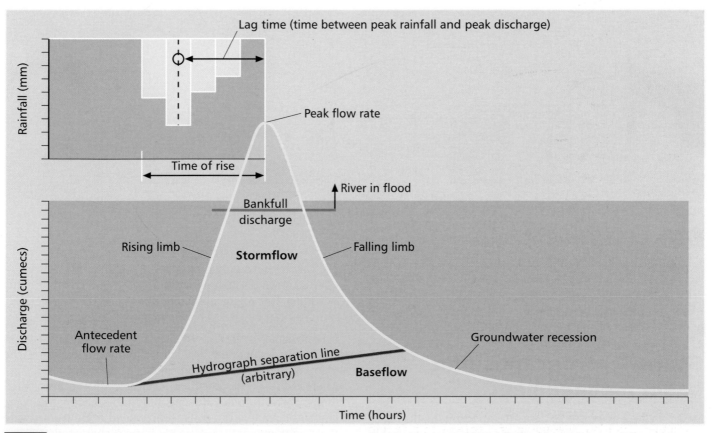

7.12 *The main characteristics of a storm hydrograph*

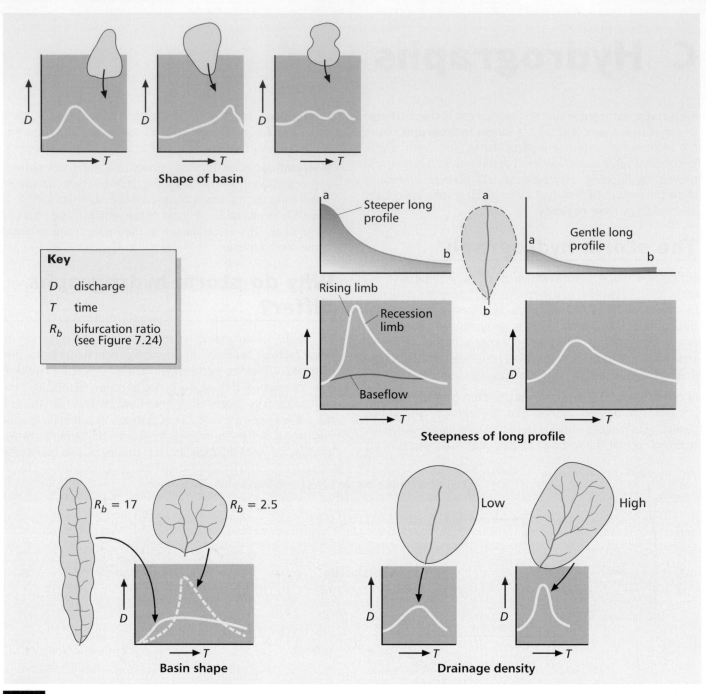

7.13 *Effect of selected basin characteristics on hydrographs*

A basin with an impermeable bedrock, such as shale, is likely to produce a much more dramatic hydrograph than a basin with a permeable bedrock, such as chalk. Look at Figure 7.13 to see how basin shape, gradient and drainage density can affect the shape of storm hydrographs.

Storm hydrographs and basin management

Storm hydrographs are extremely important in managing drainage basins. Hydrographs enable hydrologists to assess the rate of discharge increase of a stream or river, so allowing them to predict flooding. Each channel section will have a known discharge that it can contain without flooding occurring. If it seems likely that this figure will be exceeded, warnings can be issued to farmers and home-owners.

It is possible for people to alter storm hydrograph shapes too, both deliberately and inadvertently. For example, the planting of trees will often flatten a hydrograph's shape, whereas the development of suburbs, with the associated impermeable tarmac surfaces, tends to steepen the rising limb and heighten the peak. People need to be made aware that tampering with a drainage basin's characteristics can have significant effects on its 'normal' hydrograph.

NOTING ACTIVITIES

1   Draw a labelled diagram of a storm hydrograph.

2   **a**   What is meant by the term 'lag time'?

    b   Why is the lag time a key factor in determining the likelihood of flooding?

Activities 3 and 4 are best done in pairs first, to encourage discussion and the sharing of ideas.

3   How might the following conditions be expected to affect the shape of a hydrograph? For each one draw a simple hydrograph to show what it might look like. Add annotations (detailed labels) to your diagram to explain its characteristics. (For the purpose of this exercise, you have to assume that no other factors are influential.)

- a small drainage basin with very steep slopes
- a linear-shaped drainage basin
- a drainage basin that has recently been deforested
- a long period of mainly light rain in summer
- a heavy rainstorm falling on thick snow in March.

4   Figure 7.14 shows details of four drainage basins. For each one, attempt to draw an annotated hydrograph to show the likely response to a rainstorm. You should assume that similar storms occurred in all four basins.

5   In what ways can hydrographs be useful to people managing river basins?

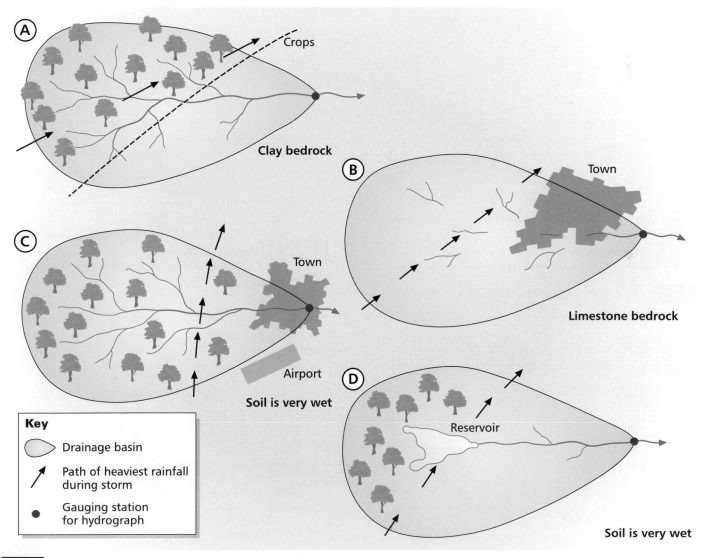

7.14   *Different characteristics in four drainage basins*

STRUCTURED QUESTION 1

Figures 7.15a and b are maps showing some of the characteristics of a small drainage basin in northern England. The basin lies on a coarse sandstone. Figure 7.16a gives details of a single rainstorm, which occurred on 18 March. Figure 7.16b is a storm hydrograph based on discharges recorded at X on Figure 7.15.

a What was the average intensity (in mm/hr) during the period of the storm? *(1)*

b Calculate the lag time in hours. *(1)*

c Describe and explain the effect on the lag time of any four characteristics of the drainage basin for which the hydrograph was produced. *(4 × 2)*

d Explain how throughflow might account for the secondary peak in discharge at about 27 hours from the beginning of the storm. *(2)*

e Describe and explain how the storm hydrograph from a storm of similar precipitation characteristics would differ from the one on 18 March, if the storm were:

- on 21 July after a dry early summer *(4)*
- on 29 January after a cold spell of three weeks with sub-zero temperatures. *(4)*

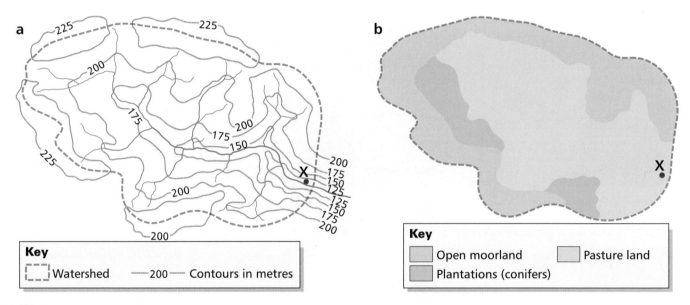

.15 *Characteristics of a basin in northern England*

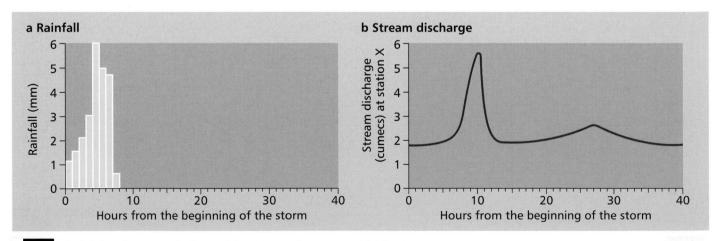

.16 *Rainfall and stream discharge following a rainstorm on 18 March*

STRUCTURED QUESTION 2

Figure 7.17 shows the River Ouse and its drainage basin. Figure 7.18 shows precipitation over this drainage basin between 13 December 1981 and 6 January 1982. Figure 7.19 shows the level of flow of some of the rivers in the drainage basin between 25 December 1981 and 7 January 1982.

a Describe the hydrograph for the River Ouse shown on Figure 7.19. *(3)*

b State two ways in which the hydrograph for the River Ouse is different from the hydrographs of the three tributaries shown. *(2)*

c With reference to Figures 7.17, 7.18 and 7.19, describe how physical factors led to the severe flooding at York between 2 and 6 January 1982. *(10)*

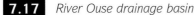

7.17 River Ouse drainage basin

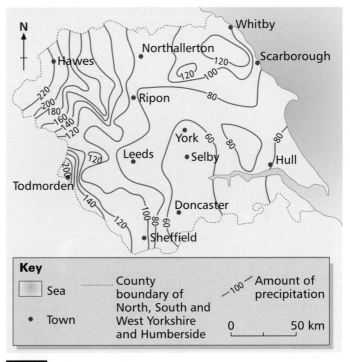

7.18 Precipitation over the Ouse basin, 13 December – 6 January 1982

In mid- to late December a series of deep low pressure systems brought snow to all parts of Yorkshire. On 2 January a depression with a marked warm front began to move across the area, bringing periods of heavy rain and a rapid rise in temperature. Thus rain and melting snow were too much for the infiltration capacity of the soil, and considerable overland flow resulted.

Following the onset of heavy rain or high river levels, a network of rainfall and river-level stations is monitored. At predetermined levels, and after consideration of the prevailing meteorological conditions, progressive flood warnings were issued by the police on the advice of the Water Authority. The only problems that arose concerned areas not covered on the maps of recorded flood events.

Source: adapted from a report by the Water Authority

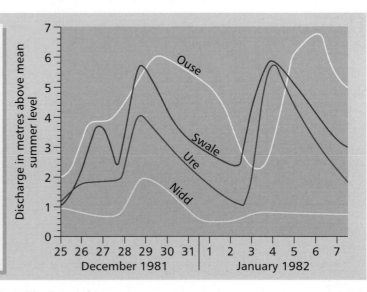

7.19 The Ouse basin's hydrological characteristics, and selected hydrographs

233

River regimes

The pattern of a river's flow over a period of time, usually a year, is called the river's **regime**. Several factors can affect a river's regime:

- **Precipitation** The amount, type and seasonality of precipitation is the primary factor affecting streamflow.

- **Temperature** Temperature has an effect on the amount of evaporation that takes place within a drainage basin. It also influences plant growth, and therefore affects rates of transpiration.

- **Water abstraction** The use of water for drinking or for irrigation reduces streamflow and therefore influences a river's regime.

- **Dams** Many rivers throughout the world have been dammed to control their flow. Water can be stored in reservoirs during times of plenty and then released during drier periods. Dams often 'flatten' a river's regime, making it less variable throughout the year.

Examples of river regimes

Figure 7.20 shows five very different river regimes:

- **River Sonjo, Tanzania** Notice that this regime has distinct wet and dry seasons, typical of a humid tropical climate. The high temperatures and heavy rainfall promote chemical weathering, which leads to the formation of deep clay soils. These quickly become

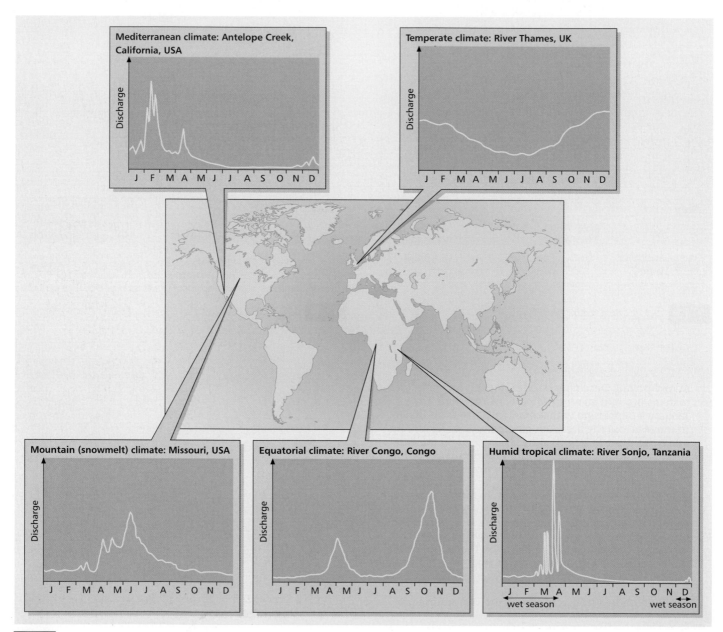

7.20 *Examples of river regimes*

saturated in the rainy season and overland flow creates sharp hydrograph peaks. In the dry season the clay quickly dries out due to high rates of evaporation.

- **River Congo, Congo** This regime is typical of an equatorial climate with its two wet seasons. Notice how there are two distinct periods of high discharge, one in April/May and the other in October/November.

- **River Thames, UK** Rainfall occurs throughout the year in this temperate climate, leading to continuous water flow. The presence of vegetation and deep soils promotes deep throughflow and groundwater flow, which maintains streamflow in times of drought. Slightly lower rainfall and higher rates of evapotranspiration in the summer cause the dip in discharge.

- **River Missouri, USA** Very high rates of discharge occur from April to June as a result of snowmelt on the High Plains and in the Rocky Mountains. Low rates of precipitation, and the fact that most of it falls as snow, explain the very low discharges during the winter.

- **Antelope Creek, California, USA** Much of the west coast of California experiences a Mediterranean climate. Dry and hot summers often cause rivers to flow at very low levels or even to dry up completely. Storms in the winter can, however, cause dramatic rises in discharge.

NOTING ACTIVITIES

1 What is meant by a river's 'regime' and how does it differ from a 'storm hydrograph'?

2 Study the river regime graph for the River Sonjo.

   a Suggest reasons for the sudden 'blip' in December.

   b What evidence is there that the wet season comprises a series of storms rather than one prolonged period of rainfall?

   c Suggest reasons why the construction of a dam might be favoured by people who live alongside the river.

   d Draw a simple graph to show how you would expect the regime of the River Sonjo to change after a dam had been constructed.

3 Study the river regime graph for the River Missouri.

   a Suggest reasons for the low flow from November through to March.

   b Why are there dramatic peaks in late spring and early summer?

4 What are the problems faced by river basin managers in Mediterranean climates?

STRUCTURED QUESTION 3

Study the graphs in Figure 7.21, which show the flow and precipitation regimes of two river catchments, A and B, in different parts of Britain.

a For catchment A:

   (i) describe the general relationship between river flow and precipitation (2)

   (ii) explain the relationship you have described in **a(i)**. (2)

b Explain the fluctuations in river flow in catchment B between January and May. (4)

c What are the likely differences between the two catchments in terms of:

   (i) infiltration (2)

   (ii) evapotranspiration? (2)

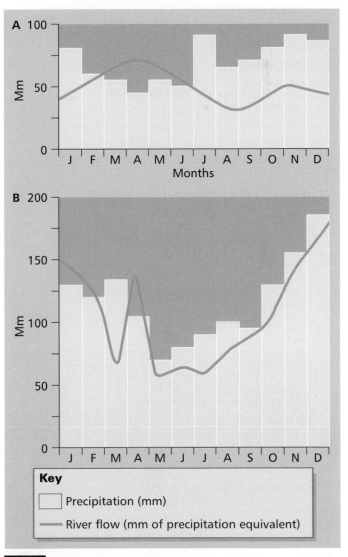

Key

☐ Precipitation (mm)

── River flow (mm of precipitation equivalent)

7.21 *Precipitation and flow regimes of two river catchments in different parts of Britain*

D Drainage basin analysis

River networks on maps often look very complex. Whilst each river network is of course unique, it is possible to find common patterns and some degree of order.

Drainage patterns

Several common types of drainage pattern can be identified (see Figure 7.22). The most common pattern of rivers in the UK is **dendritic**, which looks rather like the branches of a tree ('dendron' = treelike). It is associated with uniform drainage basin features such as bedrock and slope. Other patterns tend to be linked to particular basin characteristics, for example **parallel** drainage is common on very steep mountainsides, whereas **radial** drainage is most likely where there is a hill or a dome.

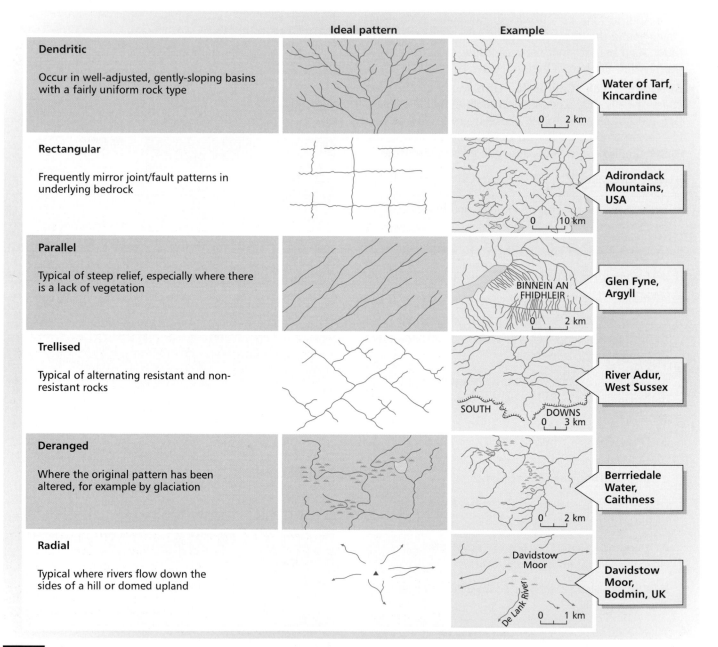

Ideal pattern   **Example**

Dendritic

Occur in well-adjusted, gently-sloping basins with a fairly uniform rock type

Water of Tarf, Kincardine

0   2 km

Rectangular

Frequently mirror joint/fault patterns in underlying bedrock

Adirondack Mountains, USA

0   10 km

Parallel

Typical of steep relief, especially where there is a lack of vegetation

BINNEIN AN FHIDHLEIR

Glen Fyne, Argyll

0   2 km

Trellised

Typical of alternating resistant and non-resistant rocks

SOUTH   DOWNS

River Adur, West Sussex

0   3 km

Deranged

Where the original pattern has been altered, for example by glaciation

Berrriedale Water, Caithness

0   2 km

Radial

Typical where rivers flow down the sides of a hill or domed upland

Davidstow Moor

De Lank River

Davidstow Moor, Bodmin, UK

0   1 km

7.22 *Drainage patterns*

Drainage density

Drainage density is one of the most readily comparable measures of drainage basins and it is one of the most important. The **drainage density** of a basin is defined as the total length of channels divided by the basin area. It is expressed as 'km per square km' (km/km²).

Study Figure 7.23, which compares different densities. High densities are characteristic of areas where:

- the bedrock is impermeable
- the soils are readily saturated
- the slopes are steep
- there is little vegetation and interception of precipitation
- the amount and intensity of precipitation is high.

If a river basin has a high drainage density, water will quickly find its way into a channel. It will then be transmitted rapidly through the basin, so increasing the flood risk. High drainage densities are also associated with areas experiencing rapid rates of denudation.

Description		Density	Geology	Location
Very coarse		2.7–3.5	Chalk	South-east England
Coarse		3–8	Carboniferous limestone	Yorkshire
Medium		15–25	Weathered igneous	California, USA
Fine		25–40	Lavas	North Wales
Ultra-fine		200–900	Loess (wind-blown deposit) in Badlands	Arizona and New Jersey, USA

7.23 *Drainage density*

Statistical analysis of river networks

In the 1930s R. E. Horton, a hydraulic engineer, carried out various studies of stream networks in which he discovered that a number of key relationships existed between certain measurable variables. The presence of these relationships, he argued, suggested that stream networks have a degree of order and logic about them and that they are not completely random. It could be suggested, perhaps, that when order appears to exist, then a state of equilibrium or balance has been reached.

Stream ordering

Networks can be compared in terms of their intricacy and complexity by using the concept of **stream ordering**. Each stream segment can be given a value relating to its 'order' in the network. The value of the highest-order stream segment becomes the **basin order**. Most river basins contain 3rd or 4th order rivers (the one in Figure 7.24 is a 4th order). For a comparison, the Mississippi is thought to be a 10th order basin.

The most widely used system of ordering was developed by the American geologist A. N. Strahler (see Figure 7.24). According to his system, all initial tributaries are 1st order segments. When two 1st order segments join, they form a 2nd order stream. Thereafter, each time two streams of the same order meet, the order increases by one. Look at Figure 7.24 to see how this works. Notice that the order does not increase if a lower-order segment joins a higher-order segment.

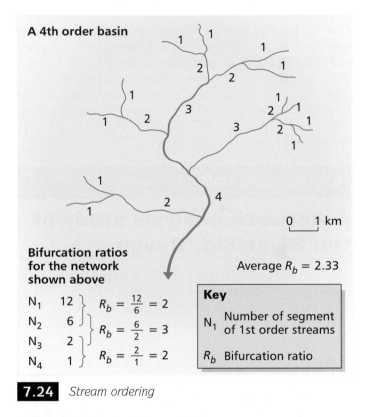

A 4th order basin

Bifurcation ratios for the network shown above

Average R_b = 2.33

N_1 12 $\left.\right\}$ $R_b = \dfrac{12}{6} = 2$

N_2 6 $\left.\right\}$ $R_b = \dfrac{6}{2} = 3$

N_3 2 $\left.\right\}$ $R_b = \dfrac{2}{1} = 2$

N_4 1

Key

N_1 Number of segment of 1st order streams

R_b Bifurcation ratio

7.24 *Stream ordering*

Stream bifurcation ratio

The **bifurcation ratio** is defined as the ratio between the number of segments of one order and the number of segments of the next highest order. The bifurcation ratio is normally between 3.0 and 5.0. If the ratios for a particular basin are similar, it suggests that the drainage basin has uniform characteristics. Research has suggested that basins with a low bifurcation ratio are more prone to flooding than those with a high ratio. Study Figure 7.24 to see how bifurcation ratios are calculated.

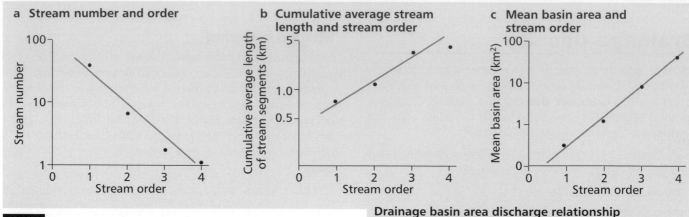

a **Stream number and order**

b **Cumulative average stream length and stream order**

c **Mean basin area and stream order**

7.25 *Stream order relationships*

Stream order relationships

Horton compared stream order with several other aspects of drainage basins. He discovered that clear relationships often existed. Some of these are shown in Figure 7.25.

One of the beauties of network analysis is that some research can be carried out without having to make a field visit. Having identified something of interest, for example many more 1st order streams than would be expected, a clear focus for further research (including fieldwork) has been established. It could be, for example, that the large number of streams are the result of deforestation or the presence of readily saturated soils.

Drainage basin area discharge relationship

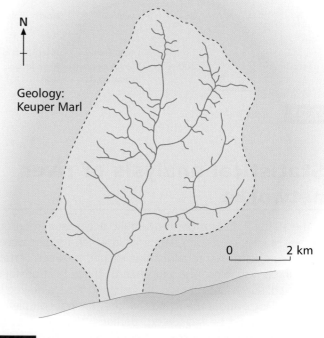

EXTENDED ACTIVITY

A network analysis study of the River Sid, Devon

The River Sid in South Devon lies in a fairly compact basin draining gently-sloping land from north to south (Figure 7.26). The rock type is mainly Triassic Keuper Marl: this consists of silty mudstones and sandstones and is generally impermeable.

In this activity you will be carrying out your own analysis of some of the River Sid's drainage network. You will need to make a copy of Figure 7.27 in order to record your results.

1 Does the River Sid conform to Horton's law of stream numbers?

a Order the streams in the network.

b Add up the total number of stream segments for each order and calculate the bifurcation ratios. Comment on the results.

c Plot the number of stream segments (*y* axis) against stream order (*x* axis) on semi-log graph paper (page 402).

d Use a ruler to see how close to a straight line the points are. Notice that three of the values form an almost perfect

7.26 *Drainage basin of the River Sid, South Devon*

Order	Number of segments	Bifurcation ratio	Total length of streams (km)	Average length of streams (km)	Cumulative average stream lengths (km)
1					
2					
3					
4					

7.27 *Stream network analysis*

straight line. Use these points to help you draw and label a best-fit line. The point for 1st order stream segments lies some way off the best fit-line. Label this as an anomaly.

e Does the overall pattern support or refute Horton's law of stream numbers?

f Why do you think there are so many 1st order streams? Use your best-fit line to suggest how many would have been expected. Try to suggest how the following factors may account for this anomaly:

- rock type
- gradient
- changes in land use.

2 Does the River Sid conform to Horton's law of stream lengths?

a Add up the total length of segments of each order. This will take some time and you need to make sure that you work out the length of all segments without repeating any!

b Calculate the average length for each order segment. Sum each successive average length to give the cumulative average lengths. It is this value that you will plot on your graph.

c Plot the cumulative average stream lengths (y axis) against stream order (x axis) on semi-log graph paper (page 402).

d Use a ruler to suggest a line of best fit. Don't be surprised if the value for 1st order lengths appears as an anomaly, given what you discovered earlier. Label the best-fit line and identify any anomalies.

e Does the overall pattern support or refute Horton's law of cumulative average stream lengths?

f Comment on the presence of any anomalies you have identified.

E River channel processes

Water in a river channel has a certain amount of available **energy**. It is greatest when there is a large amount (mass) of water and when there is a steep gradient.

Most of a river's energy is used up in overcoming friction with the bed and banks. Friction is often high in the upper reaches of a river where large boulders may protrude into the river's flow. They cause considerable **turbulence** which manifests itself as dramatic white water and swirling eddy currents (see Figure 7.28). What may appear at first sight to be rapidly moving water is, in fact, highly disturbed water that is moving down the channel relatively slowly.

How is sediment transported?

Having overcome friction, energy is then used to transport sediment. The ability of a river to transport sediment is referred to as its **competence**. The material carried by a river is called its **load**. There are three types of load:

1 **Dissolved load** This involves the invisible transport of chemicals dissolved in the water. It is particularly significant if water has flowed over (or through) limestone or chalk.

2 **Suspended load** This is sediment that has been whisked-up by the water and is then carried by the flow. It is the main form of sediment transfer and explains why most rivers look muddy or cloudy.

3 **Bedload** Material that is too heavy to be suspended may be rolled (**traction**) or bounced (**saltation**) along a river bed.

The amount and type of sediment movement depends on many factors:

• **The nature of the bed and banks** Is the river flowing over loose sediments or is it carving a channel through solid rock? Does the river flow over chalk or limestone?

• **The flow of the river** Is the river flowing fast or slow? Is its flow constant or does it fluctuate greatly?

• **Human intervention** Has a dam been constructed which might trap sediment, or the channel been dredged or lined with concrete?

What are the processes of erosion?

Any surplus energy is used by a river to carry out erosion. Erosion is the picking up and removal of material. It is possible to identify three main processes by which a river will erode its channel:

1 **Corrasion** This is where particles of rock carried by the river grind away at the bank and bed. It is the most significant type of erosion in most rivers.

2 **Hydraulic action** This is the sheer power of the water as it crashes onto the bed or against the banks. It is particularly significant at waterfalls and rapids or during times of flood. Air bubbles may burst (strictly speaking they implode) in areas of great turbulence (Figure 7.28) sending out shockwaves which may increase erosion. This is called **cavitation**.

3 **Solution** Water contains dissolved carbon dioxide from the air and this may react with limestone and chalk, causing it to dissolve.

As particles are transported down-river, they constantly crash into one another. This process is called **attrition**. It causes the particles to become increasingly rounded and smaller in size with increasing distance downstream (see Figure 7.29).

7.28 *White-water turbulence in an upland river*

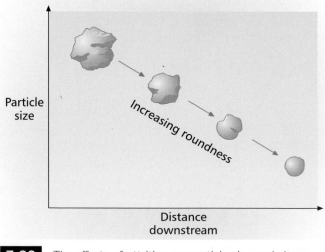

Particle size

Increasing roundness

Distance downstream

7.29 *The effects of attrition on particle size and shape*

When does deposition occur?

Deposition occurs throughout the course of a river wherever the speed of flow drops such that particles can no longer be carried. This could occur on the inside bend of a meander, on the bed of the river or close to its banks where friction is at its greatest, or where the river enters the sea or a lake and its flow is checked.

What controls river processes?

The main factor that controls transportation, erosion and deposition is the speed or **velocity** of a river. Study Figure 7.30. It is a graph showing the relationship between velocity and particle size. Known as the **Hjulstrom curve**, it resulted from research carried out in 1935. Notice that the graph has been subdivided into the three types of process.

Identify the **critical erosion velocity** line. This is the velocity required to pick up a particle of a given size. The 'picking-up' of particles is called **entrainment**. Broadly speaking, the larger the particle, the greater the velocity needed to pick it up. The main exception to the rule is for very fine clay-sized particles which require surprisingly high velocities to be entrained. This is because they have a tendency to stick together: imagine trying to separate out tiny individual particles from a mass of soggy clay!

As the graph shows, the particles requiring the lowest velocity to move are sand-sized particles. This explains why channels in sandy material are often large and frequently change their shape. Notice that the process of erosion operates above the critical erosion velocity line.

The other line on the graph is the **critical deposition velocity**. This is much more logical and straightforward: it shows that as velocity falls, so successively smaller particles are dropped. The area on the graph below this line indicates when deposition will occur.

The reason why there is a gap between the two lines is that a body of water has the ability to hold and suspend particles (as you will have experienced when swimming), even if the velocity falls below that which is required to pick them up in the first place. Notice that the area on the graph between these two critical lines represents the process of transportation.

How does velocity change with distance downstream?

However fast a river's flow appears to be in its upper reaches (Figure 7.28), and sluggish it appears in its lower reaches, measurements reveal an increase in velocity with distance downstream (Figure 7.31). There are several reasons for this:

- there is a greater volume of water as tributaries join the main river
- the 'roughness' of the channel is reduced as silt replaces pebbles and boulders on the river bed; this results in less turbulence and enables faster, uninterrupted flow
- the channel tends to adopt a more efficient profile – the term **hydraulic radius** is used to quantify efficiency (see *box* on next page).

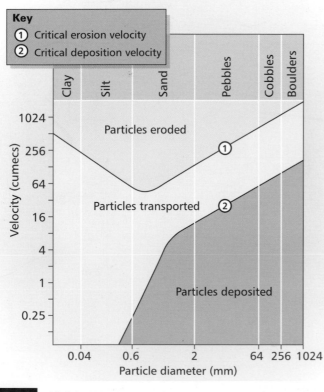

7.30 *Hjulstrom curve*

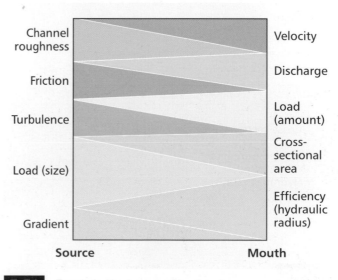

7.31 *Trends in the long profile of a river*

NOTING ACTIVITIES

1 What determines the energy of a river, and how is it used up?

2 **a** For each one of the three factors identified as affecting sediment transport (page 240), suggest how it will affect the type and amount of transportation.

   b Try to identify two other factors, and explain their role.

3 For each process of erosion (pag 240), describe its operation and suggest factors likely to maximise its effectiveness.

4 Study the Hjulstrom curve (Figure 7.30).

   a Describe the critical erosion velocity line.

   b Why do clay particles require such a high velocity to become entrained?

   c Describe the critical deposition velocity line.

   d Between which velocities will pebbles be deposited?

   e Between which velocities will boulders be transported?

   f What is the significance of sand being the easiest sized particle to erode?

Hydraulic radius

- Hydraulic radius is a measure of a stream's efficiency. It is calculated as:

$$\frac{\text{cross-sectional area (m}^2)}{\text{wetted perimeter (m)*}}$$

   * Wetted perimeter is the line of contact between the water and the channel.

- The hydraulic radius expresses the losses of energy in overcoming friction with a stream's bed and banks.

- High values indicate high efficiency, with a channel approaching a semi-circular shape.
- Low values indicate low efficiency where a channel will often be wide but shallow.
- Hydraulic radius tends to increase with distance downstream, primarily due to the increasing discharge.

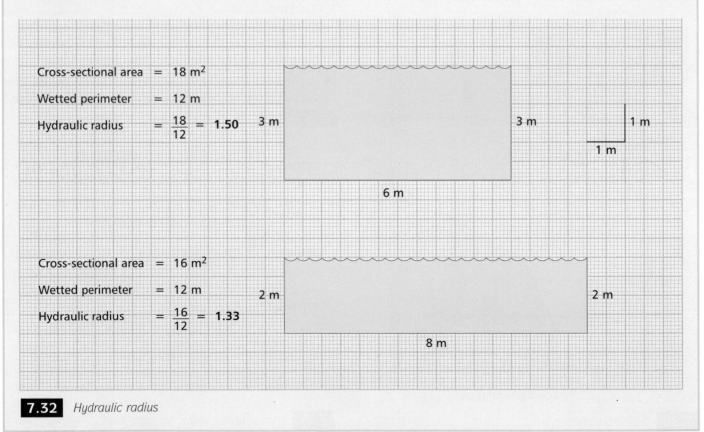

Cross-sectional area	= 18 m²
Wetted perimeter	= 12 m
Hydraulic radius	= $\frac{18}{12}$ = **1.50**

Cross-sectional area	= 16 m²
Wetted perimeter	= 12 m
Hydraulic radius	= $\frac{16}{12}$ = **1.33**

7.32 *Hydraulic radius*

STRUCTURED QUESTION 1

Study Figure 7.33, which shows the relationships between stream sediments (shape) and distance from the source, for a stream in lowland Britain.

a **(i)** Copy and complete Figure 7.34 by plotting the data given in the table. *(3)*

(ii) Draw a best-fit straight line on your copy of Figure 7.34. *(1)*

b **(i)** Describe the relationship between the best-fit lines shown on Figures 7.33 and 7.34. *(2)*

(ii) Explain the change in particle size with increasing distance from the stream source. *(2)*

(iii) Explain the change in particle roundness with increasing distance from the stream source. *(2)*

c With reference to the load of a river:

(i) state *three* ways in which material is transported *(3)*

(ii) explain how the method and speed of transport may be influenced by rock composition. *(3)*

d Suggest how human activity might add to the load of a river. *(3)*

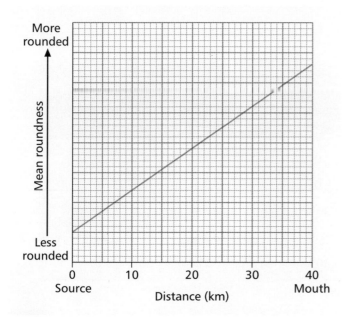

7.33 *The relationship between size of particles in a stream and its distance from the source*

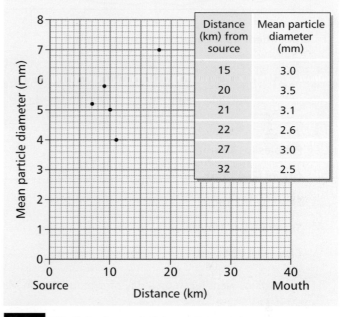

Distance (km) from source	Mean particle diameter (mm)
15	3.0
20	3.5
21	3.1
22	2.6
27	3.0
32	2.5

7.34 *Particle size and distance from a stream's source*

STRUCTURED QUESTION 2

Study the information in Figures 7.35a,b and c on the next page. The map (a) shows two adjacent river basins in an upland part of northern England. Table (b) contains information about sediment transfer, and table (c) gives fieldwork readings for four sites along the course of the main channel in basin A.

Basin A is used for cattle and sheep grazing, while basin B is partly sheep pasture and contains a reservoir.

a **(i)** Describe *two* contrasts between the bedload particle size distributions at sites A and B shown in table (b). *(2)*

(ii) Suggest reasons for the contrasts you have observed. *(2)*

b Using evidence from map (a), give *one* possible reason for the difference in the patterns of suspended load at sites A and B shown in table (b). *(2)*

c **(i)** Calculate the missing hydraulic radius values in table (c) (see *box* opposite). *(1)*

(ii) What is the meaning and significance of the downstream change in hydraulic radius from measuring point 1 to point 4 in table (c)? *(2)*

d Give *two* likely reasons why the average velocity increases downstream in table (c). *(2)*

a

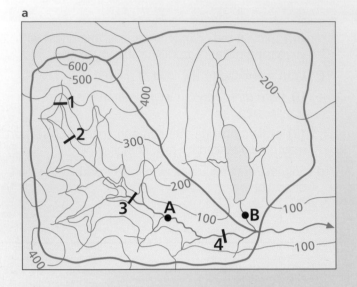

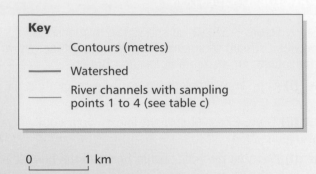

Key

——	Contours (metres)
——	Watershed
——	River channels with sampling points 1 to 4 (see table c)

0   1 km

b Sediment data collected at sites A and B

	Site A		Site B	
Particle size	**Bedload load**	**Suspended load**	**Bedload load**	**Suspended load**
Pebbles (> 4 mm)	45%	0	10%	0
Gravel (2–4 mm)	30%	0	15%	0
Sand (0.1–2 mm)	20%	30%	35%	0
Silt (0.004–0.1 mm)	5%	50%	30%	30%
Clay (< 0.004 mm)	0	20%	10%	70%

The readings were taken on a day of average flow in March 1977.

c Channel data for points 1, 2, 3 and 4 on the main channel in basin A

Measuring point	Cross-sectional area ($w \times d$) in m^2	Wetted perimeter* (m)	Hydraulic radius (m)	Average water velocity (m/sec)
1	0.1	1.5		1.2
2	1.6	4.2	0.38	1.3
3	4.8	6.8		1.6
4	10.4	10.0	1.04	1.8

All readings were taken on a single day of average flow.
* *Wetted perimeter* was measured by laying a tape-measure across the bed of the stream from the water level on one side to the water level on the other side.

7.35 *Sediment transfer in two adjacent river basins in upland northern England*

F River landforms

As water flows in rivers, it produces several distinctive landforms both within the active channel and outside it. River landforms are probably the most extensive physical features on the Earth's land surface and rivers are responsible for sculpting much of our landscape.

Potholes

A **pothole** is a circular depression on the river bed carved out of solid rock. It is formed by a kind of drilling-action as pebbles are caught in eddy currents and whisked around within a small natural crack or hollow. As time passes, the drilling action enlarges the hollow to form a pothole which may be many centimetres wide and deep. Potholes are commonly found below waterfalls or rapids, where hydraulic action is a significant process.

Waterfalls

A **waterfall** is a sudden step in a river's long profile. It is often the result of a tougher, more resistant band of rock cutting across the valley (see Figure 7.36). Unable to erode the rock at the same rate as neighbouring rocks, a step is formed and a waterfall results. Over time, the river cuts backwards into the resistant rock so causing the waterfall to retreat up-valley to form a narrow, steep-sided **gorge**. One of the most famous waterfalls in the UK is High Force on the River Tees (see *box* on the next page).

When rocks of varying resistances cut across a valley, a series of smaller steps may form **rapids**. Rapids are associated with very disturbed turbulent water.

River valleys

River valleys are often said to be V-shaped in their profile. The logic behind this is explained in Figure 7.37. As a river cuts down vertically into its bed, the valley side walls become degraded by weathering and mass movement, so that the valley itself adopts a V-shaped profile.

It is certainly true that many river valleys do manifest such a profile, but the role of geology must not be forgotten. We have already seen that steep-sided gorges can be formed when waterfalls cut back into tough rock. Look at the *box* on page 247 to see how geology is a key influence in the formation and shape of one of the USA's best-known river valleys.

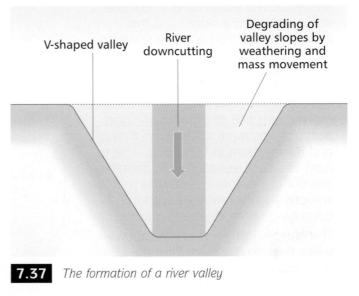

7.37 *The formation of a river valley*

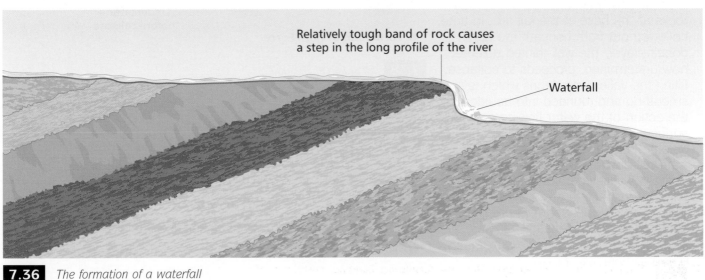

7.36 *The formation of a waterfall*

High Force waterfall, River Tees

At High Force, the River Tees falls in a single sheet some 20 m into the swirling pool below. Draining from the peat-covered Pennines to the north, the water is dark brown in colour owing to the high content of acids drawn from the peat. But tumbling over the fall it erupts into white foamy plumes, stark against the dark bands of rock framing the spectacle; the foaming waters have been likened to brown ale! Two types of rock can be seen at the fall. The lower has horizontal layers and it forms part of the Carboniferous limestone. The upper set of dark-coloured rocks has vertical joints within it and is much denser and harder than the underlying limestones. In fact it is an outcrop of igneous rock (dolerite). The tough band of dolerite is very resistant to erosion, and this explains the existence of the waterfall, for the river cannot erode this as easily as the limestone. Where there is no dolerite, the valley broadens out and waterfalls are absent, but other bands of dolerite give rise to waterfalls a short distance upstream at Cauldron Snout and downstream at Low Force. These hard bands of resistant rock, therefore, prevent the river from following a smooth sloping profile through the countryside from source to mouth.

The narrow valley and pebble-strewn valley floor downstream from the fall also owe their existence to the layer of igneous rock. As the river plunges over the waterfall, the softer Carboniferous rocks at the base of the fall are, in time, hollowed out from beneath the hard dolerite layer. The well-jointed dolerite, now undermined, proceeds to collapse, filling the valley with debris which is smoothed and rounded into boulders by the action of the water flowing over its surface. The progressive collapse of the dolerite leads to the retreat of the waterfall upstream, leaving the narrow gorge-like valley cut through the dolerite. This valley shape is very different from that 1.5 km downstream where the dolerite band is absent.

Source: A. Goudie (1985) Discovering Landscape in England and Wales, *Allen & Unwin*

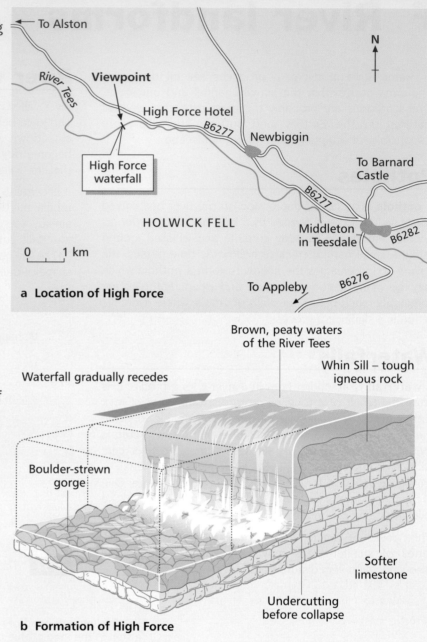

a Location of High Force

b Formation of High Force

| 7.38 | *High Force, Teesdale* |

NOTING ACTIVITY

Use the information in this box to make a case study of High Force waterfall.

a Describe the feature itself and explain how it has formed.

b Discuss the role of rock type in the waterfall's formation.

c What effect has the waterfall had on river processes?

d How may the characteristics of the waterfall change over time?

Grand Canyon, Arizona, USA

Grand Canyon, in the north-western corner of Arizona, is one of the most dramatic river valleys in the world. At its deepest, at 1829 m (well over a mile) from rim to river, it is up to 29 km wide.

The river valley that forms the Grand Canyon (Figure 7.39) is an erosional feature that owes its existence to the Colorado River, which is responsible for the depth of the canyon. Of equal importance are the forces of erosion that have shaped it and continue to shape it today – mainly running water from rain, snowmelt and tributary streams, which enter the canyon throughout its length. The climate at Grand Canyon is classified as semi-arid but, when it comes, rain falls suddenly in violent storms, particularly in the late summer, and the power of erosion is therefore more evident here than in places that receive more rain.

Grand Canyon owes its distinctive shape to the fact that the different rock layers in the canyon walls each respond to erosion in different ways: some form cliffs and some erode more quickly than others. The vivid colours of many of these layers are due mainly to small amounts of various minerals, most containing iron, which impart subtle shades of red, yellow or green to the canyon walls.

Extract from 'A quick look at Grand Canyon', Grand Canyon National Park

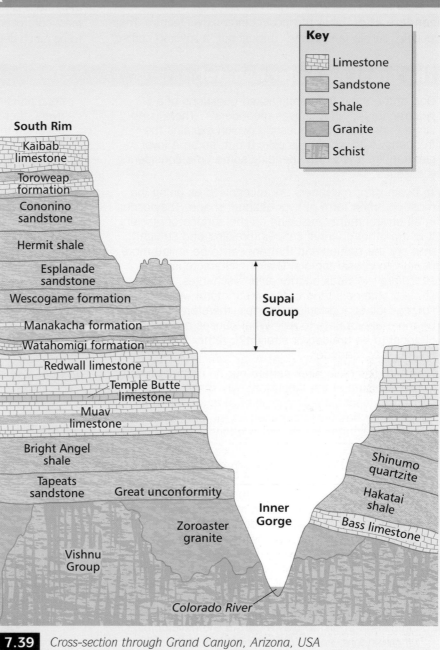

Key
- Limestone
- Sandstone
- Shale
- Granite
- Schist

7.39   *Cross-section through Grand Canyon, Arizona, USA*

NOTING ACTIVITIES

1   Study Figure 7.39.

   a   Describe the shape of the canyon.

   b   Look at the exposures of limestone on the valley sides. Do they tend to produce steep cliffs or gentle slopes? Try to account for your answer.

   c   What rock type consistently produces relatively gentle slopes on the valley sides? Suggest reasons for your answer.

   d   Most of the beds of rock are horizontal. Do you think this has affected the profile of the canyon's sides in any way?

   e   Why do you think the valley at the base of the Grand Canyon (the Inner Gorge) is narrower than the Canyon higher up?

2   How has the climate of the area encouraged the formation of the canyon?'

Meanders

Meanders are probably the most common river feature. Their sweeping curves are found throughout a river's course, although they tend to be most pronounced on the flat plains away from mountains. Meanders have long been one of the most controversial geographical features and there continues to be much debate about why they form (see *box* below).

The great meander debate!

One of the most frequently asked questions of a geographer is 'What causes meanders?'. There is no simple answer to this question (which explains the frequency with which it is asked) but, as an A level geographer, it is worth spending some time considering the issues.

To begin with, meanders are recognised as being the most common form of river channel shape. They occur in all environments including on the surface of glaciers. It is the other channel shapes (braiding and straight) that are the rarities and that appear to be linked more directly to causal factors. Braiding characterises sandy/gravelly beds and irregular discharges, whereas straight channels tend to follow structural weaknesses such as joints and faults. Perhaps, therefore, it would be more appropriate to ask 'What causes a river channel to be braided or straight?', rather than what causes it to meander.

Many studies have been carried out on meanders both in the field and in the laboratory. Whilst there is no consensus about the formation of meanders and no single factor has been identified as the primary cause, some interesting features and relationships have been discussed.

- Meandering river channels often exhibit a clear pattern of alternating shallows (called **riffles**) and deeps (called **pools**). Laboratory experiments have shown that such features do form in straight channels and that, as they develop, they can cause the water flow to begin to swing from side to side, thus initiating meandering. However, laboratory experiments are artificial and there is absolutely no evidence to suggest that, once upon a time, all river channels were straight and that they have subsequently meandered. There is little doubt that riffles and pools do play a part in the *development* of meanders, and in determining the route of the thalweg, but there is no proof that they are responsible for in some way *initiating* meanders in the first place.

- The flow of water in a meandering river is highly complex. Friction interferes with water movement at the bed and banks and this, in turn, interferes with water flow elsewhere in the channel. Research has indicated the existence of a corkscrew-like, spiral flow called **helicoidal flow** which is undoubtedly linked to meander development. However, the question remains – is meandering the result of helicoidal flow, or is helicoidal flow caused by meandering?

- Field evidence has linked meanders with gentle gradients, fine sediments and steady discharges (see Figure 7.40). Whilst these environmental conditions may well be associated with meandering rivers, there are exceptions, such as meanders that occur on the surface of glaciers where there is no sediment and discharges are far from being steady.

Returning to the question, 'What causes meanders?', the simple answer is that nobody really knows. However, perhaps it is an unfair question to be posed in the first place.

Meanders do seem to represent a state of balance or equilibrium between the amount of energy needed to transport a river's load and the need to expend any additional energy in the form of channel erosion. Meanders clearly represent a natural and efficient shape for flowing water to discharge its available energy. It may be this that accounts for flowing water's natural tendency to adopt a winding course over whatever surface it flows.

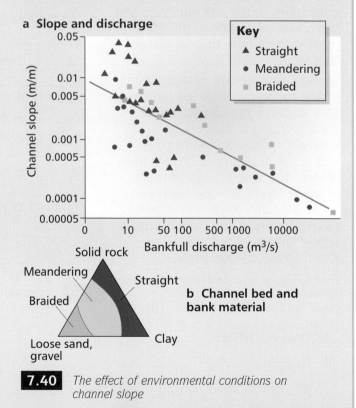

7.40 *The effect of environmental conditions on channel slope*

NOTING ACTIVITY

Having read through the information in this box, write a short account of the controversy surrounding meanders, entitled 'What causes meanders?'

Study Figure 7.41, which shows some of the main features of a meandering river. Notice that the flow of fastest water (the **thalweg**) swings from bank to bank. This causes erosion to be concentrated on the outside of the meander bends, resulting in bank undercutting and the formation of **river cliffs**. In contrast, deposition occurs on the inside bends leading to the formation of **point bars**.

Over time, the alternating zones of erosion and deposition cause meanders to migrate both across and down-valley. In addition, the meanders themselves tend to become more exaggerated and **sinuous** (winding). As opposite bends erode towards each other, the neck of a meander will get progressively narrower until, during a period of high discharge, the river will cut through to form an **oxbow lake** (Figure 7.42).

Study Figure 7.43, which shows a stretch of the River Conwy in North Wales. The old course of the river has been plotted using old political boundaries that, in the past, were drawn down the centre of the river. Notice how there used to be a much more complex pattern of meanders than there is today.

Oxbow lakes gradually silt up as they are no longer part of the river's active course. As the water becomes shallower, vegetation starts to colonise the lake. Turn to page 83 to find out about the vegetation succession (called a **hydrosere**) that occurs when an oxbow lake dries up.

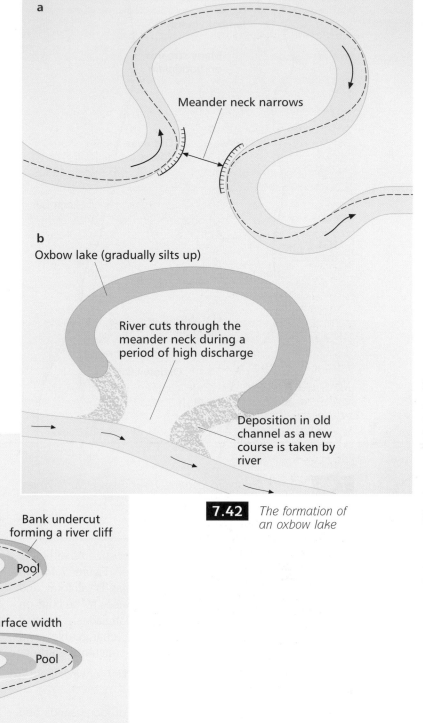

7.42 *The formation of an oxbow lake*

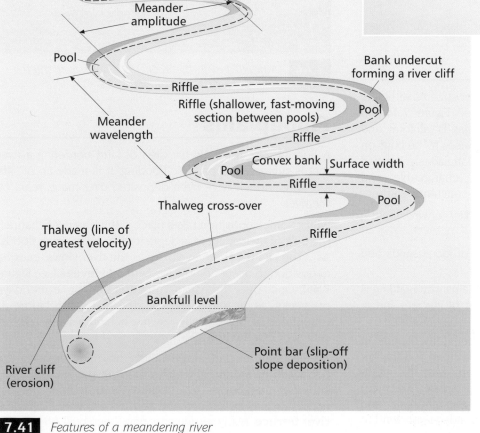

7.41 *Features of a meandering river*

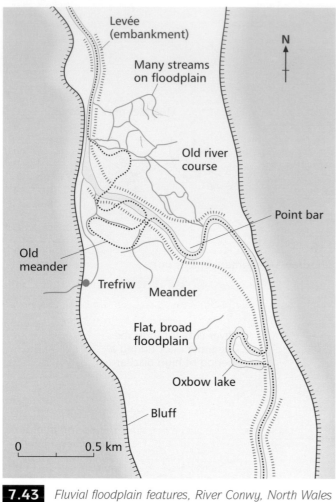

7.43 *Fluvial floodplain features, River Conwy, North Wales*

Braiding

When a river subdivides into many smaller streams within its channel, it is said to be **braided** (see Figure 7.44). In between the individual channels, small islands will form, and these may be permanent enough for houses to be built on them, as is the case with the *chars* in Bangladesh. Braided channels are, however, notoriously changeable and dangerous as the individual streams are constantly shifting.

Braiding has been linked with two key environmental conditions:

- easily erodable channel sides, made of loose sands and gravels, for example
- highly irregular discharges, which explains why braiding is common in rivers found at the snout of glaciers and in semi-arid regions.

Braided channels are regarded as being inefficient in carrying sediment because they are often broad and shallow and have a low hydraulic radius. To cope with this inefficiency, the river splits into smaller, faster streams which cut into the channel bed, steepening the gradient and increasing efficiency. This explains why braided rivers tend to have a steeper long profile than meandering rivers.

7.44 *A braided river: the Rhône as it flows from the the Rhône glacier in Switzerland*

Floodplains

A floodplain is a mostly flat area of land bordering a river that is subjected to periodic flooding. It is made up of silts and sands which have been deposited over many years by the river.

Whilst some deposition takes place within the river channel, the bulk of the sediment is deposited when the river floods. As it tops its banks, the velocity of the river slows, causing deposition to occur and the formation of **levées** (see Figure 7.45). Once over the banks, the water lies for several days or weeks as a thin sheet on the flat land before it evaporates or drains away, leaving behind a fresh layer of sediment or **alluvium**.

As meanders migrate down-valley, they may broaden the floodplain by eroding the valley sides. If the sea level falls and the river begins to cut into its own bed, the old floodplain may be left perched above the channel to form a **river terrace**. Many settlements have been built on terraces because they are less likely to be prone to flooding.

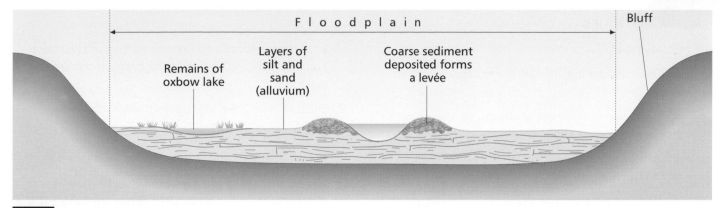

7.45 *Features of a river floodplain*

Deltas

When a river enters the sea or a lake, it loses velocity and a large amount of its sediment is deposited as a **delta**. Over time, the fan of deposited sediment breaks the surface of the water to form new land (see Figure 7.46). Rivers flowing across deltas tend to split into many smaller channels called **distributaries** as they endeavour to cope with the reduced gradient and the high sediment load.

It is possible to identify several types of delta – the **arcuate** and the **bird's-foot** are probably the most common (Figure 7.47).

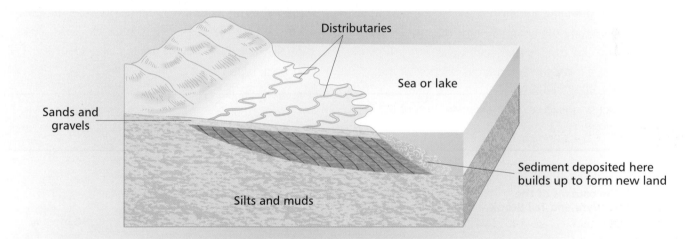

7.46 *The formation of a delta*

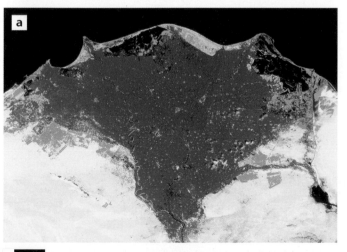

7.47 *An arcuate delta: the Nile, Egypt*

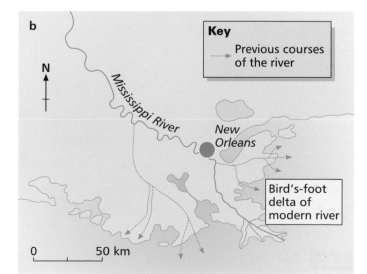

Key
⇢ Previous courses of the river

Mississippi River

New Orleans

Bird's-foot delta of modern river

0   50 km

NOTING ACTIVITIES

1   Study Figure 7.43.

a   Describe the features of the floodplain of the River Conwy.

b   Use a series of simple diagrams to show how the oxbow lake would have been formed.

c   What is a levée, and how would it have been formed?

d   Draw a fully annotated sketch of the river, describing the features of the river channel and its floodplain. Attempt to suggest the line of fastest flow (the thalweg).

e   With reference to a chosen section of the river, suggest how and why you think it might change over time.

2   Study Figure 7.44, which shows a braided river.

a   What is braiding?

b   Under what conditions is braiding likely to occur?

3   With the aid of simple sketches, describe the difference between an arcuate delta and a bird's-foot delta.

STRUCTURED QUESTION

Refer to Figures 7.48a, b and c, which describe characteristics of the upper reaches of two rivers in the Forest of Bowland, Lancashire.

a   Describe briefly the channel plans of the two rivers (b and c).   *(2 x 2)*

b   Construct a labelled sketch cross-section of the Marshaw Wyre along X–X₁.   *(2)*

c   Construct a labelled sketch cross-section of the Tail Brook along Y–Y₁.   *(3)*

d   Despite being close to one another, the two rivers have very different channel shapes. Suggest how the following factors may be responsible, at least in part, for these variations:

- discharge characteristics   *(2)*
- rock type   *(2)*
- long-profile gradient.   *(2)*

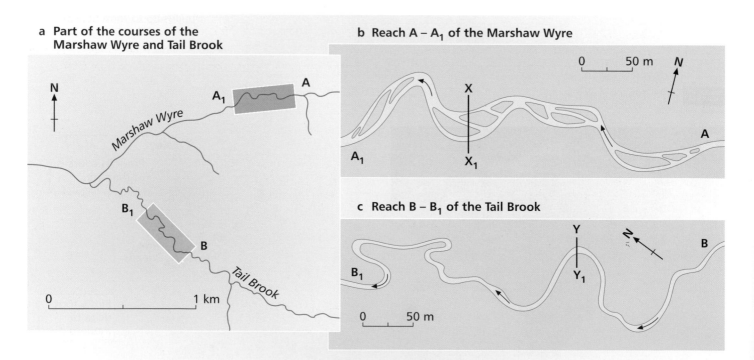

a Part of the courses of the Marshaw Wyre and Tail Brook

b Reach A – A₁ of the Marshaw Wyre

c Reach B – B₁ of the Tail Brook

7.48   *Part of two rivers in the Forest of Bowland, Lancashire*

G River channel management

Rivers that are completely natural will carve a course through the mountains and then weave their way across their marshy, lowland floodplains before discharging their water and sediment into the sea or a lake. They will, over a long period of time, tend towards an equilibrium state and be in balance with their environmental conditions.

However, very few river channels can claim to be entirely 'natural'. People have been interfering with them both directly and indirectly for years:

- building dams to create reservoirs and regulate water flow
- straightening channels to reduce the flood risk or to improve navigation
- dredging channels to increase the hydraulic radius and improve rates of flow
- cutting down bankside vegetation to increase land for farming
- abstracting water for irrigation, industrial use and domestic consumption.

In this chapter we will study the issue of riverbank erosion and then examine an increasingly popular form of sustainable river channel management.

Riverbank erosion: is it a problem?

Riverbank erosion (Figure 7.49) is, of course, a completely natural process. It is most commonly associated with lateral erosion in a meandering river.

Research has shown that a 'healthy' river (one with abundant wildlife) will have up to 10 per cent of its banks being eroded at any one time. This is because there are several advantages associated with riverbank erosion:

- a variety of habitats are created along the riverbank
- the greater the diversity of habitats, the greater will be the variety of flora and fauna
- some species have become especially adapted to collapsing riverbanks, such as sandmartins which nest in the sandy cliffs and sandpipers which nest on the exposed shingle banks often derived from collapsed banks.

However, bank erosion may be much greater than 10 per cent, particularly along short stretches of a channel. Human activity is usually to blame, and this can take several forms:

- cutting down bankside trees to increase available land for cultivation – trees help stabilise the banks by holding the soil together
- allowing farm animals to trample banks as they search for drinking water
- severe footpath erosion or general overuse for recreation.

Severe bank erosion increases the amount of sediment in a river. This can have a number of harmful effects:

- finer sediments can bury the gravel banks used as spawning grounds by trout and salmon
- larger rocks can lead to increased scouring downstream, particularly during times of high discharge
- smaller stream channels may become infilled, reducing the hydraulic radius and increasing the risk of flooding
- habitats will be destroyed

 7.49 *Riverbank erosion*

- nutrients in the soil will be released into the water where they may result in rapid growth of algae, depleting the oxygen supply in the water.

It is clear that whilst some riverbank erosion is to be welcomed, too much can be damaging to the environment.

How can riverbank erosion be reduced?

The Environment Agency has produced a guide for landowners entitled 'Understanding riverbank erosion'. It contains advice regarding management practices aimed at reducing severe erosion:

- Avoid overgrazing (too many animals in an area) and do not allow winter grazing near channels, when the vegetation cover is at its thinnest and the soil most likely to be wet and soggy.
- Fence off the river and provide alternative drinking water for animals, preferably right away from the river.
- Where there is arable farming, maintain a 'buffer zone' of vegetation alongside the river. Do not plough right up to the river's edge.
- Create a dedicated path for walkers to encourage them not to walk up and down the banks, and educate people by using posters and noticeboards.

7.50 *A riverbank restoration project*

- Plant trees that are particularly suited to riverbanks, such as willows and alders.

Study the stretch of riverbank shown in Figure 7.50 and notice the work that has been done to reduce erosion.

NOTING ACTIVITIES

1 Outline some of the advantages and disadvantages of riverbank erosion. Present your information in the form of a table.

2 Select *three* possible solutions and for each one give full reasons why it can lead to a reduction in riverbank erosion.

3 Should local landowners, for example farmers, be merely 'encouraged' to protect riverbanks, or should they be forced to do so by law?

4 Study Figure 7.50.

a What measure(s) have been taken to reduce riverbank erosion?

b Describe and comment on the measures that have been taken at this site as illustrated in the photo.

River restoration

The quality of rivers and their surrounding wetland environments have been steadily degraded as river channels have been straightened and deepened, and floodplains developed and intensively farmed. As a result, much of their natural beauty and their value to people and wildlife have been lost.

River restoration involves reversing this process. It aims to return a river and its adjacent environment to its natural state. River channels are encouraged to meander and allowed to flood as they would have done in the past.

In striving to return rivers to their natural equilibrium state, river restoration is an example of **sustainable management**, where present-day use does not threaten or degrade the long-term survival of the environment. Projects are usually locally based and frequently involve local communities in both planning and construction. By empowering local communities and giving them a degree of ownership over a particular project, there is a greater likelihood that it will become a success.

There are many examples of river restoration throughout the world. The information in the *boxes* on pages 255 and 256–57 describes two projects: one is on quite a large scale and essentially rural; the other is much smaller, and is in a large town.

Kissimmee River, Florida, USA

The Kissimmee River connects two large lakes in central Florida, USA (Figure 7.51). In the past it meandered for approximately 165 km within a 2–3 km wide floodplain. As the photographs (Figure 7.52) illustrate, the area is very low-lying and flat. It was known to have flowed very slowly (due to its gentle gradient) and flooding was common.

Wading birds, waterfowl, fisheries and other biological components were once part of this integrated and resilient river/floodplain wetland ecosystem and were supported by, and dependent on, the special mosaic of habitats, intricate food webs, and other complex physical, chemical and biological interactions and processes.

Before 1940, human habitation was sparse within the Kissimmee basin. Land use within the basin consisted primarily of farming and cattle ranching. After the Second World War, there was significant property development in the basin and concerns about flooding grew. A severe hurricane in 1947 and its associated heavy rainfall was the catalyst for major engineering works aimed at reducing the flood risk.

Between 1962 and 1971, the river channel was straightened and deepened by the US Army Corps of Engineers. The meandering river was transformed into a 90 km long, 10 m deep, and 30 m wide canal. Excavation of the canal and deposition of the resulting spoil eliminated approximately 2500 hectares of floodplain wetland habitat. In addition, deep trenches were dug across the floodplain to drain it and to enable further development to take place.

By the early 1970s several negative effects were becoming clear:

- floodplain utilisation by wintering waterfowl had declined by 92 per cent
- low flows in stretches of the canal system had led to the growth of certain types of vegetation which depleted the water of oxygen and thereby affected fish stocks
- the neighbouring wetlands had become starved of water, silt and nutrients and many habitats had been lost.

In 1990 it was decided to restore 100 km² of floodplain to wetland, and to convert 70 km of canal to meandering river. The aims of the project (which began construction in 1997 and is due to continue until 2010) are:

- to restore the ecological diversity of the region by enabling seasonal controlled flooding to take place

- to improve the quality of the water to enable traditional fish stocks to increase
- to provide increased opportunities for recreation
- to provide opportunities for environmental education.

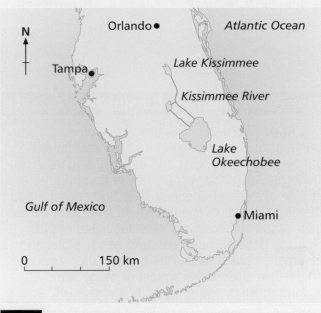

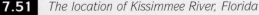

7.51 *The location of Kissimmee River, Florida*

7.52 *The Kissimmee River **a** before and **b** after it was managed*

NOTING ACTIVITY

Write a short report about the Kissimmee River restoration project. In your report you should consider the following:

- the natural context (the physical geography and ecology of the river and its floodplain)
- the post-1940 factors that led to the river being engineered

- the engineering work itself and the hydrological reasons for it
- the arguments in favour of restoration
- the restoration project details and aims
- the consequences of restoration.

(To find out how the project is progressing and to see other photographs, access the Kissimmee River website via Stanley Thornes.)

The River Skerne, Darlington, County Durham, UK

The River Skerne at Houghton-le-Skerne, Darlington, flows through an urban parkland surrounded by housing and industry (Figure 7.53). Over the past 200 years it has undergone straightening and deepening for flood control and drainage. Much of the floodplain has been raised high above the river by industrial waste-tipping. Housing development, gas and sewer pipes and electricity cables further limit restoration opportunities. The situation is typical of many rivers flowing through towns and cities in the UK, where the ecology and the visual and recreational appeal of rivers has suffered.

There were two main aims of the project:

- to restore 2 km of river in terms of physical features, flood management, habitat diversity, water quality, landscape and access for the community
- to further knowledge and understanding of river restoration by comprehensive monitoring of the effects of the project.

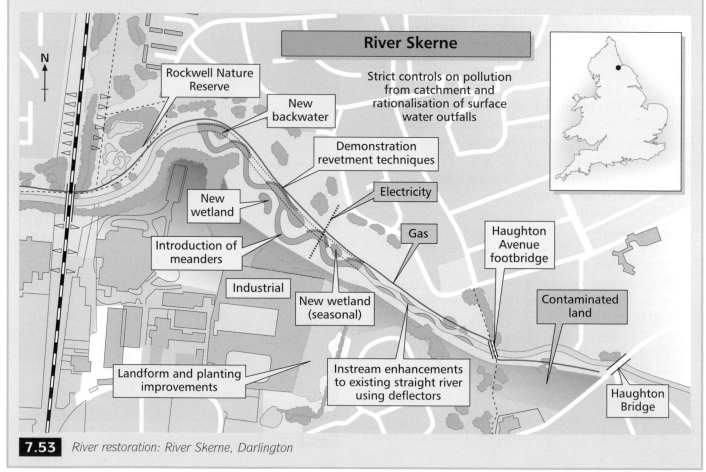

7.53 *River restoration: River Skerne, Darlington*

Construction took place between 1995 and 1996, and involved several features:

- The floodplain was lowered by removing the industrial spoil and transferring it to the valley sides – this allowed for the development of a wetland area fed by periodic flooding
- where possible, the river was 're-meandered' to produce a more natural course
- the river banks were re-profiled to create more natural shapes

- coconut-fibre rolls and willow trees were used to protect the north bank from erosion, which could expose the mains gas pipeline
- 20 000 trees and shrubs were planted by local schoolchildren.

Since 1996, the site has been carefully monitored. The river has maintained its new meanders, and many bird species have been attracted back to the area, particularly to the wetlands. The quality of the river water has improved and the whole environment is more attractive, and popular with local residents.

NOTING ACTIVITIES

1   Make a careful sketch of Figure 7.54b, which is an aerial photograph of the newly restored River Skerne in Darlington. Use the information in this box to add detailed labels (annotations) describing some of the key characteristics of the project.

2   Many rivers in the UK flow through towns such as Darlington. Assess the importance of restoration projects, similar to the River Skerne project, to local urban communities.

b After restoration

a Before restoration

7.54   *River Skerne restoration*

H River basin management

How are river basins managed?

The Environment Agency is responsible for the management of river basins in England and Wales. Formed in 1996, it oversees water resources, flood defence, pollution prevention and control, navigation, fisheries, recreation and conservation. Much of its work is carried out through its regional offices (see Figure 7.55).

It's broad aim is 'to protect or enhance the environment and to attain the objective of sustainable development'. Arguably the three most important aspects of the Environment Agency's work concern water resources, flood defence and pollution.

- **Water resources** – in attempting to balance the demands of users with the supply of water, the Environment Agency is concerned with the conservation and redistribution of surface and groundwater supplies. It is also concerned with locating new reserves.

- **Flood defence** – the primary concern is to safeguard life (through flood defences and warning systems) but the Environment Agency is also concerned with environmental protection.

- **Pollution control** – the Environment Agency monitors water quality and works with industry and local authorities in attempting to prevent pollution incidents.

Much of the land adjacent to an individual river channel is privately owned, for example by a farmer. It is the responsibility of the landowner to maintain a river channel and its banks. Landowners are encouraged to consult with the Environment Agency before planning or starting any work on the river channel. As channel alteration can lead to increased erosion and a possibly increased flood risk, flood defence consents are required from the Agency prior to any work being undertaken. The maintenance of structures crossing a channel, such as a road bridge or footpath, is the responsibility of the owner (e.g. local authority, Railtrack, etc.).

In Scotland, much of the day-to-day administration of water supply and sewerage, flood control and drainage is carried out by local authorities. The Scottish Environment Protection Agency (SEPA) monitors and advises on water quality. In Northern Ireland, the Department of the Environment administers much of water policy.

The watershed approach

One of the major problems associated with river basin management is that rivers often flow across several different planning regions. For example, rivers like the Rhine and the Danube in Europe flow through different countries, each with its own planning system and hydrological priorities. Countries close to the mouth of these rivers often have to cope with pollution dumped by countries further upstream.

In the USA, similar difficulties have arisen between individual states. Recently, the **watershed approach** has been adopted for a number of river basins. Recognising that watersheds are 'nature's boundaries', state and federal agencies work together with local community groups to ensure whole basin management planning, particularly relating to environmental matters such as pollution and wetland conservation.

Many management issues affect drainage basins. In this chapter we will study discharge extremes (floods and low flows) and pollution. We will conclude by making a whole-basin study.

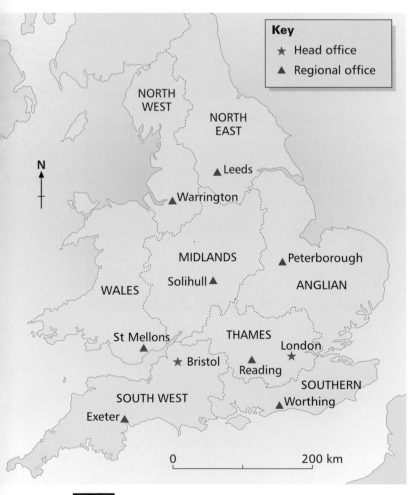

Key
★ Head office
▲ Regional office

NORTH WEST
NORTH EAST
▲ Leeds
▲ Warrington
MIDLANDS
▲ Peterborough
Solihull ▲
WALES
ANGLIAN
St Mellons ▲
THAMES
London ★
★ Bristol
▲ Reading
SOUTHERN
SOUTH WEST
▲ Worthing
Exeter ▲

N

0   200 km

7.55 *Environment Agency regional offices*

NOTING ACTIVITIES

1   The Environment Agency has responsibility for several aspects that are related to rivers and river basins. Do you think this is a good idea, or should each aspect be controlled by a separate body?

2   How might a short stretch of river, say 200 m, have several different people or organisations responsible for it?

3   Suggest some problems that might result when several different countries are responsible for an international river.

4   What is the 'watershed approach'? Do you think it is a good idea?

Managing discharge extremes

Rivers respond to changes in environmental conditions, particularly precipitation. Long periods of heavy rainfall can contribute towards high discharges and increase the risk of flooding, whereas a long period of drought can lead to abnormally low flows. Human actions, for example developing the floodplain, constructing reservoirs, or abstracting water can also have significant effects on river discharge.

1 Flooding

A **river flood** can be defined as 'the inundation of land not normally submerged by water'. In most circumstances it simply involves water flowing over the banks of a river because, for one reason or another, there is too much water for the channel to accommodate.

What causes flooding?

Flooding is encouraged when water flows quickly into a river, either by overland flow or by rapid throughflow (see pages 222–23). The following circumstances promote rapid water transfer:

- a prolonged period of heavy rain whereby the ground has become saturated and additional water is forced to flow overland

- an intense rainfall event, such as a convectional storm, where the ground cannot cope with the sheer volume of water landing on it

- snowmelt, often combined with frozen soil

- steep valley sides

- an impermeable bedrock or thin soils

- a lack of trees or grass to intercept rainfall

- urbanisation, creating impermeable surfaces and rapid water transfer through pipes and sewers

- the silting-up of river channels, often associated with soil erosion in the river basin.

It is important to realise that a good number of these factors are the result of human actions. There is no doubt that, in a number of cases, the problem of flooding has been made worse by people's activities.

River flooding is, of course, under normal circumstances a completely natural process. We should not be surprised to see water spreading out as a sheet across a floodplain (Figure 7.56). However, because we have developed floodplains for intensive farming, communication networks, housing and industry, river flooding represents a hazard that needs to be controlled, or at least monitored.

7.56   *A flooded floodplain*

How can the flood hazard be reduced?

Flood defence involves three main approaches:

1   **Basin management** This aims to reduce the amount of water reaching a river (Figure 7.57).

2   **Channel management** This involves modifying river channels so that they can cope with high discharges (Figure 7.57).

3   **Flood warnings** Throughout the UK, drainage basin instruments can be used, together with information from the Meteorological Office, to predict the likelihood of flooding, and warnings can then be issued. The Environment Agency has three levels of warning:

- *yellow* (flooding is possible) – rivers are running high and flooding is likely, especially of roads and farmland

- *amber* (flooding is likely) – flooding of a number of roads, considerable areas of agricultural land and some high-risk properties is likely

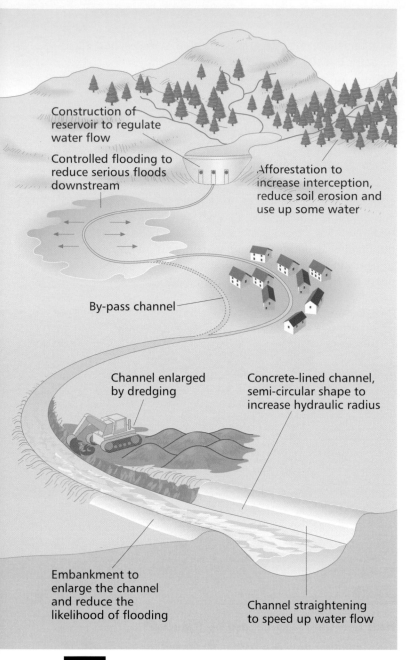

Construction of reservoir to regulate water flow

Controlled flooding to reduce serious floods downstream

Afforestation to increase interception, reduce soil erosion and use up some water

By-pass channel

Channel enlarged by dredging

Concrete-lined channel, semi-circular shape to increase hydraulic radius

Embankment to enlarge the channel and reduce the likelihood of flooding

Channel straightening to speed up water flow

 7.57 *Basin and channel flood prevention measures*

- *red* (serious flooding is likely) – flooding of a significant number of properties, roads and large areas of agricultural land is likely.

The Environment Agency's flood warning system was criticised in the official report on the Easter floods of 1998, when heavy rain triggered flooding across Wales and the Midlands. Large numbers of people who were affected by the flooding failed to receive any warning at all. The Environment Agency is now working with the Meteorological Office to establish a better system of hydrological monitoring and subsequent flood warning. One of the main problems with the Easter floods was that several places affected had no recorded history of flooding.

Despite the many schemes and flood control measures available, it is important to remember two things:

1 Flooding is a natural process. It only becomes a problem if we choose to develop land that is periodically affected by flooding. Perhaps we shouldn't have built there in the first place?

2 Flooding can be – and often is – made more serious by the actions of people, for example in deforesting hillslopes.

NOTING ACTIVITIES

1 Identify *six* factors (not including the actions of people) that encourage flooding. For each one, explain *how* it might encourage flooding.

2 In what ways can the actions of people increase the likelihood of flooding?

3 Study Figure 7.57. Identify the different methods of:

- basin management

- channel management.

For each method, suggest how it can reduce the likelihood of flooding.

4 What are the advantages and disadvantages of a flood warning system?

5 What is responsible for making flooding a 'natural disaster' rather than a 'natural event'?

CASE STUDY

Talgarth, Wales 1998

A prolonged period of heavy rain in early April 1998 led to widespread flooding in Wales and the south Midlands. Many rivers burst their banks and several thousand people were affected. Five people died directly or indirectly as a result of the floods, and damage estimated at £350 million was caused. Many people suffered massive disruption to their lives and livelihoods, and some were affected by ill health and chronic anxiety.

Talgarth, in the north-western foothills of the Black Mountains in Powys, was one of many towns and villages to be affected by the floods, which were triggered by heavy rainstorms. The ground was already saturated before a period of heavy rain and thunderstorms dumped 134 mm of rainfall during a 48-hour period between 8 and 9 April.

The flooding that occurred at Talgarth (Figure 7.58) is common to many towns and villages on the upper reaches of river systems in England and Wales. For this reason, it is an interesting case study to use.

EXTENDED ACTIVITY

The aims of this Activity are:

- to understand the causes of the flood at Talgarth
- to suggest measures aimed at reducing the risk of future flooding.

Study Figure 7.59. Notice that the catchment of the River Ennig, which flows through Talgarth, is quite small. It is roughly circular in shape and is generally steep-sided. There are many small tributaries flowing down steep valleys towards the village.

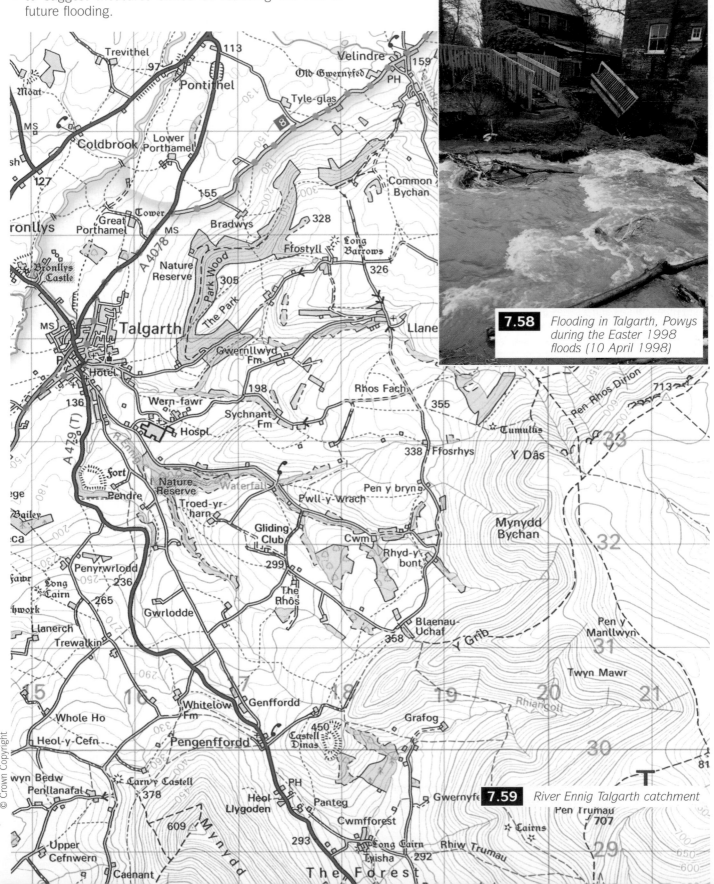

7.58 *Flooding in Talgarth, Powys during the Easter 1998 floods (10 April 1998)*

7.59 *River Ennig Talgarth catchment*

261

Read the information in Figure 7.60, which is an extract from the Environment Agency's report on the Easter floods 1998.

1 What were the effects of the flood on the village?

2 Study Figure 7.59. Suggest how the following drainage basin characteristics might have contributed towards the flooding:

- small circular-shaped drainage basin
- high drainage density (calculate the density)
- steep valley sides.

3 Read the extract in Figure 7.60.

a Describe the features of the river channels just above Talgarth. Discuss how these characteristics might have contributed to the flood.

b What caused water to overflow the channel and flow down the main road?

c Why do you think many local people were terrified by the event?

Talgarth, near Brecon, Powys, River Ennig

Background

Talgarth is a village in the north-western foothills of the Black Mountains in the county of Powys. The village has developed around the Afon Ennig and an unnamed tributary [Figure 7.59].

The catchment to Talgarth is rocky in nature, rural and hilly. The village reaches of the Ennig and its tributary have steep overall gradients with boulder-strewn pool and riffle features and a series of waterfalls. The catchment, which is 18.9 km^2 in area, rises to a level of 730 m above sea level and falls to about 130 m above sea level at Talgarth, an average gradient of 1 in 11. The valley sides above and through the village slope steeply to the watercourses. The absence of extensive floodplain areas is typical of the upper reaches of a river system. No reservoirs or other water systems which could influence flood hydrology are evident in the catchment.

Development, mainly housing, appears to have occurred over decades if not centuries, and there is little modern construction. Some properties are close to or hard against the watercourses.

The village reaches are crossed by nine bridges, four of which were blocked during the flood.

There are no flood defences at Talgarth. Only limited records of flooding of the village exist; however, recently a report of a severe flood which occurred in 1880 has been found.

Flood warning

Talgarth was not thought to be at risk of frequent flooding. The village has not been considered for coverage by the Agency's flood-warning service.

Effective warning is precluded by the rapid response of the Ennig to rainfall on the small and steep catchment to Talgarth. As a consequence, this community must rely on general forecasts of severe weather in the locality – intense convective storms in particular – for their awareness of possible flooding.

Brief description of flooding

Heavy rain over the northern slopes of the Black Mountains on 8 and 9 April resulted in flood flows in the upper reaches of this part of the Wye system.

From eyewitness descriptions it appears that fallen trees and other flood-washed debris carried down by the flood partially blocked the waterway at the road bridge on the upstream side of the village. The resulting water level caused flooding to properties alongside the bridge, and flow down the road running into the village, inundating buildings, mainly houses. Further blockages at another three footbridges within the village resulted in more overspill into the streets and buildings. Overland flow off the steep hillsides added to the discharge down the roads and the flooding in the village.

The event was particularly dangerous, and no doubt frightening for the community, because of the high velocities of flows down the steep village roads and the high flows in the watercourse channels. Also, some properties were flooded to depths in excess of one metre.

First reports that flooding of property was at risk were received by the Environment Agency at 11.30 hrs from Powys County Council Emergency Planning Office, and the first of two flood peaks occurred at about 14.30 hrs, closely followed by another at about 15.00 hrs. The floodwater had subsided about 1 m from its maximum by 16.30 hrs.

A total of 15 properties and 4 cellars were flooded, with many severely damaged. In addition, 7 properties were saved from flooding by sandbagging.

Emergency response

In the difficult and dangerous circumstances of this event, little could have been done to lessen the impact of flooding. However, it is evident that Powys County Council's Emergency Planning and Highways and Direct Services departments, which have a depot in the village, responded quickly and deployed a total of some 30 men with the support of a gang from the Environment Agency who were deployed at 11.30 hrs. The actions taken after the flood to clear obstructions from blocked bridges and otherwise restore the watercourse channels were important in reducing the risk of flooding in the event of further storms. Sandbagging of flood-prone properties was also undertaken, and seven properties were protected from flooding.

Possible courses of action

Flood alleviation schemes may involve construction to provide singly or in combination:

- channel diversion
- trap, deflector or screen devices for intercepting boulders, gravel, timber and trash swept down by the flood flows
- channel works to alter flow characteristics so that flood levels are reduced
- flood walls or embankments.

With or without flood alleviation works, flood risk is lessened by regular maintenance to control tree and bush growth on watercourse banks and to remove gravel accumulations and boulders from critical sections.

7.60 *Extract from 'Easter Floods 1998', a report produced by an Independent Review Team for the Environment Agency*

4 Examine the feasibility of each of the four flood alleviation schemes suggested in Figure 7.60.

  a For each one suggest its hydrological purpose (how it will reduce future flooding).

  b Which scheme, or combination of schemes, do you favour, and why?

  c Can you suggest any alternative schemes?

2 Low flows

During the 1990s a number of rivers suffered from abnormally low discharges or **low flows**. There was a great deal of public concern as some rivers, for example the River Ver in Hertfordshire, dried up completely (Figure 7.61). The public finger of blame was pointed at the water companies who were felt to be abstracting too much water, causing water tables to fall with the resultant drop in river levels.

Abstraction of groundwater is, however, only one possible cause of low flow. The other major cause is climatic, and relates to amounts of precipitation and the subsequent movement of water through the drainage basin. Between 1988 and 1992 there was a period of low rainfall throughout much of England and Wales which undoubtedly had a part to play in the extremely low flows recorded during that period.

7.61 *Low flow*

CASE STUDY

River Wylye, Wiltshire

The River Wylye (Figure 7.62) rises from springs in the Cretaceous Upper Greensand (a type of sandstone) in west Wiltshire. It then flows north and then eastwards mostly over chalk to join the River Nadder at Wilton. The chalk is highly porous and is an important underground freshwater reservoir (**aquifer**).

There are several boreholes along the course of the river operated by Wessex Water. Most are used to abstract water

for domestic consumption, but some are used to support (**augment**) flows in the river itself, in order to avoid any environmental damage associated with the water abstraction. The abstraction rates have increased since the 1970s, although they have remained within their licensed amounts.

Between 1988 and 1991, the River Wylye and its tributaries registered particularly low flows. People became concerned about the state of the fish stocks in the river (it is a well-known trout fishery) and about the quality of the aquatic environment in general. Some local people thought that the problem was being made worse by water abstraction.

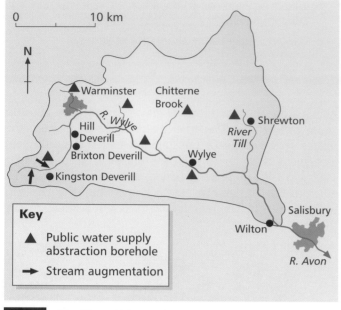

7.62 *The River Wylye drainage basin*

In 1991 a study was commissioned by the Environment Agency to examine the impact of abstraction on the Wylye and its tributaries. The main findings of the study were as follows:

- groundwater abstraction, despite being within its licensed limits, was having a detrimental effect on low flows
- whilst there was some evidence of harmful effects on fish stocks, there may be contributory causes such as algal growth (caused by nutrients from fertilisers being washed into the streams) leading to reductions in oxygen, and drought
- the most seriously affected stretch of the Upper Wylye was between Brixton Deverill and Hill Deverill (Figure 7.62) where abstraction had reduced typical summer flows by 50–60 per cent despite the augmentation upstream
- parts of the Chitterne Brook, a tributary of the River Wylye, ceased to flow at times as a result of abstraction; if abstraction were to increase to its full limits, changes to the stream would be significant.

A further study was commissioned to look at fish stocks and other environmental indicators.

As a result of the study, Wessex Water has agreed to:

- sink two new boreholes to augment flows in the Chitterne Brook and the River Till
- increase river augmentation at Brixton Deverill
- aim for a more constant level of abstraction, avoiding the seasonal (summer) peaks
- monitor the effects of abstraction on river flows and on the quality of the aquatic environment.

In addition, the Environment Agency has asked Wessex Water to stop abstracting water from the Chitterne public water

supply borehole. This would significantly improve summer flows in the Chitterne Brook and the River Till. Flow in the middle and lower parts of the River Wylye would also benefit.

NOTING ACTIVITIES

1 a What are 'low flows'?

  b Why do you think people are concerned about low flows in rivers?

  c Outline the two main causes of low flows.

2 Study the case study of the River Wylye. Write a short report including the following:

- a sketch map of the area locating the places and stretches of river mentioned in the study
- why it was decided to commission a study in 1991
- the main findings of the study
- details of the programme to reduce the problem of low flows.

Try to give reasons for the actions chosen.

River and groundwater pollution

Throughout the world and for many years, people have been polluting rivers and groundwater. These pollutants have caused disease and death in humans and animals, and they have damaged delicate wetland habitats (Figure 7.63).

What are the main causes of water pollution?

There are many sources of water pollution (Figure 7.64), although probably the most significant are associated with sewage disposal, industrial development and agriculture.

- **Sewage** Raw sewage is generally treated at a sewage works before being discharged into a river. It is, however, not unknown for raw sewage to find its way into a river, particularly during times of flood. Even treated sewage will contain some chemicals that can harm aquatic ecosystems.
- **Industrial waste** Industrial waste accounts for nearly 40 per cent of all pollution incidents. The worst culprits are industries involved with metal smelting, textiles, food processing and pulp and paper. Abandoned mines can lead to serious pollution, as has happened in the case of the old tin mines in Cornwall. Power stations discharge warm water which can harm ecosystems and, of course, nuclear power stations can discharge more sinister pollutants.

7.63   *Water pollution*

• **Agriculture** In recent years increasing amounts of nitrogen fertilisers have been applied to the land to increase yields. However, not all of the nitrogen is taken up by the plants. Some is leached into the soil and will eventually find its way into rivers and aquifers. High nitrate concentrations in water have been linked to health risks, particularly in young children. Chemicals in the water, particularly phosphates from sewage works and nitrates from fertilisers, promote the growth of algae in rivers. This process is called **eutrophication**. The algal growth depletes the oxygen in the water, causing fish and other forms of water-life to die.

NOTING ACTIVITIES

1   Study Figure 7.64. Draw up a list of the various sources of pollution, and try to identify those that affect:

   • mainly rivers

   • mainly groundwater

   • rivers and groundwater equally.

2   In what ways can industry cause water pollution?

3   Use a simple flow diagram to describe the process of eutrophication.

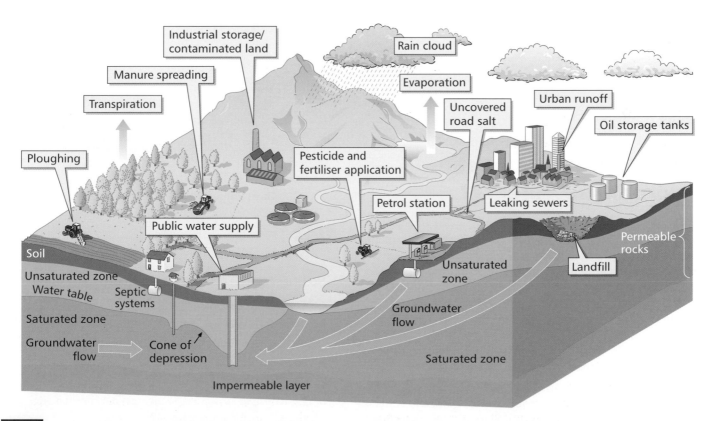

7.64   *Potential pollution sources to rivers and groundwater*

STRUCTURED QUESTION

Study Figure 7.65, which shows variations in nitrate concentration during a storm.

a   At what time does the discharge peak occur? *(1)*

b   What happens to the nitrate concentration at the time of the peak discharge? *(1)*

c   Suggest *one* reason for your answer to **(b)**. *(2)*

d   What happens to the nitrate concentration immediately after the discharge peak? *(1)*

e   Outline *one* possible reason for the change in nitrate concentration. *(2)*

f   From midnight on 27 March 1980, the nitrate concentration line broadly mirrors the trend of the throughflow line. What does this suggest about the transfer of nitrates into the river? *(2)*

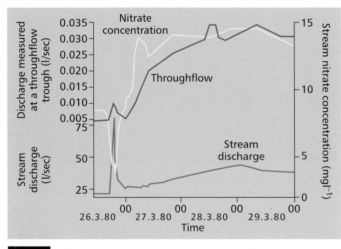

 7.65   *Variations in nitrate concentrations during a storm*

What is the role of the Environment Agency?

The Environment Agency is responsible for monitoring pollution in England and Wales. It regularly publishes maps describing the quality of water in rivers and it sets targets for future improvements. Discharges into rivers are monitored and fines are imposed if pollution incidents occur. In 1990, Shell UK were fined £1 million for spilling oil into the Mersey, and in 1991 the South West Water Authority was ordered to pay a fine of £10 000 with costs estimated at £1 million for accidentally allowing aluminium sulphate to pollute the River Camel in Cornwall.

In the mid-1990s maps were produced to show **groundwater vulnerability**. Aquifers were graded according to their potential vulnerability to pollution. The assessment of vulnerability was based on:

- the physical and chemical nature of the bedrock (its permeability, for example)
- the nature and thickness of overlying soil
- the presence and thickness of glacial drift (an impermeable layer)
- the depth of the unsaturated zone (the zone of rock above the water table).

Figure 7.66 contrasts soil/rock profiles with a high and a low groundwater pollution vulnerability. Notice, for example, that a thin, well-drained sandy soil will have a higher vulnerability than a thick impermeable clay, as water and pollutants will be transmitted much more rapidly.

In 1998 the Environment Agency identified 68 'Nitrate Vulnerable Zones'. Farmers in these areas are subject to strict regulations regarding their use and management of fertilisers and manure. Grants are available to help them improve their farm waste storage facilities.

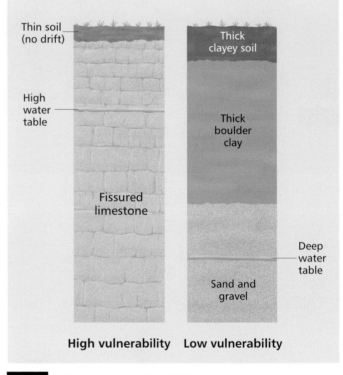

7.66   *Groundwater vulnerability*

NOTING ACTIVITIES

1   Study Figure 7.66 . Make a copy of each profile and add labels to explain the contrasts in vulnerability.

2   How can a groundwater vulnerability map be a useful tool in the decision-making process when, for example, it comes to siting a new landfill site (used for dumping refuse)?

EXTENDED ACTIVITY

The aim of this activity is to make a study of groundwater pollution in the River Thames region. Figure 7.67 shows the major aquifers and the distribution of public supply boreholes. Groundwater is a major contributor to the water supply of the region. Figure 7.68 is an extract from the Thames Region's 'Policy and practice for the protection of groundwater'.

1 Study Figure 7.67. Describe and account for the distribution of the public supply boreholes in as much detail as you can. The following questions might help you:

- Of the three aquifers listed in the key, which would appear to be the most important, and why?

- Do the boreholes appear to be particularly clustered around the major towns and cities? Try to explain your answer.
- Which river valley has a particularly high concentration of boreholes?

2 Read through the information in Figure 7.68.

- What have been the results of over-abstraction in the past?
- What are the main pollution concerns for the future, and how does the Environment Agency intend to address these issues?

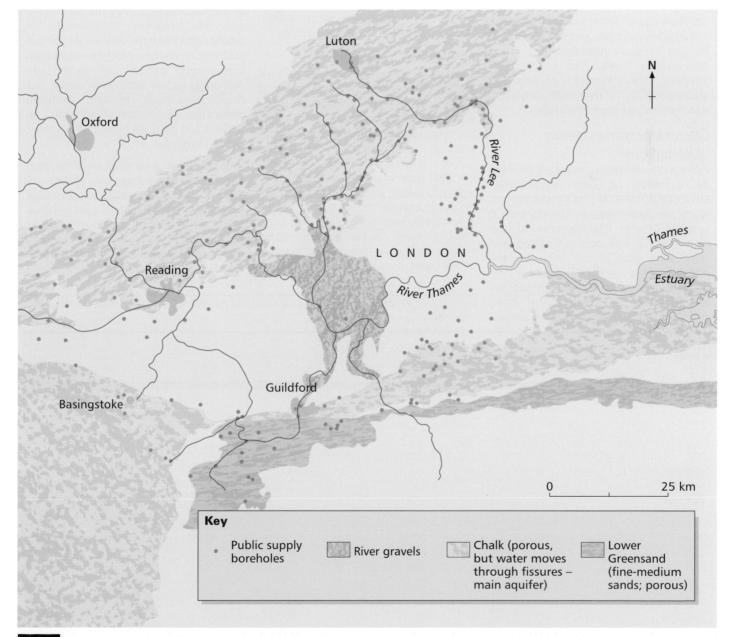

Key

- Public supply boreholes
- River gravels
- Chalk (porous, but water moves through fissures – main aquifer)
- Lower Greensand (fine-medium sands; porous)

7.67 *Public supply boreholes in part of the Thames Region*

General situation

In much of the catchment a situation has been reached where there is no remaining capacity for abstraction because of the need to protect stream flows and the valley environment. In some areas the resource has experienced historic over-abstraction leading to reduced flows and drying-up of some groundwater-fed rivers, particularly on the chalk aquifer. This also has consequences for groundwater quality since potentially polluting discharges made to such rivers would tend to soak into aquifers. Abstraction in proximity to the Thames estuary has resulted in the ingress of saline waters several kilometres inland.

A notable exception to the above trend is the chalk aquifer in the London Basin. The considerable reduction in abstractions over the last 25 years has resulted in rising groundwater levels. This is likely to increase the mineral content of groundwater as it rises into previously dewatered strata above the chalk.

Control of groundwater abstractions

Flows in several rivers have been depleted as a result of large groundwater abstractions close to the headwaters or along the river valley. In consequence a remedial scheme, 'The Alleviation of Low Flows' (ALF), has been instigated and engineering solutions are being contemplated. These include reduction of groundwater abstraction to allow baseflow recovery and the possibility of lining riverbeds with low-permeability materials.

Any proposals to increase abstraction, particularly in these river valleys, will warrant careful study.

Where groundwater has been affected by saline intrusion along the River Thames there will be limited scope for resource development in order to avoid exacerbating the existing situation.

Waste disposal to land

Most sites which have been considered suitable in the past for landfill are quarries located on aquifers, such as sand and gravel quarries overlying the chalk aquifer. However, in many of these areas, such as South Hertfordshire, groundwater is used extensively for public supply. In such circumstances there will be particularly strict limitations on landfilling activities.

Contaminated land

There is continued pressure for redevelopment of former industrial sites, many of which occupy prime locations in urban areas. The land is frequently contaminated and there is often associated groundwater pollution, with possibly considerable pollution potential remaining. These sites are often close to groundwater sources of supply. Extensive works may be necessary to decontaminate ground and remediate groundwater. Requirements are likely to be more stringent on the more important aquifers such as the chalk than for example on the Lower Thames gravel.

Diffuse pollution of groundwater

As an exercise separate from the protection policy, Nitrate Sensitive Areas have already been established for some public supply sources. In these areas farmers have been encouraged to join a scheme to change farming practice and limit the amount of nitrate leached. Rising nitrate concentrations are evident in other parts of the catchment. Consideration will be given to the establishment of further sensitive areas where concentrations are already unacceptably high or where rates of increase will lead to excessive levels.

Other chemicals, such as pesticides, are in widespread usage across the catchment, and the frequency of detection in groundwater has risen. Thames Region will continue to discuss pesticide application with relevant parties, such as Highway Authorities and the farming community. The Region will seek to limit pesticide application within sensitive areas on aquifers and pesticide types to those least harmful to groundwater.

Groundwater in some urban areas has been contaminated by leakage from sewers and through widespread usage of chemicals such as solvents. Thames Region will seek to reduce incidences of contamination by liaising with relevant parties and has instigated a programme of site visits aimed at pollution prevention.

Source: Environment Agency: from NRA, 'Policy and practice for the protection of groundwater', Regional Appendix, Thames Region

7.68 *Groundwater problems and issues in the Thames Region*

A The growth of cities

Look at Figure 8.1. What are your reactions to the picture? Is it an attractive environment? What types of people live here? What types of activities take place here? Your answers to these questions may be different from those of other people. They may depend on where you live at the moment, or on an experience earlier in your life. In any case, what you make of this picture will reflect in some way your attitude towards cities, and urban life.

The geographer John Short has argued that our perceptions of the city are based on a series of 'myths'. In Western cultures, there is a series of anti-urban myths:

- The city is unnatural and is compared unfavourably with the countryside and the wilderness.
- The city is anonymous and is contrasted with the warm community found in villages and small towns.
- The city is a place of sin, disease and moral corruption.
- The city is a threat to social order. It is a place of crime and fear.

On the other hand, there is also a series of pro-urban myths:

- The city is the height of human civilisation, compared with the backwardness of rural areas.
- Whereas some people see the anonymity of the city as a source of concern, others celebrate the freedom this offers for people to escape social constraints.
- Cities are sites of freedom. Large cities such as London and New York are truly global and are able to incorporate a range of influences from a wide range of cultures. They are less bound by tradition.
- Cities are sites of radical change in society. They are more tolerant of differences between people, and are places where people can join together to create a better world.

In 1800, only 3 per cent of the world's population lived in towns with a population of over 5000. The population of the world was overwhelmingly rural, making a living from the land. There were only a handful of cities with a population greater than half a million.

8.1 *Urban environment: part of London*

NOTING ACTIVITIES

With other people in your group, look at a range of representations of cities in films, books, television programmes and music.

1 How are cities represented? Find examples of positive and negative representations.

2 Which view of urban life best reflects your own? Where did your views come from?

3 Are there differences in the views of different members in your group?

The next 100 years saw significant changes. In 1900, 10 per cent of the population lived in towns, and there were 20 cities with more than half a million people. By 1990, 40 per cent of the population lived in towns, and there were almost 600 cities of more than half a million people. These figures indicate that **urbanisation** (the trend for an increasing proportion of the population to live in urban areas) is one of the most significant processes affecting societies at the turn of the 20th/21st centuries.

As in the case of 'rural', defining exactly what is 'urban' is a difficult task. This is because many of the functions that we find in urban areas (such as shops) are also found in rural areas. Similarly, attempts to define urban according to the population size are also difficult. For example, in Scandinavia, settlements as small as 300 people might be classed as urban, whereas in Japan a settlement must have more than 30 000 people to be classed as urban. Definitions of 'urban' tend to rely on apparent differences from 'rural' areas. These include:

- a high density of population
- the existence of an extensive area of built-up or developed land
- the lack of open space
- the dominance of manufacturing or service functions over agricultural functions.

Urbanisation refers to the process whereby an increasing proportion of the total population of regions and nations comes to live in places that are defined as urban. Often this proportion is referred to as the **level of urbanisation**. **Urban growth** refers to an absolute increase in the physical size and total population of urban areas.

In countries that urbanised many decades ago, the trend is towards lower rates of urban growth. However, countries in the economically developing world are experiencing rapid urbanisation.

Look at Figure 8.2. It shows that in the early 1920s, 24 of the world's cities had a population of over 1 million. By the early 1990s the number of **million cities** was 198. The average size of these million cities had increased to over 2.5 million inhabitants. In the early 1920s, 1 in 50 of the world's people lived in these large settlements. In the early 1990s this figure was 1 in 10. Perhaps most significant of all, the mean latitude of million cities has moved steadily towards the Equator. This means that million cities are increasingly associated with economically less developed countries.

NOTING ACTIVITIES

1 Define the following terms:

- urban
- urbanisation
- urban growth.

2 Using the data in Figure 8.2, summarise the main features of global urbanisation from the 1920s to the present.

3 Suggest reasons why data on levels of urbanisation might be unreliable.

STRUCTURED QUESTION 1

a Study Figure 8.3. Using an atlas to help you, name the cities with a population of around 20 million in 2000. *(4)*

b What do you notice about the distribution of these cities? *(3)*

c Write a sentence summarising the level of urbanisation found in each of the following regions:
- North America
- Central and South America
- Africa
- South and South-east Asia
- Europe. *(10)*

d Suggest reasons for the different levels of urbanisation found in different regions. *(5)*

	Number of million cities	Mean latitude north or south of the Equator	Mean population (millions)	Percentage of world population living in million cities
1920s	24	44° 30′	2.14	2.86
1940s	41	39° 20′	2.25	4.00
1960s	113	35° 44′	2.39	8.71
1980s	198	34° 07′	2.58	11.36

 8.2 *World million cities since the 1920s*

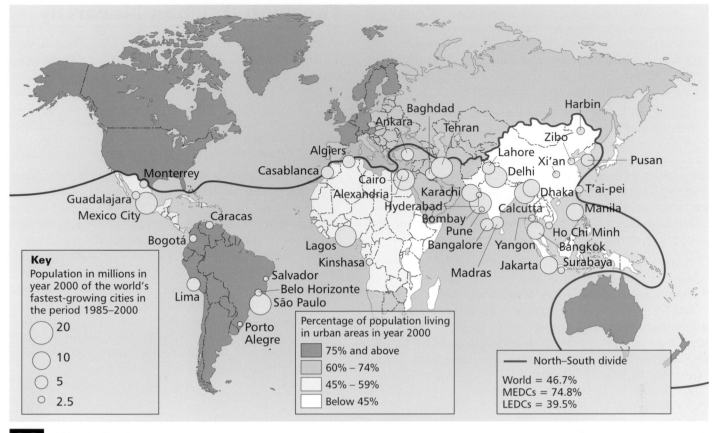

8.3   *Urbanisation today*

Key
Population in millions in year 2000 of the world's fastest-growing cities in the period 1985–2000

○ 20
○ 10
○ 5
○ 2.5

Percentage of population living in urban areas in year 2000

■ 75% and above
▨ 60% – 74%
□ 45% – 59%
□ Below 45%

— North–South divide

World = 46.7%
MEDCs = 74.8%
LEDCs = 39.5%

The growth of megacities

In many parts of the developing world, cities are growing so fast that they are merging together to form larger urban regions called **megacities**. Examples of these regions include Mexico City, São Paulo, Lagos and Cairo. Large-scale migration is fuelling this growth. In most countries, the destination for most migrants is the major cities, since these provide the greatest opportunities. This process has been called **mega-urbanisation**. Large cites grow rapidly and expand along lines of communication to envelop villages. As a result, the boundaries between urban and rural areas become blurred.

Why are cities growing?

Urban population growth results from two sets of processes:

1 **Natural population increase** Put simply, this is where the number of babies born in a given year exceeds the number of people who die.

2 **In-migration** Urban population growth occurs as a result of a net migration from rural areas.

It is difficult to make general statements about whether in-migration or natural increase is more important in explaining the rapid growth of cities. It is better to consider each city in its own right. The two are related, since the people who migrate to the city from the countryside tend to be younger.

This leads to an increase in the proportion of the total population of the city, that is in the more fertile age groups and therefore more likely to have children. This leads to an increase in the number of children born.

What prompts people to leave rural areas and move to cities?

Perhaps the most important reason why people leave rural areas is because of widespread poverty and unemployment in certain rural areas (Figure 8.4). Wage levels tend to be higher in urban areas, and social and health facilities are better in large towns and cities. The large number of people

8.4   *Rural poverty in the developing world: Tarabuco, Bolivia*

in towns and cities means that there is a wide range of opportunities for making a living there. In most developing countries, governments have focused investment in factories, roads, infrastructure and other facilities on urban areas. This means that rural dwellers are likely to perceive urban areas as offering a better quality of life.

Where regular formal work is not available, people must find other ways of making a living. This might involve selling matches or shoelaces on the street, shoe shining, craft work, collecting bottles for recycling, or rubbish collecting. These ways of making a living might be supplemented by begging, petty crime or prostitution. Such types of employment add up to an important part of the economy of many large cities in the developing world. This part of the economy is known as the **informal sector**.

This section of the book should give you some sense of the scale and importance of contemporary urbanisation. The figures suggest that the rapid growth in urban living is occurring in regions of the world where social and economic conditions are poorest and where industrial production is low. As a result, enormous pressure is being placed on the children, women and men who live in these places. These issues are discussed in Chapter 8D.

The global urban hierarchy

A global hierarchy of cities can be identified, ranging from the largest and most influential to the smallest. **World cities** are those cities that have become the command centres for the flow of information, finance and cultural products that drive the global economy. In these cities are the headquarters of transnational corporations, of international banking and finance, government offices, and the most powerful and internationally influential media organisations.

Figure 8.5 shows the location of these influential world cities. The most powerful of these are London, New York and Tokyo, because of their major role in the world financial markets, their transnational corporation headquarters, and their concentrations of financial services (stock exchanges, for example). Below these three are cities that have a strong influence over large regions of the world, and these include Brussels, Frankfurt, Los Angeles, Singapore, Paris and Zurich. A third tier consists of important international cities with more limited or specialised international functions. Cities are often keen to increase their position in the world city hierarchy, and to promote themselves as centres of importance (Figure 8.6).

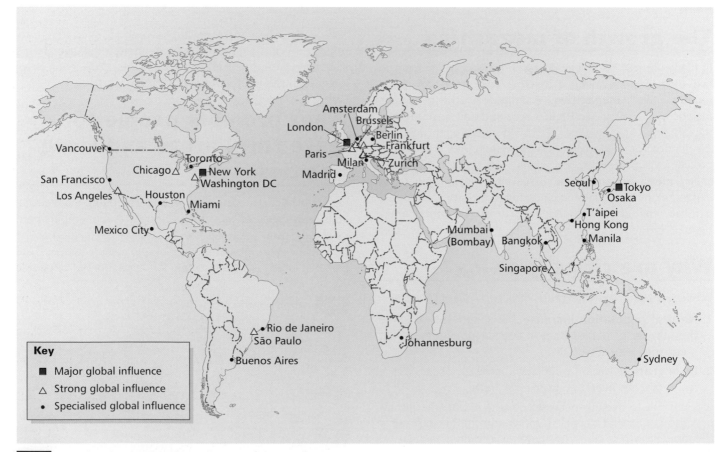

8.5 *World cities: a global hierarchy*

8.6   *Promoting a 'world city': Sydney*

Rank–size distributions

In studying urban systems within specific countries, geographers are interested in the relationship between cities of different sizes. The **rank–size rule** suggests that there is a statistical regularity in the city-size distributions of countries. The rule states that the population of the nth largest city in a country is $1/n$th of the population of the largest city in that country or region. For instance, if the largest city in a country has a population of one million, the 4th largest city should have a population one-quarter as big

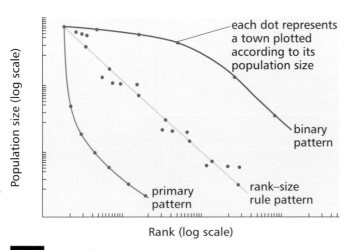

8.7   *Rank–size distribution*

(that is, 250 000), the 50th largest city should have a population one-fiftieth as big (that is, 20 000), and so on. This is called a **standard** or **normal hierarchy** (Figure 8.7).

In some urban systems, this rank–size distribution is distorted due to the disproportionate size of the largest city. In Britain, for example, London is more than nine times the size of Birmingham, the second largest city. In France, Paris is more than eight times the size of Marseilles. This condition is known as **primacy**, where the population of the largest city in an urban system is disproportionately large in relation to the second and third largest cities in that system. Cities such as London and Paris are termed **primate cities** (Figure 8.7). **Primate** distributions are those in which the urban hierarchy is dominated by one large city.

Rank–size distributions

1 Britain

In 1800, Britain was still essentially an agricultural society. Three-quarters of the population lived in places with less than 2500 people. There were some regional centres such as Norwich and Bristol, but the hierarchy was dominated by London, which played a key role in the global trading system.

The dominance of London has continued. Its population is bigger than the combined populations of the next 15 biggest cities in Britain. It is the seat of government (despite a degree of devolution of political power to Scotland and Wales), the centre for the English legal system, and the centre for banking and insurance, publishing, fashion and advertising. However, this picture of the dominance of London needs to be moderated. The view from Manchester or Cardiff is very different. In fact Britain is a country marked by intense regional rivalry. This can be seen in the way sporting events such as football and rugby generate fierce loyalties, and in the cultural distinctions between the 'north' and 'south' of Britain. The case of Britain illustrates a primate urban hierarchy in a developed country.

2 USA

The USA is an example of a **normal** rank–size distribution because no single city dominates. There is a more even spread of population. The USA is a highly urbanised country, with over 70 per cent of its population living in cities of more than 50 000. There are more than 20 cities of over 1 million. No single city dominates. Although New York is the biggest city, it is not the centre of political power, which is located in Washington D.C. Culturally New York and Los Angeles compete for dominance. A number of factors explain this:

- The country is so large that no single city can dominate. There are significant contrasts between east and west, north and south. This means that there are large and influential cities in all parts of the USA.

- The political system provides individual states with high levels of autonomy, so they can promote their own economic growth. This means that each state has its own capital city which plays an important role in economic and political life.

- Over time the economic fortunes of different regions have changed, leading to high levels of internal and external migration, which fuel population growth. For instance, Los Angeles has emerged as a major urban centre, as has Dallas in the south.

Centrality

The idea of urban primacy relates to the dominance of the largest city in terms of population size. However, size is not everything, and some cities have an economic, political and cultural importance far greater than their size would suggest. This is known as **centrality**, which refers to the functional dominance of cities within an urban system. For example,

São Paulo has only 10 per cent of the Brazilian population, but it produces about a quarter of the country's GDP and accounts for over 40 per cent of its manufacturing industry (Figure 8.8).

NOTING ACTIVITIES

1 What do you understand by the term 'world city'?

2 Choose *two* world cities. State which tier of the global hierarchy they occupy, and show how they have global importance.

3 Suggest ways in which cities might seek to increase their importance as world cities.

4 Distinguish between 'primacy' and 'centrality'.

5 Choose *one* country. State whether it displays a primate or standard distribution. Account for its distribution.

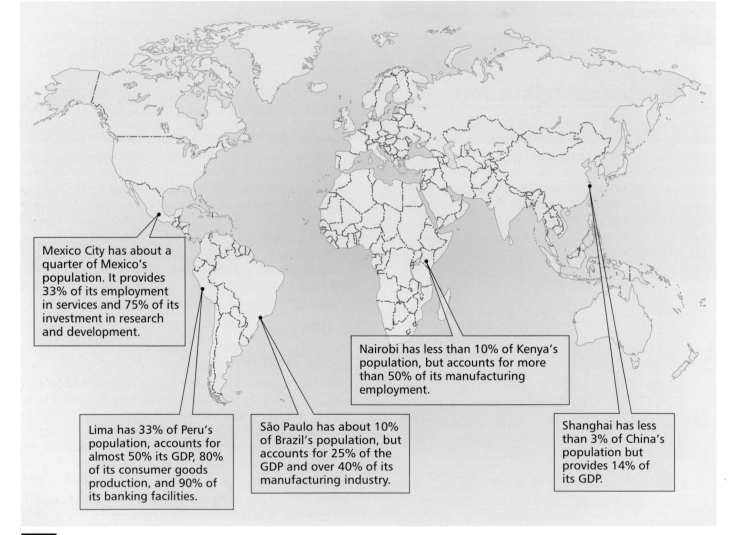

Mexico City has about a quarter of Mexico's population. It provides 33% of its employment in services and 75% of its investment in research and development.

Lima has 33% of Peru's population, accounts for almost 50% its GDP, 80% of its consumer goods production, and 90% of its banking facilities.

São Paulo has about 10% of Brazil's population, but accounts for 25% of the GDP and over 40% of its manufacturing industry.

Nairobi has less than 10% of Kenya's population, but accounts for more than 50% of its manufacturing employment.

Shanghai has less than 3% of China's population but provides 14% of its GDP.

8.8 *Urban centrality: some examples*

EXTENDED ACTIVITY

Look at Figure 8.9. It shows the population of the 10 largest cities in Brazil and the USA.

1 Calculate the expected population of each city according to the rank–size rule formula.

2 On double logarithmic graph paper, plot the actual population sizes. On the x axis plot the rank of the city, on the y axis plot the population size. Compare the curves you have drawn with the ideal typical distributions shown in Figure 8.7.

3 Comment on the difference between the distributions for Brazil and the USA. What might explain the differences you have described?

	Brazil	Population (millions)	USA	Population (millions)
1	São Paulo	16 832	New York	18 087
2	Rio de Janeiro	11 141	Los Angeles	14 532
3	Belo Horizonte	3 446	Chicago	8 066
4	Recife	2 945	San Francisco	6 253
5	Porto Alegre	2 924	Philadelphia	5 899
6	Salvador	2 362	Detroit	4 665
7	Fortaleza	2 169	Boston	4 172
8	Curitiba	1 926	Washington	3 924
9	Brasilia	1 557	Dallas	3 885
10	Nova Iguaçu	1 325	Houston	3 711

8.9   *The ten largest cities in Brazil and the USA*

STRUCTURED QUESTION 2

Study Figure 8.10, which shows data on urban populations in Africa.

a Which two African cities have a population of more than 7 million people? (2)

b Describe the distribution of countries with an urban population of over 40 per cent. (4)

c Identify two factors that lead to urban population growth. (2)

d Suggest reasons for the relatively high levels of urbanisation in (i) North African countries (ii) South Africa. (4)

e Suggest reasons for the low levels of urbanisation in much of southern Africa. (3)

f Suggest some of the problems that might be associated with a rapid rate of urban growth in African cities. (4)

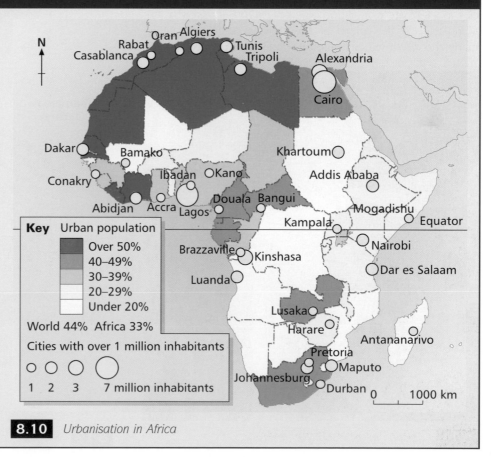

8.10   *Urbanisation in Africa*

B Land use patterns in cities

Think of a town or city with which you are familiar. Now think about its different areas. What happens in the centre of the town? How is it different from the areas further out from the centre? Are there 'old' and 'new' parts of the town, or areas of 'better' housing? Where are the concentrations of shops and leisure facilities?

We all carry around 'mental maps' of towns and cities. These are rough guides that help us to make sense of the complexity of the places in which we live (80 per cent of Britons live in towns and cities). In their study of urban areas, geographers have devised models which simplify and summarise what they think are the important features of towns and cities. These models have usually been produced through the study of actual cities, and the models have been proposed as being applicable to other cities. This chapter is an introduction to some of the main models of urban structure.

Concentric rings (Burgess)

The most famous model was developed by the sociologist E. W. Burgess, who worked in Chicago in the early part of the 20th century (Figure 8.11). When Chicago formally became a city in 1837, its population was only 4200. Its growth was stimulated by the arrival of the railways (railroads), which made the city a major transportation hub. The city became

an important centre for steelmaking, for lumber distribution and for meat processing and packing. The city's growth attracted large numbers of immigrants from both the USA and Europe, and its neighbourhoods were unique at the time for being highly segregated, with distinct cultural and social characteristics (Figure 8.12). The 1880 and 1890 censuses

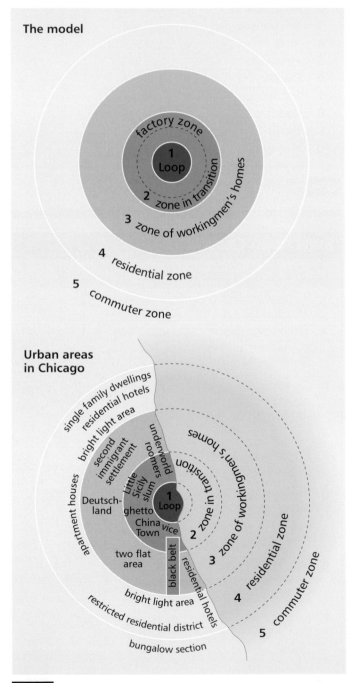

The model

factory zone
1 Loop
2 zone in transition
3 zone of workingmen's homes
4 residential zone
5 commuter zone

Urban areas in Chicago

single family dwellings
residential hotels
bright light area
second immigrant settlement
apartment houses
Deutsch-land
Little Sicily
slum
ghetto
China Town
two flat area
vice
black belt
bright light area
restricted residential district
bungalow section
residential hotels
underworld
roomers
1 Loop
2 zone in transition
3 zone of workingmen's homes
4 residential zone
5 commuter zone

8.11 *Chicago in the early 20th century*

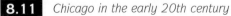

8.12 *Burgess's concentric model of urban structure*

showed that more than three-quarters of Chicago's population was made up of foreign-born immigrants.

These factors made Chicago an important place to study, and the Chicago School of Sociologists produced many studies of the city that are still influential today. Burgess studied Chicago at a time when it was experiencing massive in-migration. Most new arrivals to the city had little money and limited resources. They moved first to the inner city, where the cheapest housing could be found. As they gained a firmer foothold in the economy, they moved further out to the more expensive housing. The result of these processes of invasion and succession was a series of zones. The terms 'invasion' and 'succession' are significant because they are associated with the study of ecosystems. The Chicago School of Sociology saw human behaviour as dominated by ecological principles. Thus they believed that the most powerful groups (species) would obtain the most advantageous position in a given space through competition.

Burgess's model has been adopted by many as a model of a 'typical' city. The Burgess model represents the city as a dynamic, living system. Its growth was fuelled by the arrival of new in-migrants to the city and the desire of individuals to move 'onwards and upwards' away from the centre to better-quality housing at the edge of the city. The following passage summarises the model.

'According to this theory, the typical city could be conceptualised as consisting of five main concentric zones, the innermost of which was described as the non-residential central business district which was then circled by a "zone of transition", where factories and poorer residences intermingled, and finally by three residential zones of increasing affluence and social status. New immigrants, it was postulated, would move into the cheapest residential areas of the city and then, as they became economically established, migrate outwards. This would be a continuous process, so that the "zone in transition" would (as the name implies) have a high mobility rate.'

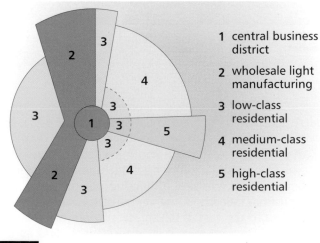

8.13   *Hoyt's sector model of urban land use*

1   central business district

2   wholesale light manufacturing

3   low-class residential

4   medium-class residential

5   high-class residential

Sector model (Hoyt)

A second model is that of the urban economist Homer Hoyt (1939). Hoyt based his model (Figure 8.13) on an analysis of land values in 25 US cities. He rejected the idea of concentric zones. Instead he suggested that residential areas took the form of a series of sectors. He identified sectors of high-status housing that developed along main transport routes radiating out from the central area of the city. These high-status areas tended to locate themselves away from areas of heavy polluting industry, and often developed on high ground, where there was less risk of flooding, and where there were panoramic views of the city. By contrast, Hoyt identified sectors of industrial development growing out from the centre along transport routes such as railways and rivers.

Multiple-nuclei model (Harris and Ullmann)

A third model of urban land use is Harris and Ullmann's multiple-nuclei model (Figure 8.14). This was based on the idea that once cities reach a certain size, the traditional downtown or CBD is no longer sufficient to serve the commercial needs of the whole city. As a result, additional nodes of shops and offices emerge in outlying districts. Once this occurs, these nodes continue to grow through processes of attraction. For

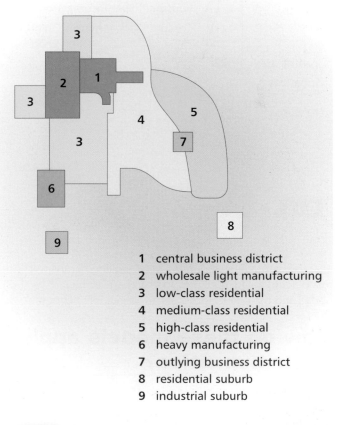

1   central business district

2   wholesale light manufacturing

3   low-class residential

4   medium-class residential

5   high-class residential

6   heavy manufacturing

7   outlying business district

8   residential suburb

9   industrial suburb

8.14   *Harris and Ullman's multiple-nuclei theory of urban structure*

Bid-rent theory

These models all share the common idea that land use in cities results from economic forces. This is based on the idea that most urban land users want to maximise the profit they gain from a particular location. The most accessible parts of cities are assumed to be the most profitable. For example:

- businesses and shops want to be accessible to one another, to markets, and to workers
- private residents want to be accessible to jobs, amenities and friends
- public institutions need to be accessible to clients.

Different land users compete for accessible sites near the city centre. The amount they are prepared to pay is called **bid-rent** (Figure 8.15).

However, some users are prepared to pay more for central sites because their utility is greater. For example, large retailers will be prepared to pay high prices for land at the city centre because this will maximise the number of potential customers they can attract (Figure 8.16). Further away from the centre, they will be willing to pay much less. As a result of this competition for land, a concentric pattern of land use develops.

According to this logic it might be expected that the poorest households will end up occupying the periphery of the city. However, in Western capitalist cities this is not generally the case. The reason for this is that the wealthier households trade-off the convenience of accessibility for the benefits of being able to consume larger amounts of space. Lower-income households trade-off living space for accessibility to places of employment, and thus find housing near to the city centre.

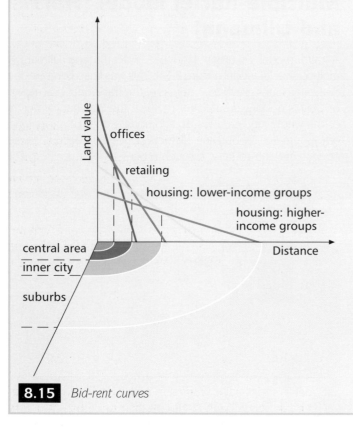

8.15 *Bid-rent curves*

8.16 *A large retail outlet in a British CBD*

example, retailing requires warehousing, transport and manufacturing. In addition, workers are attracted to these nodes and the outcome is a series of multiple centres.

How do these models apply to British cities?

A number of writers have tested the applicability of these models to British cities. For instance, Mann (1965) developed a model of the British city that drew upon Burgess's and Hoyt's models by combining the concentric and sector patterns of residential areas (Figure 8.17). His model recognised that while cities might develop outwards in patterns of concentric rings, income differences are often sectoral. Mann also built into his model the action of local authorities in the provision of housing. He suggested that pollution from industry in the centre of the city would be transported to the east of British cities by prevailing winds. This created an east–west split in residential class as the higher-income groups chose to live in the west.

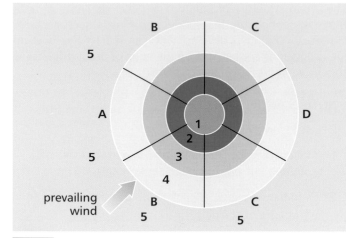

1 central business district
2 transitional zone
3 zone of small terrace houses in sectors C, D;
   larger local authority housing in sector B;
   large old houses in sector A
4 post-1918 residential areas, with post-1945
   development mainly on the periphery
5 commuting distance 'dormitory' towns
A Middle-class sector
B Lower middle-class sector
C Working-class sector
   (and main council estates)
D Industry and lowest working-class sector

8.17 *Mann's model of the structure of a hypothetical British city*

Robson (1965) produced a model of the patterns of residential segregation in Sunderland (Figure 8.18). This is how he describes the patterns he found:

- The central business district lies to the south of the river, with the industrial areas following both banks of the river and extending south along the coast to the south of the river mouth.

- The residential pattern in the north has some elements of the sectoral developments suggested by Hoyt, but in the main the northern area is in the form of a series of concentric rings of the Burgess type, progressing outwards from a poor, subdivided zone adjacent to the industrial area flanking the river, through a medium-rated area extending to the east–west railway line, and to a higher-rated zone north of this and running to the boundary of the town.

- To the south of the river, the residential areas have developed a sectoral pattern with four principal sectors: a low-class sector in the east which is highly subdivided at its northern apex; a highly rated sector next to it which reaches out from the inner areas in an expanding area to the southern boundary; a middle-class sector running out to the west; and finally, a second low-rated sector flanking the industrial areas of the river.

- A rooming-house area has developed at the townward apexes of both of the highly rated sectors, one to the north of the river, the other, more extensive, to the south.

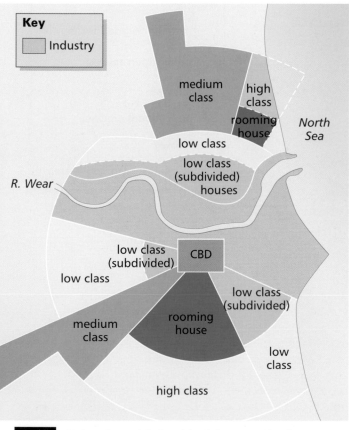

8.18 *Robson's model of residential segregation in Sunderland*

CASE STUDY

Bradford

Bradford had its origins as a small village which expanded rapidly during the Industrial Revolution. It was an important centre for the textiles industry and the development of engineering. Richardson (1976) studied the urban structure of the city. Figure 8.19 shows a number of features:

- Central business district – this was the site of the original settlement and consists of three distinct sectors: the wool warehousing and office area; the banking and shopping area; and the educational and entertainment area.

- A ring of Victorian and Edwardian terraced housing around the CBD – this grew to encompass nearby villages such as Manningham and Little Horton. This is an area of high-density back-to-back housing built prior to 1880. This zone in transition had become run-down by 1945 and afterwards became dominated by small industrial units. Bradford became a centre of immigration from the

New Commonwealth and Pakistan from the late 1950s, and this area attracted large numbers of immigrants.

- Outer areas of council and private housing – the council estates were built on the edge of town following the clearance of much of the dilapidated housing in the inner area after the First World War. The growth of the city was facilitated by the extension of public transport and the growth of car ownership. Greater amounts of open space enabled housing to be built at lower densities.

- A commuter zone – made up of outlying villages such as Allerton, Cottingley, Eccleshill, Woodhall and Wisbey.

- In addition to these broad concentric zones of housing type are sectors of industry that developed along the streams towards Frizinghall, Thornton, Clayton and Bowling. These can be identified on Figure 8.19. They include: a western sector on either side of Clayton Beck; a zone of industry running south along Bowling Beck; and an eastern sector running along the railway to Leeds.

The urban structure of Bradford displays elements of both Burgess's concentric ring model and Hoyt's sector model. However, continued development has led to the built-up areas engulfing settlements on the edge of Bradford, leading to a more multiple-nuclei structure.

NOTING ACTIVITIES

Study Figure 8.19.

1 Make a copy of the map. Label the main areas of commercial, residential and industrial land uses.

2 Summarise the development of the urban structure of the city.

3 To what extent does the model appear to fit the models of urban structure discussed in this chapter?

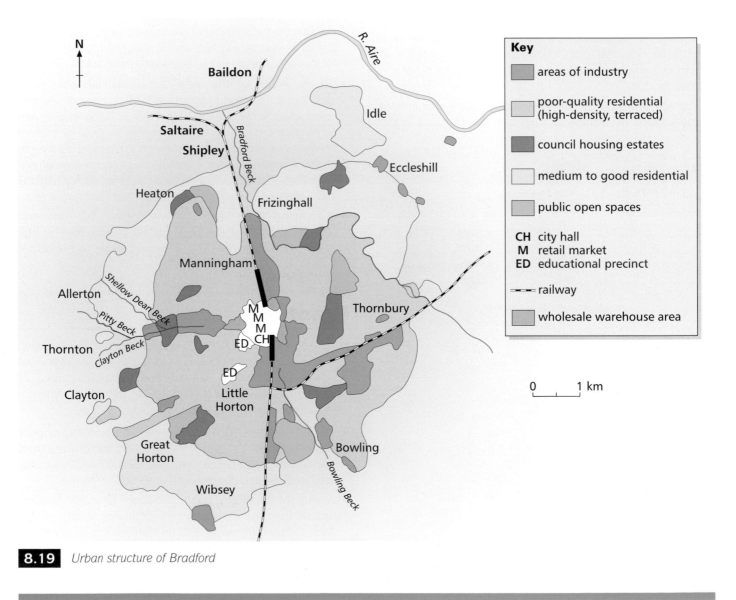

8.19 *Urban structure of Bradford*

Geographers studying urban areas have been interested to see how far these models are replicated in other cities. Much effort has gone into identifying 'zones' and 'sectors' in cities around the world. Much of this work is now criticised. The models were the product of particular economic and social conditions that rarely apply in other places. All the models discussed so far were based on a society where there was a private housing market and free economy where decisions about land use were based on the need to make profits (see *box* below on how urban structure develops in socialist societies). As cities have changed, so geographers have sought to find new models to describe and explain these changes.

Urban models in socialist cities

Many of the cities of Eastern Europe were greatly affected by socialist planning. Urban planning in socialist states reflected communist thinking (Figure 8.20). For instance, official policies sought to eliminate an excessive concentration of people in large cities in order to create a balanced urban system, and narrow the gap between urban and rural areas. The socialist city displays a series of concentric rings and sectors which are in many ways similar to those found in the capitalist city. However, there are distinct pre-socialist inner and socialist outer areas.

Most socialist cities retain at least a trace of their long history, usually in the form of a square, cathedrals and churches, and castles. Next to these historic centres there is a commercial area with pre-communist industry and housing. Away from this core is a zone of socialist transition where modern construction is gradually replacing older housing. Massive building programmes were undertaken from the 1950s to the 1970s. Usually this took the form of estates made up of multi-storey, prefabricated apartments for a number of families. These were built in clusters on the edge of cities where land was available. As a result, population densities in Eastern European cities actually increase with distance from the centre. Each building formed part of a neighbourhood unit and shared surrounding facilities such as shops, green spaces and play areas. In this case the urban form was designed to promote social interaction. The housing in these zones varies in age and reflects changes in the designs approved by the state. In comparison with capitalist cities, socialist cities display a lower level of social segregation, though higher socio-economic groups have tended to occupy the older and better housing nearer the centre. This is different from the model proposed by Burgess, which suggested that higher socio-economic groups would occupy the outer areas of the city. Socialist cities are quite well served by public transport and most have networks of buses and trams. Surrounding this, and further out, are planned open belts, followed by industrial zones. Where possible, industry is located downwind and separated from the city by green spaces and parkland.

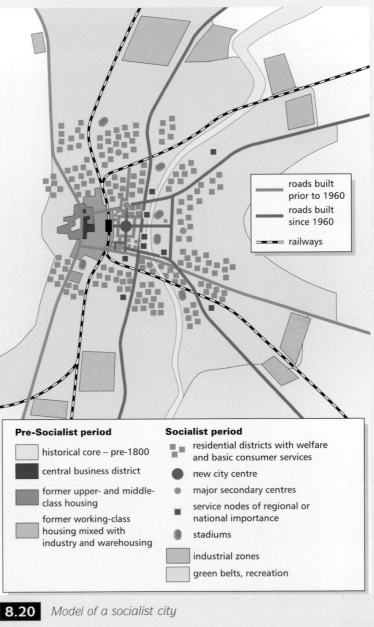

roads built prior to 1960

roads built since 1960

railways

Pre-Socialist period

☐ historical core – pre-1800

■ central business district

▨ former upper- and middle-class housing

▨ former working-class housing mixed with industry and warehousing

Socialist period

▪▪ residential districts with welfare and basic consumer services

● new city centre

● major secondary centres

■ service nodes of regional or national importance

⬭ stadiums

▨ industrial zones

▨ green belts, recreation

8.20   *Model of a socialist city*

NOTING ACTIVITIES

1 Make copies of the models of urban structure proposed by Burgess, Hoyt, and Harris and Ullmann.

2 Annotate your diagrams to show the main features of each model.

3 Summarise the ways in which the urban structure of socialist cities is both similar to and different from that in Western capitalist cities.

4 Using the idea of bid-rent, show how land use decision-making in capitalist cities is different from that in socialist cities.

5 The box on page 278 suggests that all these models assume that land use in cities results from economic forces. Explain what you understand by this.

6 Think of a town with which you are familiar. Can you identify examples where land use is not the result of economic forces? In these cases, what determines decisions about land use?

STRUCTURED QUESTION 1

Study Figure 8.21, which shows a model of urban land use for a city in the developed world.

a Suggest ways in which the urban environment of higher-income residential areas and lower-income residential areas might differ. (4)

b Why might some members of higher-income groups want to live close to the central business district (CBD)? (3)

c Suggest reasons for the location of the lower-income residential areas. (2)

d The city has a small proportion of its population from ethnic minorities. In which parts of the city might you expect these groups to live? Explain your answer. (4)

STRUCTURED QUESTION 2

Study Figure 8.22, which shows the bid-rent lines for selected land uses in one part of a city in the developed world.

a Explain the differences in bid-rents between the three land uses. (3)

b At what distance from the city centre does residential land use take over from commercial and industrial land use? (2)

c Why do the bid-rent lines decrease in value with distance from the city centre? (3)

d How would the zones of land use alter if the bid-rent for residential land use increased by £5 per m^2? (3)

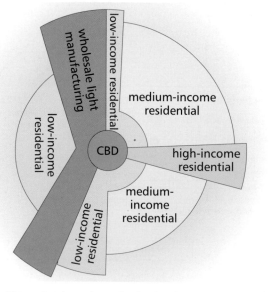

8.21 Urban land use in a city in the developed world

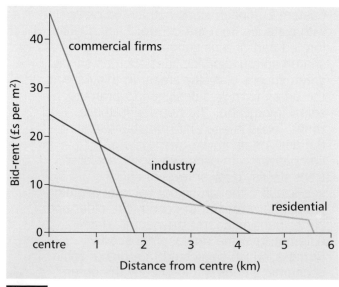

 8.22 Bid–rent lines for selected land uses in part of a city in the developed world

Peripheral model (Harris)

Harris (1997) updated the mutliple-nuclei model, suggesting that recent urban developments in the USA and other countries mean that a **peripheral model** was needed (Figure 8.23). The main feature of this model is the existence of a peripheral belt which lies within the metropolitan area but outside the central city. This peripheral belt, as the name suggests, has little to do with the central city but is linked to other developments in the periphery. It has a number of features:

- It is linked by a radial transport route.
- It has large blocks of land for development.
- It has similar social, economic and housing characteristics.
- It is free from the problems of inner city areas.
- There is land available for the development of regional shopping malls, industrial districts, theme parks, airports with motels and hotels, conference centres and parks.
- The residents of the peripheral belt have most of their ties within this sector of the city and have little to do with the central city.

Postmodern cities

In recent years urban geographers have discussed the emergence of new 'postmodern' urban forms. It has been suggested that the form, land use patterns and landscapes of cities such as Los Angeles are very different from the modern cities studied by Burgess and Hoyt. The postmodern city is seen as a response to wider changes in the operations of the world economy. Rather than being a single coherent entity, postmodern cities are thought to consist of a number of large, spectacular residential and commercial developments which are not necessarily connected. The result is a chaotic and disorganised patchwork of land uses. One attempt to represent this confusion is shown in Flusty and Dear's (1998) model of postmodern urbanism (Figure 8.24). It shows a number of features:

- The city is a patchwork of land uses. There is no clear pattern or structure to the city.
- There is no single clear centre. Instead, there is a multiplicity of centres.
- The city becomes a complex mix of unrelated land uses. High-status consumption centres are juxtaposed with ethnic ghettos, while exclusive residential communities are designed to keep unwelcome sections of the population out.

Peripheral model of metropolitan areas

1 central city
2 suburban residential area
3 circumferential highway
4 radial highway
5 shopping mall
6 industrial district
7 office park
8 service centre
9 airport complex
10 combined employment and shopping centre

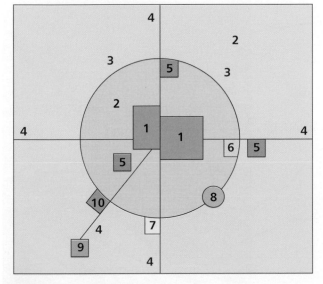

8.23 *Harris's peripheral model*

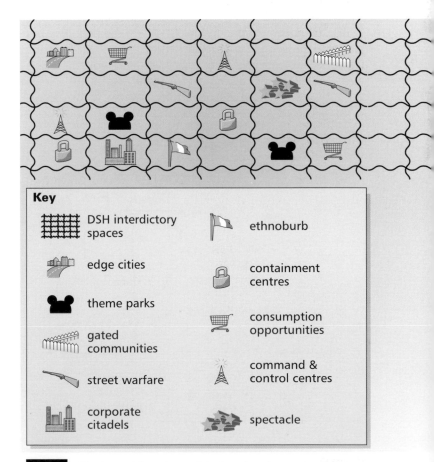

8.24 *Flusty and Dear's model of postmodern urbanism*

Urban problems in Los Angeles

Los Angeles is sometimes described as *the* 'global city'. Even if you have never been there you will know it through the media. Films such as LA *Story* portray the city as a place of sunshine, leisure and endless luxury. There are darker visions too, as in the film *Falling Down*, about a man who is at the end of his tether, crossing the city and losing his patience with anyone who happens to get in his way. Beyond fiction, you may remember the video footage showing the beating that the black motorist Rodney King received at the hands of officers of the Los Angeles Police Department.

According to the urban geographer Ed Soja, Los Angeles is the model for cities in the coming century. He argues that the trends and patterns that are emerging in Los Angeles will become apparent in other large cities. The population of Los Angeles grew from 9 million in 1970 to 14.5 million in 1996. This growth has been facilitated by the continuous expansion of the city through the development of road transport. Settlements have been incorporated into the urban area and a number of large cities have grown up on the edge of the city (Figure 8.25). The result of these developments has been a strained relationship between the centre and the periphery. Soja has described Los Angeles as the industrial city turned inside out, where the periphery has become the economic core, attracting investment from the aerospace, electronics and other high-tech industries. This process left a hole at the centre which has been filled in recent years by the headquarters of large transnational corporations. As a result the city has become a world city,

commanding investment from the Pacific Rim. This is just one side of the story, however. Alongside the lucrative new forms of employment there has been a growth in a low-wage, non-unionised economy providing employment for many of the people from minority groups that make up the population of the city.

Los Angeles has undergone a dramatic transformation in its economic and urban structure. The traditional industrial sectors that provided the basis for growth after 1945 were virtually eliminated in the long recession of the late 1970s and early 1980s. In the period 1978–83, 70 000 well-paid jobs were lost in industries making automobiles, tyres, steel and civilian aircraft. Most of these were located in the industrial belt along the Long Beach Freeway. These job losses were offset by the gaining of 1.3 million jobs between 1970 and 1990 in the five Southern Californian counties (Los Angeles, Orange, Riverside, San Fernando and Ventura). These jobs were largely in high-tech manufacturing and services, and in low-wage sectors linked to aircraft production.

There are a number of important features of the urban structure of Los Angeles that result from these developments. One of these is the social contrast between different residential environments. Mike Davis contrasts the exclusivity of areas such as Hollywood and Bel Air and their affluent suburbs where there are private security guards, neighbourhood watch programmes and surveillance cameras, with the poverty of ghetto communities such as South Central, Watts and East LA. In the east of Los Angeles, along the Santa Ana Freeway, is a narrow ribbon of cities that make up the heart of the old industrial economy of Southern California. These cities are known as the 'hub cities'. They have experienced problems of deindustriali-sation, impoverishment and housing. Between 1978 and

8.25 *Los Angeles*

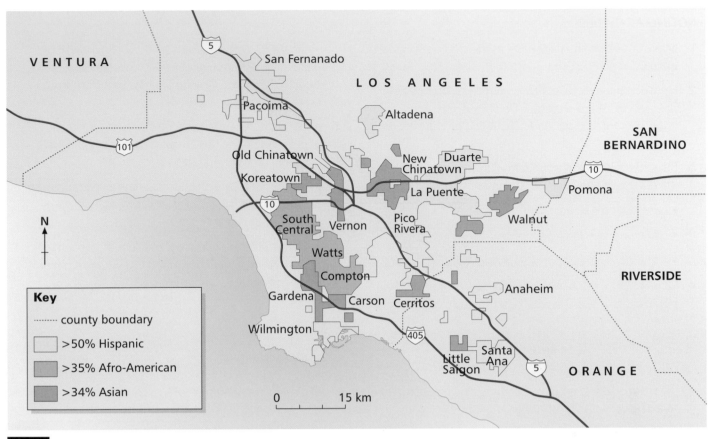

8.26   *Ethnic neighbourhoods in Los Angeles and Orange counties*

1982, 75 000 jobs were lost in the south-east of Los Angeles in plant closures in basic manufacturing industries. These cities were characterised by immigrants of Hispanic and Chicano origin. It was these groups who suffered in the plant closures of the late 1970s and early 1980s. The jobs available in these cities are now low-paid work in garment factories and services. By the late 1980s three of the hub cities – Cudahay, Bell Gardens and Huntington Park – were among the poorest suburbs in the USA.

Watts is one of the poorest districts of Los Angeles, and was affected by rioting in both 1965 and 1992. It is an area with a concentration of Hispanics (69 per cent) and Afro-Americans (24 per cent), and has high rates of mobility as immigrants settle close to the centre of Los Angeles. In 1990 one-third of households had an income below the official poverty line (Figure 8.26).

A notable feature of contemporary Los Angeles is the way in which space is divided up according to the social status of individuals. Public spaces, such as parks, are closed to all but the local people, and urban redevelopment schemes are policed by security guards in order to keep out 'undesirables'. There are sadistic street environment features, such as benches that are impossible to sleep on, and signs telling people to stop loitering. This concern with security has led to the creation of gated communities where outsiders are kept at bay by armed guards, fences, walls and security guards (Figure 8.27).

8.27   *An exclusive 'gated' community in Los Angeles*

NOTING ACTIVITIES

1   Make notes on the peripheral model and postmodern cities.

2   How do these models differ from the other models discussed in this chapter? Do they offer improvements to our understanding of the contemporary city?

3   Describe some of the main features of the urban structure of Los Angeles. Comment on:

- its pattern of growth

- the changes in the economy

- the development of social divisions within the city.

4   Study Figure 8.28, which shows the growth of the population of the five counties of Los Angeles since 1900. Draw a graph to show the growth of population for each of the five counties (Los Angeles, Ventura, Orange County, San Bernardino and Riverside).

5   a   What has happened to the growth of the population of the Los Angeles region since 1900? Suggest reasons for this.

   b   Since 1950, which counties have experienced the fastest growth in population? How might this growth be explained?

   c   What effects would the pattern of population growth have on land use in the Los Angeles region? (*Hint:* residential, transport, agriculture)

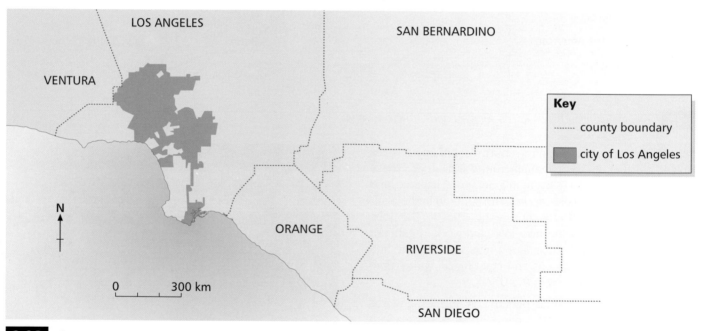

8.28   *Population growth in five counties of Los Angeles*

Census year	Los Angeles	Orange	San Bernardino	Riverside	Ventura	Five-county region
1900	170	20	28	18	14	250
1910	504	34	57	35	18	648
1920	936	61	73	50	28	1148
1930	2209	119	134	81	55	2598
1940	2786	131	161	106	70	3254
1950	4152	216	282	170	115	4935
1960	6011	709	501	303	199	7723
1970	7042	1421	682	457	378	9980
1980	7478	1932	893	664	530	11 497
1990	8863	2411	1418	1170	669	14 531

C Urban issues in British cities

In the introduction to his book *Urban Decline* (1989) the geographer David Clark states: 'British cities are in decline'. He argues that this is particularly true of large cities, at the centre of the large urban **conurbations** (Figure 8.29). Two indicators of this decline are population change and employment change. Look at Figure 8.30. These tables give an indication of the changes taking place in the largest British cities.

Key

- Cities over 200 000

Conurbations

1 Greater London
2 West Midlands (Birmingham)
3 Merseyside (Liverpool)
4 South-east Lancashire (Manchester)
5 West Yorkshire (Leeds)
6 Tyneside (Newcastle)
7 Central Clydeside (Glasgow)

8.29 *Major urban areas in Britain*

	1988 population	% change				
		1901–51	**1951–61**	**1961–71**	**1971–81**	**1981–88**
Birmingham	993 700	49.1	1.9	–7.2	–8.3	–2.4
Glasgow	703 200	24.9	–2.9	–13.8	–22.0	–9.2
Leeds	709 600	19.3	2.5	3.6	–4.6	–1.2
Liverpool	469 600	10.9	–5.5	–18.2	–16.4	–9.2
London	6 735 000	25.9	–2.5	–6.8	–9.9	–1.0
Manchester	445 900	8.3	–5.9	–17.9	–17.5	–3.6
Newcastle	279 600	26.1	–2.3	–8.4	–9.9	–1.6
Sheffield	528 300	23.0	0.4	–2.1	–6.1	–3.6

a *Population and percentage change*

	Birmingham	Glasgow	Leeds	Liverpool	Manchester	Newcastle	Sheffield
Primary	–18.6	–16.3	–39.3	–64.8	–17.3	–38.3	–52.0
Chemicals etc.	–48.3	–67.4	–14.8	–41.9	–44.5	–46.3	–73.0
Engineering etc.	–37.8	–45.4	–26.2	–49.4	–32.8	–49.0	–31.7
Other manufacturing	–19.8	–40.8	–26.9	–60.4	–48.6	–37.2	–18.7
Construction	–10.0	–15.0	–14.0	–24.7	–26.2	–41.1	–13.5
Producer services	33.0	22.0	54.1	10.3	16.2	46.2	32.2
Distributive services	–6.4	–25.3	9.8	–36.9	–16.0	–20.0	11.1
Personal services	22.2	4.6	19.3	3.3	–3.6	16.7	32.5
Non-market services	5.3	11.1	2.6	–11.2	–0.7	26.3	7.5
Total	**–10.8**	**–13.1**	**–0.6**	**–25.7**	**–12.9**	**–4.5**	**–15.1**

b *Employment change (%) by sector, 1981–91*

8.30 *Population and employment change in some large British cities*

Geographers have identified a **spiral of decline** that affects many parts of large cities. Population decline is linked to declining employment opportunities. As a result, investment is withdrawn. Unemployment means that people have less money to spend, and this has a multiplier effect on other businesses, such as shops and other services. Those who are able to do so move to find work and a better environment, leaving behind a residual population. The area is perceived as a 'problem area', and it becomes difficult to attract staff for schools, hospitals and social services. These problems add up to what has been called an **'urban crisis'**. Figure 8.31 outlines some aspects of the urban crisis.

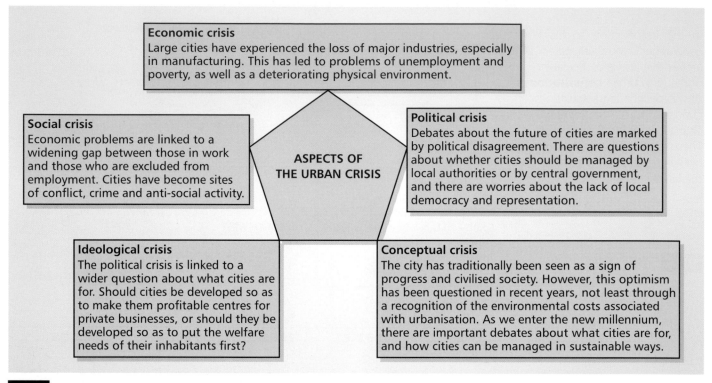

Economic crisis
Large cities have experienced the loss of major industries, especially in manufacturing. This has led to problems of unemployment and poverty, as well as a deteriorating physical environment.

Social crisis
Economic problems are linked to a widening gap between those in work and those who are excluded from employment. Cities have become sites of conflict, crime and anti-social activity.

ASPECTS OF THE URBAN CRISIS

Political crisis
Debates about the future of cities are marked by political disagreement. There are questions about whether cities should be managed by local authorities or by central government, and there are worries about the lack of local democracy and representation.

Ideological crisis
The political crisis is linked to a wider question about what cities are for. Should cities be developed so as to make them profitable centres for private businesses, or should they be developed so as to put the welfare needs of their inhabitants first?

Conceptual crisis
The city has traditionally been seen as a sign of progress and civilised society. However, this optimism has been questioned in recent years, not least through a recognition of the environmental costs associated with urbanisation. As we enter the new millennium, there are important debates about what cities are for, and how cities can be managed in sustainable ways.

8.31 *Aspects of the urban crisis*

NOTING ACTIVITIES

1 What do you understand by the term 'conurbation'?

2 Study Figure 8.30.

   a Which city experienced the most rapid population growth between 1901 and 1951?

   b What has happened to the populations of the largest cities since 1951?

   c In which decade did the most dramatic changes occur?

   d Which types of employment declined most rapidly between 1981 and 1991?

   e Which types of employment grew in the same period?

   f How might employment change and population change be linked?

3 Draw a diagram to illustrate the 'spiral of decline' in parts of Britain's cities.

Housing issues

Figure 8.32 is a map showing the quality of life within different wards (small areas) within London. It divides the wards in London into five groups, or quintiles. If you study the map, you will notice that there is a broad pattern:

- The wards with the highest quality of life tend to be found in the outer parts of London, whilst the wards with the lowest quality of life are clustered in inner London, especially to the east.

- The map provides evidence of a marked contrast between inner and outer London, with the most intense deprivation in inner London being towards the east.

- This simple pattern is complicated by the fact that there are pockets of affluence in inner London and pockets of deprivation within outer London (Figure 8.33).

'Deprivation' is a difficult concept to define. This map is based on four indicators of deprivation. These are:

- % unemployed
- % households that are overcrowded
- % households not owning their own home
- % households without a car.

There are other measures, and your own studies could make use of census data to analyse deprivation in your locality.

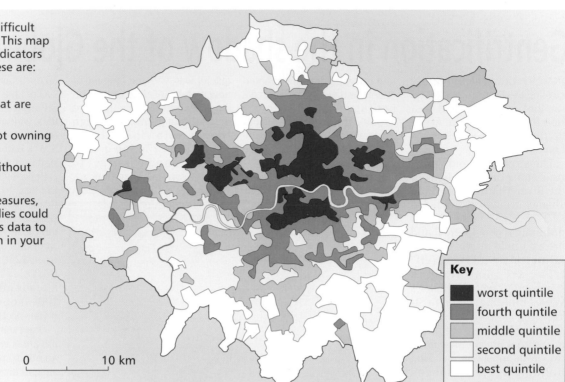

Key
- worst quintile
- fourth quintile
- middle quintile
- second quintile
- best quintile

0   10 km

8.32 *Scores of deprivation in Greater London*

8.33 *London: a city of contrasts…*

What is gentrification?

Read the newspaper article Figure 8.34. It outlines how a 'run-down' area of inner London is set to become a fashionable place to live. The process is fuelled by developers and estate agents who have renovated many of the older buildings, thus raising their value, and are looking to sell them to people who can afford to buy them. The article outlines the attractions of the location, stressing its proximity to the City, its cultural attractions, and the growth of expensive restaurants and wine bars. This process, whereby predominantly working-class areas of cities are colonised by middle-class residents, is called **gentrification** (Figure 8.35).

The causes of gentrification

The process described in Figure 8.34 goes against the predictions of Burgess's model (see Figure 8.12), which suggests that higher-income groups will seek to move away from the central areas of cities. Instead, some members of

Gentrification in the shadow of the Globe Theatre

INVESTORS are constantly circling the London market on the lookout for the next area ripe for gentrification. At the moment all eyes are on London SE1.

The stretch of land which runs along the south bank of the River Thames from Shad Thames in the east to Waterloo in the west is seeing a dramatic rise in interest and prices. Could this be the next Islington or Clerkenwell?

The area has become synonymous with the Butler's Wharf Building, home to Sir Terence Conran's string of restaurants, where the prime ministerial Blairs entertained the presidential Clintons. That building has just changed hands.

The new owner, Prestbury, plans a wholesale transformation of what is essentially a tired rental block, to bring the interiors up to the highest London standards. Many of the other buildings which make up the Butler's Wharf estate are being finished this year, with the Sold Out signs already up.

Nicholson Estates opened its offices in January expecting a cold winter chill, and instead was trampled by a stampede of first-time buyers and investors looking for reasonably cheap central London property. It sold 105 out of 118 homes in less than two months.

Galliard Homes has followed suit, selling all 62 of its flats in the converted warehouse development at Tamarind Court.

Tom Marshall, of Cluttons Daniel Smith, who has worked in the area for more than 10 years, says Shad Thames has shifted from part of the Docklands to a trendy Soho-style alternative for the West End crowd. It has the cobbled streets and authentic industrial heritage which most other areas along the river lack. What West End buyers come for is that authentic warehouse look.

Prices for flats are rising faster in Southwark than in any

Location of Southwark within Greater London

other part of London. Flats that were selling for £175 a sq. ft last year are selling for £250 a sq. ft now. On the river, prices at developments such as Horseshoe Wharf in Clink Street are at Covent Garden levels of around £400 a sq. ft.

One key to Borough's popularity is transport. The arrival of the new Jubilee Line Underground stations – which serve the river stretch of Borough far better than Shad Thames – and the new footbridge across the Thames to St Paul's Cathedral, will transform the area's accessibility.

Other projects raising Borough's status include the Globe Theatre, the new Tate Gallery of Modern Art and the new office of the Mayor of London.

At the same time gentrification is creeping in in a quieter way, with the opening of new restaurants and pubs. The glass-walled café Fish! is the new talking point in Borough Market, with Vinopolis, an eating and drinking centre based around wine, set to follow. Town Hall Chambers is a 150-year-old building on Borough High Street, where the Slug and Lettuce chain plans to open one of its café-style bars on the ground floor.

On the upper floors, Silverstreet is developing 10 flats, with air conditioning, cast-iron radiators and Philippe Starck bathroom fittings. Prices start from £170 000.

Berkeley Homes is likely to be the next to reap the benefit of Borough's popularity with its scheme Benbow House, next to the Globe Theatre. Prices on the riverside development range from £250 000 to £1 million. Berkeley's hoardings have already attracted more than 1000 inquiries, mostly from owner-occupiers.

One estate agent says: 'There are areas where you can create an environment like Islington, such as around Borough Market and along the river. The area is so well located, with the new bridge and the tube. It feels like it's going to go places. But there are parts of it that are ugly and inconvenient. The danger is that developers will try to push prices too fast. You cannot create a good new area overnight.'

From *The Financial Times*, 19/20 June 1999

8.34 *Gentrification in progress*

8.35 *Gentrified London*

higher-income groups are seeking to move back into the city. There are a number of possible explanations for this:

- The continuing existence of high-income employment in the central city: the changing nature of the economy, with the growth of professional, managerial and other white-collar occupations, has led to changes in the population of large cities, and property in some parts of inner cities provides an attractive investment opportunity for these groups.

- New patterns of household structure, with an increasing number of single-person and non-children households: studies of gentrification in North American cities have found that inner areas of cities are attractive to professional women because they are close to city centres, they have well-developed social and informal networks, property here is modestly priced, and there is a wide range of community services.

- New patterns of consumption and new housing preferences amongst younger, affluent households: gentrification is well suited to households where both partners work and there is a need to reorganise childcare arrangements.

All these reasons stress the importance of the choices made by individuals and couples about where to live. An alternative explanation of gentrification focuses on the process whereby land values become so cheap that developers are willing to invest in improving properties in order to sell them. Neil Smith has argued that many inner city neighbourhoods have experienced a lack of investment. Instead, capital has been invested in suburban locations. This creates a **rent gap**, or the difference between the actual land value and potential land value of an accessible inner city location. When land values fall so low, investors see the potential to make profits from redeveloping an area. Gentrification is one result of this process.

There is a tendency to think about gentrification as a process of in-migration. However, gentrification also involves out-migration, as people who can neither afford the increased rents of the gentrified neighbourhood nor live with the area's new cultural demands (expensive restaurants and wine bars replace cheap cafés and pubs) are displaced. The cartoons in Figure 8.36 offer an alternative view of the process of gentrification to balance the more optimistic accounts of gentrification.

Dr. Dan explains gentrification (Doonesbury © 1980)

8.36   *A view of gentrification*

Gentrification

The term 'gentrification' was coined in 1963 by the sociologist Ruth Glass to describe the social and geographical expansion of a new 'gentry' into the old working-class residential areas of inner London. She linked this to the trend towards smaller households, which meant that the number of middle-class households increased, as did their demand for housing:

> 'One by one, many of the working-class quarters of London have been invaded by the middle class – upper and lower – and shabby modest mews and cottages have been taken over when their leases expired and have become elegant, expensive residences... Once this process of "gentrification" starts in a district it goes on rapidly until all or most of the original working-class occupiers are displaced and the whole social character of the district is changed.'

Central London has always had its upper- and middle-class areas, but Glass argued that the increase in the number of middle-class households meant that the demand for housing spread into other areas.

By the mid-1960s the opening of the Victoria tube line led to the gentrification of the Barnsbury area of Islington. Other areas such as Primrose Hill, Camden Town, Notting Hill and Holland Park, all close to central London with good tube links, quickly became gentrified in the 1970s. In the mid-1970s gentrification spread further afield, down the Kings Road into Fulham, into Ealing and North Kensington and across north and east Islington into the more attractive parts of Hackney. The most dramatic example of gentrification occurred east and south of the city, in London's Docklands. Gentrification has also spread into other areas of the old East End such as Mile End Road, and south of the Thames into Battersea, Vauxhall and Southwark, long seen as the home of the traditional working-class.

NOTING ACTIVITIES

1 Define the term 'gentrification'.

2 Describe the way in which gentrification has spread through London.

3 Outline two explanations for why gentrification has occurred.

4 Study Figure 8.35. Suggest reasons why certain groups might find this environment attractive.

5 Make brief notes to explain why gentrification is considered to be a controversial process.

STRUCTURED QUESTION

Study Figure 8.37, which shows **(a)** the distribution of housing types and **(b)** accessibility to doctors' surgeries in Edinburgh.

a Describe the distribution of **(i)** low-status, 19th-century apartment blocks **(ii)** 20th-century local authority housing. *(2)*

b Explain the patterns you have described. *(4)*

c Why are high-status residential areas sometimes found close to city centres in the developed world? *(2)*

d Describe the relationship shown in Figure 8.37 between housing type and accessibility to doctors' surgeries. *(4)*

e Outline two possible reasons for the relationships you have described. *(4)*

a Housing areas and neighbourhood types

b Access to doctors' surgeries

Key
- 18th-century apartments (high status)
- 19th-century apartments (low status)
- 20th-century owner occupied housing
- 20th-century local authority housing
- mixed 20th-century owner-occupied and local authority housing
- open spaces
- ◆ CBD
- .... city boundary

N

0   3 km

Key
- High
- Low

8.37 *Housing areas, neighbourhood types and access to doctors' surgeries in Edinburgh*

Homelessness

A major feature of urban areas in the 1990s has been the return of the homeless as a visible part of city life. It is difficult to estimate the number of people who are homeless, because there is no single agreed definition of 'homeless'. In fact there are varying degrees of homelessness, ranging from people living in insecure, unsafe or unaffordable housing who are at risk of homelessness to people living on the street, in parks or squats.

In 1997, 165 790 households were officially recognised as homeless by local authorities in England. The housing charity Shelter estimates that this represents 400 000 individuals.

Local authorities have a duty to house homeless people who are in priority need. These include families with children, pregnant women, and those who are vulnerable due to age, disability, and mental or physical illness. In Britain, the rate of homelessness is 7 per 1000 people across the country. However, there are significant variations, with the highest being London which has a rate of 25 per 1000.

Why are some people homeless?

The make-up of the homeless population has much to do with methods of recording. In Britain the Housing (Homeless Persons) Act of 1977 gives housing priority to pregnant women and to families with children. As a result, 71 per cent of those certified as homeless are women and children, and another 12 per cent are pregnant women without children.

Homelessness does not strike at random. Some groups are more vulnerable than others. For example, people from ethnic minorities in Britain are four times more likely to be homeless or living in poor accommodation. The average age of people living in shelters for the homeless is 35 years.

8.38   *A homeless person selling* The Big Issue

The 'typical' homeless person is a school-leaver with few or no educational qualifications and poor employment prospects. Around 180 000 children become homeless in England each year. Most of these are forced out of their homes along with their mothers, who are often fleeing an abusive partner. The vast majority of these children are in single-parent families living on a low income. Many of the children have behavioural problems and show significant delays in social or language development. These problems are compounded by the fact that these children are often excluded from or unable to attend school.

Most homeless people are unemployable because of ill-health, disability, old age or a lack of skills. Many may work on a casual basis, but such work provides neither security nor benefits and pay is generally too low to lift them out of poverty (Figure 8.38).

A range of factors contribute to homelessness, but the most common is when people lack sufficient income to meet their housing needs. Other factors contributing to homelessness include:

- violence (particularly domestic violence) that forces people to leave their homes
- unemployment, which causes people to lose their income and so become unable to maintain rent or mortgage repayments
- drug and alcohol abuse
- disability
- family breakdown
- loss of social support networks.

Social issues

The 1991 British census defined **ethnicity** as being 'of non-European origin'. In 1991, the number of people in Britain belonging to this category totalled just over 3 million, or 5.5 per cent of the population. The largest group was the Indian population (840 255), followed by the Black Caribbean population (just under half a million) and Pakistanis at 476 555.

Although these groups make up a relatively small proportion of the British population, they are spread unevenly (Figure 8.39). At the national scale, ethnic minorities are concentrated in England rather than in Scotland or Wales. At the county level, it is clear that the ethnic population is concentrated in large urban areas to a greater extent than the white population, although the figures vary according to ethnic group.

In his analysis of 1991 census data, the geographer Ceri Peach (1996) studied the degree of segregation amongst Britain's ethnic groups. He concluded that Bangladeshi segregation is very high, not only from the white population but also from other ethnic groups. Pakistani segregation levels, although high, are significantly lower than the Bangladeshis, while Indian segregation is significantly lower than for the Pakistanis. Levels of segregation for the Black groups are generally much less marked than for the South Asian groups.

Study Figure 8.40, which shows indices of segregation for different ethnic groups in British cities. The values of these indices range from 0 to 100. The higher the index, the greater the degree of segregation. Thus a score of 0 would indicate complete integration, and a score of 100 would indicate complete segregation. An examination of the table reveals that the Bangladeshi population shows the highest degree of segregation. The least segregation is found among Black Caribbeans.

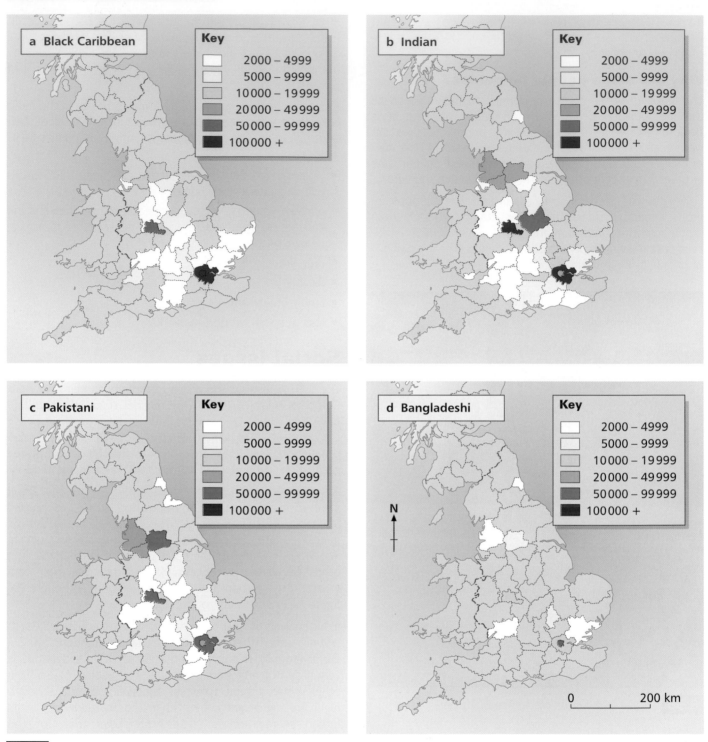

8.39 *Numerical distribution of ethnic groups, 1991*

		White	Black Caribbean	Black African	Black Other	Indian	Pakistani	Chinese	Average IS
a	**Bangladeshi**	73	57	55	59	58	46	65	69
b	**Pakistani**	61	46	42	45	39	46	53	58
c	**Black Caribbean**	45	54	28	20	39	45	39	41

8.40 *Indices of segregation (IS) for different ethnic groups in British cities with over 100 000 people in that group (1991)*

Figure 8.41 shows some of the main reasons for poverty among ethnic minorities. Historically, migration to Britain was associated with labour shortages in key industries such as manufacturing (iron and steel, textiles, carmaking) and public services (transport and health), largely in manual occupations with low pay and minimum employment rights. The restructuring of the UK economy has led to large-scale unemployment, and these problems have been exacerbated by problems of discrimination in the health, education and housing sectors. The outcome is that people from the New Commonwealth and Pakistan are disproportionately found in poor housing, in low-income households, and suffer from ill-health.

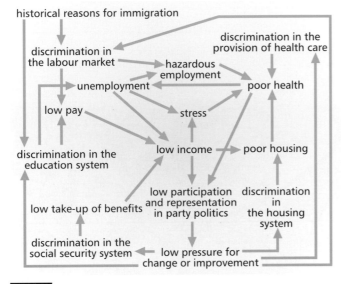

8.41   *Factors affecting poverty amongst ethnic minorities in British cities*

There is thus considerable evidence of a range of factors that constrain the choice of where people live. On the other hand, it has been suggested that residential segregation reflects choices made by members of ethnic groups for particular reasons, for example:

- **Defence** – when discrimination is widespread and intense, leading members of the ethnic group withdraw from the hostility of the wider society.

- **Support** – this ranges from formal ethnic-oriented institutions and businesses to informal friendship and kinship ties. For example, Sikh temples and Muslim mosques in British cities have become the focus of Sikh and Pakistani local welfare systems. There may be a desire to avoid outside contact in order to ensure community cohesion. There are clusters of ethnic enterprises such as banks, butchers, grocery stores, travel agencies and clothing shops (Figure 8.42).

- **Preservation** – groups may seek to preserve a distinctive cultural heritage. This may be related to the demands of religion, such as diet, preparation of food, attendance for prayer and religious assembly.

8.42   *Shops for ethnic minorities in a British city*

- **Attack** – it may be that a concentration of group members allows for electoral power and the gaining of official representation within the urban political system.

NOTING ACTIVITIES

1   Study Figure 8.39. Describe the distribution of Britain's ethnic minority groups at national and county scales.

2   Study Figure 8.40, which shows indices of segregation for different ethnic groups.

   a   Explain what the index of segregation measures.

   b   Which group has the highest average index of segregation?

   c   Comment on the degree of segregation of each of the Black Caribbean, Bangladeshi and Pakistani groups.

3   Write a paragraph in response to the question 'Does Britain have ghettos?'.

EXTENDED ACTIVITY

Study Figure 8.43, which shows particular features of four districts in Preston, Lancashire in 1991. Figure 8.44 shows the location of these districts.

1   Describe how the age structure of the population differs between the districts A–D.

2   Choose *one* of the districts A–D. Suggest some of the ways in which the age structure of the population of that district presents a challenge to the providers of welfare services.

	District A	District B	District C	District D
Age structure %				
0–15	37.0	35.3	3.9	20.8
16–29	25.5	21.4	39.8	17.2
30–60	35.0	35.6	56.3	57.7
> 60	2.5	7.7	0	4.3
Housing tenure %				
Owner-occupied	79.6	20.3	73.4	98.2
Private rented	4.8	0	23.3	1.8
Public rented	14.3	78.1	0	0
Other	1.3	1.6	3.3	0
Ethnicity %				
White	22.8	99.5	98.1	95.1
Non-white	77.2	0.5	1.9	4.9

8.43 *Age structure, housing tenure and ethnic composition of four enumeration districts in Preston, 1991*

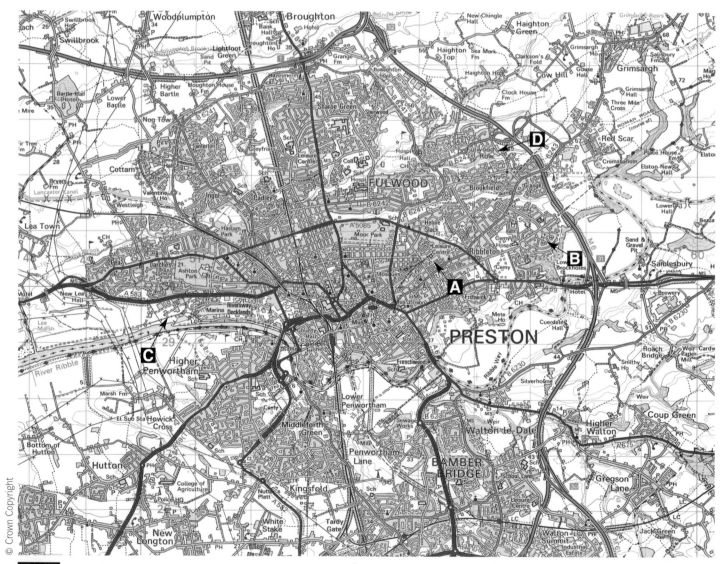

8.44 *Location of four enumeration districts in Preston*

3 Suggest possible reasons for the concentration of ethnic minority groups in district A.

4 With reference to Figures 8.43 and 8.44, suggest the possible income level and family status of the people living in district D.

5 Suggest reasons for the location of this group.

Retailing issues

In recent years the central business districts of cities and towns have declined in importance as the centre of retailing activity, and have experienced competition from out-of-town shopping and suburban locations. This process is known as **decentralisation**. This is especially true of North American cities, but also holds for British cities. There have been three waves of decentralisation in Britain in the post-war period:

- From the late 1960s, food-based superstores began to be established in suburban locations on the edge of cities.

- In the 1980s there was a trend towards large discount retail warehouses selling non-food bulky goods such as furniture, carpets, DIY and electrical equipment.

- Several very large regional shopping centres selling goods traditionally sold in high street retailers – clothing, jewellery, shoes and so on – were opened, such as the MetroCentre Gateshead and Lakeside, Thurrock.

There are several explanations for these changes:

1 Changing shopping habits There has been a reduction in the frequency of short, daily shopping trips for small quantities of convenience goods, and an increase in weekly bulk-purchase trips. This is linked to increased personal mobility, the growth of female employment, and greater use of freezers. These changes have encouraged the growth of family weekend and evening shopping and increased demand for one-stop shopping centres.

2 The shift in population to the suburbs Younger, better-off and more mobile sections of the population to the suburbs created a large source of demand that was under-served by retailing outlets.

3 High cost of land in city centres As the average size of stores has increased, the demand for more space is more and more difficult to meet. The developers of large out-of-town centres have sought to attract large retailers in order to 'anchor' their developments.

4 Access to labour The suburbanisation of the population means that sources of labour have changed. Retailing is largely reliant on a part-time female workforce, and access to a primarily suburban labour pool benefits out-of-town retailers.

Economic factors

The argument is often made that the development of out-of-town retailing leads to the death of the city centre. As a result planning authorities are concerned about their proliferation. However, it may be that the impact is exaggerated. Thomas and Bromley found that the development of a suburban retail park in the Swansea Enterprise Zone 7 km from the city centre did not have a strong effect on the city centre but did affect nearby middle-order shopping centres. Many large chain stores maintained their city centre branches after setting up units in the suburbs.

Social factors

There is concern that the decentralisation of retailing widens inequalities between the middle-class car-oriented suburban centres and the traditional shopping centres which may become impoverished and run-down.

Environmental factors

Planners are now more concerned about the development of urban sprawl. Retail development is looked at critically for its effects on the Green Belt. However, many town centres suffer from too much traffic, not enough car parking, and development demands that conflict with conservation policies.

NOTING ACTIVITIES

1   What do you understand by the term 'decentralisation of retailing'?

2   Suggest reasons for the growth of retailing in non-central locations.

3   For a town with which you are familiar, what changes have taken place in the location and type of retail facilities in recent years?

4   Suggest some of the advantages and disadvantages of out-of-town retail developments.

Transport issues

In towns and cities of the developed world, the principal transport problems are caused by the widespread use of the private car, which creates severe congestion on road networks and leads to a decline in the use of public transport systems. Transport issues have assumed greater importance in political debates, with politicians proclaiming the need to reduce the use of private cars in town centres.

Congestion

Congestion occurs when transport networks are unable to accommodate the volume of traffic that uses them. This is a particular problem at peak times. Congestion in urban areas has been caused by the increased volumes of private cars, public transport and commercial traffic. The situation is worse in the older city centres where street patterns have survived from the 19th century and earlier.

Decline in public transport

The growth of private car ownership has led to a decline in the use of public transport networks in many cities. The reduced demand for services often leads to a decline in revenues which then leads to lower frequencies of services and higher fares.

Car parking

Car parking is a problem for workers and shoppers in the central business districts of urban areas. The extension of pedestrian precincts and retail malls in city centres is intended to provide a more attractive environment, but such 'traffic-free' zones may cause problems since they create new patterns of access for car-borne passengers and users of public transport.

Peripheral developments

The development of out-of-town shopping centres and leisure complexes has a mixed effect on transport patterns. Whilst they may divert traffic away from city centres, these new developments may not be served by existing bus routes (some large retailers have introduced their own bus services to transport people to their stores), and car-borne shoppers using these centres often create traffic problems in suburban areas.

NOTING ACTIVITIES

1 Study Figure 8.45. Draw graphs to show the changes in employment and population in the seven European cities.

2 Summarise the changes shown by your graphs.

3 What transport problems might result from these trends?

4 Suggest possible solutions to these problems. Why might it prove difficult to implement these solutions?

Transport deprivation

Certain groups suffer from problems of access to transport. The elderly, the sick and disabled, and the young, are groups that are generally 'transport poor'.

Environmental problems

The effects of increased levels of private car use in urban areas are well publicised. They include:

- atmospheric pollution
- high noise levels and vibration
- the construction of new routes can lead to disruptions to communities and inflict excessive noise.

Managing urban problems

The years after 1945 were characterised by an optimism about the ability of society to provide a higher standard of living and better quality of life for all its members. Much of the older housing in Britain's large cities was considered unfit for human habitation, and a programme of slum clearance was started in many cities, along with the development of New Towns to house the displaced population.

However, the optimism of the immediate post-war period was gradually replaced by the realisation of the seriousness of many urban problems. Economic decline, leading to unemployment, and the continued existence of poverty in large cities, led to an awareness that the quality of life for many people was not improving. By the 1980s, the major issues that faced large cities were an increase in unemployment and widening social polarisation. Unemployment was felt most heavily in the inner areas of older industrial cities among groups such as the young, the poorly qualified, semi-skilled male workers, and ethnic minorities. As the decade wore on, the gap between the rich and the poor was increasingly visible, and commentators began to talk of the 'divided city' or 'dual city'. In this context, unrest developed, and several inner city areas such as Handsworth (Birmingham), Brixton (London) and Toxteth (Liverpool) experienced civil disturbances or 'riots'. In addition there was much concern in the early 1990s about drug abuse, car theft and ram-raiding on peripheral housing

City	Population			Employment		
	Years	Core city	Suburbs	Years	Core city	Suburbs
Antwerp	1970–81	–0.8	+1.2	1974–84	–0.7	+0.4
Copenhagen	1970–85	–1.5	+1.0	1970–83	–0.3	+3.2
Hamburg	1970–81	–0.8	+1.9	1961–83	–0.8	+1.9
Liverpool	1971–80	–1.6	–0.4	1978–84	–2.6	–3.1
Milan	1968–80	–0.6	+1.3	1971–81	–0.9	+1.9
Paris	1968–80	–1.1	+1.1	1975–82	–1.1	+0.9
Rotterdam	1970–80	–1.6	+2.2	1975–84	–1.1	+1.5

 8.45 *Population and employment change in selected European cities*

8.46 *Inner city disturbance*

estates in cities like Newcastle upon Tyne, Glasgow and Coventry (Figure 8.46).

The ways in which governments attempt to solve problems reflect their underlying beliefs or ideologies. So in order to understand attempts to resolve urban problems in the 1980s and 1990s, we need to understand how the Conservative governments saw those problems and the remedies they proposed. The Conservative governments from 1979 to 1997 implemented a number of measures aimed at improving run-down inner city areas. These measures had several things in common, for example:

- they attempted to bypass local councils, which were seen as too bureaucratic
- they looked to regenerate the local economy by attracting private investment into the inner cities.

Two of the main programmes developed by the Conservative governments are considered here: Enterprise Zones, and Urban Development Corporations.

Enterprise Zones

Enterprise Zones were relatively small areas of land in which special incentives were offered to encourage firms to locate there. These incentives included exemptions from certain local and central taxes, relaxed planning controls and reduced government interference.

Enterprise Zones were set up in 1981 and were located in large urban areas which had suffered the loss of manufacturing employment. Examples were the Isle of Dogs in inner London, Gateshead in Newcastle, and Clydebank in Scotland. Enterprise Zones looked to attract small and medium-sized firms in growth sectors of the economy such as computing and high-tech manufacturing. The wealth created would then 'trickle down' to other sectors and boost the local economy. Geographers have debated the question of how successful the policy was. They created new jobs, but the numbers were small in comparison with job losses in the economy as a whole. In addition, there is some suspicion

that these jobs would have been created anyway. There is also the problem that the establishment of Enterprise Zones in one part of a city had a negative effect on surrounding areas, as these automatically became less attractive sites for investment.

Urban Development Corporations

The most important of the urban policies developed in the Conservative era was the introduction of Urban Development Corporations (UDCs). These were government-appointed agencies set up with the aim of regenerating designated parts of cities. They were run by bodies made up mainly of representatives from the local business communities. The UDCs aimed to encourage private firms and businesses into urban areas that had lost their manufacturing base. The idea behind this was that firms would seek to locate in attractive areas where there were prospects for making profits.

UDCs were set up to act as catalysts for redevelopment. In practice, this meant that they took control of areas of land, reclaimed it and sold it to developers. They would provide infrastructure such as roads, water and gas supplies and sites, and could provide grants and financial aid to developers. UDCs had a large amount of freedom. They could do anything they felt would lead to 'regeneration'. However, 'regeneration' itself was not defined.

Figure 8.47 shows the location of the UDCs. It was envisaged that each UDC would have a limited life of between 10 and 15 years. In this time they would set the scene for the regeneration of the area and then be dismantled.

8.47 *Urban Development Corporations in England and Wales*

Bristol Urban Development Corporation

Bristol's UDC was set up in December 1987. Bristol is located in the south-west of England, which in the 1980s was one of the fastest-growing regions in the UK with a number of growing sectors such as insurance, banking and finance, medical services and hotels and catering. These new jobs were created at the same time as jobs were being lost in the paper and packaging, food, drink and tobacco, and aerospace industries.

The decline of these traditional manufacturing industries brought with it a number of problems, including growing polarisation in the labour market, a divide between the prosperous parts of the city and areas of high deprivation, and the ongoing problem of social order. The UDC sought to address the problems of derelict land and premises, an ageing and inadequate road system, land contamination and fragmented patterns of land ownership.

From the start the UDC drew a number of criticisms:

- outside intrusion into key areas of planning and development
- the use of public funding to subsidise private development
- the lack of the UDC's local accountability
- potential clashes with existing local government policies
- arguments about the definition of 'regeneration'.

The Bristol UDC adopted a view of regeneration that stressed the importance of allowing the market to decide. It sought to improve the physical image of Bristol through flagship architectural projects and marketing campaigns. This meant that 'social aspects' of regeneration were played down, and there were no guarantees of direct employment benefits for local people. The Bristol experience is an example of where a UDC tended to consult more with local businesses and land and property interests than with the local people and local authority (Figure 8.48).

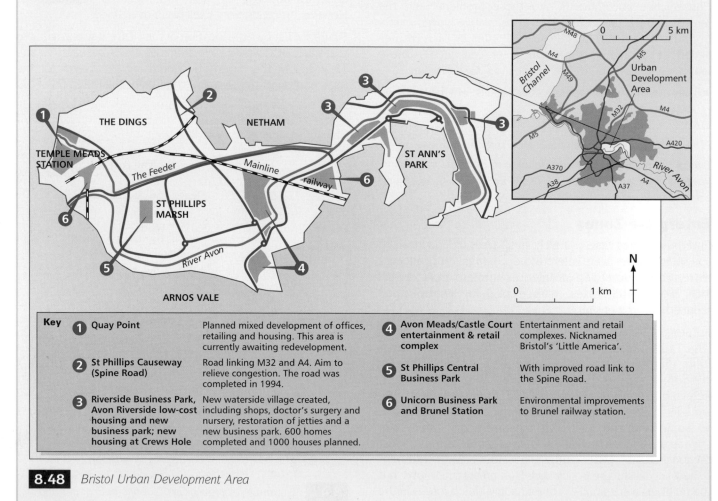

Key

1	Quay Point	Planned mixed development of offices, retailing and housing. This area is currently awaiting redevelopment.
2	St Phillips Causeway (Spine Road)	Road linking M32 and A4. Aim to relieve congestion. The road was completed in 1994.
3	Riverside Business Park, Avon Riverside low-cost housing and new business park; new housing at Crews Hole	New waterside village created, including shops, doctor's surgery and nursery, restoration of jetties and a new business park. 600 homes completed and 1000 houses planned.
4	Avon Meads/Castle Court entertainment & retail complex	Entertainment and retail complexes. Nicknamed Bristol's 'Little America'.
5	St Phillips Central Business Park	With improved road link to the Spine Road.
6	Unicorn Business Park and Brunel Station	Environmental improvements to Brunel railway station.

8.48 *Bristol Urban Development Area*

Recent developments in urban policy

Attempts to manage urban areas are closely tied up with politics. The election of a new Labour government in May 1997 led to a whole range of new initiatives designed to focus on areas of greatest need and break the cycle of economic and social problems that lead to deprivation.

This government set up eight **Regional Development Agencies**. These are responsible for administering the Single Regeneration Budget and European regeneration money. In addition, the government set up the Social Exclusion Unit which is responsible for ensuring that all policies that may have an impact on social exclusion are part of an integrated strategy incorporating health, education and crime policy.

The election of this new Labour government heralded a different approach to the inner cities. Labour's approach reflects the new political consensus that neither government intervention through the welfare state nor private enterprise based on free-market policies have created a society free from poverty, crime and violence. The new approach focuses on the importance of community, and ways of tackling problems of social exclusion.

This approach is based on the idea that communities which have declined and disintegrated should be restored in order to create a new moral, political and social order. Schools and family life are seen as crucial in order to build up a sense of citizenship and social responsibility. The intention is that all disadvantaged groups will be integrated into communities and not excluded to form a dispossessed and threatening 'underclass'.

CASE STUDY

Glasgow

Throughout the 1800s and early 1900s Glasgow grew to become one of the leading heavy engineering centres of the British Empire. In 1801 the population of the city was 77 385; by 1901 this had risen to 761 709, a tenfold increase over the century. By 1931 the population was 1 088 417, and the city was the second biggest city in Britain. At the beginning of the 20th century, Glasgow was one of the leading industrial centres of the world. Its concentration of heavy industries such as coal, iron and steel, engineering and shipbuilding assured it an important position in the national and international economic system. In the peak year of 1913 shipyards on the banks of the River Clyde produced one-third of Britain's output, or 18 per cent of world output (Figure 8.49).

During the inter-war period (1918–39) the region experienced decline in its key industries. By the late 1930s coal and shipbuilding operated at about half their 1913 levels, pig-iron production was down by two-thirds and steel output was stagnant. Throughout the inter-war period unemployment averaged around 20 per cent.

Glasgow never really recovered from this decline in its manufacturing base. The economy of the city has been restructured. Over the period 1961–81 the numbers employed in services in Glasgow rose from 48 per cent of the workforce to 68 per cent. By 1991 this had increased to 77 per cent. This transformation challenged the long-standing image of Glasgow as an area of heavy manual work. By 1985 Glasgow was the principal business centre in Scotland and the seventh largest commercial office centre in the UK. These new jobs were less attractive to male manual workers (Figure 8.50).

Problems of population contraction

One of the effects of the loss of population from inner areas is the pressure it puts on essential services such as schools and health services. The loss of population means that the provision of such services can be financially unviable. To make matters worse, the people who are left behind in inner areas are often those most in need of such services. Schools

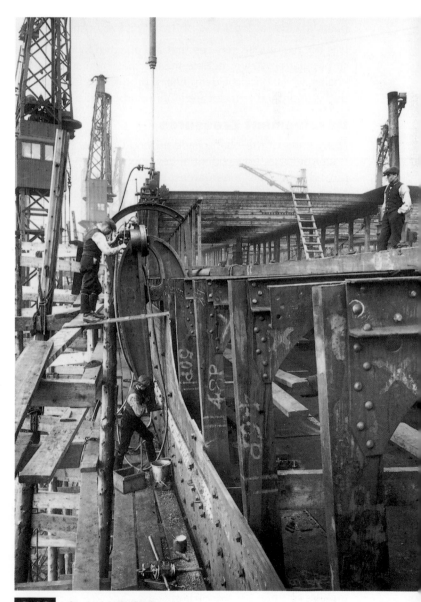

8.49 *Clydeside in its heyday*

	1981		1991			
	Number	**%**	**Number**	**%**	**Actual change**	**Percentage change**
Primary	6 217	1.6	5 140	1.6	−1 077	−17.3
Secondary	87 651	23.2	48 782	14.5	−38 869	−44.3
Construction	27 144	7.2	23 008	6.8	−4 136	−15.2
Services	256 819	68.0	259 614	77.1	2 795	1.1
Total	377 831	100	336 544	100	−41 287	−10.9

8.50 *Industrial structure and change in Glasgow, 1981–91*

and hospitals in these areas find it hard to attract quality staff, and local people have to travel further to obtain services. In Glasgow, falling school rolls have meant that the local education authorities have come under pressure to achieve cost savings by closing schools. In 1988 a plan was put forward to close seven schools in Glasgow. Local people have resisted these plans. The same argument applies for health services. In 1992 the Health Authority announced plans to close at least one of the city's five hospitals.

It is not just public services that are affected by the decline of population. Fewer people means that large supermarket chains, which offer a wider range of products at lower prices, find it unprofitable to locate in inner urban areas.

Development pressures

Many parts of the countryside around Glasgow have come under pressure for residential development as people have moved from the inner areas of the city. This pressure has been particularly important since 1980, when planning controls on new developments were relaxed. The pressure for new development has been especially placed on the Strathkelvin district on the northern edge of Glasgow. New developments affect the social mix of such areas. For example, the proportion of owner-occupied housing grew from 49 per cent to 62 per cent between 1975 and 1998, whilst the proportion of council rented housing fell from 49 per cent to 36 per cent in the same period.

Villages on the outskirts of Glasgow, such as Lennoxtown, Milton of Campsie and Torrance, were subject to intense development pressure in the 1990s. The issue of whether new housing development should be focused on brownfield or greenfield sites is important here.

Conservation

In Glasgow the period of Victorian expansion destroyed or transformed much of the medieval and Georgian heritage of the city. After the Second World War, much of the Victorian housing which was in slum condition was cleared and replaced by concrete high-rise housing. Since the 1960s there has been an increasing awareness of the need to conserve Glasgow's built environment. A series of Conservation Areas has been designated and a programme of stone cleaning has been undertaken. These have contributed to an improved image for the city. However, only a small part of the built-up area is within a Conservation Area. A large number of the city's inhabitants live in unsatisfactory housing, much of which was built within the last 50 years.

Transport

There has been a major shift in transport policy within Glasgow over the past 30 years, away from plans to build large roads and motorways towards improving public transport (Figure 8.51). Glasgow has the largest local rail

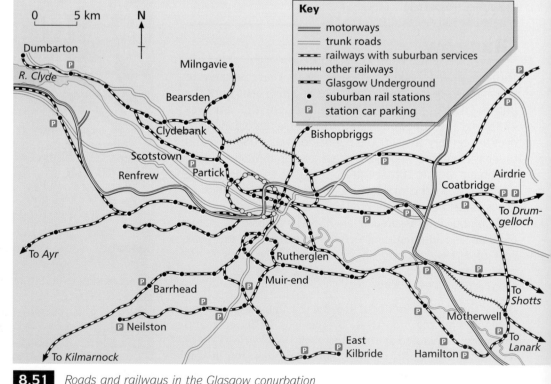

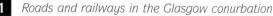

8.51 *Roads and railways in the Glasgow conurbation*

service outside of London. Its central area has an underground circle line that carries 14 million passengers per year. There are also plans to develop a rail link to Glasgow Airport, and a light railway system to serve the outer parts of the city which were developed from the 1960s without adequate transport links. In the city centre there is discussion on introducing measures to restrict parking and imposing charges on cars entering the central area.

The loss of population from inner city areas to new housing estates on the edge of the city has led to a decline in bus and rail travel. The effect has been a reduction in the frequency of services and higher fares. Attempts to halt the decline of public transport include:

- the electrification of the suburban railways north and south of the River Clyde
- the refurbishment of the underground rail loop
- the provision of small-capacity buses in the outlying housing estates.

Attempts to solve Glasgow's problems

Various attempts have been made to deal with the impacts of economic decline in those parts of the city that have been hardest hit by the loss of manufacturing. The most significant of these is the Glasgow Eastern Area Renewal (GEAR) project, which was established in 1976.

The GEAR project covered an area of 1600 ha (8 per cent of the area of Glasgow) in the city's East End (Figure 8.52). This was an area of heavy manufacturing industry which was in decline. It had experienced plant closures and the loss of skilled and semi-skilled employment. To add to the economic problems, slum clearance and piecemeal redevelopment had broken up communities, and many families had moved to outer city council estates. Between 1951 and 1976 half the population had left the area, leaving behind a population that was disproportionately elderly, disabled, on a low income, and suffering from ill-health and high mortality rates.

GEAR had six main objectives:

- to increase residents' competitiveness in securing employment
- to arrest economic decline and establish the GEAR area as a major employment centre in the city
- to alleviate the social disadvantages experienced by residents
- to improve and maintain the environment
- to stem population decline and achieve a better age balance and social structure
- to foster residents' commitment and confidence.

The success of the project can be assessed with regards to housing, economic development and welfare.

Housing

Over half the public expenditure in GEAR was on housing. By 1987 roughly two-thirds of residents were living in new or

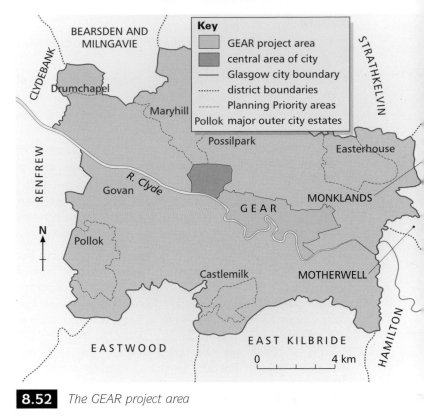

8.52   *The GEAR project area*

modernised housing. Three agencies were responsible for this work: Glasgow District Council, the Scottish Special Housing Association, and the Housing Corporation.

Economic development

The Scottish Development Agency (SDA) played a key role in making the area more attractive to businesses. Landscaping, shop-front renewal and stone cleaning improved the physical appearance of the area. By 1987 the SDA had assembled 190 ha of industrial land including the Cambuslang Investment Park, a greenfield site.

Overall, 80 000 m² of industrial floorspace was provided, including 159 new advanced factory units and the refurbishment of several buildings. The SDA's work was supplemented by the provision of light industrial units by the Regional Council and the Clyde Workshop Project, sponsored by British Steel Corporation Industry Limited.

The overall impact of these developments was reduced by a further decline in the local economy. It is estimated that the policies pursued by GEAR created over 2000 additional jobs between 1976 and 1985, but some 16 000–17 000 jobs were lost through economic decline in the same period. Indeed, the GEAR area's rate of decline was more rapid than the city as a whole. The position of local people in the labour market remained marginal. In 1985 the official unemployment rate was 25 per cent (male 33 per cent, female 16 per cent), and over half the unemployed were long-term unemployed. In effect the rate of unemployment in the GEAR project doubled over the course of the project.

Welfare

In terms of welfare, the picture is one of considerable progress:

- Three health centres were built and a broad range of social services was provided.
- A major campaign to increase the take-up of welfare benefits was launched in the 1980s.
- There was an improvement in the provision of community meeting places, and outdoor leisure facilities improved.

However, the provision of shopping facilities and transport remains poor, reflecting the weak state of the local economy and low income level.

GEAR began as a partnership of public sector agencies, but as the 1980s wore on, attempts were made to attract private-sector investment. The goal was to encourage wealth creation that could 'trickle down' to local people. The main successes in this area were in private housebuilding, with companies like Barratt, Bovis and Wimpey investing £87 million in 1000 owner-occupied homes. The area was attractive to housebuilders because planning restrictions were less strict than in the Green Belt. Overall, the attraction of private-sector investment in other activities was limited.

GEAR ended in 1987. It did not lead to a substantial change in the structure and composition of the population in the area, although some younger, dual-income, higher-status people were attracted to the new private housing. The population has stabilised. Social disadvantage, low incomes and high unemployment remain. The environment has been improved, leisure and community facilities provided and the standard of housing has improved.

At the end of the GEAR project, one of Glasgow's leading councillors commented that GEAR's prime achievement was the provision of 'prettier street corners for the unemployed to stand on'. Figure 8.53 gives some statistics for the GEAR area and Glasgow.

Since 1980, Glasgow District Council has tried a number of initiatives to overcome the negative image of the city. It was hoped this would attract both potential investors and tourists. The main initiative was the 'Glasgow's miles better' slogan (try saying it quickly to get the double meaning of the title), which ran from 1983 to 1990. In terms of the number of visitors to the city, the scheme was a great success. Prior to the campaign Glasgow attracted 700 000 visitors per year, but by the time of the Glasgow Garden Festival in 1988, this had risen to 2.2 million. In 1990 Glasgow was designated the European City of Culture, which focused international attention on the positive attributes of the city.

These attempts to improve the image of Glasgow were questioned, however, by those who argued that the version of the city being promoted failed to reflect its 'real' culture – the working-class experience. The celebrations promoted only a shallow version of culture that was designed to appeal to international visitors to the city, rather than to its own working-class residents. The celebrations cost local authorities large amounts of money at a time when funds were desperately required for the improvement of the city's working-class housing, especially the large housing estates on the edge of the city.

	1971		1981	
	GEAR area	Glasgow	GEAR area	Glasgow
Population per district	533	403	247	306
Working age (%)	54.7	53.6	55.1	56.8
Pension age (%)	17.4	15.4	22.8	19.3
Male unemployment (%)	13.7	8.6	27.3	19.4
Female unemployment (%)	6.5	4.3	13.6	10.5
Owner-occupied housing (%)	20.7	22.5	17.2	24.2
Council housing (%)	49.0	52.2	60.8	56.2
Privately rented property (%)	28.7	18.9	7.1	8.4
Over 1.5 persons per room (%)	19.3	11.9	8.2	6.4
1–1.5 persons per room (%)	13.9	14.1	12.7	13.0
No car (%)	84.0	68.5	77.5	67.0
Lack bath and WC (%)	15.1	9.5	8.8	4.4
Share bath and WC (%)	0.3	1.6	0.2	0.8
Two or more cars (%)	1.1	2.8	2.0	4.3
One or two rooms (%)	50.2	22.0	34.8	19.4
Six or more rooms (%)	1.4	5.0	1.3	3.7
Single-parent families (%)	6.2	4.5	6.4	6.8
Permanently sick/disabled (%)	2.1	1.2	4.4	2.7

8.53 *Data for GEAR project area and Glasgow*

NOTING ACTIVITIES

1 a Study Figure 8.50. Draw pie-charts to show the proportion of the workforce employed in each sector in 1981 and 1991.

b Draw graphs to show the actual number of workers employed in each sector in 1981 and 1991.

c Comment on the results shown by your graphs.

2 a Describe how the economic fortunes of Glasgow have fluctuated over the last 100 years.

b Identify some of the social and environmental problems resulting from economic decline.

3 a Describe the aims of the Glasgow Eastern Area Renewal (GEAR) project.

b Outline the achievements of GEAR. Use the following subheadings to help you:

- housing
- welfare
- economic development.

4 Use the information in this case study to make some evaluative comments on the success of the GEAR project.

CASE STUDY

London's Docklands

The decline of older heavy industries in parts of large cities in recent decades has led to controversy about how they should be redeveloped. In many places, the solution has been to clean up the area and attract new groups of people and new types of economic activity to the area. Such developments are often fiercely opposed by local residents, who feel that their locality is being taken over by outsiders. Nowhere is this process more clearly seen than in the case of London's Docklands. The following quote offers some idea of the way of life that grew up around London's Docklands in the past.

'London's docklands cover sizeable parts of five Thameside boroughs. The area has strong traditions which die hard. To those who grew up downstream of London Bridge and within easy reach of the river the docks were a fact of life. The sight of the line of working cranes seen over the endless terraced housing of Bermondsey, the smell of leather tanning, soap and biscuits, and the sounds of foghorns on murky nights were an unforgettable part of childhood. It is true that those born south of the river and owing allegiance to Millwall and the New Cross speedway team had little love for the aliens on the north shore who supported West Ham. But there was more to unite the two than to divide them — as became apparent in a thousand ways during the 1940–41 London blitz.'

P. Ambrose (1986) *Whatever Happened to Planning?*

The West India Dock was the first, opened in London in 1802. The opening of docks continued throughout the 1800s. The docks were at the centre of a collection of related industries such as refining and processing which made use of the imported products, and shipbuilding and repair. These combined with other manufacturing industries and the small workshops and trades that served the consumer markets of the capital city, especially clothing, furniture and printing (Figure 8.54).

A distinct way of life developed in the communities around the docks. The residential areas were quite isolated, and most people worked, went to school, and shopped without leaving the locality. As a result, a strong local identity developed. By 1870 the middle and upper classes had left the East End, leaving it almost wholly a working-class area. In this community, there was a strict division between the worlds of men and women. Dock work was almost entirely male. Women worked in the clothing and furniture trades or in the factories (in matchmaking, packaging and food processing). East End communities had a long history of immigration, as work was readily available and many people settled close to the docks where they had entered Britain.

8.54   *London Docks in their prime*

The Chinese settled in Limehouse, for example, and Somalis in Wapping.

Housing conditions were often squalid and poverty was widespread. This was aggravated by the casual labour systems operated by dock-owners and manufacturers. Workers responded by organising for better pay and conditions, and the area developed a strong tradition of trade unionism and support for the Labour Party.

The decline of Docklands

The area suffered substantial bomb damage in the Second World War and this led to the need for a substantial rebuilding programme. In the first two decades after the Second World War, many buildings in Docklands came to the end of their usefulness. A number of factors contributed to the decline in the importance of the docks:

- The role of London as the centre of world trade had declined, so much of the dock space and warehousing was no longer needed.

- Some manufacturing activities were attracted to the growing New Towns and other out-of-town sites where costs were lower. As a result, much of the canal and railway land had fallen derelict.

- The economic viability of the docks was further reduced by changes in transport technology. Containerisation meant there was a need for deepwater docking facilities, and these could only be found downstream.

As a result of these factors, the docks closed. East India Dock was closed in 1967, and trade in other docks began to fall. Between 1961 and 1971, almost 83 000 jobs were lost in the five boroughs in the Docklands area (Greenwich, Lewisham, Newham, Tower Hamlets and Southwark). Many of these jobs were from large transnational corporations. The decline was heightened by government policies that favoured the growth of industry outside of London. Unemployment was accompanied by population decline. Whilst inner London lost 10 per cent of its population between 1961 and 1971, the figures for Tower Hamlets and Southwark were 18 per cent and 14 per cent respectively.

The London Docklands Development Corporation

London's Docklands had come to be recognised as 'the largest redevelopment opportunity in Europe' by the 1970s. The London Docklands Development Corporation (LDDC) was established by the Conservative government in 1981. It took responsibility for the development of the area (Figure 8.55). These responsibilities included the reclamation of land, the provision of transport infrastructure, environmental improvement and the attraction of private-sector investment.

The LDDC acquired land and promoted transport development in the form of the Docklands Light Railway (DLR) and London City Airport. It sought to attract development and fill up the Isle of Dogs. New roads and office buildings, such as Canary Wharf, were built (Figure 8.56). From 1981 Docklands saw the largest concentration of new house-building in inner London. The tenure of this new housing was predominantly owner-occupied. Docklands became synonymous with the London housing boom, with prices spiralling rapidly. However, the policy of housebuilding was controversial, since some argued that it failed to meet the local housing needs. The increase in house prices put the new homes beyond the reach of local residents, and local authorities and housing associations were unable to purchase land because of the high land costs. The Docklands residents feared that the policies of the LDDC would replace them with a new community. Prior to the LDDC, Docklands was a predominantly working-class area with 50–60 per cent of households falling in the manual category. The cost of housing meant that the incoming population had higher incomes and were from a different socio-economic group. Their patterns and places of work differed: 64 per cent of new residents worked in the City of London and only 10 per cent worked in Docklands. Among the existing population 46 per cent of local residents worked in the boroughs in which they lived. The new residents tended to be concentrated in the service industries. They were more affluent, with lower rates of unemployment and higher rates of car ownership. They were more transient – half of the new residents planned to move within two years – and had fewer children. This transient feeling was reinforced by the lack of community, social facilities and shops.

The LDDC caused a great deal of controversy, and geographers have debated the economic, social and cultural impacts of its activities. In her assessment of the impact of the LDDC, Susan Brownill (1999) raises some questions about how the success of Docklands regeneration should be assessed:

- It was argued that Docklands represented an example of how 'market forces' could change inner city areas. On the other hand, it might be suggested that everything that happened in Docklands was a result of spending large sums of taxpayers' money.

- To what extent did local people benefit from the injection of money into their area?

- Is the redevelopment of parts of cities best left to democratically elected local councils or to government-appointed bodies?

- How should the costs and benefits of development be measured?

The LDDC finished its work in Docklands in 1998. The effectiveness of its work is still debated, though it has undoubtedly left its mark on the landscape. In its place there is a whole series of initiatives designed to continue the regeneration of the area. These include the Millennium Dome, the Channel Tunnel terminal at Stratford, the Thames Gateway Project and the Greenwich Waterfront Partnership.

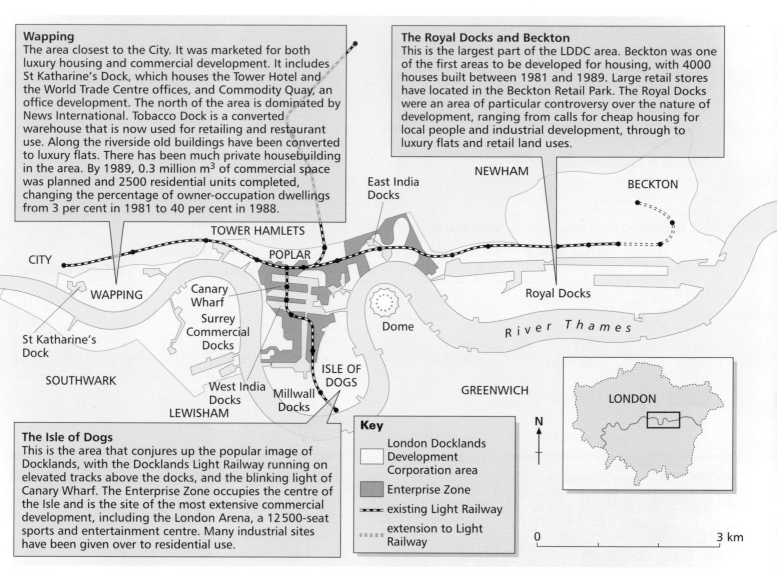

Wapping

The area closest to the City. It was marketed for both luxury housing and commercial development. It includes St Katharine's Dock, which houses the Tower Hotel and the World Trade Centre offices, and Commodity Quay, an office development. The north of the area is dominated by News International. Tobacco Dock is a converted warehouse that is now used for retailing and restaurant use. Along the riverside old buildings have been converted to luxury flats. There has been much private housebuilding in the area. By 1989, 0.3 million m^3 of commercial space was planned and 2500 residential units completed, changing the percentage of owner-occupation dwellings from 3 per cent in 1981 to 40 per cent in 1988.

The Royal Docks and Beckton

This is the largest part of the LDDC area. Beckton was one of the first areas to be developed for housing, with 4000 houses built between 1981 and 1989. Large retail stores have located in the Beckton Retail Park. The Royal Docks were an area of particular controversy over the nature of development, ranging from calls for cheap housing for local people and industrial development, through to luxury flats and retail land uses.

The Isle of Dogs

This is the area that conjures up the popular image of Docklands, with the Docklands Light Railway running on elevated tracks above the docks, and the blinking light of Canary Wharf. The Enterprise Zone occupies the centre of the Isle and is the site of the most extensive commercial development, including the London Arena, a 12 500-seat sports and entertainment centre. Many industrial sites have been given over to residential use.

Key

- London Docklands Development Corporation area
- Enterprise Zone
- existing Light Railway
- extension to Light Railway

LONDON

0 —————— 3 km

8.55 *London Docklands Development Corporation area*

8.56 *Property development in London Docklands*

NOTING ACTIVITIES

1 What were the aims and objectives of Urban Development Corporations?

2 Summarise the activities and achievements of the LDDC.

3 With other members of your group, draw up a list of criteria that might be used to evaluate the success or failure of the work of the LDDC.

4 Using Figure 8.57 and the information in this section, evaluate the activities of the LDDC.

5 Look at the cartoon, Figure 8.58. To what extent do you think this is a fair assessment of the work of the LDDC?

a Docklands before and after the LDDC

	1981	1997
Population	39 429	81 231
Employees	27 213	72 000
Home ownership	5%	43%
Dwelling stock	15 000	35 665
Service sector employment	31%	70%
Financial services employment	5%	42%

Commercial floor space since 1981 = 2.3 million m²
Housing units since 1981 = 21 615

b Regeneration of Docklands by area, March 1997

	Office space (m²)	Total commercial (m²)	Housing units	Social housing units	% private housing
Isle of Dogs	1 million	1.4 million	4178	754	82
Wapping	155 192	303 717	3874	653	84
Surrey Docks	169 374	370 172	7654	1843	75
Royal Docks	12 673	234 348	5909	2638	55
UDA	1.38 million	2.23 million	21 615	5968	72

8.57 *The impact of the LDDC*

8.58 *A view of the LDDC*

EXTENDED ACTIVITY

Study Figure 8.59, which shows some data for the achievements of the Urban Development Corporations.

1 What do you understand by the term 'Urban Development Corporation'?

2 Use the data in the table to comment on the achievements of the Urban Development Corporations with regards to:
   - housing
   - the environment
   - employment
   - attracting investment.

3 **a** For each UDC:
   - calculate the ratio of private investment to public subsidy (this is called the **leverage ratio**)
   - draw graphs to represent this data.

   b Given that a leverage ratio of 1:3 is generally considered to be the minimum acceptable, comment on the success of the UDCs.

4 Why might the data for housing and employment not tell the whole story about the achievements of the UDCs?

Urban Development Corporation	Land reclaimed (ha)	Housing units completed	Non-housing completed (thousand m^2)	Roads built or improved (km)	Gross gain jobs	Private-sector investment (£m)
London	776	2165	2300	282	70 484	6505
Merseyside	382	3135	589	97	19 105	548
Black Country	363	3441	982	33	18 480	987
Cardiff Bay	310	2260	379	27	9 387	774
Teesside	492	1306	432	28	12 226	1004
Trafford Park	176	283	636	42	23 199	1513
Tyne and Wear	507	4009	982	39	28 111	1115
Bristol	69	676	121	7	4 825	235
Sheffield	247	0	495	15	18 812	686
Central Manchester	35	2583	138	2	4 944	373
Leeds	68	571	374	12	9 066	357
Birmingham Heartlands	115	669	217	41	3 526	217
Plymouth	11	28	3	4	29	4
Totals	3551	21 126	7648	629	222 194	14 318

8.59   *Achievements of Urban Development Corporations*

D Urban issues in LEDCs

Look at Figure 8.60. It shows the total urban populations of more and less developed regions since 1950. Whilst the increase in the total urban population of more economically developed countries (MEDCs) is growing relatively slowly, the less economically developed countries (LEDCs) are experiencing a rapid growth in their urban population. It is often argued that there is an 'urban explosion' in the LEDCs, and that this raises a number of issues concerned with how to accommodate this growth (Figure 8.61). This chapter looks in more detail at the issues raised by this growth.

However, it is important to take a critical look at the figures. There are a number of problems with the type of data provided in Figure 8.60:

- There is a lack of reliable census data with which to make comparisons of the urban population between countries.

- Even where data does exist, comparisons are difficult because there are large differences in the ways that governments define urban centres.

- Dividing national populations into 'rural' and 'urban' assumes that there are clear divisions in the way people live in rural and urban areas. In many urban centres, a large proportion of the population may be engaged in agriculture or agriculture-related activities.

- Even if the data is assumed to be accurate, there is a risk of assuming that the growth rates experienced today will continue into the future. In fact, over the past two decades many countries have not grown at the rates predicted in 1980.

Despite the difficulties of using data about the urban population, it is possible to make some generalisations about urban processes in LEDCs in recent years:

- Most nations experienced a more rapid growth in urban population than in rural population, which means that an increasing proportion of their populations live in urban centres (the level of urbanisation).

- In most LEDCs there has been an increasing concentration of population and economic activity in one or two major cities.

- Overall, natural population increase has contributed more to the growth of urban population than net rural-to-urban migration. However, for many cities and some nations, this is not the case.

Levels of urbanisation and rates of urban growth differ widely between different countries.

8.61 *La Paz, Bolivia: a city in the developing world*

NOTING ACTIVITIES

Study Figure 8.3 on page 271. It shows the level of urbanisation for different countries.

1 Complete a copy of Figure 8.62 to show the level of urbanisation within each of the countries listed. Use an atlas to help you locate the countries.

2 Why is it difficult to make comparisons between countries on the basis of this data alone?

3 Using your completed table to help you, write paragraphs commenting on the levels of urbanisation in Central and South America, Africa and Asia.

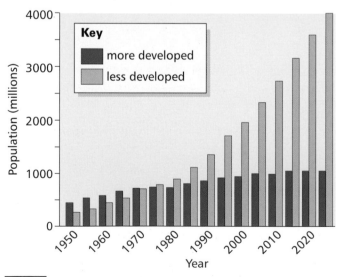

8.60 *Total urban populations of more and less developed regions, 1950–2025*

	Below 45%	45–59%	60–74%	75% and above
Argentina				
Brazil				
Chile				
China				
Egypt				
Ethiopia				
India				
Indonesia				
Mexico				
Myanmar (Burma)				
Nicaragua				
Nigeria				
Philippines				
Republic of South Africa				
Tanzania				

8.62 *Levels of urbanisation in selected countries*

The problems of rapid urban growth

Economic

In theory, there should be no reason why increased levels of urbanisation and rapid urban growth should pose problems for nations. After all, urbanisation usually goes hand in hand with industrialisation. As a country develops, it attracts investment by transnational corporations, and new employment opportunities are available in urban centres. This attracts people from the rural areas. However, transnational corporations have been very selective in where they choose to invest, and few developing countries have experienced this form of industrialisation. Mexico, Brazil and the 'Tiger' economies of South-east Asia have experienced such growth, but even in these countries the factories are not large.

8.63 *Informal sector activity: Goa, India*

In practice, then, many migrants who come to the city in search of work find it difficult to secure employment. Many are **underemployed**, which means that they work less than full-time even though they would prefer to work more hours. As a result, most find work in what is often called the **informal sector** (Figure 8.63). The informal sector is made up of jobs such as shoe shining, ice-cream vending, car washing or taxi driving. Many children are involved in such work. Urban geographers recognise that the informal sector may have some benefits. For example, it provides a vast range of cheap goods and services that would otherwise be out of reach for many people, and it allows average wages to be kept low, which may prevent some transnational corporations from seeking to invest in cheaper locations.

Despite these benefits, it is widely recognised that one of the effects of the lack of formal employment in urban areas is increasing poverty. It is difficult to assess the real situation. The World Bank in 1993 claimed that poverty was being rapidly eliminated by the spread of the market economy. However, other institutions were observing a growth in urban poverty. Part of the problem lies in the difficulty of defining poverty, and the unreliability of data.

Another effect of rapid economic growth is linked with the ideas of primacy or centrality. This can lead to a concentration of investment in one or two large centres or regions and the loss of people and resources from other parts of the country. Rangoon (in Burma, or Myanmar) is located at the centre of the national transport and communications network. It is the economic and political heart of Burma. It is the main centre for services, and virtually all the import and export trade passes through its port. Nairobi (Kenya) accounts for about two-thirds of Kenya's manufacturing employment and industrial plants. It contains about 10 per cent of the national population and acts as a major magnet for migrants from other parts of Kenya.

NOTING ACTIVITIES

1 Study Figure 8.64. It gives data about the urban growth rates and economic growth rates of 14 countries.

   a Plot this data on a scattergraph.

   b Attempt to draw a line of 'best fit' on your graph.

   c Calculate Spearman's rank correlation coefficient to analyse the relationship between urban growth and economic growth.

   d To what extent do your results indicate a relationship between urban growth and economic growth? Suggest reasons for your answer.

2 Distinguish between the 'formal sector' and the 'informal sector'.

3 What role does the informal sector play in many cities?

4 What problems are associated with urban primacy?

	Average urban growth rate per annum 1980–92 (%)	Average economic growth rate per annum 1985–94 (%)
Australia	1.5	1.2
China	4.3	7.8
Egypt	2.5	1.3
France	0.4	1.6
Ghana	4.3	1.4
India	3.1	2.9
Italy	0.6	1.8
Jamaica	2.1	3.9
Kenya	7.7	0.0
Mexico	2.9	0.9
Pakistan	4.5	1.3
Peru	2.9	–2.0
Poland	1.3	0.8
Thailand	4.5	8.6
UK	0.3	1.3
USA	1.2	1.3

8.64 *Urban and economic growth rates in selected countries*

Housing

The lack of employment and poverty in many developing world cities leads to problems of overcrowding. The rapid influx of migrants to cities means that the supply of urban housing is scarce. One response to this shortage is the appearance of makeshift shanty housing. This housing is often built on the cheapest and least desirable sites. This may include low-lying land prone to flooding, derelict land, or steep slopes that may be at risk from landslides. These settlements are known as **shanty towns** or **squatter settlements**. It is estimated that one billion people worldwide live in inadequate housing (Figure 8.65).

One response of governments to these squatter settlements is to eradicate them. People are evicted from their sites and their homes are bulldozed. Often this is done to provide space for luxury housing or urban renewal. In Seoul, South Korea, the run-up to the 1988 Olympic Games saw the forced removal of nearly 750 000 people in an effort to 'beautify' or clean up the city. Destroying squatter settlements is generally an ineffective policy, since their occupants have little choice but to build new settlements elsewhere.

In recent years, the governments of some countries have revised their attitude to squatter settlements. It is recognised that they play an important role in solving immediate housing problems, acting as reception areas for recently arrived migrants, and offering informal networks of support for people. In some cases, there have been moves to formally recognise the existence of these settlements and provide resources for improving the conditions of buildings and developing their infrastructure by providing roads, piped water, electricity and waste disposal services.

NOTING ACTIVITIES

Look at Figure 8.66. It shows some of the challenges that face migrants from rural areas in their search for shelter in large urban centres.

1 Identify the barriers that faced the Ramirez family in their search for shelter. Classify these into economic, environmental and political barriers.

2 What is meant by the term 'squatter settlement'?

3 Suggest why squatter settlements are a common feature of urban areas in the developing world.

4 'Slums of hope, or slums of despair.' Which of these best describes squatter settlements?

8.65 *A squatter settlement in Mumbai (Bombay), India*

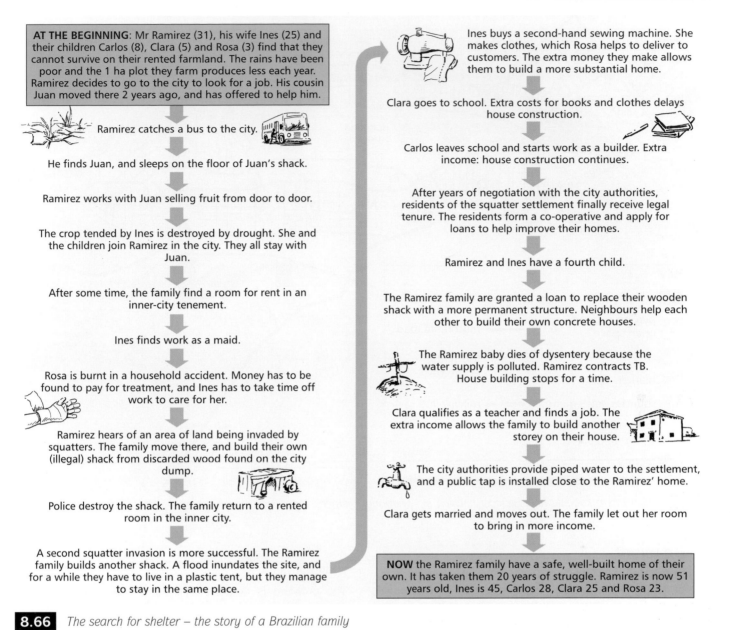

AT THE BEGINNING: Mr Ramirez (31), his wife Ines (25) and their children Carlos (8), Clara (5) and Rosa (3) find that they cannot survive on their rented farmland. The rains have been poor and the 1 ha plot they farm produces less each year. Ramirez decides to go to the city to look for a job. His cousin Juan moved there 2 years ago, and has offered to help him.

Ramirez catches a bus to the city.

He finds Juan, and sleeps on the floor of Juan's shack.

Ramirez works with Juan selling fruit from door to door.

The crop tended by Ines is destroyed by drought. She and the children join Ramirez in the city. They all stay with Juan.

After some time, the family find a room for rent in an inner-city tenement.

Ines finds work as a maid.

Rosa is burnt in a household accident. Money has to be found to pay for treatment, and Ines has to take time off work to care for her.

Ramirez hears of an area of land being invaded by squatters. The family move there, and build their own (illegal) shack from discarded wood found on the city dump.

Police destroy the shack. The family return to a rented room in the inner city.

A second squatter invasion is more successful. The Ramirez family builds another shack. A flood inundates the site, and for a while they have to live in a plastic tent, but they manage to stay in the same place.

Ines buys a second-hand sewing machine. She makes clothes, which Rosa helps to deliver to customers. The extra money they make allows them to build a more substantial home.

Clara goes to school. Extra costs for books and clothes delays house construction.

Carlos leaves school and starts work as a builder. Extra income: house construction continues.

After years of negotiation with the city authorities, residents of the squatter settlement finally receive legal tenure. The residents form a co-operative and apply for loans to help improve their homes.

Ramirez and Ines have a fourth child.

The Ramirez family are granted a loan to replace their wooden shack with a more permanent structure. Neighbours help each other to build their own concrete houses.

The Ramirez baby dies of dysentery because the water supply is polluted. Ramirez contracts TB. House building stops for a time.

Clara qualifies as a teacher and finds a job. The extra income allows the family to build another storey on their house.

The city authorities provide piped water to the settlement, and a public tap is installed close to the Ramirez' home.

Clara gets married and moves out. The family let out her room to bring in more income.

NOW the Ramirez family have a safe, well-built home of their own. It has taken them 20 years of struggle. Ramirez is now 51 years old, Ines is 45, Carlos 28, Clara 25 and Rosa 23.

8.66   *The search for shelter – the story of a Brazilian family*

STRUCTURED QUESTION 1

Study Figure 8.67, which shows the distribution of squatter settlements and low-cost housing schemes in Kariobangi District, on the eastern edge of Nairobi. Nairobi is the capital of Kenya. Its population has grown rapidly since 1950. In 1993, 55 per cent of Nairobi's population lived in squatter settlements which occupied just 6 per cent of the city's residential area.

a What factors might explain the rapid population growth of Nairobi since 1950?   *(2)*

b What does the statistic for the large proportion of the population living in squatter settlements suggest about levels of economic development in Nairobi? Give reasons for your answer.   *(4)*

c Using evidence from Figure 8.67, suggest three possible reasons for the location of squatter settlements in Kariobangi District. Explain your answers.   *(6)*

d Dandora (see Figure 8.67) is a government-sponsored low-cost housing scheme for low-income households in Nairobi.

(i) Why might the government have sponsored this type of development?

(ii) What might be the advantages and disadvantages for local people of living in such a scheme?   *(4)*

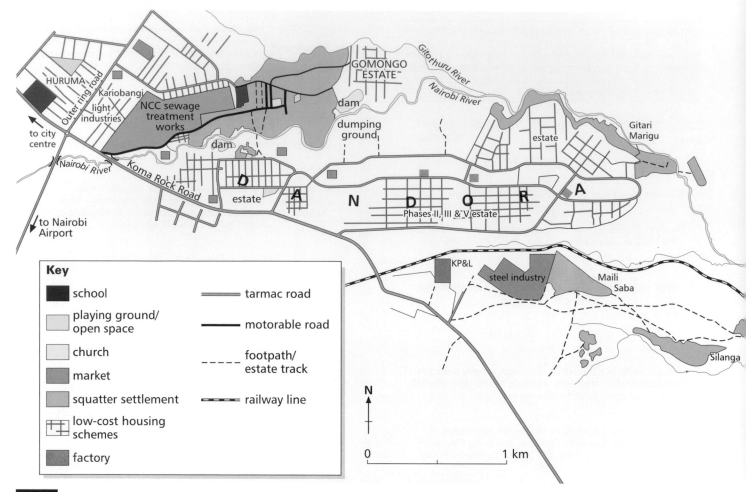

8.67 *Squatter settlements and low-cost housing in Kariobangi District, Nairobi*

Environment

Environmental problems are significant in many cities in LEDCs. Many of these cities have large concentrations of industries. For example, countries such as China, India, Mexico, Brazil and South Korea are among the world's largest producers of industrial goods. Industrial production has increased rapidly in many LEDCs and this has often occurred without any strong controls on pollution.

The problem is compounded by the fact that in many countries industrial production is heavily concentrated in one or two core cities or regions. For example, in Thailand a large proportion of all industry is located in the city of Bangkok and its neighbouring provinces.

Cubatão

The city of Cubatão is close to São Paulo and the major port of Santos. It is known as the 'Valley of Death'. Here there is a high concentration of heavy industry which developed with little or no attempt by the government to control pollution. The industries include both foreign-owned and Brazilian companies in fertilisers, oil, chemicals and steel. People here suffer from ill-health associated with air pollution, including high levels of stillborn or deformed babies, tuberculosis, bronchitis, and asthma.

The Cubatão River was once an important source of fish, but this is no longer the case. Toxic wastes have been dumped in the surrounding forests and have contaminated surface and groundwater that is used for drinking and cooking. Vegetation has suffered from air pollution, and on some slopes the loss of vegetation has weakened the structure of the soil, leading to landslides.

Life for the inhabitants of Cubatão is hazardous. There has been little or no protection for workers in many of the industries, and many suffer from serious diseases or disabilities due to their exposure to chemicals or waste products while at work.

Environmental problems are not confined to industrial pollution. Where large proportions of the population do not have access to sewerage services and waste disposal there are problems of land and water pollution. Traffic congestion and poorly maintained engines add greatly to air pollution. It is estimated that 600 million urban residents live in conditions that continually threaten their health.

Water

Some 170 million urban residents lack access to potable (drinkable) water near their homes. In Indonesia only one-third of the urban population has access to safe drinking-water. Many are forced to use water from contaminated sources. Population growth in cities such as Bangkok and Mexico City has led to an overuse of underwater supplies, leading to the problem of subsidence. Communal taps may operate for only a few hours a day, and many people have to carry water in small quantities over long distances.

Waste disposal

During the 1980s, the number of urban residents without access to adequate sanitation increased by 25 per cent. Human waste lies untreated, increasing health risks, and is eventually washed into streams, lakes or seas, and into groundwater supplies. A survey of 3000 towns in India revealed that only 8 had full sewerage facilities, whilst 209 had partial treatment facilities. Along the Ganga River, 114 towns and cities dump untreated sewage into the river each day.

Air pollution

Governments may be reluctant to enforce controls on industrial production for fear of discouraging investment. This has led to the contamination of industrial air and water pollution by some companies. The most notorious example was the Union Carbide plant in Bhopal, India where poisonous gases killed 3300 people and seriously injured another 150 000. Most of these people were from poor households living adjacent to the plant.

Many cities are located in areas that are prone to hazards such as floods, earthquakes, and landslides. It is often the poorest households who live in these areas. Lack of resources and unenforced building standards often mean that people in these areas suffer most from hazards.

Transport

Urban dwellers in the LEDCs often face transport difficulties which are far worse than those experienced by urban dwellers in the MEDCs. Though the situation varies in different cities, it is possible to identify a number of problems:

- Rapid traffic growth – in many cities in LEDCs there have been sharp increases in the availability and use of cars. For example, in Taipei, Taiwan the number of cars increased from 11 000 in 1960 to over 1 million in 1990. Not only are there more people and more traffic, but there have also been important changes in the organi-

sation of cities, as new industries and services have been established in the CBD. The result is an increase in home-to-work commuting and congestion.

- Poorly maintained infrastructure – transport networks are expensive to build and maintain, and many governments lack the funds to improve dated transport networks. For example, foreign currency may not be available to import new buses or trains, and as transport fleets get older, buses and trains may break down. The growth of squatter settlements on the edge of cities exerts increasing pressure on already inadequate transport networks. In Mexico City there are traffic queues of more than 90 km (60 miles) per day on average. In Bangkok the 24 km (15 mile) trip into the city from Don Muang Airport can take three hours.

- Transport problems of low-income groups – the cost of using public transport is relatively high, and many urban dwellers may not be able to afford to use it.

- High accident rates.

The inadequacies of transport have encouraged the growth of **paratransit** modes. These are usually part of the informal sector of the economy and take the form of minibuses, hand-drawn and motorised rickshaws, scooters and pedicabs (tricycles used as taxis). These paratransit modes do meet a demand for transport, but frequently add to the problems of congestion on already busy roads (Figure 8.68).

8.68 *'Informal transport' in Dhaka, Bangladesh*

STRUCTURED QUESTION 2

Study Figure 8.69, which shows how the quality of the environment in urban areas is related to levels of economic development.

a For each of the six indicators, describe how the quality of the urban environment is related to levels of economic development. *(6)*

b Explain the patterns you described in **a**. *(6)*

c Discuss the likely consequences of these patterns for people living in large urban areas in LEDCs. *(5)*

d Suggest ways in which the environmental quality in urban areas in LEDCs might be improved. *(5)*

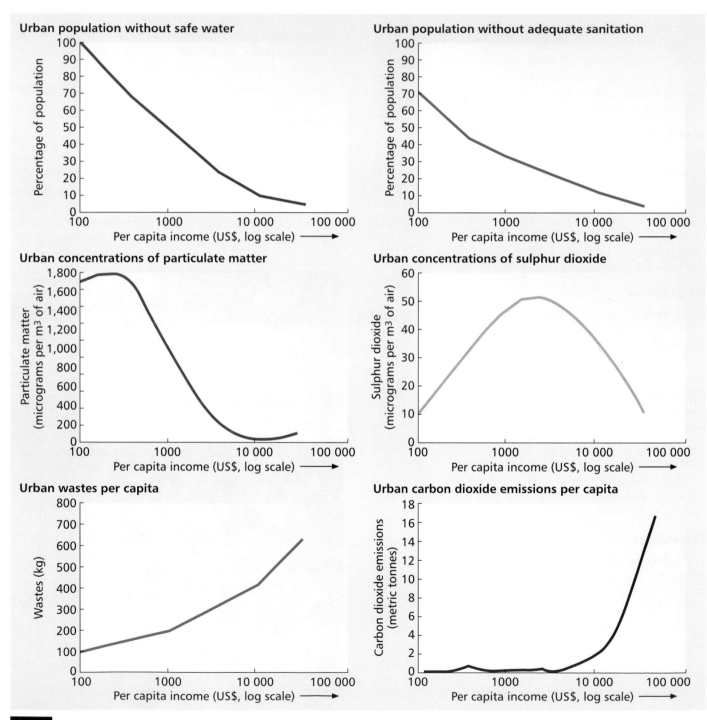

8.69 *How urban environmental indicators relate to economic development*

Land use in cities in LEDCs

Cities in developing countries are numerous and varied. In general they have experienced rapid growth rates as a result of rural-to-urban migration, without high rates of economic growth to accommodate this population increase. It is difficult to generalise about land uses in the 'typical' city, but there are generally three important common features:

- a central concentration of modern commerce, retailing and industry – this is the central business district
- a distinctive zone or sector of elite residential neighbourhoods
- squatter settlements or shanty towns on the periphery of the city or in spaces of unused land within the city.

Figure 8.70 is a model of a Latin American city. This model has several distinct zones.

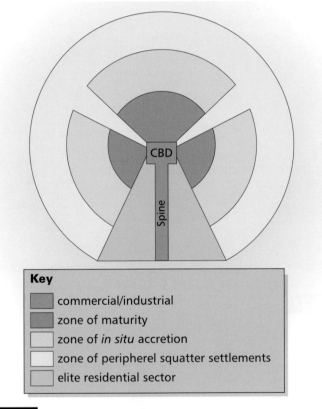

Key

■ commercial/industrial
■ zone of maturity
□ zone of *in situ* accretion
□ zone of peripherel squatter settlements
■ elite residential sector

8.70 *Model of land use in a Latin American city*

1   Central business district

This is characterised by a highly dynamic commercial and administrative core. It is likely to have undergone modernisation in recent decades, with streets widened, old mansions demolished, parking lots created, skyscrapers built and shopping malls constructed.

2   Spine and elite residential sector

Outside the CBD the key feature of Latin American cities is a commercial spine surrounded by an elite residential sector. This is lined by the city's most important amenities, and contains nearly all the professionally built upper-class and upper-middle-class housing. In this sector are found the major tree-lined boulevards, golf courses, museums and zoos. It houses the best theatres and restaurants, and the wealthiest neighbourhoods. The elite nature of the area is maintained by strict zoning which controls development. The spine and elite residential sector represents a dominant aspect of most Latin American cites, despite the fact that only a small proportion of the city's total population lives there.

3   Zones

Away from the spine/elite sector the city's structure consists of a series of concentric zones similar to those suggested by Burgess for Western cities. In general terms, socio-economic levels and housing quality decrease with distance from the central core:

- *Zone of maturity* The zone of maturity is the area of 'better residences' within a city. Whilst the best houses tend to be the older colonial houses, many of the residences in this zone are self-built houses that have been gradually upgraded. There have been gradual improvements in amenities so that the zone of maturity has paved streets, lighting, good public transport, schools and sewerage.

- *Zone of* in situ *accretion* This zone has a great variety of housing types, sizes and qualities. Some of it is 'completed', while on the same block there may be huts and shacks. Only the main thoroughfares are paved, but small shops and schools are present. Many neighbourhoods are without electricity.

- *Zone of peripheral squatter settlements* Recent in-migration from the countryside brings large numbers of people who are not easily absorbed into the urban economy. Many lack regular employment and are unable to afford the relatively high rents or mortgage repayments to secure housing. People build their own homes with available materials such as cardboard, plastic, tin, scrap wood and so on. The makeshift homes are gradually improved as and when money and time allow. Often these settlements are illegal, and lack basic services such as paved roads, piped water and sewerage facilities.

NOTING ACTIVITIES

1   Make a copy of Figure 8.70. Annotate it to show the main features of the Latin American city.

2   In what ways is the model similar to and different from the models of typical Western cities described by Burgess and Hoyt?

3   Describe the processes that have led to the growth of the Latin American city.

4   Explain the role of the zone of *in situ* accretion in the development of the typical Latin American city.

Lagos (Nigeria)

Lagos is spread out over a large area on the mainland and four small islands to the south (Figure 8.71). The city developed from an initial settlement at Iddo and on the northern shore of Lagos Island. Lagos grew quite slowly up until the 1950s, when it experienced a rapid growth in population. This was the result of the demographic transition and the discovery of oil reserves in south-eastern Nigeria, which triggered an economic boom.

This is a difficult site of sandspits and lagoons, which means that the growth of Lagos has been irregular and sprawling, with extremely high densities in the central area. Most of the city's growth has been unplanned and irregular, with swamps, coves and canals impeding efficient development (Figure 8.72).

The rapid growth of the population is linked with a rapid growth of urbanisation. The economy is unable to provide sufficient regularly paid employment and the government lacks sufficient funds to support the population. Lagos thus displays symptoms of overurbanisation. The result has been the development of squatter settlements.

Typical of such developments is the settlement of Olaleye. This developed as a farming settlement outside Lagos but as the city grew it was engulfed and became part of the city. It has experienced rapid population growth. In the mid-1960s about 2500 people lived there compared with today's population of about 25 000. This population growth has been brought about through the rapid increase in the population and built-up area of Lagos, and a railway line constructed in the 1960s.

Population growth has outstripped the city's capacity to deal with the daily movement of people. The central city is an island site, with limited access to road bridges.

Lagos is a primate city. There is a strong contrast between the elite groups and the mass of the population. Rapid population growth means that there are insufficient employment opportunities. This means that incomes are low and people are unable to afford housing.

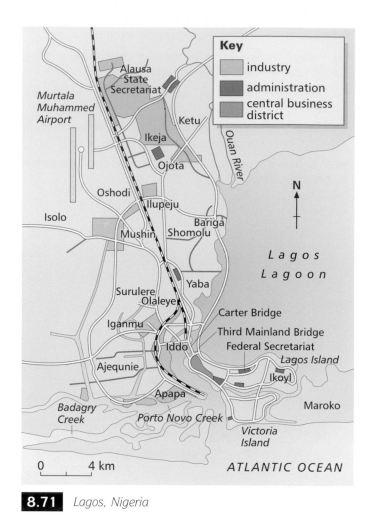

8.71   *Lagos, Nigeria*

8.72   *View of Lagos*

NOTING ACTIVITIES

1   Study Figure 8.73. What was the population of Lagos in **a** 1950, **b** 1990?

2   Suggest reasons why the population of Lagos has grown so rapidly since 1950.

3   Why has the population density of Lagos increased?

4   Study Figure 8.74, which is a simplified map of housing quality in Lagos. Describe and explain the distribution of in-migrant housing areas in Lagos.

5   Suggest why area B has a low population density.

6   What problems have resulted from rapid population growth in Lagos?

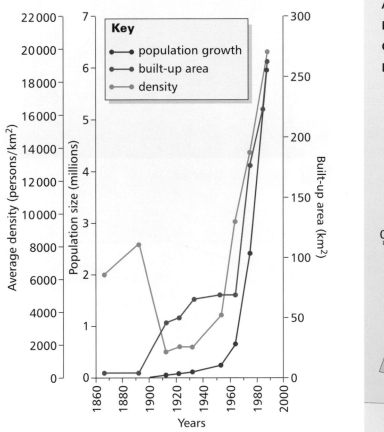

 Growth of population and the built-up area of Lagos, 1860–1990

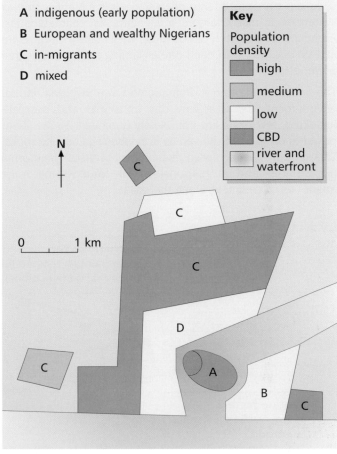

 *Housing areas in Lagos*

A Managing the coastal system

The coast is the narrow strip of shoreline that separates the land from the sea. It is an ever-changing environment and one that requires careful management.

The **coastal system** (see Figure 9.1) is an example of an open system in that it has inputs from outside (for example, river sediment) and outputs into other systems (for example, sediment transported into deep seas). It is a highly complex system with many factors influencing its characteristics. Along any one stretch of coastline, these factors interrelate in a unique way, with each one assuming a different degree of importance.

The coastline is often a dramatic and awe-inspiring natural environment. On a stormy day at a wild and desolate headland, few of us are completely unmoved by the sheer power and might of the sea as it crashes against the rocks. This power has sculpted magnificent coastal landforms (Figure 9.2) and is responsible for transporting huge quantities of sediment along the coast.

9.2 *The coastal landscape*

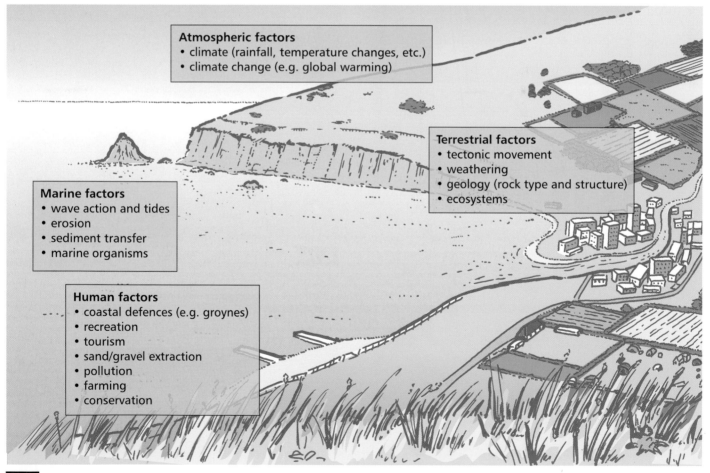

Atmospheric factors
• climate (rainfall, temperature changes, etc.)
• climate change (e.g. global warming)

Terrestrial factors
• tectonic movement
• weathering
• geology (rock type and structure)
• ecosystems

Marine factors
• wave action and tides
• erosion
• sediment transfer
• marine organisms

Human factors
• coastal defences (e.g. groynes)
• recreation
• tourism
• sand/gravel extraction
• pollution
• farming
• conservation

9.1 *The coastal system*

The coastline provides many important and often unique natural habitats and ecosystems. Many species of plant, animal and bird rely upon the sand dunes and the saltmarshes that are commonly associated with coastlines.

Why does the coastline need to be managed?

The single most important reason why the coastline needs to be managed is that it provides many opportunities for human actions and activities. In order for the natural systems associated with the coastline to be maintained and, at the same time, provision made for human exploitation in a sustainable way, the coastline needs to be carefully managed.

There are many individual demands on the coastline and reasons for management:

- A high proportion of the world's population live in towns and cities located on the coast. Many people depend on the sea for their livelihood, e.g. fishing and port activities. Property close to the coast needs to be protected from dangers of flooding or cliff collapse (see the *box* below).
- The coastline is an important tourist destination. Uncontrolled development (hotels, roads, etc.) can readily lead to unsustainable tourism where ecosystems are damaged, for example marshes and coral reefs.

- The construction industry extracts huge quantities of sand and gravel from the sea floor just off the coast in order to produce cement and concrete. The extraction sites need to fit into the natural sediment transfer system to avoid possible harmful knock-on effects elsewhere along the coast.
- The natural coastal environments (e.g. saltmarshes and sand dunes) and landforms need protection and conservation if they are not to be damaged.
- Recent climate change is causing sea levels to rise in some parts of the world. Flooding of low-level land adjacent to the coast is now more likely, resulting in the need for coastal management to plan well inland from the land/sea margin. With increased storminess predicted, coastal erosion may well increase in certain localities, thereby increasing the threat of cliff collapse.

How is the coastline managed?

Perhaps the main problem with managing the coast is that there is no single body or authority in charge of planning. Much of the 'land' is under the control of the local authorities, but this involves all levels from borough councils through to county councils. Coordination and cooperation are essential because, whilst actions may be very localised (for example, the construction of a sea wall), the

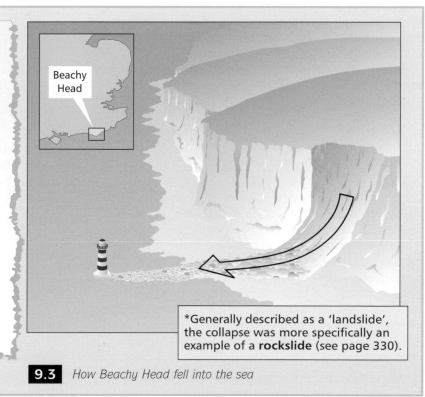

Beachy Head cliff crashes into the sea
by Helen Johnstone

The unmanned Beachy Head lighthouse was effectively rejoined to the Sussex coast by thousands of tonnes of chalk which fell 150 m into the sea at the weekend. Experts believe the collapse – probably the biggest single loss of coastline in living memory – to have been caused by water entering the chalk rock and expanding on freezing, forcing the cliff to crumble. The landslide* was spotted more than 3 miles out to sea by lifeboat crew members who estimate that a 15 m slab of cliff face fell away along a 200 m stretch.

Ray Kemp of the Environment Agency commented, 'One minute we are in a drought situation and then the chalk is sodden. In eight months out of twelve we have had above-average rainfall. The coastline is increasingly vulnerable as climate change starts to bite.'

The Times, 12 January 1999

Beachy Head

*Generally described as a 'landslide', the collapse was more specifically an example of a **rockslide** (see page 330).

9.3   *How Beachy Head fell into the sea*

consequences may be far-reaching, well beyond the boundary of the acting authority.

The 'sea' part of the coastline is managed mostly by government authorities and organisations concerned with, for example, fisheries, mineral extraction, pollution and maritime safety. Again, coordination and cooperation are essential if conflicts are to be avoided. The Environment Agency has a key role to play in England and Wales, particularly regarding pollution and flood defence.

NOTING ACTIVITIES

1   Study Figure 9.1.

   a   Why is the coastline an example of an 'open' system?

   b   List some inputs and outputs to the system.

   c   Try to suggest some additional factors not mentioned in the diagram.

2   Why does the coastline need to be carefully managed?

3   Outline some of the difficulties of coastal management.

What is a Shoreline Management Plan?

In an attempt to establish greater coordination in the management of the coastline of England and Wales, in 1993 the Ministry of Agriculture, Fisheries and Food (MAFF) introduced the concept of the **Shoreline Management Plan**.

A Shoreline Management Plan is defined as:

> 'a document which sets out a strategy for coastal defence for a specified length of coast taking account of natural coastal processes and human and other environmental influences and needs.' (MAFF)

The coastline of England and Wales has been divided into 11 major **sediment cells** (Figure 9.4). These are lengths of coastline which, research has indicated, can be considered to be relatively self-contained as far as the movement of sand or shingle is concerned. In other words, actions taking place in one sediment cell would not be expected to have a significant effect on adjacent cells. Notice on Figure 9.4 that the boundaries of these sediment cells tend to correspond with large estuaries or headlands. Each sediment cell has been split into minor cells to create more manageable-sized units for planning.

Having established the geographical extent of the sediment cells, MAFF has encouraged the setting up of voluntary

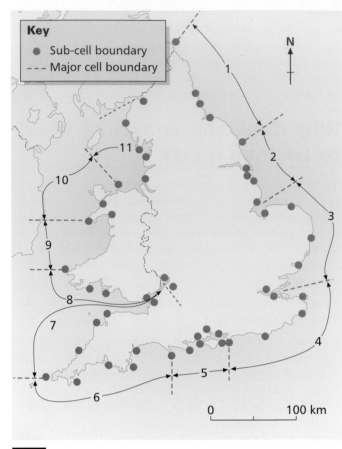

 9.4 *Major sediment cells in England and Wales*

coastal groups comprising the local authorities and other interested parties. The groups are charged with developing an integrated management strategy for their particular cell, taking into account the following aspects:

- natural coastal processes
- coastal defence needs
- environmental considerations
- planning issues
- current and future land use.

Whilst the Shoreline Management Plans have no statutory status, they provide valuable information for the local authorities whose responsibility it is to implement strategies. The main objectives of a Shoreline Management Plan are to:

- assess a range of strategic coastal defence options and agree a preferred approach
- outline future requirements for monitoring, management of data and research related to the shoreline
- inform the statutory planning process and related coastal zone planning
- identify opportunities for maintaining and enhancing the natural coastal environment, taking account of any specific targets set by legislation or any locally set targets
- set out arrangements for continued consultation with interested parties.

The great value of a Shoreline Management Plan is that it helps to inform and provide the overall context for an individual local authority charged with the management of a short stretch of coastline. No longer will piecemeal, isolationist policies be followed. In the future, the strategy of each individual authority should fit in with those of neighbouring authorities for the benefit of the whole sediment cell.

NOTING ACTIVITIES

1   **a**   What is a Shoreline Management Plan?

b   How are Shoreline Management Plans prepared?

c   What are the main purposes of a Shoreline Management Plan?

d   Should Shoreline Management Plans have statutory powers?

2   Make a copy of the sediment cells around England and Wales (Figure 9.4). The following headlands and river estuaries form the boundaries of the major sediment cells. Use an atlas to help you label them on your map.

> Great Orme
>
> Thames
>
> Solway Firth
>
> Land's End
>
> Flamborough Head
>
> Bardsey Sound
>
> Selsey Bill
>
> The Wash
>
> Portland Bill
>
> St David's Head
>
> St Abb's Head
>
> The Severn

3   **a**   Define the term 'sediment cell'.

b   Why do you think headlands and estuaries form the boundaries between major sediment cells?

c   Why have sediment cells been chosen as the basic coastal planning unit? Do you think it is a good idea?

d   How realistic do you think it is to divide the coastline into supposedly self-contained units?

What are the options for coastal defence?

Some stretches of coast are vulnerable to flooding during storms. Other sections suffer from cliff recession which threatens property and farmland. It is the responsibility of the local authority to implement a strategy of coastal defence.

There are four accepted options for defending the coast:

1   Do nothing. This involves no defence activity other than ensuring safety.

2   Hold the existing line. This involves defending the coastline using structures such as sea walls or enlarging beaches (beach nourishment).

3   Advance the existing line. This involves moving defences seaward.

4   Retreat the existing line. This involves **managed retreat**, whereby, through planned intervention, the coastline is allowed to retreat in a controlled way.

In recent years, a number of local authorities have begun to follow the strategy of managed retreat. It is a relatively new concept as previously it was felt that coastlines had to be defended at all costs. Managed retreat allows for the natural processes of change to occur, albeit in a controlled and managed way. It can be considered a **sustainable** form of management.

Managed retreat is well suited to low-lying, saltmarsh environments, for example the estuaries of Essex and Suffolk. The construction of sea walls in the past along this coast has starved saltmarshes of seawater thereby damaging their fragile ecosystems. By allowing controlled breaching of the defences, the saltmarshes can be restored. Over time, the broad intertidal marsh that will be created will help reduce wave energy, thus providing a low-cost form of coastal defence. It is this dual purpose – ecosystem enhancement and low-cost coastal defence – that is beginning to make managed retreat an attractive alternative option.

NOTING ACTIVITIES

1   Briefly outline the four options for coastal defence.

2   What are the twin benefits of 'managed retreat'?

3   How can the option of 'managed retreat' be considered sustainable?

B The Dorset coast: an overview

We have already identified that the coastline is a highly complex environment with many interrelating factors. For this reason, we are going to continue our study of the coastal environment with reference to a single stretch of coastline, the Dorset coast (Figure 9.5). This will enable you to appreciate the whole picture and see how the different aspects of the coastal environment are linked together. Whilst we shall study coastal processes and issues at a number of different localities, they will all be set in the context of the Dorset coast.

The geological background

The Dorset coast has some of the most magnificent coastal scenery in the whole of the UK (Figure 9.6). One of the main reasons for this is the arrangement and nature of the rocks that are exposed at the coast.

Study Figure 9.7. It is a simplified geological map showing the main rock types that outcrop along the Dorset coast. Look at the key and notice that many types of rock are represented, including resistant rocks such as chalk and limestone, and weaker rocks such as clays and shales. It is this variety of rock types that is largely responsible for the varied nature of Dorset's coastal landforms.

During the Alpine **orogeny** (mountain-building period), some 7–50 million years ago, the rocks of southern England, including those of Dorset, were severely crumpled and **folded**. Figure 9.8 shows the effect of this folding to the east

9.6 *The Dorset coast looking west from Durdle Door*

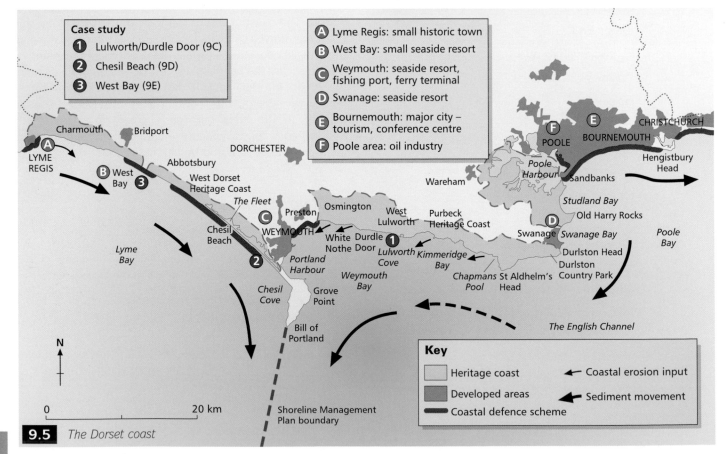

Case study
1. Lulworth/Durdle Door (9C)
2. Chesil Beach (9D)
3. West Bay (9E)

A Lyme Regis: small historic town
B West Bay: small seaside resort
C Weymouth: seaside resort, fishing port, ferry terminal
D Swanage: seaside resort
E Bournemouth: major city – tourism, conference centre
F Poole area: oil industry

Key
- Heritage coast
- Developed areas
- Coastal defence scheme
- ← Coastal erosion input
- ← Sediment movement

9.5 *The Dorset coast*

of Weymouth. Notice that the rocks at the coast have been so intensively folded that they are almost vertical.

Having been subjected to such enormous stresses, it is hardly surprising that the rocks actually 'snapped' in places, forming **faults**. There are many faults trending at right-angles to the line of the coast in southern Dorset, and these have been readily exploited by marine erosion to form inlets and bays.

As Figure 9.5 shows, several stretches of the Dorset coast have coastal defences. There are some schemes aimed at reducing cliff erosion (for example at Lyme Regis and West Bay) and others aimed at reducing the possibility of flooding (for example along parts of Chesil Beach). We will study some of these schemes in detail in subsequent chapters.

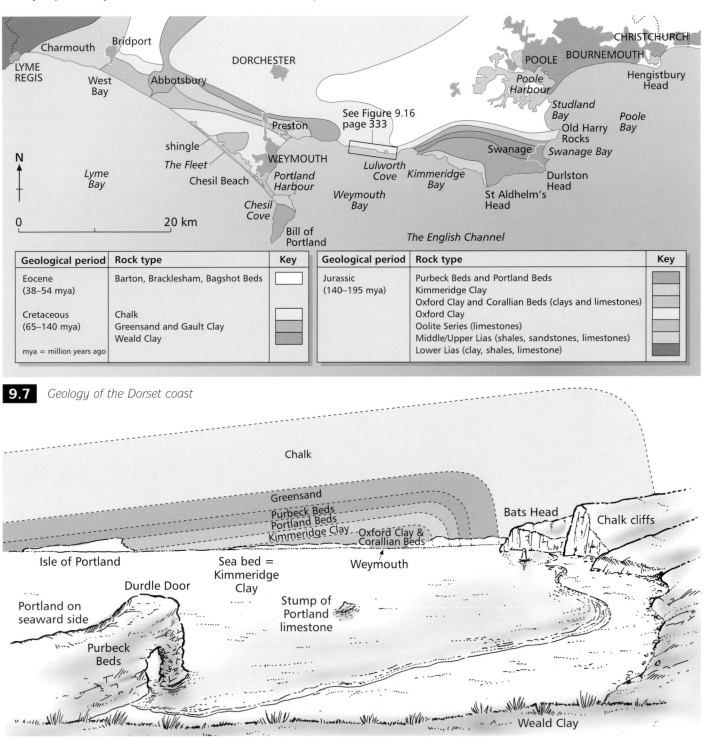

Geological period	Rock type	Key
Eocene (38–54 mya)	Barton, Bracklesham, Bagshot Beds	
Cretaceous (65–140 mya)	Chalk	
	Greensand and Gault Clay	
	Weald Clay	
mya = million years ago		

Geological period	Rock type	Key
Jurassic (140–195 mya)	Purbeck Beds and Portland Beds	
	Kimmeridge Clay	
	Oxford Clay and Corallian Beds (clays and limestones)	
	Oxford Clay	
	Oolite Series (limestones)	
	Middle/Upper Lias (shales, sandstones, limestones)	
	Lower Lias (clay, shales, limestone)	

9.7 *Geology of the Dorset coast*

9.8 *Folding of rocks, looking west towards Weymouth*

The ecological background

There is a great diversity of coastal wildlife, and several stretches of the Dorset coast have been designated as **Sites of Special Scientific Interest** (SSSIs). The Fleet – the lagoon sheltered by Chesil Beach (Figure 9.5) – is an important locality for birds, especially terns and mute swans. The limestone cliffs, for example on the Isle of Portland, are home to wild orchids and rare butterflies, and there are several rare species of plant on the sand dunes at Studland. The headlands of Portland, Durlston and St Aldhelm's (see Figure 9.5), due to their relative proximity to continental Europe, are important landmarks for migratory birds. Figure 9.9 identifies some of the ecologically important habitats in Dorset.

The human background

Much of the Dorset coast is inhospitable as far as development is concerned. There are few natural harbours (such as Poole Harbour) and only Poole and Weymouth have developed as ports. Several settlements have developed as seaside resorts (e.g. Swanage, West Bay, Lyme Regis and Bournemouth) and tourism is extremely important to the local economy. Some 16 million people visit the coast every year and an estimated 37 500 jobs are tourism-related.

Whilst the area is not especially industrial, there are significant oil reserves in the Poole Harbour area, and oil is extracted on land at Wytch Farm, near Poole.

Managing the Dorset coast

Approximately 25 per cent of the coastline has been developed. Much of this took place in a rather haphazard and uncontrolled manner prior to the Town and Country Planning Act of 1947. Since then, the emphasis has been on protection and conservation. In 1972 two stretches of the coastline were designated as a **Heritage Coast** (Figure 9.5). This designation, with an emphasis on conservation, imposes very strict planning restrictions and requires the drawing up of Heritage Coast Plans.

Much of the large-scale strategic planning for the coastline is undertaken by Dorset County Council. At a more local scale, planning is the responsibility of the district and borough councils. Offshore, various government authorities, e.g. the Department of the Environment and the Ministry of Agriculture, Fisheries and Food, are responsible for planning and enforcing regulations. In addition, there are several other bodies and organisations that have a consultative role in the planning process. Two Shoreline Management Plans include reference to the Dorset coast. The boundary between them is Portland Bill (see Figure 9.5). The item in the *box* opposite describes how the Dorset coast is managed.

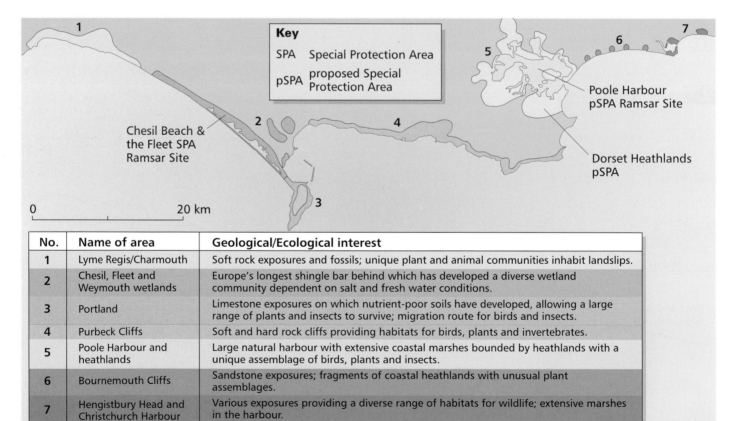

No.	Name of area	Geological/Ecological interest
1	Lyme Regis/Charmouth	Soft rock exposures and fossils; unique plant and animal communities inhabit landslips.
2	Chesil, Fleet and Weymouth wetlands	Europe's longest shingle bar behind which has developed a diverse wetland community dependent on salt and fresh water conditions.
3	Portland	Limestone exposures on which nutrient-poor soils have developed, allowing a large range of plants and insects to survive; migration route for birds and insects.
4	Purbeck Cliffs	Soft and hard rock cliffs providing habitats for birds, plants and invertebrates.
5	Poole Harbour and heathlands	Large natural harbour with extensive coastal marshes bounded by heathlands with a unique assemblage of birds, plants and insects.
6	Bournemouth Cliffs	Sandstone exposures; fragments of coastal heathlands with unusual plant assemblages.
7	Hengistbury Head and Christchurch Harbour	Various exposures providing a diverse range of habitats for wildlife; extensive marshes in the harbour.

9.9 *Important coastal habitats in Dorset*

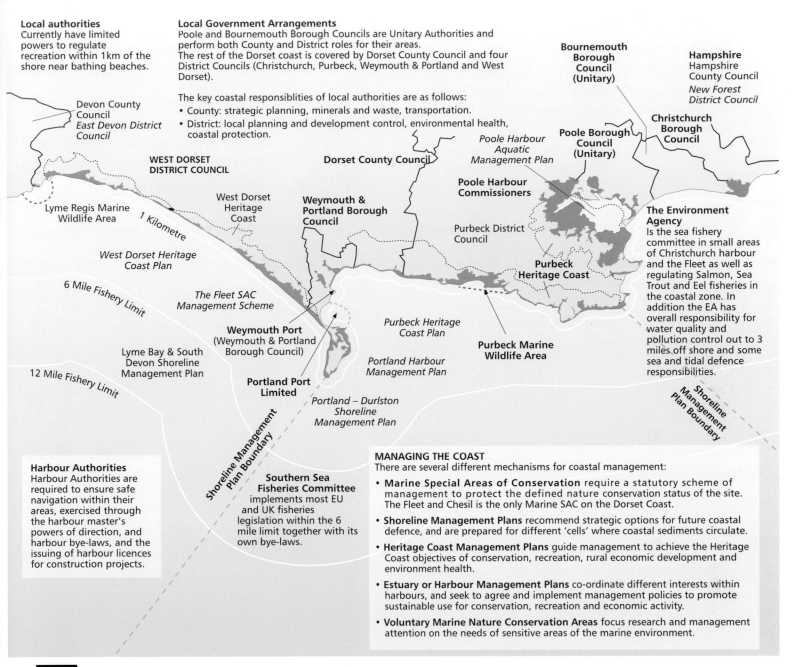

Local authorities
Currently have limited powers to regulate recreation within 1km of the shore near bathing beaches.

Local Government Arrangements
Poole and Bournemouth Borough Councils are Unitary Authorities and perform both County and District roles for their areas.
The rest of the Dorset coast is covered by Dorset County Council and four District Councils (Christchurch, Purbeck, Weymouth & Portland and West Dorset).

The key coastal responsiblities of local authorities are as follows:
• County: strategic planning, minerals and waste, transportation.
• District: local planning and development control, environmental health, coastal protection.

Bournemouth Borough Council (Unitary)

Hampshire
Hampshire County Council
New Forest District Council

Devon County Council
East Devon District Council

Poole Borough Council (Unitary)

Christchurch Borough Council

WEST DORSET DISTRICT COUNCIL

Dorset County Council

Poole Harbour Aquatic Management Plan

West Dorset Heritage Coast

Weymouth & Portland Borough Council

Poole Harbour Commissioners

Lyme Regis Marine Wildlife Area

1 Kilometre

West Dorset Heritage Coast Plan

Purbeck District Council

The Environment Agency
Is the sea fishery committee in small areas of Christchurch harbour and the Fleet as well as regulating Salmon, Sea Trout and Eel fisheries in the coastal zone. In addition the EA has overall responsibility for water quality and pollution control out to 3 miles off shore and some sea and tidal defence responsibilities.

6 Mile Fishery Limit

Purbeck Heritage Coast

The Fleet SAC Management Scheme

Weymouth Port (Weymouth & Portland Borough Council)

Purbeck Heritage Coast Plan

Purbeck Marine Wildlife Area

Lyme Bay & South Devon Shoreline Management Plan

Portland Harbour Management Plan

12 Mile Fishery Limit

Portland Port Limited

Portland – Durlston Shoreline Management Plan

Shoreline Management Plan Boundary

Shoreline Management Plan Boundary

Harbour Authorities
Harbour Authorities are required to ensure safe navigation within their areas, exercised through the harbour master's powers of direction, and harbour bye-laws, and the issuing of harbour licences for construction projects.

Southern Sea Fisheries Committee implements most EU and UK fisheries legislation within the 6 mile limit together with its own bye-laws.

MANAGING THE COAST
There are several different mechanisms for coastal management:

• **Marine Special Areas of Conservation** require a statutory scheme of management to protect the defined nature conservation status of the site. The Fleet and Chesil is the only Marine SAC on the Dorset Coast.

• **Shoreline Management Plans** recommend strategic options for future coastal defence, and are prepared for different 'cells' where coastal sediments circulate.

• **Heritage Coast Management Plans** guide management to achieve the Heritage Coast objectives of conservation, recreation, rural economic development and environment health.

• **Estuary or Harbour Management Plans** co-ordinate different interests within harbours, and seek to agree and implement management policies to promote sustainable use for conservation, recreation and economic activity.

• **Voluntary Marine Nature Conservation Areas** focus research and management attention on the needs of sensitive areas of the marine environment.

9.10 *Who manages the Dorset coast?*

NOTING ACTIVITIES

1   Make a list of the authorities and organisations that are involved in management of the Dorset coast. Present your list in the form of a Venn diagram, comprising two overlapping circles, one circle involving the land and the other the sea. The overlapping segment covers both land and sea.

2   Having completed your diagram, write a paragraph explaining why coastal management is a very complex process.

3   Do you think coastal management should be made simpler? Work in small groups to outline a proposal for a more straightforward form of coastal management. Consider which authorities would have statutory powers (the ability to pass and implement laws) and which would be purely advisory. Should the land and the sea be managed separately?

NOTING ACTIVITIES

1 Study Figure 9.5.

   a Describe, with the aid of a simple sketch map, the movement of sediment along the Dorset coast. Identify on your map the inputs of sediment from cliff erosion.

   b Why do you think Portland Bill was chosen as a boundary between two Shoreline Management Plans?

2 Study Figure 9.7. Look closely at the Swanage area. Notice that Swanage itself is in a bay, Swanage Bay, and that on either side of the bay there are two headlands, Old Harry Rocks to the north and Durlston Head to the south. This is a very good example of the way in which rock type can affect the shape of a coastline. The rocks that form the headlands here are relatively resistant to erosion, whereas the rocks that form Swanage Bay are weaker and more easily eroded.

   Draw a simple sketch map of this area, including the geological details, and write a paragraph accounting for the shape of the coastline.

3 Describe the effects of folding and faulting on the Dorset coast.

4 Use the information in Figure 9.9 to draw a simple sketch map to show the main habitats along the Dorset coast.

5 Having read through this chapter, explain the need for the careful management of the Dorset coast.

C Coastal erosion: the Lulworth Cove area

Kimmeridge Bay

Lulworth Cove

Stair Hole

9.11 *View of part of the Dorset coast – Durdle Door is about 1 mile west of Lulworth (off the bottom of this photograph)*

The length of the Dorset coastline, from Durdle Door in the west to Lulworth Cove in the east, is one of the most dramatic and well-known stretches in the UK. There are many classic coastal landforms here (see Figure 9.11), which is why it makes an excellent case study of coastal erosion. Before looking at the landforms in detail, we first need to consider the processes in operation at the coast.

Coastal processes

Several groups of processes interact to create coastal erosional landforms:

- weathering processes
- mass movement
- marine erosion processes
- human activity.

Weathering processes

Weathering processes operate on both the cliffs and the shoreline platform, which is periodically covered by seawater. Whilst the processes themselves are the same as elsewhere, there are some differences in the way they operate and in their relative importance:

- **Frost-shattering** is promoted by the presence of large amounts of water (although saltwater freezes at a lower temperature than freshwater) and the tendency for coastal rocks to be severely cracked by the processes of erosion. However, coastal locations tend to be warmer than inland localities, reducing the likelihood of frosts. One rock that is thought, though, to be vulnerable to frost-shattering in the Lulworth area is chalk, which is porous and has many cracks and joints.

- **Salt crystallisation** is active at the coast due to the presence of saltwater. In the tidal or wave splash zone, saltwater is readily available and evaporation leads to the formation of salt crystals. As these crystals grow, they exert pressures within the rock, causing it to gradually break apart.

- **Wetting and drying** is also active in the tidal/splash zone. Clays and shales are common along the Dorset coast and they are particularly prone to the expansion and contraction associated with wetting and drying. One of the results of this process is the formation of cracks which further weakens the rock and encourages erosion.

- **Solution,** involving the dissolving of soluble minerals, is an active process at the coast. It particularly affects limestone and chalk, both of which are present along the Dorset coast. The effect of solution or, more specifically, carbonation, is to enlarge the joints in the rocks, creating pitted and jagged rocky surfaces.

- **Biological action,** involving marine organisms, is important. Shellfish, for example piddocks, have a specially adapted shell to enable them to literally drill into rock. They are particularly active in chalk where their activity produces a sponge-like honeycombed rock pitted with holes (Figure 9.12). Seaweed attaches itself to rocks and the action of the sea can be enough to cause the swaying seaweed to prise away loose rocks from the seafloor. Blue-green algae, often associated with pollution, are thought to secrete chemicals capable of promoting processes such as solution.

9.12 *Honeycombed chalk fragment showing piddock holes*

NOTING ACTIVITIES

1 **a** What coastal factors promote the process of frost-shattering?

  b Despite these factors, why is frost-shattering often less effective here than inland?

2 Outline briefly the operation of salt crystallisation, wetting and drying, and solution weathering.

3 Study Figure 9.12.

  a Describe the characteristics of the rock.

  b What has happened to the rock to cause these characteristics?

  c Discuss whether this is truly weathering, or erosion.

  d What other forms of biological weathering operate at the coast?

Mass movement

Many of the processes described in Chapter 2E (pages 56–62) are active at the coast. This is largely because the rocks at the coast are being constantly undercut by the sea, making them potentially unstable. In addition, the presence of large amounts of water encourages slippage. Some coasts, such as the Dorset coast, have been developed for tourism, and the additional weight of buildings, together with alterations to the natural drainage of the slopes, have contributed to their potential instability.

Perhaps the most common forms of mass movement at the coast are rockfalls and landslips (rotational slips). **Rockfalls** are most commonly associated with cliffed coastlines and relatively resistant rocks, such as limestone and chalk. The undercutting by the sea combines with weathering to loosen chunks of rock on the cliff face. Eventually they fall, either as separate fragments or as part of a much larger section of cliff in the form of a **rockslide**. Look back to the *box* on page 321 to read about a recent rockslide at Beachy Head.

Landslips (rotational slips) tend to be associated with weaker rocks, such as shales and clays. They result from the saturation of the rocks and their subsequent slumping, often triggered by undercutting or a heavy rainstorm. The saturated material commonly flows out from the base of the cliff to form a tongue of mud (see Figure 9.13). The role of landslips (rotational slips) is examined in more detail in Chapter 2E.

9.13 *A mudflow tongue*

NOTING ACTIVITIES

1 Suggest the factors that promote mass movement at the coast.

2 Suggest which processes are most likely to operate with the following rock types. Give reasons for your suggestions.

   • A relatively tough limestone, well-bedded and dipping seawards

   • Horizontal, thin beds of alternating shales, clays and sands.

   (Look back to Chapter 2E to remind you of the mass movement processes.)

3 Discuss the importance of mass movement as part of the coastal system. Consider, for example, its role in cliff retreat, sediment supply, etc.

Marine erosion processes

The dominant force behind the processes of coastal erosion is the action of waves (see *box* below).

Waves and tides

Waves

Waves are created by the frictional drag of air as it moves over the water. This explains why high waves are usually associated with windy, stormy conditions, and still seas are associated with calm weather conditions.

In open water, the bobbing of the waves results from an orbital motion (Figure 9.14). However, close to land this orbital motion is interrupted by friction with the sea bed, forcing it to become more elliptical in shape. This causes the wave to extend upwards and break. Water rushes up the beach as **swash** and then draws back towards the sea as **backwash**.

The strength of a wave is determined by wind speed, wind direction and **fetch** (distance of open water over which the wind has blown). The most powerful waves are generated by storms, or by seismic events such as earthquakes, which result in huge waves called **tsunamis**.

At Lulworth, the longest fetch is several thousand kilometres across the Atlantic Ocean in a south-westerly direction. This is also the direction of the prevailing winds, and explains why powerful waves often pound the Lulworth area.

At the coast, two types of wave can been identified (Figure 9.15).

• A **constructive wave** has a low frequency and a low height and it is associated with a gentle offshore profile. Constructive waves gradually build up a beach, as the swash of each wave extends a long way up the beach before being interrupted by the backwash from the previous wave. Backwash is frequently less powerful because of water percolating into the beach itself.

• A **destructive wave** is associated with stormier conditions. It is a frequent wave and grows tall before crashing down onto the beach. There is little forward movement of water and most of the energy is used in scouring the beach. Backwash is more significant than swash. Destructive waves are commonly associated with steeper beach profiles.

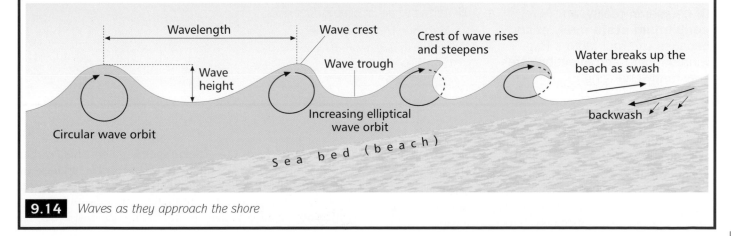

9.14 *Waves as they approach the shore*

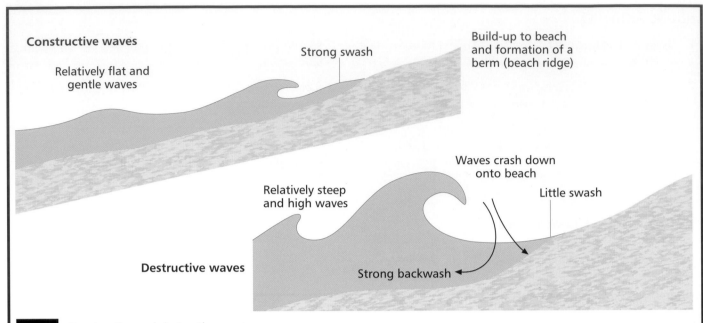

Constructive waves

Relatively flat and gentle waves

Strong swash

Build-up to beach and formation of a berm (beach ridge)

Relatively steep and high waves

Waves crash down onto beach

Little swash

Destructive waves

Strong backwash

9.15 *Constructive and destructive waves*

Most beaches are subjected to the alternating action of constructive and destructive waves. Constructive waves build up a beach and result in a steeper beach profile. This encourages waves to become more destructive (destructive waves tend to be associated with steeper profiles). However, as time goes by, the destructive waves transport sediment in a seaward direction, thus reducing the beach angle once more and encouraging the formation of constructive waves. This is a good example of a **negative feedback** in geography, where a process becomes less effective the longer it operates. It encourages the existence of an equilibrium state. Of course, in reality, an **equilibrium state** rarely exists because external factors (e.g. wind strength and direction) are constantly changing!

Tides

Tides are another important factor. They are caused by the gravitational pull of the moon as it orbits the Earth and, to a lesser extent, by the sun as it is orbited by the Earth. Most coastal locations experience two tides each day. The strongest tides occur twice a month and are called **spring tides**. They occur when the sun,

moon and Earth are all in alignment, causing a strong gravitational pull on the sea.

Variations in the frequency and height of tides are influenced by the presence of land masses and shallow water. Coastal areas that are affected by a narrow tidal height-range experience more concentrated zones of erosion and deposition.

NOTING ACTIVITIES

1 What causes waves to form?

2 Describe the causes and consequences of a breaking wave. Use the correct terminology in your description.

3 With the aid of simple diagrams, describe the differences between a constructive wave and a destructive wave.

4 **a** What is the meaning of a 'negative feedback' in geography?

b How can this concept be applied to constructive and destructive waves?

There are four processes of erosion that operate at the coast:

1 Quarrying When a wave breaks against a cliff face, the sheer force of the wave is capable of detaching loose fragments of rock. Air may be blasted into cracks (cavitation) to further loosen and then detach rock fragments.

2 Abrasion This is the effect of rock fragments held within the water, bashing or grinding against rocky surfaces. In stormy conditions, a lot of sediment may be flung directly at the base of a cliff, contributing towards its erosion. Attrition is also active on intertidal rock platforms, as sediment is drawn back and forth, grinding away at the platform as it does so.

3 **Attrition** Individual fragments are themselves eroded when they crash into each other. Sharp edges are gradually worn down to produce rounded pebbles typical of many shingle beaches.

4 **Solution** Seawater flowing over soluble rocks such as limestone and chalk will cause some dissolving to occur.

Human activity

People are a significant agent of erosion in the Lulworth area. The coastline is popular with walkers who wander about on the cliffs causing vegetation and soil to be worn away. Footpaths are heavily used and footpath erosion is serious in several places. Once vegetation has been destroyed by trampling, the exposed soil and rock is more readily weathered and eroded.

NOTING ACTIVITIES

1 Write brief definitions of the processes of coastal erosion.

2 Suggest how the following factors affect the rate of coastal erosion:

- wind speed and direction

- fetch

- nature of the waves

- rock type

- geological structure (folding/faulting)

- tides

- human activity.

CASE STUDY

The Lulworth Cove area

One of the most important controls in the formation of the landforms in the Lulworth Cove area (Figure 9.11) is geology. Figure 9.16 shows the detailed geology of the area.

There are two important controls on the development of the coastal landforms along this stretch of coast:

- the rocks run broadly parallel to the coast
- the rocks vary greatly in their resistance to erosion.

The most resistant rock in the area is the Portland Stone (limestone). Along much of the coast this thin band of rock (Figure 9.16) forms a barrier to marine erosion because it is so strong. Its very steep dip (remember that the rocks here have been intensely folded so they are almost vertical) gives the limestone a wall-like appearance. However, the intense folding created cracks within the limestone which have subsequently been exploited by the sea to form **caves**.

At Stair Hole (Figure 9.16) the sea has broken through the wall of Portland Stone, gaining access to the weaker Purbeck Beds behind. These shales and clays have been readily

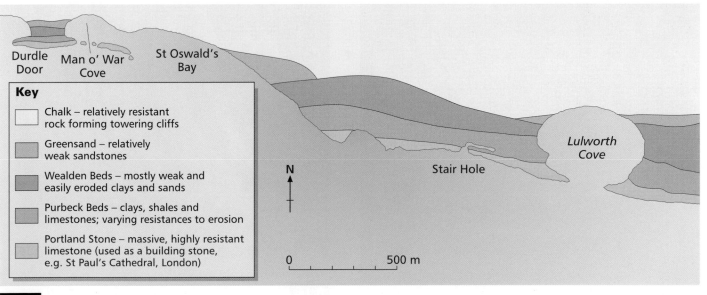

Key

- ☐ Chalk – relatively resistant rock forming towering cliffs
- ▨ Greensand – relatively weak sandstones
- ▨ Wealden Beds – mostly weak and easily eroded clays and sands
- ▨ Purbeck Beds – clays, shales and limestones; varying resistances to erosion
- ▨ Portland Stone – massive, highly resistant limestone (used as a building stone, e.g. St Paul's Cathedral, London)

Durdle Door   Man o' War Cove   St Oswald's Bay

Lulworth Cove

Stair Hole

N

0   500 m

9.16 *Geology of the Lulworth Cove area*

eroded to form Stair Hole (Figure 9.17). Look carefully at Figure 9.17 and notice the following features:

- the wall of steeply dipping Portland Stone on the right-hand side
- the crumpled nature of the Purbeck Beds
- the sea, which enters Stair Hole through two arches
- Lulworth Cove in the distance beyond the crumpled rocks.

Once the sea has broken through the outer wall of Portland Stone, erosion of the Purbeck Beds is rapid (Figure 9.18). Look again at Figure 9.16 and locate Durdle Door to the far west. Notice that the Portland Stone forms isolated rocky outcrops, remnants of a once continuous cliff wall.

Durdle Door is an impressive **arch** (Figure 9.19) and shows what the smaller arches at Stair Hole will probably look like at some time in the future. Other arches would have existed in the past linking the now separate outcrops of limestone that mark the entrance to Man o' War Cove (see Figure 9.16). These isolated outcrops of rock are called **stacks**. Over time, the stacks are eroded by the sea to form rocky outcrops exposed only at low tide, called **stumps** (Figure 9.20). As a cliffline retreats, all that is left is a bare rock surface called a **wave-cut platform** which is exposed at low tide but covered at high tide.

Continued erosion of the weaker rocks, once protected by the Portland Stone but now exposed to the full fury of the sea, results in the formation of a **bay** or cove. The clearest example of a bay is Lulworth Cove.

Lulworth Cove is a small, almost circular bay with a narrow opening to the sea. Whilst it is convenient to assume that the sea broke through the Portland Stone, rather as it has done elsewhere along the coast, there is some evidence that the break in the limestone was formed by fluvial rather than marine action.

During periods of ice advance in the Pleistocene period, southern England was affected by **periglacial** conditions (it was at *no time* covered by glaciers!). During these times, the chalk would have been frozen and rendered impermeable. When the surface layers of rock and soil thawed in the summer, water would have flowed over the surface forming river networks. At the same time, the sea level would have been much lower than it is today, as water was locked-up as ice on the land. It is thought that, during these times, the break in the Portland Stone was formed as seasonal rivers

9.17 *Stair Hole*

9.18 *Differential erosion of Purbeck Beds, looking west in Man o' War Cove*

flowed south to the sea, cutting through the tougher band of rock (Figure 9.20).

As sea levels have recovered to their present levels, marine erosion has exploited the weaker rocks behind the Portland Stone to form Lulworth Cove. As Figure 9.16 shows, the back of the bay is made of chalk. This is a relatively tough rock and forms a steep backdrop to Lulworth Cove and the

other bays along this stretch of coast (Figure 9.11). At the foot of the cliffs is a beach.

The Dorset coast from Durdle Door to Lulworth Cove is significantly indented. There are several headlands and rocky outcrops as well as a number of bays (Figure 9.21). The shape of the coastline has an important effect on the waves as they approach the coast, causing them to be bent or refracted. **Wave refraction** (see *box* on next page) is a very important concept because it affects the precise locations of high-energy and low-energy waves and, as a result, erosion and deposition.

9.19 *Durdle Door*

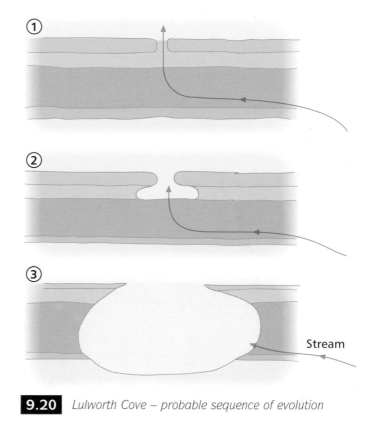

9.20 *Lulworth Cove – probable sequence of evolution*

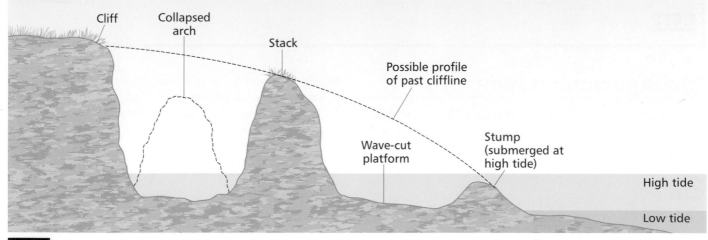

9.21 *Formation of stacks and stumps*

Wave refraction

If waves approach a coastline at an angle, or if the coast itself is indented, wave fronts may become distorted (Figure 9.22). This in turn distorts the spread of energy, with energy being concentrated at the headlands and dissipated in the bays. The high-energy waves carry out erosion at the headlands, accounting for the presence of cliffs, stacks and wave-cut platforms. In the bays, the low-energy waves tend to deposit sediment to form a beach.

The concept of negative feedback can be seen to operate here. Firstly, variations in rock strength lead to the formation of headlands and bays. This causes wave refraction which in turn encourages erosion of the headlands but deposition in the bays. If conditions remain unaltered for a long period of time (which they don't!), it is possible to envisage a state of equilibrium whereby the shape of the coastline remains static due to a balance between the potential erodability of the rocks and the effect of wave refraction.

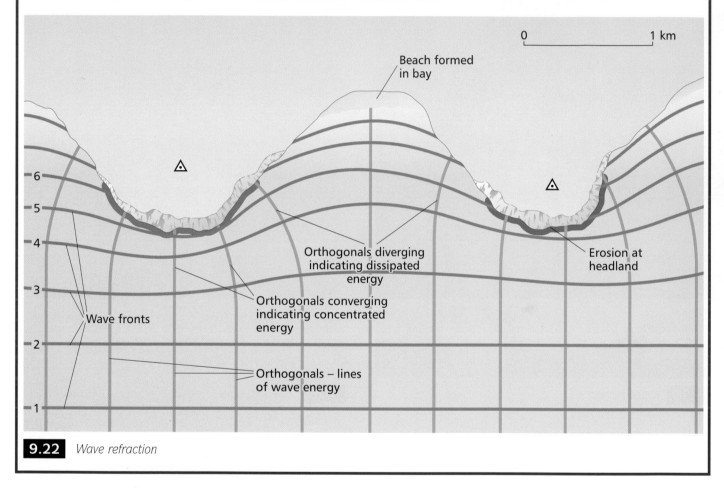

9.22 *Wave refraction*

Management issues

The cliffs in the Lulworth area are prone to sudden collapses and rockfalls. In 1977 a school party studying geology was buried by a cliff fall in Lulworth Cove itself, killing a teacher and one of the students. Two years later a women was killed when a 3 m rock overhang collapsed onto the beach near Durdle Door.

Cliff collapse is intermittent, which makes it hard to plan for. It most commonly occurs after a prolonged period (several years) of frost weathering and wetting and drying, which progressively weakens the rock. The trigger for an individual collapse is frequently a period of heavy rain.

People are warned about the dangers of possible cliff collapse by notices posted at regular intervals, and footpaths are kept well back from the cliff edge. Fences are also erected in potentially dangerous localities. However, people do take risks and it is impossible to prevent them from doing so, short of erecting a barbed-wire fence all along the clifftop.

If sea levels rise as a result of global warming, and storms become more frequent, cliff collapse is likely to become a more serious management issue in the future. It may well be that large sections of potentially active cliffs need to be completely sealed off from public use. What do *you* think?

EXTENDED ACTIVITY

The aim of this activity is for you to produce your own report about the Lulworth Cove area case study. In addition to the information presented in this chapter, other resources can be found at the Stanley Thornes website.

A comprehensive report should include the following elements:

- a map of the area showing the geology and the location of the main coastal features and places
- sketches of photographs (e.g. Figure 9.17) with full labels (annotations)
- written accounts of various features, explaining their formation and detailing the processes responsible
- an attempt to show the sequential nature of the coastal landforms in the area (Figure 9.23). This could form the basis of a plan for your report.

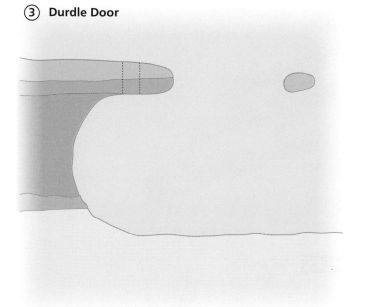

①

Portland Stone

Purbeck Beds

Wealden Beds

Greensand

Chalk

② **Stair Hole**

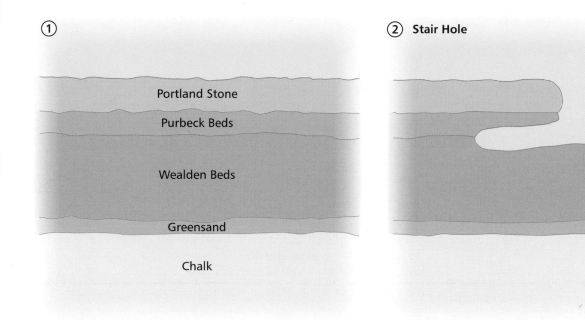

③ **Durdle Door**

④ **Man o' War Cove and St Oswald's Bay**

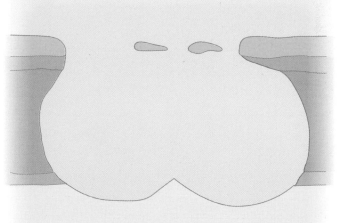

9.23   *Probable sequence of coastal evolution west of Lulworth*

STRUCTURED QUESTION

a Look at Figure 9.24.

(i) What is the wave feature marked X? *(1)*

(ii) What is the wave feature marked Y? *(1)*

(iii) What name is given to the distance marked Z? *(1)*

(iv) What causes waves to break as they approach the beach? *(3)*

(v) What are the differences between constructive waves and destructive waves? Make reference to both the causes and the effects of the waves. *(4)*

b Look at Figure 9.25. For each diagram, suggest how geology, subaerial processes (weathering, mass movement) and marine processes may have combined to produce the distinctive cliff profile. *(3 x 3)*

c **(i)** Choose *one* feature produced by coastal erosion and name an example that you have studied. *(1)*

(ii) Describe the nature and appearance of the feature. You may use a diagram to help your description. *(4)*

(iii) How have marine and other processes combined to produce the characteristics of your chosen feature? *(4)*

d Outline the management issues for a stretch of cliffed coastline that you have studied. *(4)*

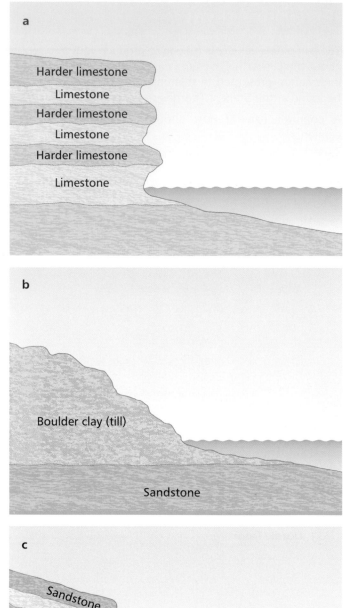

a

Harder limestone
Limestone
Harder limestone
Limestone
Harder limestone
Limestone

b

Boulder clay (till)

Sandstone

c

Sandstone
Clay
Sandstone
Clay
Sandstone

Clay

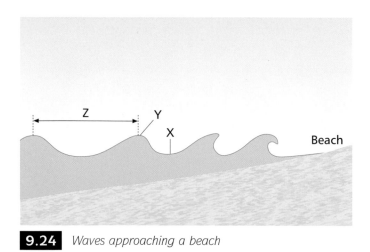

9.24 *Waves approaching a beach*

9.25 *Cliff profiles*

D Coastal deposition: Chesil Beach

Coastal deposition takes place when wave velocity drops so that the water is unable to hold sediment in suspension. Material is dropped and then transported from one place to another. There are three main sources of sediment:

- **Cliff erosion** – this is very localised but is important along several stretches of the Dorset coast.
- **Fluvial sediment** is brought to the coastal system by rivers flowing off the land. This is the main source of coastal sediment.
- **Offshore sediments** are brought onshore by tidal currents or wave action.

Landforms of coastal deposition

1 Beaches

Beaches are formed by constructive waves piling up material ahead of them. They can vary enormously in terms of their size, morphology (features) and material. Along the Dorset coast, whilst there are a few sandy beaches in sheltered bays (e.g. Swanage), most of the beaches are made of shingle. The most common beach feature is the **berm** which is a ridge about 1–3 m in height. There may be several berms at different positions on a beach, as Figure 9.26 illustrates. Notice that the berm furthest up the beach results either from a storm or from a particularly high tide. In common with other beach features, berms are dynamic features and are subject to change on a daily basis.

Another common feature, found between the high and low tide marks, is the **beach cusp** (Figure 9.26). A cusp is a semicircular-shaped accumulation of sand or shingle enclosing a small hollow or depression. They vary in size from a few centimetres to several metres across and are formed when waves break directly onto a beach and when the actions of swash and backwash are powerful.

2 Spits

A **spit** is a bank of sand or shingle protruding from the coast into the sea, or partially across a river estuary. It results from the movement of sediment along the coast by the process of **longshore drift** (Figure 9.27). Unprotected by solid land, spits are vulnerable features and are frequently breached

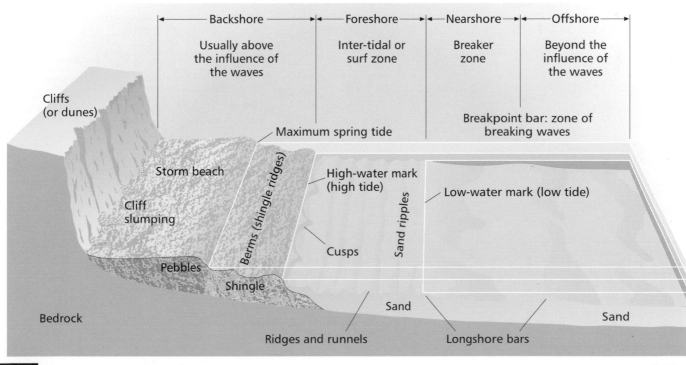

Backshore	Foreshore	Nearshore	Offshore
Usually above the influence of the waves	Inter-tidal or surf zone	Breaker zone	Beyond the influence of the waves

Breakpoint bar: zone of breaking waves

Cliffs (or dunes)

Storm beach

Cliff slumping

Berms (shingle ridges)

Maximum spring tide

High-water mark (high tide)

Low-water mark (low tide)

Sand ripples

Cusps

Pebbles

Shingle

Sand

Bedrock

Ridges and runnels

Longshore bars

Sand

9.26 *The features and landforms of a beach*

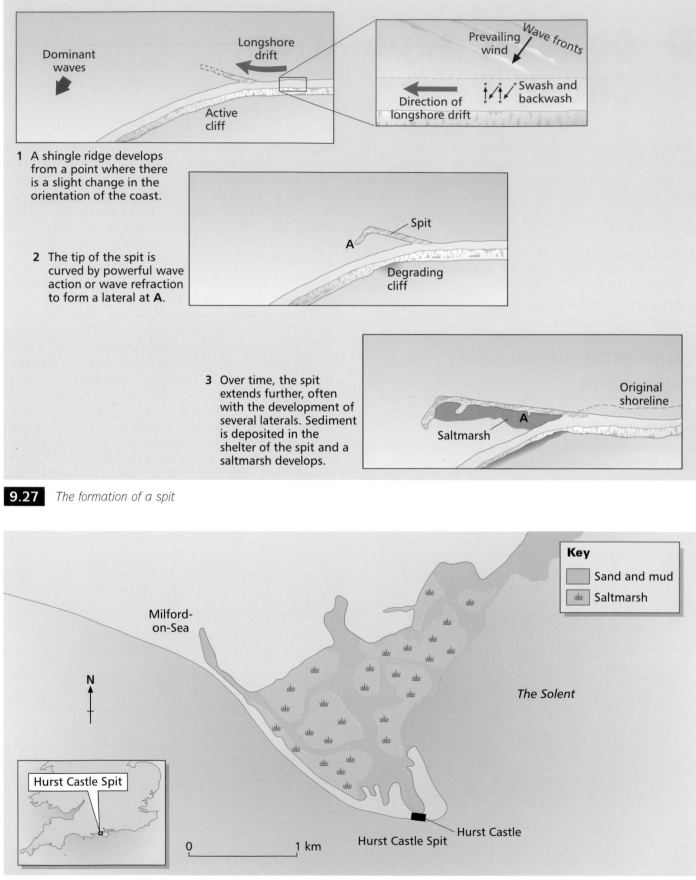

1 A shingle ridge develops from a point where there is a slight change in the orientation of the coast.

2 The tip of the spit is curved by powerful wave action or wave refraction to form a lateral at **A**.

3 Over time, the spit extends further, often with the development of several laterals. Sediment is deposited in the shelter of the spit and a saltmarsh develops.

9.27 *The formation of a spit*

9.28 *Hurst Castle Spit, Hampshire*

during storms. Figure 9.28 describes the characteristics of Hurst Castle spit, at the eastern end of Christchurch Bay and about 20 km from Bournemouth.

In places, spits can grow to extend right across an estuary joining two headlands. This feature is called a **bay bar** and, were it not for constant dredging, a bay bar might well have formed across the entrance to Poole Harbour (Figure 9.5 page 324). Where a spit links the mainland to an island, it is known as a **tombolo** (see case study below).

3 Forelands

There are places along the coast where two sediment cells converge to cause an accumulation of sediment. A triangular feature called a **foreland** may result (see Figure 9.29).

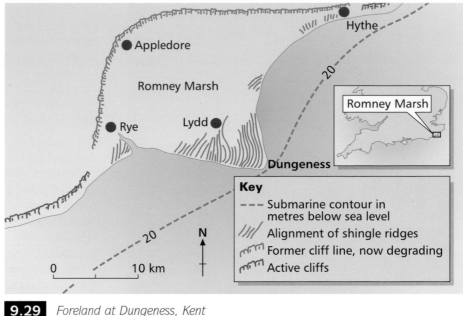

9.29 *Foreland at Dungeness, Kent*

NOTING ACTIVITIES

1 Study Figure 9.26.

   a What is a berm?

   b Why are berms found at different heights on a beach?

2 Study Figure 9.27.

   a Describe, with the aid of simple diagrams, the formation of a spit. Use labels to explain why a spit forms at a particular point on the coastline.

   b What is meant by the term 'degrading cliff'? Describe its likely appearance and explain its position.

   c Why are laterals formed?

   d Explain the presence of a saltmarsh in the lee of the spit.

3 Draw a sketch of Hurst Castle spit (Figure 9.28) and add labels describing the processes and features associated with the formation of the spit.

4 Study Figure 9.29.

   a Describe the shape of the foreland at Dungeness. Use the scale to help you give an indication of the feature's dimensions.

   b What evidence is there that sediment is accumulating in this locality?

   c Describe the alignment of the shingle ridges. What do they indicate about the growth of the foreland?

   d Draw a simple sketch of the foreland. Mark on the main features and add labels to describe the processes at work, including the likely directions of longshore drift in the area.

CASE STUDY

Chesil Beach

Chesil Beach (Figure 9.30) is an extensive shingle ridge stretching for 29 km from West Bay to Chiswell on the Isle of Portland (Figure 9.31). It is a good example of a tombolo. In the lee of Chesil Beach is an important wetland environment called The Fleet (Figure 9.31).

Chesil Beach is an imposing ridge, rising in height in an easterly direction from 7 m above sea level at Abbotsbury, to 15 m above sea level at Chiswell. Another interesting fact is that the pebble size increases in the same direction, from an average of 2.5–3.0 cm diameter at Abbotsbury, to an average of 7 cm diameter at Chiswell.

How was Chesil Beach formed?

Looking at a map, it seems that the most obvious explanation involves the process of longshore drift from west to east. However, it is now generally considered that Chesil Beach started life as an offshore shingle bank. During a period of rapid sea-level rise some 14 000–7000 years ago,

as glaciers and ice sheets melted following the last ice advance, the shingle bank was pushed towards the land by powerful waves and currents. Today, there is still some landward movement as the beach rolls over onto itself, but as a feature it is fairly static. There is some research which suggests that Chesil Beach may form its own closed sediment cell, as there is little evidence of material entering the cell from outside.

The increase in pebble size from west to east is attributed to the process of longshore drift (see Figure 9.31). The dominant direction of longshore drift transports pebbles of all sizes from west to east. Occasionally, when the wind comes from a south-easterly direction, longshore drift operates in reverse, from east to west. In this direction, longshore drift is less effective and is only capable of transporting the smaller pebbles. Over time, therefore, the pebbles become sorted so that the larger ones are found to the east and the smaller ones to the west. The increase in height of the beach also reflects the dominant west to east movement of material.

9.30 *Chesil Beach, from the northern end of Portland, with Chiswell in the foreground*

NOTING ACTIVITY

With the aid of a simple sketch map, describe the main characteristics of Chesil Beach and discuss its likely mode of formation.

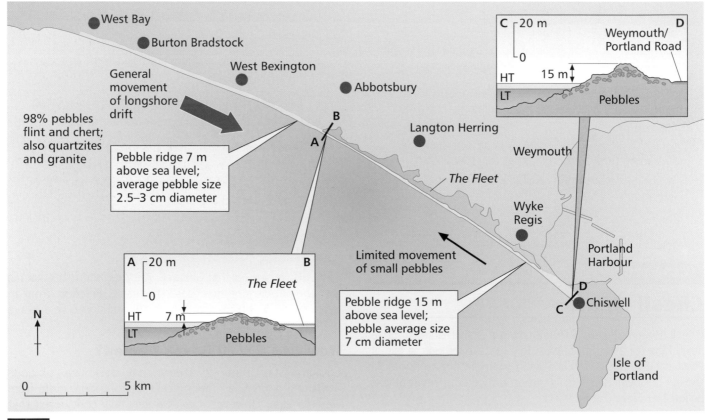

9.31 *Chesil Beach*

The Chesil Beach sea defence scheme

At the most easterly end of Chesil Beach is the small village of Chiswell. Look at Figure 9.30 and notice that there is only the ridge between the village and the open sea. Chiswell has a long history of flooding due to storms. In 1824, 26 people perished, and in two recent events in 1978 and 1979 there was significant flooding of property and communications.

In response to the two floods, Wessex Water Authority (now part of the Environment Agency) and the Weymouth and Portland Borough Council commissioned a firm of consultant engineers to devise a scheme to protect the people of Chiswell from future flooding. Read the information in the *box* below to see what action was taken.

Flooding on Chesil Beach

Flooding experienced by the 134 inhabitants of Chiswell can be attributed to both physical and human causes. Storm surges, coinciding with a high tide and a depression from the Atlantic, lead to highly unstable conditions on the beach, whose crest becomes lower and more susceptible to percolation flooding (Figure 9.32).

Every 5–10 years part of Chiswell is flooded, and families have been evacuated for up to five days. Damage caused to the beach is, however, repaired by natural processes, as constructive waves build back the beach. Thus the main impact of a lack of management here is the stress caused to the local residents and the cost to them and the local authority of recovering from fairly regular flooding.

Responses have been varied. A cost benefit assessment was undertaken in relation to keeping the A354 Weymouth road open. A Chesil Residents Action Group (CRAG) was formed to try to ensure that residents' wishes were taken into consideration. This time the conflict arose between the cost of protecting the area, providing security for the residents, and still enabling the landscape to retain its original appearance in this important tourist destination.

- In 1983 the 300 m long esplanade wall was modified and a concrete apron added with steel toe piles, to prevent undermining.
- A new wave wall was added to reduce the amount of overtopping. At the same time ramps were added to improve access for pedestrians, and facilities were provided for fishermen to remove their boats through gates.
- The most expensive part of the scheme, a culvert with openings in the seaward side and top, to allow water to enter, was built along the landward side of the beach. Long steel pipes were added in the underlying Kimmeridge Clay to drain water away. The culvert discharges into an open channel leading into Portland Harbour, via culverts under Weymouth Road.
- The final stage was completed when the A354 was raised above previous flood levels.

The success of this scheme has been tested several times and, since its completion, the area has not suffered any serious flooding. In 1989 some overtopping did occur, yet flooding was minimal, the drains proved adequate and the £5 million spent was felt to be justified.

'The scheme is meant to cope with normal storm conditions, but this was one of the worst this century … a once every 50 years storm, and it was always known the sea would come over the defences if this happened.'

Borough Council Leader David Hall

A compromise has been reached between the amount of money that could justifiably be spent and the amount of protection that could be given. As a reflection of their confidence in the scheme, the council have built new houses within the former flood-risk zone. So far, the effectiveness of this particular investment has not been brought into question.

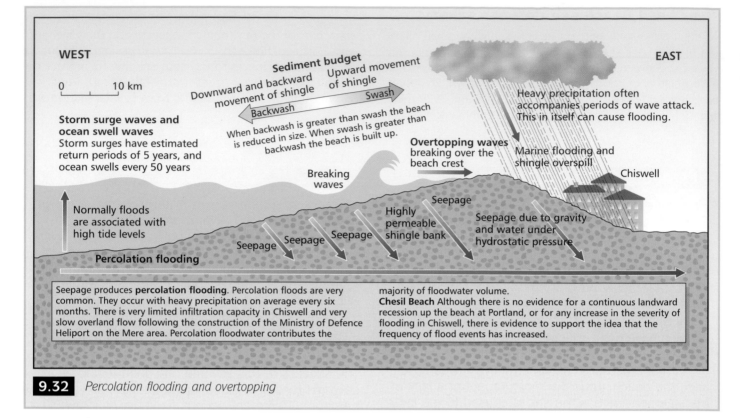

WEST

0 ____ 10 km

Sediment budget

Downward and backward movement of shingle | Upward movement of shingle

Swash

Backwash

When backwash is greater than swash the beach is reduced in size. When swash is greater than backwash the beach is built up.

Storm surge waves and ocean swell waves
Storm surges have estimated return periods of 5 years, and ocean swells every 50 years

Breaking waves

Overtopping waves breaking over the beach crest

EAST

Heavy precipitation often accompanies periods of wave attack. This in itself can cause flooding.

Marine flooding and shingle overspill

Chiswell

Normally floods are associated with high tide levels

Seepage

Highly permeable shingle bank

Seepage due to gravity and water under hydrostatic pressure

Percolation flooding

Seepage   Seepage   Seepage

Seepage produces **percolation flooding**. Percolation floods are very common. They occur with heavy precipitation on average every six months. There is very limited infiltration capacity in Chiswell and very slow overland flow following the construction of the Ministry of Defence Heliport on the Mere area. Percolation floodwater contributes the

majority of floodwater volume.
Chesil Beach Although there is no evidence for a continuous landward recession up the beach at Portland, or for any increase in the severity of flooding in Chiswell, there is evidence to support the idea that the frequency of flood events has increased.

9.32 *Percolation flooding and overtopping*

NOTING ACTIVITIES

1 Study Figure 9.30. Describe the site of Chiswell and suggest why it is vulnerable to coastal flooding.

2 With the aid of a diagram, describe the causes of flooding at Chiswell.

3 Apart from reducing the flood risk for the residents of Chiswell, why else was it deemed necessary to create a sea defence here?

4 Describe the defence scheme that was implemented.

5 How successful have the sea defences been to date?

6 Comment on the wisdom of allowing new houses to be constructed in the former flood-risk zone.

STRUCTURED QUESTION 1

Study Figure 9.33, which shows an idealised beach profile at low tide.

a Name the beach features labelled X. *(1)*

b Explain the presence of more than one feature marked X. *(3)*

c Suggest *three* possible sources for the sediment on this beach. *(3)*

d Outline the physical factors likely to influence the distribution of beach material on the beach profile. *(4)*

e How might marine processes cause the beach profile to change over time? *(4)*

f Study Figure 9.34, which shows the location of a marked tracer along a beach on selected dates in January.

   (i) State the total distance moved by the tracer over the month as shown on the map. *(1)*

   (ii) Between which two dates was the tracer moving fastest? *(1)*

   (iii) Suggest an explanation for the variations in direction and rate of movement of the tracer. *(4)*

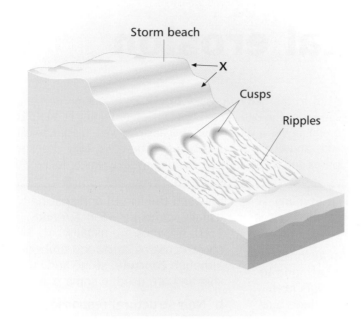

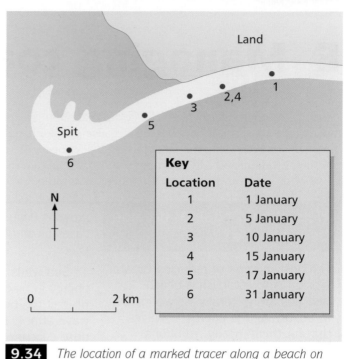

9.33 *An idealised beach profile at low tide*

9.34 *The location of a marked tracer along a beach on selected dates in January*

STRUCTURED QUESTION 2

Study Figure 9.35, which shows the coastal sediment budget for East Anglia.

a (i) State the percentage of the input of sediment at Cromer that moves:

   A west

   B east. *(2)*

(ii) Suggest where the rest of the input of sediment has gone. *(2)*

b Suggest why the north Norfolk cliffs produce more sediment than those at Dunwich. *(2)*

c With the aid of a diagram, explain the process of longshore drift. *(3)*

d Why might the figures for the amount of sediment moved by longshore drift vary from year to year? *(3)*

Key

All values in thousand m³/yr

60 → Direction and amount of sediment moved by longshore drift

⬅ Input of sediment

⬅ Removal of sediment

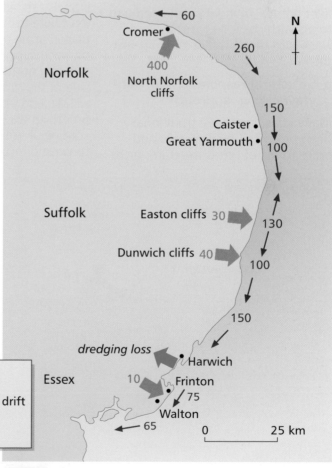

9.35 *The coastal sediment budget for East Anglia*

E Managing coastal erosion

Coastal erosion usually involves cliff collapse and retreat. For much of the coast this is not an issue, because people are not involved; however, where the coast has been developed, there is a threat to farmland and to property and communications. In the past, some developments were undertaken without due regard to the natural coastal processes in operation. Some sections of the British coast have been retreating at a rate of several metres a year, usually because they are made of weak sands and clays. This has brought the coast ever closer to houses and roads, thereby creating a management issue.

How should coastal erosion be managed?

The best method of reducing coastal erosion is to establish a beach. Beaches are generally very effective at absorbing the energy of waves that hit them. Many traditional methods of coastal management rely upon the creation or the simulation of a beach. Increasingly, however, there is a belief that we should not fight nature; that we should come to terms with coastal processes and work with them, rather than against them.

a Structural responses: the 'hard engineering' approach

Many coastlines have traditionally been protected by sea walls and other forms of basal structure, or by groynes (Figure 9.37). In many cases a combination of methods is used:

1 Sea walls *Aim:* to absorb wave energy in place of a beach and to protect the bottom of cliffs from wave attack. Designs include bull-nose concrete walls, slatted revetments and stone and concrete blocks.

2 Groynes *Aim:* to trap and then stabilise shingle by slowing down longshore drift. Groynes are usually placed at an angle of 5–10° to the perpendicular in order to prevent scouring on the leeward (downdrift) side of the groyne. The exact angle will depend upon the direction of the prevailing waves. The length and spacing of groynes are also related to wave direction; the relationship between the length and spacing usually varies between 1:4 and 1:10. Groynes are usually constructed of hardwood timber, although concrete, stone and steel sheeting are used in some places.

b Non-structural response: the 'soft engineering' approach

This approach is based largely upon the philosophy that human interference should be minimised, and that natural processes of beach renewal should be maximised.

(i) Beach nourishment This is the replacement of beach material that has been removed by longshore drift and by attrition. Successful schemes have been in operation since the 1930s. The Copacabana Beach in Rio de Janeiro, Brazil, was rebuilt in 1970 with beach sand from a nearby area. In Bexhill, Sussex, erosion accelerated after extensive sea defences were constructed at Eastbourne in the early 1940s. Groynes alone did not solve the problem, so between 1975 and 1985, in addition to the construction of over 100 new groynes, over 150 000 m³ of shingle was placed on the beach. The shingle came from three main sources: nearby land-based gravel pits, dredged from offshore, and recycled from nearby Hastings beach, where gravel accumulation was becoming a problem. At nearby Seaford, major engineering works were completed in 1987 at a cost of £13 million. Every year about 160 000 tonnes of shingle is moved from the eastern part of the beach

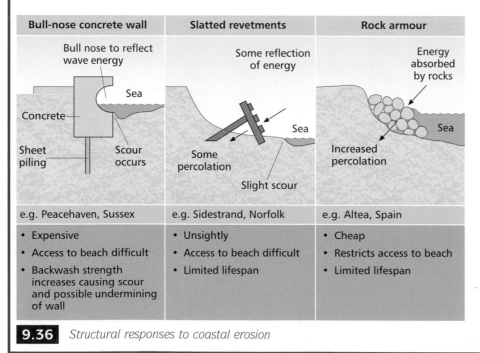

Bull-nose concrete wall	Slatted revetments	Rock armour
Bull nose to reflect wave energy · Sea · Concrete · Sheet piling · Scour occurs	Some reflection of energy · Sea · Some percolation · Slight scour	Energy absorbed by rocks · Sea · Increased percolation
e.g. Peacehaven, Sussex	e.g. Sidestrand, Norfolk	e.g. Altea, Spain
• Expensive • Access to beach difficult • Backwash strength increases causing scour and possible undermining of wall	• Unsightly • Access to beach difficult • Limited lifespan	• Cheap • Restricts access to beach • Limited lifespan

9.36 *Structural responses to coastal erosion*

In Chapter 9A we discussed the options of coastal defence. When property and communications are involved, the most common option has been to 'hold the line'. The options of 'do nothing' and 'managed retreat' have, understandably, been less popular (particularly with people whose houses are perched on a clifftop!).

Measures aimed at reducing coastal erosion have traditionally involved so-called 'hard' engineering structures, such as sea walls and groynes. Recently, there has been a move towards 'soft' engineering solutions, for example **beach nourishment**. Here, sand and shingle are added to a beach to increase its height and width. A beach forms a natural buffer to the sea and so an enlarged beach helps to dissipate wave energy, therefore reducing coastal erosion. Read through the article in the *box* below to find out more about this and other strategies of coastal erosion management.

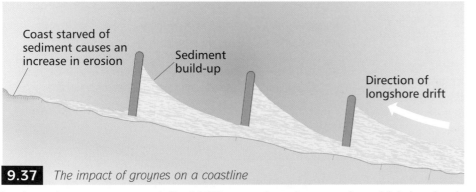

9.37   *The impact of groynes on a coastline*

currently over 1 m per year, but reaches far higher levels at a local scale. In parts of the USA, 'retreat' policies have been in operation for some time. In northern California, small properties have been required to be built at least 30 times the rate of erosion away from the shore since legislation was passed in 1979. Large properties must be at more than double the distance.

Source: Stanley Thornes, Geofile 201, *September 1992*

back to the west, at a cost (in 1987 prices) of £60 000 per annum. The defences were completed just in time to save Seaford from severe flooding during the 'hurricane' of October 1987.

(ii) Beach stabilisation by reducing slope angle, providing drainage of cliffs and revegetating vulnerable coastal areas. This has been carried out successfully at Whitby in Yorkshire, and in several locations in Cornwall.

(iii) Wave alteration This involves trying to reduce the erosional power of the waves. Plans were considered to build a submerged barrier off the Norfolk coast. The aim is to change the rhythm of the waves and thus reduce their ability to erode the cliffs.

c   Other recent initiatives

In 1991 plans were submitted to build a £30 million scheme off the Holderness coast. This involved building an 'armour-plated' reef of mine and quarry waste, bound together by recycled fuel waste from nearby furnaces. In early 1991 Lincolnshire tried a scheme to combat dune erosion which involved the planting of discarded, rooted Christmas trees!

d   Do nothing!

Instead of asking the question 'how should we protect the coast from eroding?', perhaps we should ask ourselves whether we should protect the coast at all. Millions of pounds of taxpayers' money are spent annually on protecting coastal land from natural forces (in 1991 over £26 million, and in 1992 an estimated £35 million, was spent by central government alone on coastal defence work). At Barton on Sea a £1.3 million project was destroyed within four years. Sea defences need constant attention and lots of money; perhaps it would be cheaper to let nature take its course, and instead pay compensation to those affected.

e   Retreat from the coast!

Norfolk became the first British county to propose a ban on development upon coastal areas. In their current structure plan they propose a 75 m 'setback line' from the sea. Erosion in Norfolk is

NOTING ACTIVITIES

1   With the aid of diagrams, describe the following 'hard engineering' measures of coastal defence. Explain their design, their purpose, the effect on the environment, and any negative side-effects associated with them.

  a   sea walls

  b   groynes

  c   rock armour.

2   What is meant by 'beach nourishment', and how does it reduce erosion?

3   Why do you think 'soft engineering' solutions are proving more popular nowadays?

4   Is the 'do nothing' option a realistic option? Examine the arguments for and against this policy.

The problem of coastal erosion in Dorset

The rocks that are exposed along the Dorset coast vary enormously. Some of the tougher rocks, such as chalk, are vulnerable to rockfalls and rockslides (see page 330), whereas the weaker shales and clays are more prone to slumping and mudslides.

The cliffs between Lyme Regis and West Bay (Figure 9.5 page 324) consist of Lower Cretaceous and Jurassic rocks. These rocks comprise a mixture of thinly bedded and alternating sandstones, clays, shales and limestones. Whilst the sandstones readily transmit water, the clays and shales do not. They become heavy and unstable when saturated, and contract and crack during periods of dry weather. The presence of alternating beds often leads to water accumulating between beds of different permeability and, with a slight seaward dip, slippage along bedding planes is encouraged. The cliffs are made yet more unstable by the constant pounding and undercutting of the sea.

Look at Figure 9.38. It shows a cliff profile typical of the stretch of coast from Lyme Regis to West Bay and illustrates the processes operating on it. Notice the variety of rock layers and see how they dip gently towards the sea. Several terrestrial processes are in operation, for example rotational slips, mudslides and mudflows (see Chapter 2E for details of the mechanisms involved). These forms of mass movement all require large amounts of water. Notice on Figure 9.38 the absence of the drier processes such as rockfalls and rockslides. Wave attack at the base of the cliff forms a secondary cliffline broken only by the lobe of a mudslide. Once sediment enters the sea, it becomes part of the sediment cell and may move along the coast by longshore drift, or offshore, depending upon the nature of the waves and the currents.

NOTING ACTIVITY

Study Figure 9.38.

a Suggest how the following factors promote instability on the cliff profile:

   (i) the nature of the rock types

   (ii) the fact that several different rocks are exposed

   (iii) the seaward dip of the rocks

   (iv) undercutting by the sea.

b At some localities, properties have been built on the clifftop. How can this increase cliff instability?

c Look back to Chapter 2E and describe the operation of the following processes that are active on this cliff profile:

   (i) rotational slips

   (ii) mudflows

   (iii) mudslides.

d What happens to the sediment eroded from the cliff?

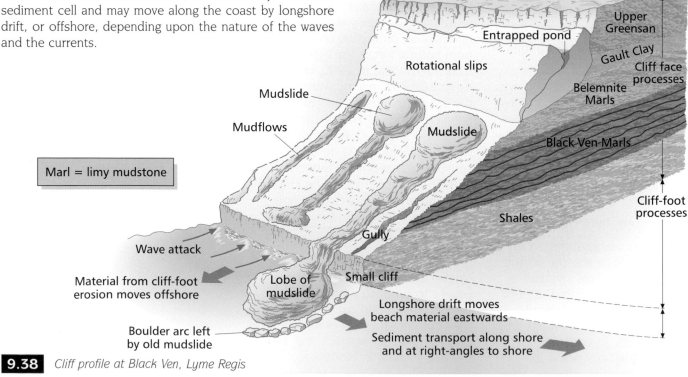

Marl = limy mudstone

9.38 *Cliff profile at Black Ven, Lyme Regis*

Managing the coast at West Bay

West Bay is a small seaside resort and fishing village at the mouth of the River Brit, a couple of kilometres south of the town of Bridport (Figure 9.5 page 324). Two piers extend into the sea from the river mouth to provide shelter and a clear passage for boats using the small harbour (Figure 9.39).

What is the geology at West Bay?

There are great contrasts in the geology of West Bay (Figure 9.40). The cliffs to the west of the bay, known as West Cliff, are mostly made up of a Middle Jurassic rock called Frome Clay, which comprises clays with thin beds of limestone. The overlying layer of Forest Marble is also made up of clays and limestones. At West Cliff there are three faults which run through the rocks (see Figure 9.41). Not only have these faults weakened the structure of the rocks, but they have enabled water to seep into them, further reducing their cohesion. It is the combination of weak rocks together with the faults that make the cliffs at West Cliff extremely unstable. To the east of the bay, most of East Cliff is made of a sandstone called the Bridport Sands. Overlying this is a relatively resistant band of Inferior Oolite (limestone).

9.39 *View of West Bay from East Cliff*

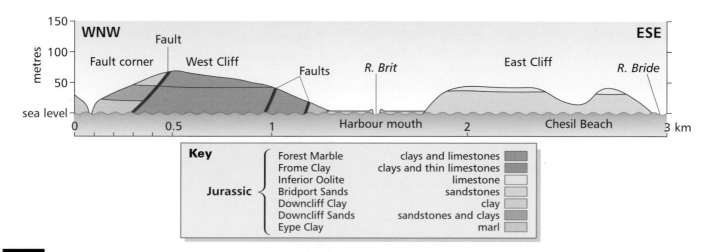

9.40 *Geology of West Bay*

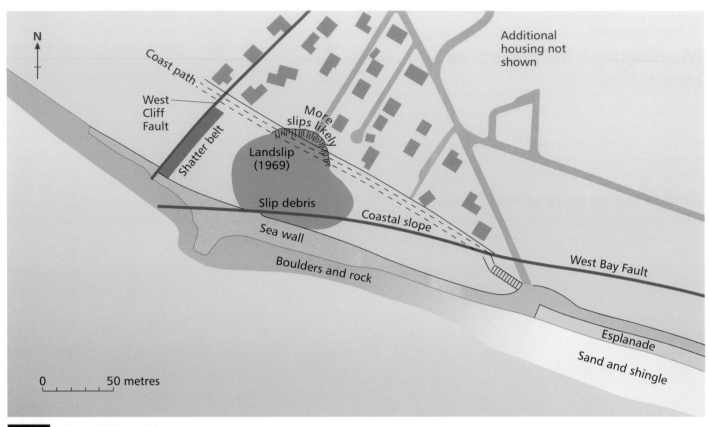

9.41 *West Cliff landslip*

What are the management issues?

There are several management issues at West Bay:

- West Cliff has been prone to landslips for many years, and a recent large landslip has threatened an area of housing (see Figure 9.41). Photographs and old maps have revealed that clifftop recession at West Cliff has been in the order of 25 m since 1903. This works out as an average of 0.44 m per year.

- Immediately to the east of the bay, the main threat is from flooding. As Figure 9.39 shows, the land here is flatter, and several buildings are situated close to the seafront, along with the harbour and harbour buildings. Serious flooding occurred in this part of West Bay as recently as 1990. The shingle beach here is the start of Chesil Beach, but the general easterly movement of sediment by longshore drift has gradually depleted it. In the past, some of the shingle was also extracted for commercial purposes. A beach is the best form of sea defence (see *box* on pages 346-47) because it acts as a barrier and, in causing waves to break, it absorbs much of the potentially destructive wave energy.

- The piers and harbour area need to be managed to maintain access to the harbour for fishing and pleasure boats.

- The village depends heavily on tourism, so the area needs to be managed sympathetically to provide public access whilst ensuring safety. Aesthetic consideration needs to be given to management schemes.

What has been done so far?

Various forms of coastal protection have been implemented over the years (see Figure 9.42). However, they have been piecemeal and have concentrated on individual problems affecting small stretches of the coast. There has not yet been an integrated management plan.

The following schemes have been implemented to date:

- In 1887 a sea wall (The Esplanade on Figure 9.42) was constructed to protect the cliffs at West Cliff. There were

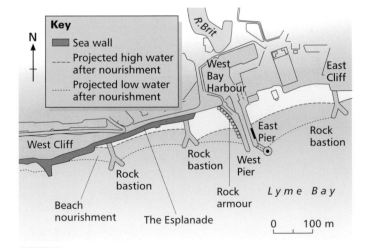

9.42 *Current coastal defences at West Bay and those proposed by scheme A*

9.43   *West Bay Esplanade, regraded cliff and coast footpath*

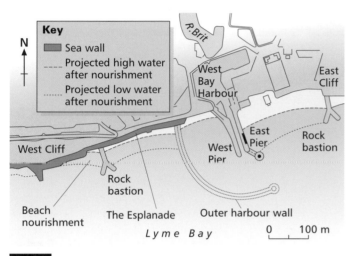

9.44   *Proposed coastal defences at West Bay: schemes B and C*

further extensions in 1959 and 1971, and strengthening work has been carried out periodically.

- Following damage to the sea wall, rock armour has been placed along the western side of West Pier.

- A new rock bastion (a groyne made of huge boulders) has been constructed to add stability to the beach and promote the build-up of a beach (the middle one in Figure 9.42).

- The cliffs behind the Esplanade at West Cliff have been regraded and drained (Figure 9.43) in recent decades to improve stability; however, this has not prevented further landslips from occurring. The most recent engineering work took place in 1995–96 and involved protecting a clifftop property that would otherwise have had to be demolished.

- To the east of the bay, beach nourishment has taken place to maintain the beach.

What are the options for the future?

The 'do nothing' option is not considered to be an option at all at West Bay. The beach to the east would become increasingly reduced in size and the risk of flooding in the harbour area would increase. The harbour piers are old and need maintaining to prevent their collapse. If this happened, the harbour itself would be forced to close. The sea wall to the west is in need of repair. Already storm waves overtop the wall and, if the projected rise in sea level occurs as a result of global warming, damage to property is likely. Cliff collapses at West Cliff will become more likely if the sea wall breaks up, threatening property on the clifftop.

Assuming that action needs to be taken to 'hold the line', three schemes have been proposed:

Scheme A This is the simplest and cheapest scheme and is estimated to take only a year to complete. It involves strengthening the existing defences, extending the rock bastions and constructing two new ones, increasing the rock armour at West Pier, and nourishment of the beach on both sides of the bay (Figure 9.42). This scheme is likely to reduce the risk of flooding around the harbour area, although storm waves will still be able to enter the harbour directly between the two piers. An increase in the size of the beaches may well attract more tourists to the area, but the increase in physical structures may have a negative visual impact.

Scheme B This involves all the improvements in scheme A but with the extension of one of the rock bastions to form an outer harbour wall (see Figure 9.44). This would help protect the piers from further erosion, reduce the flood risk to the harbour, and improve harbour access. However, it will have a significant visual effect on the coastline and may lead to increased erosion elsewhere along the coast, particularly at East Cliff and beyond.

Scheme C This is much as scheme B but, instead of a rock bastion outer harbour wall, it would involve the construction of a more solid stone wall together with a walkway. Mooring facilities would be improved with the possible development of a marina. This is the most expensive and ambitious scheme and would take at least two years to construct. As with scheme B, the introduction of a large artificial feature protruding into the sea is bound to have a destabilising effect on the coastal system, at least in the short term.

Before deciding on which scheme to adopt, a **cost–benefit analysis** needs to be made. This involves attempting to quantify the 'costs' (e.g. the cost of materials, the impact on the environment, the possible negative knock-on effects elsewhere) and the 'benefits' (e.g. the reduction of costs associated with damage from flooding and landslips, increased tourism, and an increase in the use of the harbour). Some of these 'costs' are relatively easy to quantify, for example the actual physical building costs, but others are more difficult. If the benefits outweigh the costs, then the scheme is deemed to be a worthwhile option.

EXTENDED ACTIVITY

West Bay forms an excellent case study of coastal management. There is an interesting range of problems, several issues to be considered, and different proposed solutions.

The aim of this activity is for you to write a report about coastal management at West Bay. In your report you should consider the following questions:

1 What are the physical problems affecting the coastline?

a Why are landslips common at West Cliff?

b Why is the harbour area prone to flooding?

2 What are the management issues?

a What are the various demands on the coast at West Bay?

b Are there any potential conflicts?

3 a What measures of coastal defence have been implemented so far, and how successful have they been?

b What is the purpose of each coastal defence measure?

c Suggest how the dangers of landslips and flooding are reduced.

4 a What are the advantages and disadvantages of the three proposed management schemes?

b Which of the three schemes do you favour, and why?

c Why is the 'do nothing' option not considered to be an option at all here? Do you agree?

d Can you suggest an alternative scheme? (What about 'managed retreat'?)

Use annotated sketches and maps to illustrate your report, and look up the Stanley Thornes website for additional material.

F Changing sea levels

The position of high tide and low tide on a coastline varies considerably from day to day, depending upon sea conditions (storms, wind direction, etc.) and the gravitational pull of the moon and the sun. The presence of berms at different levels on a beach is testament to this variation. However, over geological time (thousands of years, say), sea levels have changed considerably, rising or falling by many metres.

Sea level is the relative position of the sea as it comes into contact with the land. Its position will fluctuate if the amount of water in the oceans increases or decreases, or if the land rises or falls relative to the sea.

Changes in the amount of water are called **eustatic** changes. They are often associated with periods of glaciation. When water is locked-up as ice and snow, there is less liquid water in the oceans, and sea levels fall as a result. When the ice melts, the water turns to a liquid state and the sea level rises.

Ice on land is very heavy and, during a glacial period, land may actually sink relative to the sea. This movement of the land is an **isostatic** change. When the ice melts and the weight is removed, the land slowly rises and recovers (**isostatic recovery**), causing a relative fall in sea level.

This concept of rising and falling sea levels may sound simple but, in reality, the situation is highly complex. For a start, isostatic change tends to be much slower than eustatic change. In the UK, isostatic recovery is still taking place after the end of the last glacial period, some 10 000 years ago. Furthermore, isostatic change tends to affect different parts of the coastline in different ways. For example, in the UK there is a tendency for the south-east to be sinking whilst the north-west is rising!

What are the implications of global warming?

There is little doubt that the climate is getting progressively warmer (see page 219), and there is a strong probability that this will cause sea levels to rise in some parts of the world. Whilst it is easy to sensationalise the possible effects of sea-level rise (e.g. the wiping-out of entire island chains, the total alteration of the geography of the UK, etc.), authorities with a responsibility for managing the coast are taking the threat seriously.

A report by the Proudman Oceanographic Laboratory (1999) stated that, with few exceptions, sea levels around the coast of Britain were already rising at a rate of 1–2 mm/year. Whilst some of this increase may be due to longer-term adjustments, there is evidence to suggest an increasing rate of change since the second half of the 19th century, possibly as a result of global warming. The report urged the need for a more extensive programme of sea-level monitoring, together with research into isostatic change using GPS (the Global Positioning System).

There seems little doubt that sea levels will continue to rise, but it will not be a uniform rise all around the coast. Arguably, a more significant effect of global warming could be an increase in storminess (the frequency and magnitude of storms). This could have significant effects on rates of coastal erosion and sediment transfer.

Coastal authorities are taking account of the possible effects of global warming as they continue to develop plans for coastal defence. Where engineering works are being considered, the implications are that costs will increase. Indeed, the implications of global warming may encourage some authorities to move towards a 'do nothing' approach, or one of 'managed retreat'.

Landforms associated with rising sea levels

The effect of a relative rise in sea level is to flood the coast. Deltas, spits and beaches all disappear under water. Broad river estuaries called **rias** (Figure 9.45) are formed as floodplains become inundated. Rias frequently have a number of branches as tributary valleys have become flooded. The land often rises steeply directly from the water's edge, which may even be marked by a small cliff feature. If the valley has been glaciated, a **fjord** is formed.

9.45 *A drowned river estuary (ria): Salcombe, Devon*

Landforms associated with falling sea levels

When the sea level falls, more of the coastline is revealed. Beaches, no longer combed by the waves, are left stranded and exposed above the new sea level (Figure 9.46). These features, called **raised beaches**, are very common around the coast of western and northern Britain. The lower part of a raised beach will often show signs of marine erosion, and a small cliff may be formed (Figure 9.46).

Any former cliffline is also left stranded high and dry. No longer undercut and eroded by the sea, its slope gradually declines and it becomes a **degraded cliff** (Figure 9.46). Over time it will often become clothed in shrubs and bushes, a sure sign of physical inactivity.

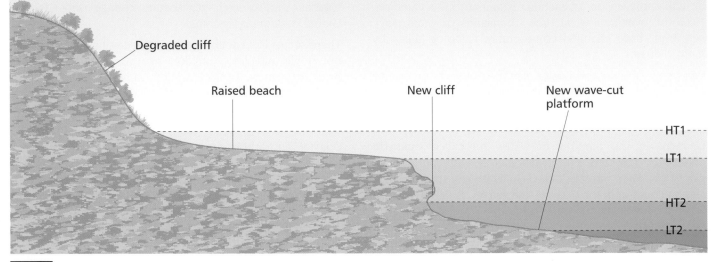

Degraded cliff

Raised beach

New cliff

New wave-cut platform

HT1
LT1
HT2
LT2

9.46 *Features associated with falling sea levels*

The Dorset coast

There are few features associated with sea-level change along the Dorset coast. There are, however, some good examples of raised beaches on the southern tip of the Isle of Portland. The best example, just to the west of Portland Bill, is at a height of about 16 m above the present sea level (Figure 9.47). It is thought to have been formed during the last Ice Age in a warm period (an **interglacial**) some 210 000 years ago when sea levels were much higher than they are today. To the north-east of Portland Bill, a second raised beach exists at between 7 and 11 m. This is also thought to

9.47 *Raised beach and degraded cliff on the Isle of Portland*

have been formed during an interglacial period, but probably a more recent one, about 125 000 years ago.

A future rise in sea levels resulting from global warming will have a considerable effect on the landforms of the Dorset coast, many of which are extremely vulnerable. The weak rocks at Lyme Regis and West Bay will be more prone to erosion, Chesil Beach may be more vulnerable to breaching, and Chiswell's flood problems may increase again. However, coastal managers in Dorset are more concerned about the effect of the predicted increase in storminess. There is already some evidence of an increase in mean wave height. Visual and shipborne wave records have shown a 28 per cent increase in mean wave height between 1952 and 1989. Certainly in the short term, an increase in the frequency and magnitude of storms will have a far greater effect on the Dorset coast than a slow rise in sea levels.

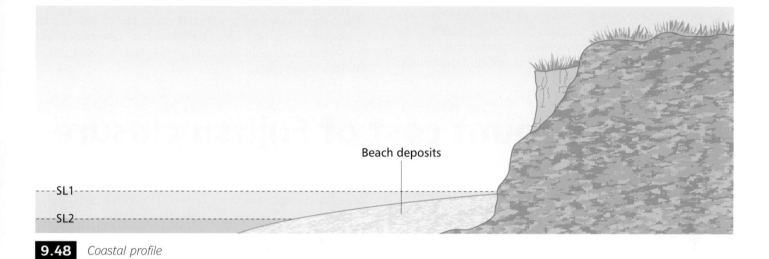

Beach deposits

SL1

SL2

9.48   *Coastal profile*

NOTING ACTIVITIES

1   Outline the differences between isostatic and eustatic change.

2   What evidence would you look for to indicate that a stretch of coastline had been affected by a relative rise in sea level?

3   The coastline in Figure 9.47 has been affected by falling sea levels. Draw a simple sketch of the coastline, and use Figure 9.46 to help you add labels to identify the features.

4   Figure 9.48 shows a coastal profile being actively eroded by the sea. Assume that sea level drops to SL2. Describe, with the aid of an annotated diagram similar to Figure 9.48, how you would expect the coastal profile to change.

5   In this section you have learned a good deal about the processes and features of the Dorset coast, and you have considered some of the issues facing coastal managers. Write a short essay describing the possible impact of global warming on the Dorset coast. Consider both the effects of sea-level rise and increased storminess, and refer to actual places/landforms.

A The world of work

Read the newspaper article (Figure 10.1). It describes the impact on workers of the closure of the Fujitsu Microelectronics plant in Newton Aycliffe, north-east England, in 1998. As soon as the announcement of the closure was made, the production of semiconductors was halted and workers were given 90 days' notice. The Japanese-owned company blamed overcapacity in the global market for the decision. The closure had wider meaning, though, because it followed the closure of the Siemens plant nearby, on north Tyneside. It raised doubts about the ability of the North East region to maintain economic growth and provide employment for its workers.

This type of news story is not uncommon, and simply by reading the financial pages of a broadsheet newspaper for a few weeks, you could build up a collection of examples of job losses and gains.

Workers count cost of Fujitsu closure

Unemployment returns to haunt the North-east, reports **Peter Hetherington**

'We believed it offered a high degree of job security … but if there is no security at a place like this, where is there?'
Gary Carney, redundant worker

NO ONE was prepared for such a sudden death after being assured only days ago that new life was being pumped into their gleaming white factory.

Rob Lothian, and his girlfriend, Jacqui Milford, organised their five-day honeymoon in Paris this month on the strength of it.

Gary Carney booked his holiday in Ibiza next week after being told by bosses that Fujitsu Microelectronics Ltd was committed to its seven-year-old Newton Aycliffe plant.

Last night, as workers, who were earning around £17,000 a year, trickled out of a microchip plant that was supposed to herald a bright new future for the North-east – particularly for Tony Blair's Sedgefield constituency – they all agreed: 'It's like a bereavement, losing a job after being told days ago it was safe.'

This was no ordinary closure. No time for the 600 workers to grieve. Although they were served with 90-day redundancy notices, production of semiconductors ended last night. Many will get out as quickly as possible, insecure in the knowledge that someone with seven years' service at Fujitsu will get seven months' pay; those with two years' service will get three months.

That does little to reassure Mr Lothian, aged 30, a process worker, who, along with Miss Milford, has a £57,000 mortgage. 'If I don't get another job I'll lose my house – simple as that.'

The couple, who live in a new semi on the outskirts of Newton Aycliffe, are getting married in three weeks. They have invited 75 guests to the wedding, and booked a honeymoon in Paris at a cost of £1,000, which they are now trying to cancel.

Jacqui, also 30, a customer services manager with a mobile phone company, added: 'It was to have been a dream holiday but we'll now have to make strict economies. I just hope we can get our money back for the honeymoon.'

Mr Carney, although he said he was devastated by the news, was trying to be upbeat. 'I'm still going to Ibiza, although I'm worried about debts when I return.'

He had given up a job with an electrical engineering company for Fujitsu on the grounds that 'it offered a high degree of job security'.

Outside the £500 million plant for some final goodbyes, many others spoke of being lumbered with hefty mortgages and big debts. They were – maybe still are – the upwardly mobile of the North-east; young men and women in search of a better life, from families which had only known unemployment.

Newton Aycliffe is a new town, born 50 years ago to provide a fresh start for a county dependent on coal mining, and subsequently devastated by its rundown.

'The Fujitsu workers are now experiencing what their parents went through,' said local county councillor Tony Moore.

'I'm extremely worried about the future with all these young people, mortgages and big debts who felt they were settled and had a great future. The Government have to do something.'

Although an admirer of Tony Blair, the local MP, Mr Moore, felt the Government was too obsessed with an economic policy geared to the service industries of the South-east rather than the manufacturing heartlands of the North.

Fifty per cent of Sedgefield's workers are employed in manufacturing – double the national average.

Waving his redundancy notice last night outside the plant, David Evans, another young worker, said he still could not take in the impact of closure.

'My girlfriend first told me and I didn't believe her.'

Outside the main entrance, John Evans, the plant's external relations manager, acknowledged that staff had been told only recently that the plant was secure following the closure of the Siemens complex 20 miles north on Tyneside.

'They were accurate and valid at the time. This is all very sad, but we are living in the real world.'

10.1 *The closure of Fujitsu's plant in north-east England*

In 1991, Fujitsu invested large amounts of **capital** (money and plant) in the new factory at Newton Aycliffe. It combined this capital with labour in order to produce a **commodity** (semiconductors). These commodities were then sold on the market for a greater value than they cost to produce. The difference between the price the semiconductors sell for and the cost of production is called **profit** (or surplus value). As long as a profit is being made, the process continues, with the company continuing to invest in the production of the commodity. Economists call this the **circuit of capital** (Figure 10.2).

Within the global economy, much economic activity is capitalist in nature, which means that goods and services are produced and sold with the aim of making a profit. In Figure 10.2, money (M) is placed into the circuit at the top of the circle by those who wish to invest. This money is then used to purchase commodities C in the form of labour power (LP) and the means of production (MP). These are then combined in a production process (P) which produces further commodities (C^1). The new commodity is then sold for more money (M^1) than was originally invested. The difference between M and M^1 is called **surplus value** (profit). This amount is ready to be re-invested in a new round of production. In a capitalist economy, it is the search for surplus value that is the rationale behind the circuit of capital.

The circuit of capital is tied up with geographical spaces – these are spaces of production, spaces of investment, and spaces of consumption. These geographies are highly dynamic and mobile, since both capital and labour are (at least in theory) free to move. The production taking place at the Fujitsu plant is only a moment in a wider set of circuits. These include investment; the hiring and firing of labour; the organisation of processes of production at particular locations; and the sale and distribution of the finished product. All these activities are geared to the generation of profit which may be re-invested to begin further rounds of accumulation or profit-making.

Why did the Fujitsu factory close?

Fujitsu decided it could no longer continue to invest in the factory at Newton Aycliffe because the demand for semiconductors had fallen, leading to a fall in the price of the product. Production at Newton Aycliffe was unprofitable.

The effects of factory closures such as Fujitsu can be dramatic. The wages earned by employees are spent on goods and services, and this serves to boost the local economy. Obviously, when companies such as Fujitsu withdraw their investment, there is a downturn in the fortunes of a place. It may have occurred to you that it is the company – Fujitsu – that is making all the decisions. This is because *capital* is mobile: it can be moved from place to place relatively easily. Conversely, *labour* is relatively immobile: people do move in order to find work, but this is usually a serious undertaking. This inequality in the ability to move creates problems for places where investment is being withdrawn, since it leads to unemployment, loss of income and relative deprivation. These were the issues that featured in the debates after Fujitsu announced its decision.

As Figure 10.1 states, the announcement was a source of embarrassment for the Prime Minister Tony Blair because the factory was in his constituency. Pressure was put upon the government to intervene and save the jobs that were threatened. In fact the Prime Minister visited the workers at the factory. His advice to them was interesting. He suggested that the old days of a 'job for life' were no longer possible, and workers had to be flexible, prepared to accept the fact that they may have to change jobs. The government could not tell companies such as Fujitsu how to run their business and if it did, such companies would simply avoid investing in Britain.

A year after Fujitsu announced the closure of its factory at Newton Aycliffe, the telecommunications company Orange had bought the site and were using it as a **call centre**, where telephone operators are on hand to answer calls from the public. This is another example of the circuit of capital in operation, as another company sees an opportunity to realise a profit from investment in capital and labour.

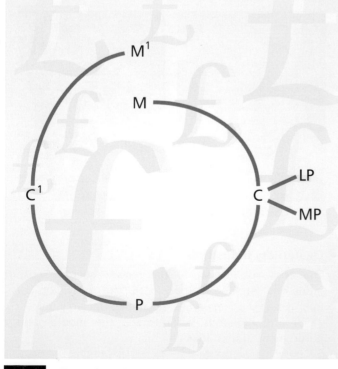

10.2 *The circuit of capital*

How is economic activity classified?

Geographers use a number of different ways to classify economic activity. The most common is classification of economic activity into four groups (Figure 10.3).

Secondary industry (manufacturing)

- **Primary industries**

    These are concerned with extracting raw material directly from the Earth. Examples of these industries include agriculture, fishing, forestry and mining. The vast majority of the world's population, concentrated in China, India, South-east Asia, and Africa, is engaged in primary economic activities. For instance, in China, primary activities account for more than 70 per cent of the labour force. However, in the developed economies primary activities generally account for less than 10 per cent of the labour force, and often the figure is less than 5 per cent.

- **Tertiary industries**

    These are concerned with the provision of a service. As a result, they tend to be located near their markets. Examples include transport and distribution, retailing, banking and insurance.

Primary industry (mining)

Tertiary industry (teaching)

Quaternary industry (a call centre)

- **Secondary industries**

    These are concerned with the manufacture or processing of raw materials into finished products, for example the processing of iron ore to make steel, the processing of food, or furniture making from timber. Figure 10.4 shows some of the main centres of manufacturing industry in the world economy.

- **Quaternary industries**

    This is a subdivision of the tertiary sector. These industries are concerned with the provision of information and expertise. Examples include universities and research laboratories, the media, and government 'think-tanks'.

10.3 *Industrial sectors*

The **São Paulo region**, with less than 25% of Brazil's population, accounts for more than 40% of the country's industrial production. The region's industrial growth took off in the 1950s and 1960s with the arrival of transnational vehicle and chemical companies. The vehicle industry accounts for 20% of São Paulo's industrial production.

Taipei has become the world's leading producer of image scanners, computer mice and monitors, as well as an important manufacturer of semiconductors.

South Africa's manufacturing sector employs 1.4 million people – about 11% of the country's workforce – and accounts for nearly 25% of the country's GDP.

Shanghai and the Pudong district attracted billions of dollars of foreign investment in the 1990s, contributing to a regional growth rate of 14% per year between 1993 and 1995.

The **Kansai region**, around Osaka, has become Japan's principal centre of pharmaceutical production, attracting significant amounts of investment from transnational corporations.

The city of **Bangalore**, in the state of Karnataka in south-central India, has become the centre of a booming high-tech sector, including telecommunications, computers and defence research.

The **upper Midwest**, part of the US manufacturing belt that grew to prominence in the early 20th century, has experienced a painful episode of economic restructuring. As a result, it is now once again the fastest-growing exporter of manufactured goods in the USA.

Stimulated by the presence of Japanese vehicle manufacturers such as Toyota, Nissan, Honda and Isuzu, **Bangkok** is set to become Asia's vehicle assembly centre.

Europe's largest single industrial region is the **Ruhr**, in north-west Germany, which was built on top of scores of coal mines in the 19th century, with dozens of steel mills, landmarks for which the region became famous through the world.

10.4   *Some major centres of manufacturing industry*

NOTING ACTIVITIES

1   Distinguish between the terms 'primary', 'secondary', 'tertiary' and 'quaternary'. Give several examples of each type of industry.

2   Figure 10.4 gives information about some major centres of manufacturing industry.

   a   On a blank outline map of the world, label these centres of manufacturing.

   b   From your completed map, identify:

   • two regions that emerged as important centres of manufacturing in the 19th century

   • two regions that emerged as important centres of manufacturing in the late 20th century.

   c   How might the types of manufacturing industry in these regions differ?

Economic structure

The balance of employment in the different sectors or types of industry is called the **employment structure**. The employment structure of more economically developed economies (MEDCs) is generally different from that of less economically developed economies (LEDCs). Developed economies employ a smaller proportion in the primary sector and higher proportions in the secondary and tertiary sectors.

The balance of employment changes over time, and it has been suggested that there are different stages of economic development. A simple model of economic change is the development-stage or the 'three-sector' model of economic development (Figure 10.5). This model was first proposed in the 1930s and suggests that, as a country develops, the

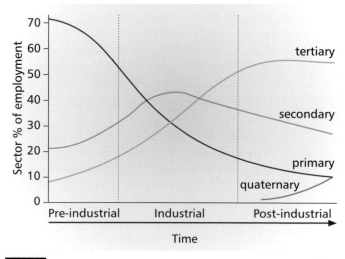

10.5   *The development-stage model*

three sectors of the economy grow and then decline. In the long term, employment in the economy gradually shifts from the primary sector, into the secondary sector, and then to the tertiary sector. There are two main reasons for this:

1 As economies grow, rising productivity levels enable technological advances which allow workers to move into the next sector. Thus, just as in the 19th century the Industrial Revolution led to the movement of workers from the primary sector to the secondary sector, so in the second half of the 20th century rising productivity in the secondary sector allowed a shift into the tertiary sector.

2 As national incomes rise, the increase in demand is channelled first into the secondary sector and then into the tertiary sector. This is based on the idea that as people have more money to spend, they spend a smaller proportion of their income on basic goods such as food, and more on consumer goods and services.

The development-stage model is useful for thinking about changes in employment patterns. However, some doubt its validity. Does such a neat progression through the sectors actually occur? For example, in Britain the tertiary sector was an important provider of jobs as long ago as the 19th century, and it has been suggested that the long-term trend is for employment to shift from the primary sector into both the secondary *and* tertiary sectors at the same time. This is certainly the case in countries such as Japan, Canada and the USA.

Long waves of economic development

A second way of thinking about employment changes over time is based on the long waves of production, or **Kondratieff waves** (named after the Russian economist who first detected them). Kondratieff observed the tendency for economic output to rise and fall in cycles of approximately 50 years (Figure 10.6). Each cycle had an upswing and a downswing, and Kondratieff detected four such waves, each coinciding with the development of key industries. For example, the first cycle (1781–41) coincided with the development of cotton manufacturing and iron and steel making, whilst the second cycle (1842–94) was linked with the development of shipbuilding and machine tools based on new methods of steel production.

From a geographical point of view these cycles are interesting because each wave of new industries led to the emergence of new regions and to significant social changes. For example, the rise of cotton manufacture and iron and steel led to increased urbanisation and the growth of factory towns in the north and west of Britain.

Kondratieff waves provide a useful way of thinking about the development of the economy over time. They are particularly helpful in understanding the rise and fall of industrial regions in Britain. One of the problems with such a model is that it simply describes changes without offering explanations for the reasons behind these changes. In addition, some economic historians dispute the existence of these cycles of economic activity.

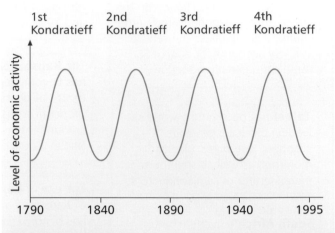

1st Kondratieff wave
1781–1841
Key technological development was the smelting of iron ore with coal, and the development of water- and steam-powered machinery which allowed the development of textile production. There was a decline in the percentage of the population working in agriculture as a result of enclosure of land. The urban population increased from 30% in 1790 to 46% in 1841.

2nd Kondratieff wave
1842–94
Key developments were based around the use of coal to power machinery. This led to the growth of coalfield towns around the Pennines in England, in central Scotland, South Wales and the Midlands. London experienced rapid growth. By 1881, 70% of Britain's population was urban, and this period saw the emergence of the towns of the Industrial Revolution – Glasgow, Liverpool and Manchester, for example.

3rd Kondratieff wave
1895–1939
The key developments were the new consumer-based industries involving car manufacture, chemicals and electrical goods, plus the food-processing industries. Services continued to grow in importance. There was a shift away from the older industrial regions towards the south and the Midlands.

4th Kondratieff wave
The *post-war period* saw the rapid development of the economy based around developments in aircraft technology and electronics. Motorways eased accessibility and made industrial location more 'footloose'. The continued decline of traditional manufacturing in the north and west of Britain was matched by the growth of consumer services (banking, finance, insurance) in the densely populated south and east.

10.6 *The Kondratieff cycle in Britain*

NOTING ACTIVITIES

1   What do you understand by the term 'Kondratieff wave'?

2   Identify and date four Kondratieff waves for Britain's economy.

3   Using Figure 10.6 to help you, annotate a blank outline map of Britain to show how different regions have experienced periods of economic growth and decline.

STRUCTURED QUESTION

1   Study Figure 10.7, which gives some information about the economic structure of four countries.

  a Make a copy of the table below, and use the graph to complete your table.   *(3)*

  b Look at Figure 10.5. Which stage of development do you think each country has reached?   *(4)*

  c Why is secondary employment decreasing as a proportion in European countries?   *(2)*

  d Why is secondary employment increasing as a proportion in Asian countries?   *(2)*

2   Comment on the validity of the development-stage model of economic development.   *(3)*

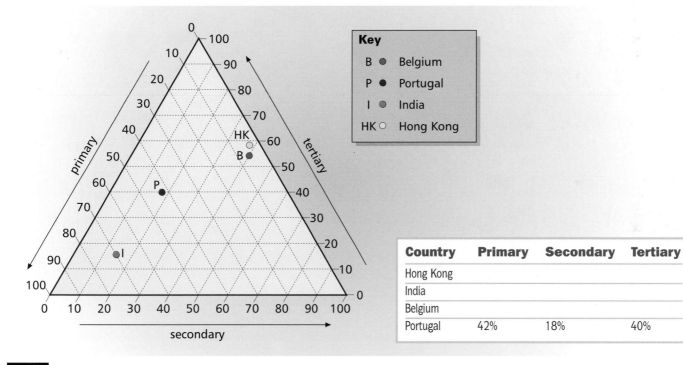

Country	Primary	Secondary	Tertiary
Hong Kong			
India			
Belgium			
Portugal	42%	18%	40%

10.7   *The economic structure of selected countries*

B Employment sectors

Primary industries

Primary activities are concerned with natural resources. They extract both renewable and non-renewable resources. **Renewable resources** are those that can be used without being permanently depleted, such as forests, water, fishing grounds and agricultural land. **Non-renewable resources** are depleted when used, as for example, minerals. In many countries, primary activities dominate the economy. For example, in many African and Asian countries, between 50 per cent and 75 per cent of the labour force is involved in primary sector activities. Generally, primary industries develop close to manufacturing districts, since these make use of the materials extracted by primary industries (Figure 10.8).

Primary industries tend to require large amounts of space. They can have a dramatic effect on the landscape. For instance, mining may create slag-heaps, gaping open-pit mines, and forests may be 'stripped' (Figure 10.9a) Other primary industrial landscapes are valued for their beauty, for instance the fishing villages of Portugal (Figure 10.9b).

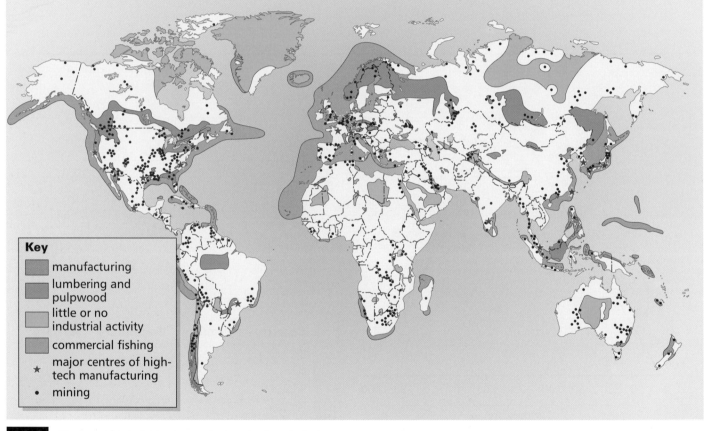

Key
- manufacturing
- lumbering and pulpwood
- little or no industrial activity
- commercial fishing
- ★ major centres of high-tech manufacturing
- • mining

10.8 *Regions of selected primary and secondary industry*

10.9 *Primary industries*   *a A strip (opencast) mine*   *b A Portuguese fishing village*

Study Figure 10.8, which shows regions of selected primary and secondary industries.

1   Identify *three* regions where mining is important. Using an atlas, for each of these state which resources are extracted.

2   Identify *two* regions that are dominated by tropical timber extraction.

3   Describe the location of regions where there is little or no industrial activity. Suggest reasons for this distribution.

4   Why do you think large centres of manufacturing are often found close to important regions of primary industries? Can you find exceptions to this?

CASE STUDY

Coal in Britain

Coal has been mined in Britain for thousands of years. However, large-scale extraction from the coalfields (Figure 10.10) developed during the Industrial Revolution. The introduction of coke smelting in the iron industry increased the demand for coal and the introduction of the steam engine allowed the exploitation of deeper reserves to meet the increased demand for coal by industrial users. The development of large cities increased the demand for coal for heating purposes.

The year of peak production in the British coal industry was 1913, when 287 million tonnes were produced. Since that time, the level of production has declined. After the First World War (1914–18), the demand for British coal from abroad declined. British coal became less competitive in world markets. Productivity was higher in continental mines, which tended to be newer, larger and more mechanised. To increase productivity and reduce costs, new mines using modern equipment would have been required. However, this was not possible due to the dominance of the industry by small independent mining companies which prevented the rationalisation of production, and the poor industrial relations record of the industry.

These problems were recognised by governments in the inter-war period and the post-war Labour government nationalised the coal industry in 1947, forming the National Coal Board. The NCB became the largest employer in Britain. The immediate concern of the NCB was to expand output at all costs. The 1950 Plan for Coal proposed a massive investment

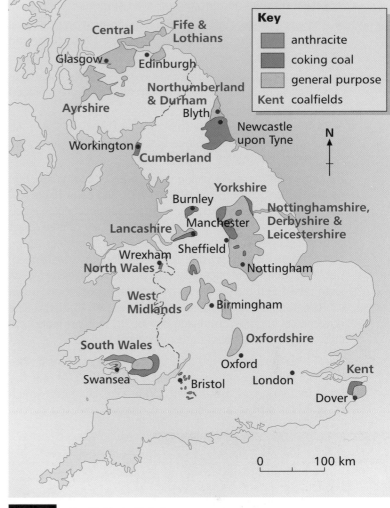

10.10   *Coalfields in Britain*

over a 15-year period to raise output to 240 million tonnes per annum. However, throughout the 1950s the NCB could not meet the demand for coal. Much production took place in inefficient pits with high costs and low productivity.

The period of expansion in the 1950s was followed by a period of energy surplus and the availability of cheap energy supplies. The situation had changed from the 1950s, when Britain was still dependent on coal for over 80 per cent of its energy supplies. The development of nuclear power from the 1950s, and the development of North Sea oil and gas from the 1960s, meant that coal declined in importance as a source of energy. Many pits were closed on the grounds that they were 'uneconomic'. The pits that were closed were those with high costs and low productivity, and tended to be found in coalfields such as South Wales, Northumberland and Durham. The fall in the demand for coal was due to a number of factors:

- Two of the main consumers of coal – the electricity and iron and steel industries – improved their fuel-burning efficiency. Larger plants increased efficiency and improvements in blast furnace design reduced coke requirements. The electricity industry generated more power through nuclear and oil-fired stations.

- The gas industry shifted from a dependence on coal to oil during the 1950s, and then from the late 1960s converted to natural gas.

- The modernisation of the railways meant that main-line routes were electrified and steam engines were replaced by diesel.

- Domestic markets were lost due to the 1956 Clean Air Act. Modern flats and office blocks were built which were suited to alternative fuels.

It is important to avoid seeing the demise of coal production as inevitable. Governments of the 1960s and early 1970s saw the decline of the coal industry as politically unacceptable and introduced measures to offset its decline. They offered direct financial support for the industry and sought to protect it from competition. Throughout the 1960s coal imports were banned, public buildings were encouraged to burn coal, taxes on fuel oil were raised and the electricity generating board slowed down its programme of building oil-fired power stations. The government required the electricity board to burn more coal and the government paid the extra costs involved.

The 1970s is remembered as a period of 'energy crisis'. The price of imported oil increased sharply in 1973/74, but the coal industry was beset with industrial disputes which resulted in strikes and energy shortages that led to the introduction of a three-day week. Whilst new reserves were proved to exist, the problem faced by the NCB was the extraction of coal at a competitive price. The crisis in the coal industry finally came to a head in the 1984/85 miners' strike (Figure 10.11). The strike was about the decision of the NCB to close some of its 172 pits. The year-long dispute resulted in the loss of power of the National Union of Miners and was followed by a series of pit closures. However, the issue of pit closures returned to the political agenda in 1992, when the announcement of the closure of 31 of the 50 pits being operated by British Coal caused a widespread public outcry. The scale of contraction in the coal industry can be seen by the fact that in 1984, the NCB operated 172 pits and employed 174 000 workers. By mid-1994 there were just 15 pits in operation and producing coal. In total these employed 6000 administrative staff and 8000 miners.

What comes after coal?

The decline of the coal industry in particular towns and villages has important effects on other economic activities. There is a loss of consumer demand for locally traded goods and services, and a physical run-down of the area. An example is the case of Grimethorpe (South Yorkshire), a small town almost wholly reliant on coal. Its power stations closed in 1991 and the colliery closed in 1993. After that, the physical structure and appearance of the place worsened. Petty crime increased dramatically. Employers who remained talked of moving out.

Certain localities associated with the coal industry were one-industry economies. Places such as Blidworth, Welbeck, Ollerton and Harworth in Nottinghamshire were 'company towns', built by the owners of the mines in the 1920s. These communities were steeped in mining, and had little need to diversify their economy. The economic and social culture was built on hard, manual labour. When mines closed, there were very few jobs for the workforce to do. The coal industry tended to have a semi-rural setting, with many pits remote from the major centres of population and therefore far from alternative sources of employment.

For example, South Kirkby was a pit-town from 1880 when the shaft was sunk. It is located between Wakefield and Barnsley in West Yorkshire and has a population of around 11 000. In 1984, over half of the jobs in the town were in coal mining. However, South Kirkby pit was closed in 1988 on the grounds that it was 'uneconomic'. In order to offset the impact of decline, the government established an Enterprise Zone in the form of an industrial estate in South Kirkby. By August 1990, an additional 905 jobs had been created. However, many of these jobs had been relocated from nearby, which meant that only about 300 were new jobs, compared with the 1350 that had been lost in the mining industry. The largest employer in the Enterprise Zone was a men's outfitter employing 370 people. On moving into the Enterprise Zone, the company closed one existing factory in South Kirkby and another three miles away. There was therefore no stimulus to the local economy. This example suggests that for places such as South Kirkby, the problem is how to diversify the economy after being dependent on coal for so long.

10.11 *The 1984/85 miners' strike*

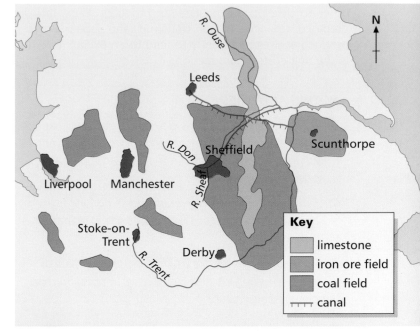

NOTING ACTIVITIES

1   Study Figure 10.12, which shows the output of coal from deep-mined and opencast pits between 1960 and 1996. How much coal was produced in deep mines in **a** 1960 **b** 1996?

2   Describe the changes in the output of coal between 1960 and 1996.

3   Suggest reasons for the sharp decline in output of coal in 1984/85.

4   Suggest the likely effects of the closure of coal mines on local communities.

5   How might the increase in opencast coal production affect the environment?

6   Why do you think past governments have sought to protect the coal industry from rapid decline?

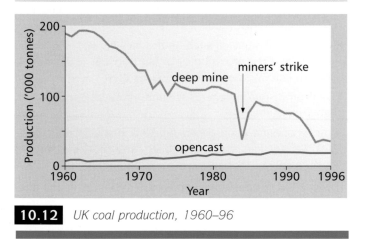

10.12   *UK coal production, 1960–96*

Secondary industries

Secondary industries add value. They can either utilise the raw materials produced by primary industries to manufacture materials (for instance the production of foodstuffs), or they may combine components already produced by other secondary industries (the production of cars, for example).

Localised sources of raw materials and energy sources were particularly important in the early development of manufacturing industry. For instance, when water was a major source of power, mills needed to be located next to rivers. As steam power took over, coalfield locations were favoured. For example, the growth of Sheffield as a manufacturing region was largely the result of the location 'pull' of the weight-losing inputs needed for iron and steel manufacture: limestone, coal, iron ore and water (Figure 10.13). At first, iron ore came from the deposits of ironstone found alongside the nearby coal, water came from the River Sheaf, and limestone from outcrops a few miles to the east. As the

steel industry grew, the flat and wide valley of the River Don provided sites for larger steelworks, which got their coal from the newer coal mines in the large coalfield to the east and high-grade iron ore deposits found in the area near Scunthorpe. In Sheffield itself a large number of firms sprang up to make use of the high-quality iron and steel, making everything from anchors, cutlery, files and nails to needles, pins and wire. Today, the location is no longer suited to large-scale steel manufacture. However, there is still some activity in this area, even though the original reasons for the location of the industry have disappeared. This is an example of **industrial inertia**.

A key element in the decision of a company to locate in a particular place is the availability and characteristics of the workforce. Labour costs can vary both between countries and within countries, and where labour costs are a major part of a firm's total costs, this may be an important factor. It is now thought that firms are seeking to locate away from large sources of labour, where workers have been able to organise themselves through unions to secure relatively high wages, and instead seek out sources of 'green' labour – often female and non-unionised, in semi-rural and rural locations. The transfer of manufacturing operations from Europe and the USA to countries such as the Philippines, Indonesia, Thailand and Mexico during the 1980s and '90s exemplifies this. Some companies are now even relocating from Malaysia to the Philippines because the difference in labour costs between the two countries has made the move viable.

Agglomeration economies are savings that result from sharing services and from making linkages between firms. For example, a clothing manufacturer might benefit from locating close to a design company or advertising agency in order to develop and market its products. Firms may be prepared to incur higher transport and labour costs if production rises sufficiently to lead to an overall reduction in the unit costs of production.

NOTING ACTIVITIES

1 Make a copy of Figure 10.13. Annotate your map to show the locational factors that were important in the development of Sheffield as an important steel-producing centre.

2 Suggest reasons why Sheffield is no longer a suitable location for large-scale steel manufacturing.

3 For each of the factors affecting the location of industry, suggest how its relative importance has changed over time.

4 Study Figure 10.14, which shows the factors affecting the choice of location for different types of business.

   a Why do you think 'labour supply' was the most important factor for both office and factory enterprises?

   b Why do you think 'proximity to markets' was the most important factor for retailing?

   c How might 'quality of life' affect the location of an industry?

Factors	Ranking by type of business			
	Office	Factory	Warehouse	Retail
Labour supply	1	1	4	4=
Proximity to markets	3	3	2	1
Access to transportation networks	5	2	1	2=
Property	2	4	3	2=
Quality of life	4	6	8	4=
Proximity to suppliers	6=	5	5=	6=
Availability of grants	6=	7	5=	6=
Other miscellaneous factors	8	8	5=	8

1 is highest, 8 is lowest.   An = symbol indicates that factors were ranked as being equally important.

10.14 *Factors influencing choice of business location*

Tertiary industries

Britain's employment structure is dominated by services. Service industries now account for around 80 per cent of employees. But what exactly is meant by a 'service industry'? There is no agreed way to define services. However, we can identify a number of components of the service sector:

- F.I.R.E. – or *finance, insurance* and *real estate* – including banking, securities, insurance, brokers and estate agents
- business services such as legal services, advertising, engineering and architecture, public relations, research and development, and consulting
- transport and communications, including the media (newspapers, television, radio), trucking, shipping, railways and local transport (taxis, buses etc.)
- wholesaling and retailing
- personal services and entertainment, including cafés and restaurants, hairdressers, repair and maintenance, gyms, private education, and all forms of amusement (e.g. tourism, film distribution, etc.)
- government services such as the civil service, the armed forces and the provision of public services (e.g. schools, health care, police and fire services etc.)
- non-profit-making agencies such as charities, churches, and museums.

Some economic geographers make a distinction between consumer services such as retailing, and producer services (which mainly provide services for other firms).

The location of **consumer services** is explained by the need to be close to customers, and so the distribution of retail outlets reflects the distribution of the population (think about Christaller's central place theory here). Population shifts can affect the location of services – in the case of retailing, the gradual loss of population from large inner urban areas, coupled with the growth of the suburbs and increased car ownership, has led to changes in the nature of retail provision.

Quaternary, or 'knowledge-based' industries

Quaternary industries are those services mainly used by producers. These include insurance, legal services, banking, advertising and consultancy. These services are concerned with the collection, generation, storage, retrieval and processing of computerised knowledge and information. In the case of **producer services**, the most important thing to note is their tendency to agglomerate in large urban areas. There are a number of reasons for this.

- To maximise their accessibility to clients and suppliers. For instance, advertising firms must have access to their clients. Often these services rely on face-to-face meetings, which can only be managed through close proximity.

- To maximise access to information – in highly competitive industries, knowledge is power, and being close to the 'action' reduces uncertainty and risk.
- Many producer services require a highly skilled, well educated and creative workforce. They tend to be located in the largest cities in the developed economies. On a smaller scale, they tend to cluster around major universities and research centres. For example, the presence of Stanford and the University of California at Berkeley have helped make the San Francisco Bay area a major centre of such industry.

In Britain, employment in producer services is concentrated in the South East region. The growth of financial services has been especially strong in the South East. This has had considerable impacts on the physical landscape of the City of London, with new office building designed to cope with the requirements of technological change, and the famous Canary Wharf development in London's Docklands. The rapid increase in activity also led to skill shortages in the 1980s and the rise in salaries, which were reflected in the rise of a 'yuppie' and 'loadsamoney' culture.

While there has been a tendency for the highest level of decision-making functions to concentrate in the large metropolitan centres of Europe, there is also evidence of decentralisation of more routine tasks. These have moved to fresh locations within large cities, to peripheral locations on the edge of cities, and occasionally to more distant locations in rural areas.

An example of these trends is the growth of **call centres**, where staff wearing headphones sit at desks, with a computer terminal in front of them, and answer calls, selling anything from kitchenware to insurance. These centres have been enabled by technological developments such as Automated Call Distribution, which allows incoming calls to be sent out in orderly queues to waiting operators, and if necessary transferred to less busy centres in other cities or even other countries.

Call centres

The growth of call centres has been rapid. In 1996 there were 123 000 workstations in call centres in the UK, and this had grown to 198 000 by 1998. It is estimated that the number of call centre workstations in the UK in 2000 will be 243 000 or 3 per cent of the workforce.

The call centre phenomenon started when the first centres were set up near London in the 1970s, and then spread outwards in search of lower costs, leading to growth in cities such as Glasgow and Leeds. London and the South East currently account for 26 per cent of all call centre jobs in Britain, followed by the North West which accounts for 20 per cent.

The location of call centres is linked to a range of factors:

- evidence of existing successful call centres in a city or region
- a labour force of sufficient number and with the required skill levels
- the availability of good-quality business networks
- financial incentives provided by local authorities or development agencies
- good-quality location or property, premises ready to occupy
- good place to live – this is important for attracting foreign nationals
- telecoms support
- place marketing by local or national development agencies.

Examples of call centre locations

London

As call centres are relatively 'footloose', it is reasonable to ask whether London, which has relatively high land prices, is the best place to locate a call centre. However, 15 per cent of call centre jobs are in London, and the South East region has a further 11 per cent. These outer South East locations include Milton Keynes, Brighton and Kent. An important factor that makes London attractive is language skills. For example, Delta Airlines set up its European reservations sales centre at Park Royal in west London in 1996. It saw London as the only European capital capable of providing the multilingual workforce required to serve customers from all over Europe. Employees at AT & T Solutions' global customer services centre in the City of London each speak an average of three or four languages, and 80 per cent are graduates.

Some organisations see a London location as adding status to their operations. London provides access to a large pool of native European speakers and also has a number of universities and other higher educational institutions; for instance, Westminster University teaches 26 languages – the most of any institution in the UK. In addition, the sheer size of London ensures that there is a wide range of suitable properties and working environments.

North West

In the North West region of England alone, more than 30 000 people are employed in call centres, accounting for 20 per cent of call centre jobs in the UK. An example is British Telecom which has one of its five call centres at Warrington, employing more than 2000 people. Customers are called on a regular basis to tell them about various savings schemes. The centre was originally a warehouse but was converted in 1996. Successful locations in Liverpool, Warrington and Manchester have access to a good mix of available skilled labour, appropriate sites and telecommunications infrastructure. Grants and investment packages provided by local development agencies also help to reduce initial costs and encourage investment. An example is Ventura (a subsidiary of the clothing company Next) which gained a package of incentives to set up its call centre in an area of high unemployment in South Yorkshire.

The rapid growth in call centres has led to labour shortages in some regions. As a result, companies are looking to places such as Wales, Scotland, the north-east of England or the Republic of Ireland because of concerns that the North West is saturated and there is an insufficient supply of labour. The number of call centres on Merseyside meant that IBM chose Scotland for its new multilingual call centre, with an expected 600 employees. However, not all towns are benefiting from call centre growth. Places such as Carlisle and Furness in Cumbria have not proved attractive sites for inward investment due to their lack of public transport, poor road infrastructure and long travel distances for potential staff.

The competition for call centre business is increasingly global in nature. The Netherlands and the Republic of Ireland are attractive sites for call centre investment. There are an estimated 4000 people working in call centres in the Republic of Ireland. In 1996/97, 16 call centres were established in Dublin alone. The reasons for this are the availability of multilingual employees, reasonable labour costs, and incentives provided by the Irish Development Agency. However, pressure for space and the availability of suitable sites has led to increased rents, and the Irish Development Agency is seeking to attract call centre investment to places such as Cork, Limerick and Galway. For example, in March 1998 Merchants Group (a UK-based customer-management company) opened a call centre in Cork employing 600 people.

An evaluation of the impact of call centres must consider the type of work involved. Whilst employment in call centres has grown rapidly in the UK in recent years, doubts have been cast on the real impact of these jobs. It has been argued that jobs have been lost elsewhere as fast as they have been created, since they replace activities previously done elsewhere in separate branches. In response it is argued that while this may have been the case in the early days of call centres, companies are increasingly selling new services by telephone.

More controversial is the quality of the jobs. It has been suggested that the calls employees make are subject to constant monitoring for their quality and length, and workers have to follow a monotonous compulsory formula when making calls. Some people have labelled call centres as the new 'sweatshops'. However, it is probable that working conditions vary widely between call centres. Indeed, motivation of the workforce is an important factor, because labour costs account for 70–80 per cent of total costs. Some companies, such as the mail order company Freeman's, have designed their telephone area for comfort, and provide a restaurant, gym and rest facilities for employees.

The global shift in producer services

As we have seen, banking, finance and business services are important for the development of manufacturing. For example, they lend money for firms to finance investment, and provide insurance in case things go wrong. These services tend to locate near to their customers. However, producer services increasingly operate on a global scale. This has been facilitated by a number of factors including advances in telecommunications and data processing.

Figure 10.15 shows that the location of the largest banks is highly concentrated in just a few countries. They are all located in highly specialised financial districts of large cities. The reasons for this are linked with the benefits of agglomeration. Cities such as New York, London, Paris and Tokyo have a specialised infrastructure – specialised office space,

stock markets, communications networks – which allow them to deliver services to clients both nationally and internationally. They have large numbers of professionals who are able to provide expert knowledge and advice.

Whilst the control functions of these business services are found in large urban areas, there has also been a trend towards decentralisation as 'back office' functions have been relocated to small-town and suburban locations. Back office functions are record-keeping and administrative functions that do not require frequent personal contact with clients. Developments in technology allow much of this work to be relocated to office space in cheaper locations. A recent trend has been the development of offshore back offices. For example, in the airline industry it is now common for tickets to be processed offshore.

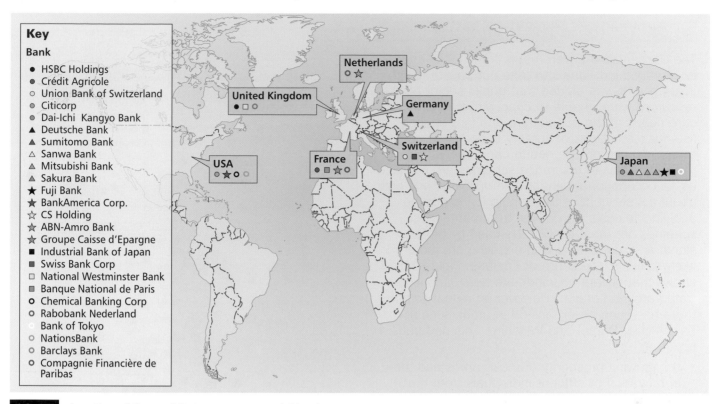

Key

Bank

- HSBC Holdings
- Crédit Agricole
- ○ Union Bank of Switzerland
- Citicorp
- Dai-Ichi Kangyo Bank
- ▲ Deutsche Bank
- ▲ Sumitomo Bank
- △ Sanwa Bank
- △ Mitsubishi Bank
- △ Sakura Bank
- ★ Fuji Bank
- ★ BankAmerica Corp.
- ☆ CS Holding
- ☆ ABN-Amro Bank
- ☆ Groupe Caisse d'Epargne
- ■ Industrial Bank of Japan
- ■ Swiss Bank Corp
- □ National Westminster Bank
- ■ Banque National de Paris
- ○ Chemical Banking Corp
- ○ Rabobank Nederland
- ○ Bank of Tokyo
- ○ NationsBank
- ○ Barclays Bank
- ○ Compagnie Financière de Paribas

10.15 *Location of the world's largest commercial banks*

American Airlines sends its accounts and ticket coupons from Dallas in Texas to Barbados. There, details of the bookings are entered into computers, and the data is returned via satellite to the data centre in the USA. Insurance clients of another company, New York Life, send their health insurance claims to Kennedy Airport in New York. These are then sent to Shannon Airport in Dublin, where they are taken by road to the firm's processing centre. The claims are processed and the results are sent by satellite to the company's data processing centre in New Jersey (Figure 10.16).

Quinary industries

Quinary industries involve consumer-related services such as education, government, recreation, tourism and health. In Britain since 1945 many of these services have been provided by the **public sector** – that is, funded by the government. Between 1971 and the late 1990s, the number of employees in the public sector (health, welfare, public administration, education and defence) increased from 4.1 million (18.7 per cent of total employment) to 5.3 million (24 per cent of total employment). This increase has been linked to the growth of female employment, especially in the areas of health and education.

Much of this employment is located in line with population distribution, especially in health and education where proximity to clients is essential. However, governments have

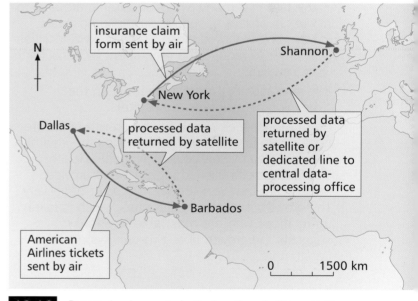

10.16 *Recent developments in the location of office activities*

attempted to disperse the pattern of public sector employment. There are two main reasons for this policy:

- first, to improve the employment prospects of people in areas which have suffered deindustrialisation and high rates of unemployment
- second, to take advantage of lower office rents and cheaper sources of labour.

Examples of major relocations include the Department of Health to Leeds and the expansion of the Employment Department in Sheffield.

The growth of private health care

The Conservative governments from 1979 to 1997 encouraged individuals to take out private health insurance. It was argued that public sector health resources were unable to cope with the health care needs of the population, and long waiting lists for many types of surgery encouraged people to seek to use private health care. The vast majority of health insurance schemes are provided as an occupational benefit by employers, and only about 13 per cent of the population is insured against the costs of medical treatment. These schemes are most likely to be used by professional and managerial workers, who are more likely to live in the south of England. Figure 10.17 shows that private hospitals are heavily concentrated around London. However, in recent years there has been an expansion in areas such as East Anglia, Dorset and Wiltshire. This pattern is a result of:

- the growing prosperity of these areas and the in-migration of affluent individuals from the London area

- the fact that many NHS consultants are able to work in private hospitals, so these hospitals have been located close to major NHS facilities in order to reduce the time spent by consultants in travelling

- a relaxation of planning controls on the development of private hospitals.

One of the issues raised by the growth of private health care is whether it undermines the idea of a National Health Service. The National Health Service was established in 1948 and provided health care that was 'free at the point of delivery'. The growth of private health care, which is taken up by those who are most able to afford it, suggests the growth of a 'two-tier' system of health care.

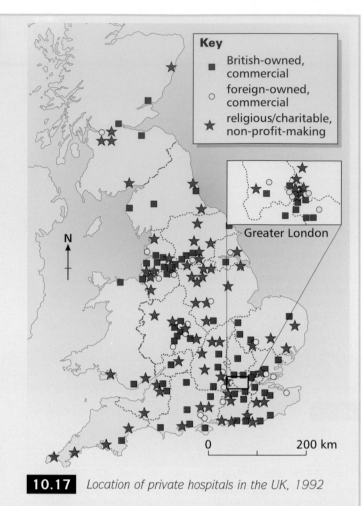

10.17 *Location of private hospitals in the UK, 1992*

NOTING ACTIVITIES

1 With other members of your group, investigate the provision of health care in your locality. What types of facilities are provided? Are there plans to open or close facilities in your area?

2 Discuss the arguments for and against the provision of private health care.

C The global economy

A 'global shift'?

Some economic geographers talk of a 'global shift' in the world economy. By this they mean that more and more manufacturing production and employment is found in the cities of developing economies, while the older industrial cities of developed economies have experienced a contraction of industrial production and employment. In the 19th century, when transport costs were high, industry was tied to sources of power or proximity to markets. As a result, factory towns grew in Europe and North America close to water power, coalfields or large urban centres. There was a clear distinction in the world economy between the industrial **core**, which concentrated on producing manufactured goods, and the **periphery**, which exported raw materials and imported manufactured goods (Figure 10.18). In the 20th century this pattern changed as the periphery became industrialised. Since 1945, mass production of goods, improvements in transport, and the global flow of capital for investment has meant that

the world, rather than a single nation, is the arena of decision-making for modern transnational corporations.

The importance of labour

Geographers have argued that an important factor influencing the location decisions of transnational corporations is the cost and militancy of labour. Workers in the older industrial countries tend to be more expensive, better organised and able to resist the introduction of new working methods. As a result, transnational corporations have shifted production from the core to the periphery of the world economy. The net effect has been the industrialisation of parts of the periphery and the de-industrialisation of the core. This process has been selective. Not all places have been affected in the same way. There are several examples, including South Korea and the city-states of Hong Kong, Singapore and Taiwan; the growth of industries in cities such as Lima and Caracas; and, more locally, the new

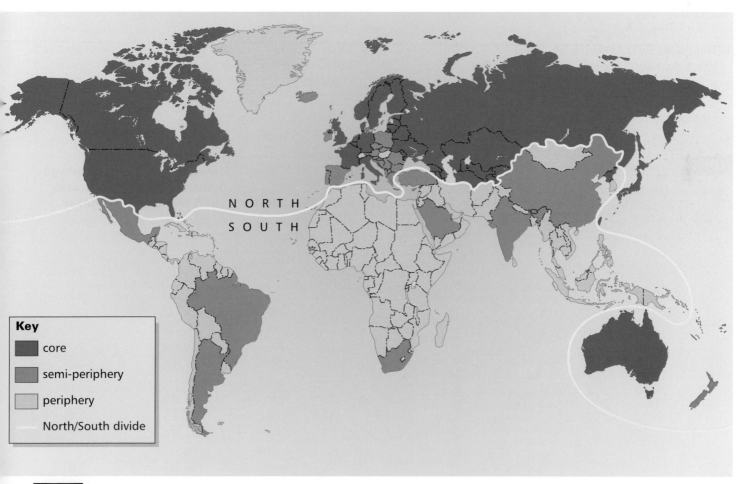

Key

- core
- semi-periphery
- periphery
- North/South divide

10.18 *Core and periphery in the world economy*

industries of cities such as Tijuana in Mexico, close to the Mexican–US border, where wage rates are one-tenth of those across the border, and there are less stringent environmental and safety regulations than in the USA (Figure 10.19). The flip-side of this process is the loss of manufacturing employment in the older cities of the core, as for example in the car cities of Detroit (USA) and Birmingham (UK).

0.19 *A factory in Tijuana, on the Mexican–US border*

NOTING ACTIVITIES

1 Study Figure 10.18.

   a Describe the distribution of the **core** economies.

   b What types of economic activity are characteristic of these countries?

   c Describe the distribution of the **peripheral** economies.

   d What types of economic activity are characteristic of these countries?

2 What is meant by the term 'global shift'?

3 Suggest reasons for the increasing tendency for manufacturing activities to relocate in the semi-periphery and periphery.

EXTENDED ACTIVITY

Study Figure 10.20, which shows the global distribution of manufacturing in 1994, and the world 'league table' of manufacturing.

1 Identify the two countries with more than $500 billion of manufacturing output (value added) in 1994.

2 Figure 10.18 shows the 'core', 'semi-periphery' and 'periphery' of the global economy. Comment on the levels of manufacturing output in each of these regions.

3 On an outline map of the world, shade the 15 largest countries in terms of manufacturing output. Describe their distribution.

4 Study Figure 10.21. Suggest some of the advantages and disadvantages of the increasing globalisation of the world economy. Present your findings in the form of a table.

0.20 *Global distribution of manufacturing, 1994*

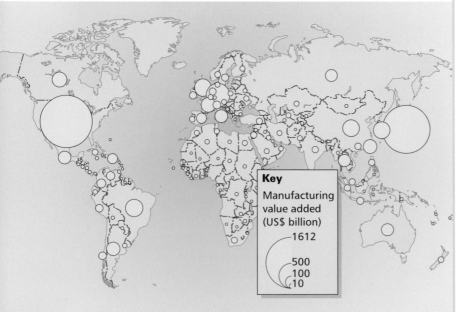

Key
Manufacturing value added (US$ billion)
1612
500
100
10

Rank	Country	Manufacturing value added (US$ million)	Percentage of world total
1	USA	1 611 763	26.9
2	Japan	1 257 761	21.0
3	Germany	692 191	11.6
4	France	268 611	4.5
5	UK	243 653	4.1
6	South Korea	159 172	2.7
7	Brazil	154 425	2.6
8	China	139 031	2.3
9	Italy	128 486	2.2
10	Canada	100 322	1.7
11	Argentina	88 366	1.5
12	Spain	81 196	1.4
13	Taiwan	73 295	1.2
14	Australia	64 417	1.1
15	Switzerland	60 111	1.0
			Total 85.8

UN attacks growing gulf between rich and poor

Charlotte Denny and **Victoria Brittain**

The combined wealth of the world's three richest families is greater than the annual income of 600 million people in the least developed countries, according to a UN report out today, and a 'grotesque' gap between rich and poor is widening.

Economic globalisation is creating a dangerous polarisation between multi-billionaires like Microsoft's Bill Gates, the Walton family who own the Wal-Mart empire, and the Sultan of Brunei – who have a combined worth of $135 billion – and the millions who have been left behind, the UN's annual human development report states.

The UN is calling for a rewriting of global economic rules to avoid inequalities between poor countries and wealthy individuals. It also wants a more representative system of global governance to buffer the effects of a 'boom and bust' economy.

'Global inequalities in income and living standards have reached grotesque proportions,' the report says.

Thirty years ago, the gap between the richest fifth of the world's people and the poorest stood at 30 to 1. By 1990 it had widened to 60 to 1 and today it stands at 74 to 1.

In terms of consumption, the richest fifth account for 86% while the bottom fifth account for just 1%. Almost 75% of the world's telephone lines – essential for new technologies like the net – are in the west, yet it has just 17% of the world's population.

'The world is rushing headlong into greater integration, driven mostly by a philosophy of market profitability and economic efficiency,' says the report's main author, Richard Jolly. 'We must bring human development and social protection into the equation.'

Breakthroughs like the internet can offer a fast track to growth, but at present only the rich and educated benefit.

Of the net's users, 88% live in the west, says the report, adding: 'The literally well connected have an overpowering advantage over the unconnected poor, whose voices and concerns are being left out of the global conversation.'

To counter the downside of globalisation, the UN makes a number of recommendations, including an international forum of business, trade unions and environmental and development groups to counter the dominance of the world's largest economies in global decision-making; a code of conduct for multinationals; and the creation of an international legal centre to help poor countries conduct global trade negotiations.

10.21   *The costs of globalisation*

The different parts of the global economy

In his book *Global Shift*, the geographer Peter Dicken (1998) argues that over the past 40 years some important changes have occurred in the pattern of world manufacturing output. He notes that between 1953 and 1995 the older industrialised economies' share of world manufacturing output declined by 95 to 80 per cent, while that of the developing economies quadrupled to 20 per cent. Despite the continued dominance of the world economy by a small number of highly developed (some would say 'over-developed') nations, Dicken argues that there have been some important changes in recent decades. He divides the global economy into three main groups of countries.

1 Older industrialised economies

These core industrial nations still dominate the world economy in terms of their output. However, significant changes have occurred. Most important is the substantial decline in the USA's relative share of world manufacturing production (Figure 10.22). In 1963 this was 40 per cent of world output; by 1994 its share had declined to 27 per cent. Other important changes are the uneven manufacturing performance of Western European countries, the decline in the relative importance of the USA, and the spectacular rise of Japan. In 1963 Japan accounted for 5.5 per cent of world manufacturing output; by 1994 it was the second largest manufacturer and accounted for 21 per cent of world output.

10.22   *City dereliction: St Louis in the USA*

2 Transitional economies of Eastern Europe and the former USSR

After 1989, the collapse of the USSR-led group of countries (the communist bloc) produced a group of so-called 'transitional economies'. These are formerly centrally planned economies which are now in various stages of transition to a capitalist market economy. They have fared badly in terms of manufacturing. In 1985, for example, the USSR accounted for around 10 per cent of world manufacturing output; by 1994 its share had declined to 1.5 per cent. Figure 10.23 shows the growth and, more commonly, the decline in the rates of economic growth in these economies.

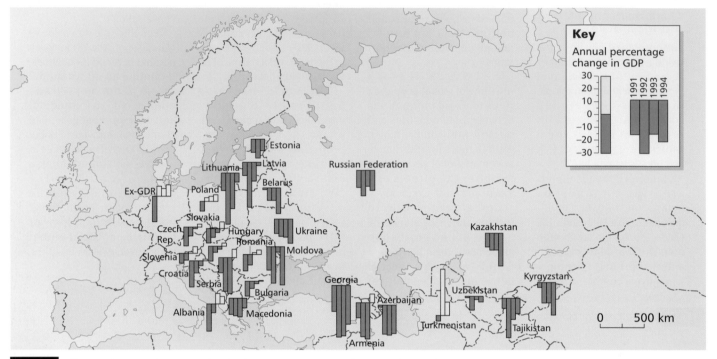

10.23 *Economic growth and decline in Central and Eastern Europe, 1991–94*

3 Developing economies

An important feature of the global shift is the growth of manufacturing in developing economies. These economies experienced a 3.5 per cent growth rate in manufacturing between 1938 and 1950, and a 6.6 per cent growth rate between 1950 and 1970. An important feature of this process has been the increasing importance of the **newly industrialising countries (NICs)**. The industrialisation of developing economies is very uneven, and many of the least developed economies have seen a decline in their share of total manufacturing production. The most rapidly growing economies in the developing world have been the NICs such as China, Brazil, India, South Korea and Taiwan (Figure 10.24). The so-called 'Tiger' economies of South-east Asia, such as South Korea, Singapore, Taiwan and Hong Kong, experienced rapid growth throughout the 1970s and 1980s. A number of reasons have been suggested for their success:

- Political stability – these countries all enjoyed a long period of political stability during their rapid growth stage. The argument is that it is difficult to encourage rapid economic development if there are frequent changes of government.

- Culture – it has been suggested that a philosophy which stresses respect for authority provides the best conditions

10.24 *East Asia and the 'Tiger' economies*

for rapid economic growth. These countries have been based on the idea that authority should be placed in the hands of the educated, and those in authority have a responsibility to act in the interests of society as a whole.

- The role of the state – governments play an important role in allocating resources, encouraging enterprise and trade, and influencing the direction of the economy.

- Human capital – given the lack of natural resources, the use of 'people power' has been important. Investment has been made in education and health care. Low population growth has allowed women to participate fully in the economy.

- Types of activity – it is argued that these economies have flourished by encouraging appropriate manufacturing activities, beginning with labour-intensive manufactures such as textiles and then diversifying into higher-value-added activities as labour became more skilled and capital more easily available.

- Openness to trade – the lack of domestic markets made it difficult for firms to take advantage of **economies of scale** (bulk production that lowers costs), so governments in these countries encouraged or promoted export activity.

- Willingness to use foreign capital – these countries welcomed the presence of foreign transnational corporations. Although there are arguments about whether such inward investment is beneficial, these countries generally benefited from the technology used in production.

- Flexibility – these economies were able to adapt quickly to changing patterns of international demand.

However, claims about the economic success of NICs need to be examined critically in the light of the events of 1997/98 when the economies of many of these countries suffered severe economic setbacks as a result of a downturn in the global economy.

NOTING ACTIVITIES

1   Label a blank outline map of the world to show the main areas of the global economy. Identify the following areas:

- older industrialised economies

- transitional economies of Eastern Europe and the former USSR

- developing economies.

2   On your map, summarise the key features of each type of economy, and its role in the global economy.

3   For each of these types of economy, suggest how it is changing, and what are its future prospects.

CASE STUDY

USA

By the end of the 19th century the USA had become a major power in the world economy. Its development was linked to the vast natural resources of land and minerals within its territory, which provided the raw materials for a wide range of industries. These industries served a large and growing market created by waves of immigrants, who also provided a cheap and industrious labour force. The USA also had links with Europe, which provided it with technological know-how and, importantly, capital. The USA's industrial strength was based on technologies that included the internal combustion engine, oil and plastics, electrical engineering, and radio and telecommunications. The outcome was a distinctive manufacturing region, known as the Manufacturing Belt, that stretched from Boston and Baltimore in the east to Milwaukee and St Louis in the west. Within this region, individual cities became famous for their own specialist industries. For example, Detroit – popularly known as Motown – was famous for its cars, Pittsburgh for steel, and Milwaukee and St Louis for brewing.

After the Second World War the USA enjoyed an unrivalled position as the largest economy in the world. It produced around 40 per cent of the world's manufacturing output, leading the world in steel and car manufacturing. However, from the 1960s it began to experience mounting competition. The 1970s saw plant closures, rising unemployment and a deteriorating trade balance. Whilst the rate of decline was gradual, the overall effect was to transform the Manufacturing Belt, where the bulk of the job losses were concentrated, into a Rustbelt. There were three main reasons for the decline of manufacturing.

- Rates of productivity growth were low compared with those of other economies.

- From the 1950s and 1960s, large numbers of US transnational companies invested abroad, first in Europe and then in the NICs. By 1980, US transnational companies earned at least one-third of their profits from overseas operations.

- US-based manufacturing was unable to compete with countries where labour was cheaper.

In 1950, 45 per cent of employment in the USA was in manufacturing, but by 1990 this figure was just 18 per cent.

The decline of the Manufacturing Belt has been accompanied by the rise of the Sunbelt, or the southern and western regions. New manufacturing industries have developed in states such as California, New Mexico, Texas, Georgia and Florida. Economic growth in these regions has been accompanied by population growth, as people migrate from the older manufacturing belt to the prosperous south and west.

NOTING ACTIVITIES

Study Figure 10.25.

1 Using an atlas to help you, locate each of the following states:

California	Pennsylvania
Florida	Texas
Michigan	Virginia
Mississippi	Washington
North Carolina	

Then make a large copy of this table:

	Decline (negative change)	Slow growth (less than 20%)	Rapid growth (more than 60%)
1960–70			
1970–80			
1980–90			

2 From your analysis, summarise the regional pattern of manufacturing change in each decade:

   a 1960–70

   b 1970–80

   c 1980–90.

3 Suggest reasons for the patterns you have identified.

4 What would be the likely economic, social and environment impacts of these changes on:

   a the Manufacturing Belt of the north and east

   b the Sunbelt of the south and west?

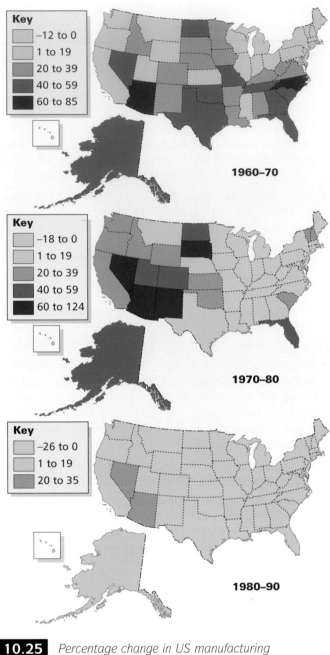

10.25 *Percentage change in US manufacturing employment, 1960–90*

CASE STUDY

Poland

Poland was the first communist country to become non-communist in 1989. It adopted a programme of economic reform. The Balcerowicz Plan (named after Poland's finance minister) was based on the privatisation of state-owned industries and the pursuit of free market policies. The economy suffered throughout the early 1990s as trade with the Eastern bloc collapsed. However, Poland was the first post-communist state to recover its 1989 level of GDP in 1996. There have been important changes in the nature of its industry. The service sector was undeveloped under communism but is now growing, and the share of GDP provided by the private sector has doubled from 30 per cent to 60 per cent since 1989. The patterns of trade show a dramatic shift. In 1985, 22.5 per cent of exports were to EU countries compared with 49 per cent to Eastern bloc countries. In 1995, the figures were 69 per cent and 16 per cent respectively. However, this situation is likely to become more balanced in the future.

Poland has attracted substantial amounts of foreign investment. US-owned transnational corporations are the largest investors. Germany is the most important trading partner, accounting for 38.5 per cent of exports and 27 per cent of imports in 1995. (German and Korean companies are important investors in Poland.) The fortunes of different sectors of Poland's economy vary. Shipbuilding is successful, especially at Szczecin in the north-west. Steelmaking remains old-fashioned, and coal mining is making heavy losses. The restructuring of the coal industry is likely to lead to the loss of 90 000 jobs over the next few years. In car manufacturing, Fiat has been long established in Poland and is planning to make its new 'world car' in Poland. Other investors are the Korean company Daewoo, which bought the Polish car-maker FSO; and GM, which plans a greenfield investment at a new factory in Gliwice in Silesia making 70 000–100 000 cars a year.

Other examples of inward investment include the Korean LG Group, which is investing in various civil engineering, telecommunications and other projects. Other investors include Michelin, Goodyear, ING, Coca-Cola, PepsiCo, Procter and Gamble, Siemens, Matsushita, Unilever, and Cadbury Schweppes (Figure 10.26).

A major challenge to be overcome if Poland is to realise its potential as a 'business centre' between East and West is the state of the country's infrastructure. Poland has a large road network, but much of it is in poor repair. A new motorway programme has been launched with the aim of building 2600 km of multi-lane highway in the next 10–15 years. The three international airports at Warsaw, Gdansk and Kracow are being upgraded. Special Economic Zones are now being established in which investors will benefit from reduced taxes and less bureaucracy (Figure 10.27).

10.26 *A US transnational in Poland*

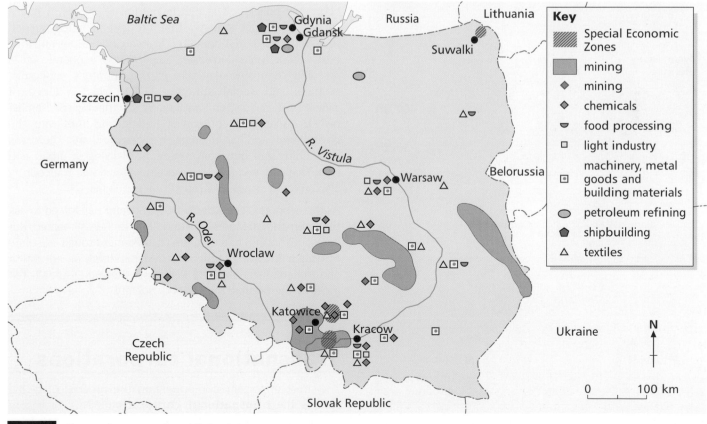

10.27 *Economic geography of Poland*

NOTING ACTIVITIES

1 Use Figure 10.27 to help you draw an annotated map showing the main features of Poland's economy.

2 What is the difference between a communist economy and a free market economy?

3 How and why have Poland's patterns of trade changed since 1989?

4 Why might a US or European-owned transnational company seek to establish production in Poland?

5 What factors might hinder Poland's rate of economic development?

CASE STUDY

Thailand

In the last two decades Thailand has had one of the world's fastest-growing economies. It is an example of a newly industrialising economy. Between 1971 and 1994 its GDP grew at an average of 7.4 per cent per annum. Thailand's

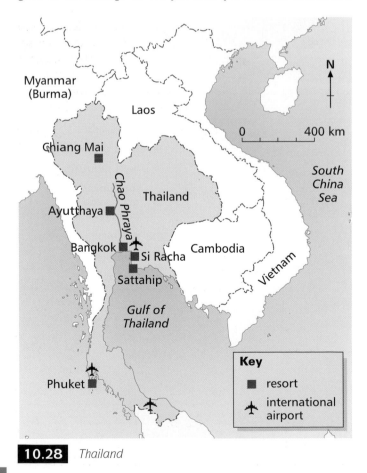

10.28 *Thailand*

rapid industrialisation had brought many changes to its economic and social structure (Figure 10.28).

Manufacturing employs 9 per cent of the labour force and contributes 37 per cent of GDP. Manufacturing has developed through a series of stages as the economy has diversified. Traditional reliance on the textile industry has given way to petrochemicals, electronics and motor vehicles. These sectors have been boosted by foreign investment because transnational corporations perceived Thailand as a country marked by political stability, sound economic management and a large and willing labour force.

Thailand's rapid economic growth has brought material benefits to its people. Average incomes have risen, and many more people are able to purchase consumer goods. However, there have been problems in providing a suitable infrastructure in order to support further economic development. Thailand's port facilities are limited, the capital city – Bangkok – suffers from overcrowding and traffic congestion, and environmental problems include traffic pollution and rapid deforestation due to logging and agricultural development. The overdependence on Bangkok as a primate city means that many people have left the countryside and the country faces a significant imbalance between its rural and urban sectors.

Alongside these problems are concerns about the stability of the economy. Thailand in many ways reflected the strengths and weaknesses of the rapidly growing Asian 'Tiger' economies. Its economic growth was based on sound economic policies, it is rich in natural resources, it had a successful tourist industry that brought in much needed foreign capital, and it benefited from an industrious and low-cost labour force. However, the 1997/98 economic crisis revealed serious weaknesses in the country's economic, political and social structure. There was evidence of widespread corruption, and many of the biggest family-owned banks and companies had strong links with the government, and had been given excessive loans. There was a widening gulf between standards of living in urban and rural Thailand, and this led to political tensions. The country also suffered severe environmental problems.

As the economic crisis deepened, Thailand was forced to ask the International Monetary Fund (IMF) for help. It agreed, but in return Thailand had to agree to implement tough 'austerity measures' involving reductions in expenditure on education and welfare. Invariably, the impact of these measures is felt most by the poorest members of society.

Transnational corporations

The most important component in the process of global shift is the **transnational corporation** (TNC). As Figure 10.29 shows, the total sales of many TNCs exceeds the gross national product (GNP) of many small nation-states. The

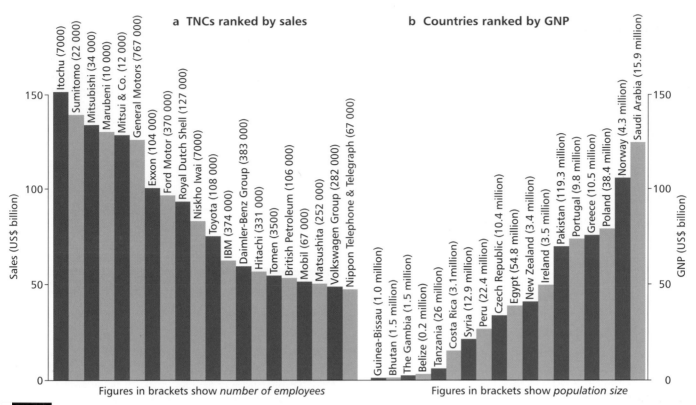

a TNCs ranked by sales

b Countries ranked by GNP

Figures in brackets show *number of employees*

Figures in brackets show *population size*

0.29   *A comparison between some TNC sales and the GNP of selected countries*

	Location	Location requirement	Changes
Company headquarters	Metropolitan areas	• Need for face-to-face contact • Close to business services • Close to government agencies	• Beginnings of suburbanisation • Developments in telecommunications dispensing with need for physical proximity
Research and development sections	Suburban areas, small cities	• Good environment to attract workers • Low taxes	• Movement to and growth in smaller towns in amenity-rich areas
Routine assembly plants	Small cities, rural areas	• Cheap labour • Low taxes	• Growth in Sunbelt (USA) and developing world

0.30   *Generalised model of a TNC*

sheer size of many TNCs means that they have considerable influence on the economy and politics of those countries in which they operate. Most TNC investment is in highly industrialised countries (most US TNC investments are concentrated in Europe, for example).

Figure 10.30 is a general model of how a TNC can divide up its operations in order to take advantage of different types of labour at different stages of its operations.

In the model, the headquarters are usually located in large metropolitan areas for reasons of prestige and status, and agglomeration economies can also be achieved through proximity to business services such as banks, advertising agencies and government functions. Research and development activities tend to be located in less central locations or in smaller urban centres in order to take advantage of sources of specialist labour (such as a university engaged in high-level research) or to avoid high taxes. Routine assembly is more footloose and its location is linked to the availability of cheap sources of labour and places where taxes are lower.

STRUCTURED QUESTION 1

Study Figure 10.31, which shows the location of the different functions of a TNC.

a Why might a TNC wish to have its headquarters in a major city?   (2)

b What might be the advantages for the TNC of locating production in:

(i)   a large urban centre   (2)

(ii) a peripheral region?   (2)

c Suggest why a TNC might locate different functions in different types of places.   (3)

d What effects might the relocation of production from a peripheral region to a large urban centre have on:

(i)   the peripheral region   (2)

(ii) the urban centre?   (2)

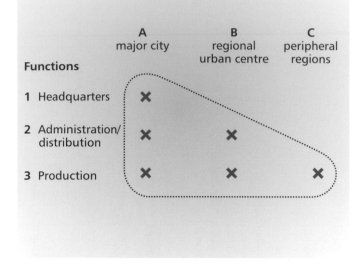

10.31 *Location of different functions in a TNC*

The costs and benefits of TNCs

Transnational corporations enter a host country's economy in a number of ways. These include the setting up of new plants, the acquisition of existing firms, or joint ventures with local companies. Many governments are suspicious of the activities of TNCs and may place limits on their activities through controls on the amount of foreign ownership permitted, or through agreements about the conditions of labour and taxes. TNC investment in a country has both costs and benefits. The benefits include:

- the generation of employment and skills
- increased incomes and lower unemployment
- increased levels of productivity
- the transfer of technology that can lead to an upgrading of the skills and aptitudes of the workforce.

The costs include:

- by competing with less efficient local firms, TNCs may force them into bankruptcy
- some TNCs may seek to avoid environmental restrictions and release pollutants into the local environment
- TNCs may seek to take advantage of cheap, flexible, non-unionised labour
- the levels of commitment to a country may be low, with the prospect of **runaway plants** where investment is quickly removed once economic conditions are less favourable. This was thought to be one of the major factors in the economic crisis that affected countries in South-east Asia in 1997 and 1998.

STRUCTURED QUESTION 2

Study Figure 10.32, which shows information about the world's 10 largest non-financial TNCs.

a How many of the top 10 companies have more than half their assets outside the country of their HQ? *(1)*

b Suggest why the motor vehicle TNCs have the majority of their assets in the country of their HQ. *(2)*

c Give an example of a TNC that has a high proportion of its production outside its country of origin. Suggest possible reasons for this. *(4)*

d With reference to a TNC that you have studied, suggest what benefits and disadvantages the company might bring to the countries in which they operate. *(6)*

Rank	Corporation	Country of HQ	Industry	% of assets held outside country	% of sales outside country of HQ
1	Royal Dutch Shell	UK/Netherlands	petroleum refining/chemicals	65	39
2	Ford	USA	motor vehicles	32	48
3	General Motors	USA	motor vehicles	29	31
4	Exxon	USA	petroleum refining/chemicals	60	78
5	IBM	USA	computers	52	61
6	British Petroleum	UK	petroleum refining/chemicals	53	73
7	Asea Brown Boveri	Switzerland	industrial and farm equipment	89	96
8	Nestlé	Switzerland	food	70	98
9	Philips Electronics	Netherlands	electronics	76	94
10	Mobil	USA	petroleum refining/chemicals	53	77

'Assets' refers to factories, machinery, land and companies owned by a TNC.

10.32 *The world's top 10 TNCs, ranked by value of foreign assets, 1990*

Nike

The US sports company Nike provides an example of the types of strategy employed by large TNCs in order to maximise their profits. The key to success is summed up in one word – *flexibility*.

Nike originated in the 1960s as an importer of shoes. The founder, Phil Knight, imported shoes from Japan and sold them at athletic meetings in the north-west USA. Today the company has its headquarters at Beaverton, near Portland in Oregon. It employs over 6000 people at its advertising, research and development centre, which is a landscaped park crossed with jogging trails and buildings such as the Michael Jordan Building and the Bo Jackson fitness centre.

No Nike shoes are produced in the USA. Instead, Nike seeks to develop flexibility in its production facilities, its location and its labour force by **outsourcing** (or **subcontracting**) its work to other firms. All of Nike's products are manufactured by contract suppliers operating throughout Asia. Over time, production has shifted from Japan to South Korea and Taiwan, and then to lower-wage regions in Indonesia, China and Vietnam (Figure 10.33). The shoe is seen as being made up of large numbers of component parts and the company shifts its production in order to seek out the lowest costs (Figure 10.34).

The reason behind this strategy is the intense competition within the industry, which constantly seeks to drive down costs and increase the profit margin on the product. Firms find it very hard to compete if their labour costs are higher than those of their competitors. It is suggested that many less developed economies in East Asia have large supplies of cheap labour that make them attractive to TNCs such as Nike. In fact the same factories are often involved in making shoes for Reebok, Adidas, Puma, LA Gear, and others.

This strategy is challenged by those who see it as exploitative of the workers involved in the production of footwear. Nike maintains that relative to other workers in these countries the Nike jobs are prized and valued. Nike's human rights critics argue that Nike's strategy of outsourcing its production to low-wage regions is exploitative (Figure 10.35). They present figures for workers in Indonesia to show that 45 workers shared just over $1.60 for making a $70 pair of Nike Air Pegasus shoes. A Nike spokeswoman stated that an $80 pair of shoes contains $2.60 in labour costs. The industry is a classic example of the de-skilling of production. The movement of shoe assembly jobs from one Asian nation to another is only possible because the task has been broken down into highly specialised gluing and stitching tasks.

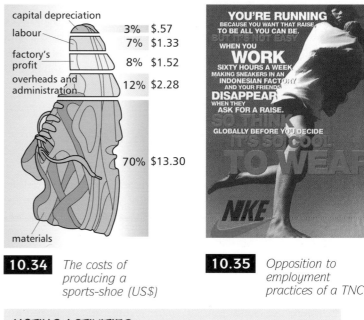

capital depreciation	3%	$.57
labour	7%	$1.33
factory's profit	8%	$1.52
overheads and administration	12%	$2.28
	70%	$13.30
materials		

10.34 *The costs of producing a sports-shoe (US$)*

10.35 *Opposition to employment practices of a TNC*

NOTING ACTIVITIES

1 Construct a table to compare the positive and negative impacts of TNCs.

2 On an outline map of the world, show the global nature of Nike's operations.

3 To what extent do Nike's operations match the general model of a TNC shown in Figure 10.30?

4 Figure 10.33 shows the changes in the source of Nike's suppliers over time.

   a Describe how Nike's suppliers have changed since 1989.

   b Suggest reasons why Nike might change its suppliers.

5 Nike's activities in south and east Asia have been a source of controversy in recent years. Why do you think this is?

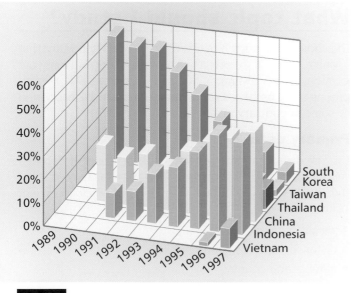

10.33 *Contract suppliers for Nike footwear, 1989–97*

A Planning a field investigation

The majority of the geography you have studied at school will have been conveyed to you through books (with information about real places) and through theories (like those of Christaller, Burgess and Hjulstrom). Whilst these approaches have made the subject interesting (you've got this far!), geography can only ever really be about the real world. As geographers, what we should really be doing is going out to measure the real world and attempting to explain why patterns conform to what we expect, or if they don't conform, explaining exactly why they don't. Fieldwork can be a lot of fun and very revealing. It can also be a useful way of generating good marks for your AS qualification. This section shows you how to approach fieldwork enquiries and how an actual enquiry progresses through the use of a running fieldwork example of a river investigation.

What do I need to do?

The precise requirements will depend upon your Examination Board, and you should read the syllabus details carefully and discuss options with your teacher before starting. This information is available in both published syllabuses or through the internet.

The main points to consider:

- Select a topic from a syllabus area that you will cover in the AS or A2 examinations.

- Ideally, choose a local site where return visits are possible.

- Ensure that there is an issue or a discussion at the heart of your study.

- Try to select a topic with supporting geographical theory or concepts.

- Check whether data can be pooled in a group or must be collected individually.

- Ensure that sufficient primary data will be available.

- Most important of all, the topic should hold some interest for *you* – make sure you have a genuine desire to find out more about it.

Reduced to its bare essentials, a geographical investigation can be considered as a three-part process:

EXPECT – what you should find according to geographical theory.

FIND – the patterns you found during your research.

EVALUATE – a comparison of the results with the expected pattern.

THE SEQUENCE

- Idea
- Topic
- Background theory and research
  - the formation of hypotheses or questions
  - what is expected in theory

THE DATA

- Method
  - ways in which the data is collected, and an evaluation of the reliability of the data
  - sampling and techniques
- Results
  - presentation of all data collected
  - visual representation of data, graphs and tables

EVALUATION

- Results analysis
  - the use of statistics to assess whether the results are significant
  - testing the hypotheses
- Discussion
  - comparison of results with the expected pattern (introduction)

What topic should I study?

The choice of topic is a demanding one. It is important that a viable topic is selected and this will depend on your location, time, commitments and resources. All investigations must be based on **primary** data and be supported by some **secondary** materials.

Primary data

This is data that you directly collect yourself through measuring, asking or recording. This data must be original and involve direct contact with the real world. It might include:

- weight or mass
- distance, width or depth
- number and counts
- questionnaire results
- sketches or photographs (taken by you).

Secondary data

This is obtained indirectly from books, notes, articles or maps. It has not, as such, been collected by you but is used in your investigation. It could include:

- maps
- books
- records
- census data, photographs (taken by others)
- internet material.

Human or physical?

It can be difficult sometimes deciding whether to attempt a 'human' or a 'physical' topic, and there are advantages and disadvantages to both (Figure 11.1). It is ideal, of course, if you can at least have a small element of cross-over: a 'human' project with some 'physical' slant to it, or vice versa. But take care: while a project that looks at footpath erosion *is* a possible cross-over project, one that looks at human perceptions of heavy metal and cyanide pollution of water could be very difficult, unless you have access to specialist laboratory facilities. Don't be too ambitious unless you know that you do have access to the hardware/expertise involved.

11.1 *Human – or physical?*

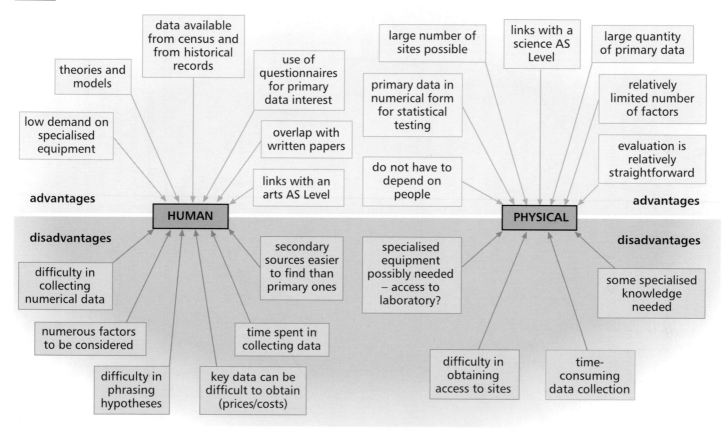

Checklist

A great deal of discussion and thought should go into planning your project. Before you make a final decision, come back to this book and look at the following checklist. As long as you are confident that these questions are answered to your satisfaction, you can submit your idea formally and get started.

- Is the topic included in your syllabus?
- Is there a theory to compare your data with?
- Do you have access to sufficient primary data?
- Can you obtain the required equipment?
- Is this a topic you are *really* interested in?

Possible themes

Your choice of topic must be based on your own requirements, interests and situation. Ideally the choice should be made with reference to your local area so that return visits are possible if they are required to complete the database, to check results or to collect follow-up data. For local studies start with an Ordnance Survey map and identify the possible topics in your own area. Figure 11.2 shows some of the possibilities in the area to the east of Oxford.

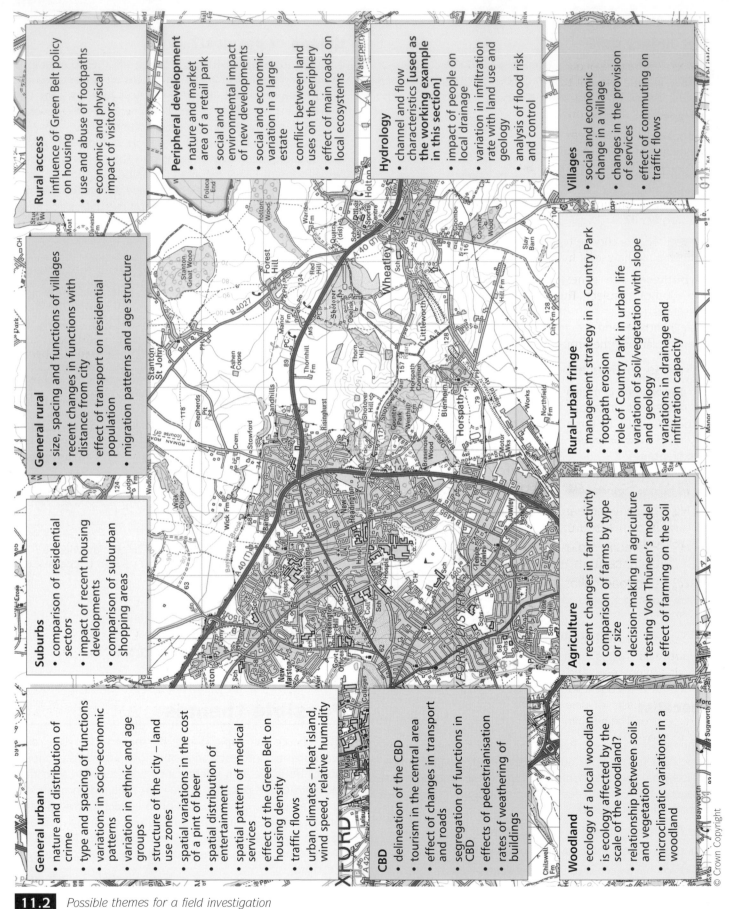

Rural access
- influence of Green Belt policy on housing
- use and abuse of footpaths
- economic and physical impact of visitors

Peripheral development
- nature and market area of a retail park
- social and environmental impact of new developments
- social and economic variation in a large estate
- conflict between land uses on the periphery
- effect of main roads on local ecosystems

Hydrology
- channel and flow characteristics [used as the working example in this section]
- impact of people on local drainage
- variation in infiltration rate with land use and geology
- analysis of flood risk and control

Villages
- social and economic change in a village
- changes in the provision of services
- effect of commuting on traffic flows

General rural
- size, spacing and functions of villages
- recent changes in functions with distance from city
- effect of transport on residential population
- migration patterns and age structure

Rural–urban fringe
- management strategy in a Country Park
- footpath erosion
- role of Country Park in urban life
- variation of soil/vegetation with slope and geology
- variations in drainage and infiltration capacity

Suburbs
- comparison of residential sectors
- impact of recent housing developments
- comparison of suburban shopping areas

Agriculture
- recent changes in farm activity
- comparison of farms by type or size
- decision-making in agriculture
- testing Von Thünen's model
- effect of farming on the soil

General urban
- nature and distribution of crime
- type and spacing of functions
- variations in socio-economic patterns
- variation in ethnic and age groups
- structure of the city – land use zones
- spatial variations in the cost of a pint of beer
- spatial distribution of entertainment
- spatial pattern of medical services
- effect of the Green Belt on housing density
- traffic flows
- urban climates – heat island, wind speed, relative humidity

CBD
- delineation of the CBD
- tourism in the central area
- effect of changes in transport and roads
- segregation of functions in CBD
- effects of pedestrianisation
- rates of weathering of buildings

Woodland
- ecology of a local woodland
- is ecology affected by the scale of the woodland?
- relationship between soils and vegetation
- microclimatic variations in a woodland

11.2 *Possible themes for a field investigation*

What am I studying?

Once the topic has been selected, research is required to determine the type of data you will need to collect. Research should include relevant theory so that you have a clear idea of the issues involved. The quality and quantity of data collected will depend on why it is being collected.

Many students find it easier to look at their project from the viewpoint of a simple question which can be answered by the evaluation of the data collected; for instance:

'To what extent does the microclimate change between summer and winter within a deciduous woodland?'

Other students choose to follow a more scientific approach and employ the use of hypotheses.

Hypotheses are generally made to help sharpen the direction of a project. These provide the specific targets of the investigation and act as a spine for the whole work. Framing the hypotheses is probably the most important step in planning the project.

What is an hypothesis?

An **hypothesis** is a clear statement which gives a specific aim to the investigation. There can be a number of working hypotheses (up to three), and they focus on what you are trying to show or prove.

River Study: hypotheses

H1   Velocity increases with increasing distance from the source.

H2   Hydraulic radius increases with increasing distance from the source.

H3   Bed load calibre decreases with increasing distance from the source.

In testing an hypothesis, you need to establish a **null hypothesis**. This is a statement that is the opposite of the working hypothesis, and it will be tested through statistics to arrive at a conclusion. Initially this sounds odd, but it forms the basis of scientific enquiry and is intended to reduce error. The null hypothesis rests on the idea that if you try to prove that a statement is incorrect and *fail*, then it must be correct. On the other hand, if you set out to prove a statement correct you are likely to select and interpret data to fit your case and end up with a false conclusion (this is called **experimenter bias**).

River Study: null hypotheses

H^o1 Velocity does not change with distance from the source.

H^o2 Hydraulic radius does not change with distance from the source.

H^o3 Bed load does not change with distance from the source.

River Study: getting started

River studies are popular because they lend themselves well to the collection of data and the testing of various theories and hypotheses. To start with you need to conduct some research from textbooks to find an aspect that interests you.

Research

Your research will enable you to write the **introduction**. The content of the introduction will directly depend on the hypotheses selected and will explain why you are investigating these aspects of the topic, including what you expect to find. There are several possible aspects worthy of study, including:

- the hydrological cycle and surface runoff
- channel formation through fluvial erosion and the processes involved
- the load characteristics of the river (Hjulstrom curve)
- the energy balance within a channel – gradient, mass, external friction, internal friction, hydraulic radius
- increase in mass downstream (stream order)
- channel shape as a balance between energy and resistance.

B Collecting data

1 Primary data

Data collection is your main contribution to the investigation and is what makes it unique. Focus on data that is directly relevant to the title or hypotheses, and aim for as high a **quality** and as large a **quantity** as possible.

For statistical testing of data, a minimum number of sites or subjects is required. As a rule, the greater the size of the sample then the more detailed and accurate the results become, making the discussion an easier and richer task. If in doubt collect more than seems necessary and discard later.

Before starting your data collection you need to plan very carefully indeed.

- **What** do I need to collect, and **why?**
- **Where** am I going to collect it, and **why?**
- **When** am I going to collect it, and **why?**

It is essential that you have good reasons for collecting the data in the manner that you propose to. Examiners will appreciate your reasoning when reading your final submission.

River Study: data collection

There are several sets of data that can be collected, although what *you* choose to collect will depend upon your hypotheses. Some examples include:

- measurements of depth, width and cross-sectional area of the channel
- bed load characteristics
- measurements of channel gradient and river velocity
- observations and mapping of bank configuration and adjacent land uses.

How should I collect my data?

There are several forms of data collection, perhaps the most common being measurements involving sampling and seeking responses using questionnaires.

Sampling

It is generally not possible to measure or question all the 'population' in your investigation, and some selection will be necessary. This process is known as **sampling**. It is important that you include a sufficient population for the results to be accurate but there is also an upper limit imposed by time, energy and availability. Similarly it is not possible to measure

---

River Study: why sample?

The river channel constantly changes and no two sections are the same, but data collection at each river site is slow and demanding. A number of sites (12 in this worked example) should be chosen as this balances time and effort collecting data with accuracy. This also gives a sufficient sample to allow variations to be shown and to use the statistical test known as Spearman's rank correlation coefficient.

---

all the variables along the entire course of a beach or river. Sampling is therefore a crucial aspect of data collection.

How many should there be in a sample?

This depends on time and resources. When physical measurements have to be taken, 10–12 sites are generally sufficient. Questionnaires are more rapid and should generally exceed 50. When pooling data as part of a group, a far larger sample is possible. It is always worth bearing in mind that the bigger the sample the more accurate your findings are likely to be. How many samples you take comes down in the end to a balance between maximising the size of your sample, and common sense and practicality.

Types of sampling

There are four main ways of selecting the location of data collection points.

- **Random sampling** – each point selected has the same chance of selection as all other sites. No rules or sequences are used in the selection process. Use random number tables, select from a hat, or ask the next person once an interview has finished.

- **Systematic sampling** – uses a rule or a set procedure to determine the place or person. This could be to ask every tenth person or divide a transect into 12 equal sections.

- **Stratified sampling** – the most accurate method involves an initial analysis of the whole population and its division into relevant categories. The sample is allocated according to the size of these categories, and specific sites are then selected by random techniques. This technique ensures that the eventual sample is a true reflection of the whole population.

- **Pragmatic sampling** – the use of one, or more, of the above methods to suit field conditions. Sampling is an

River Study: sampling techniques

a Random sampling

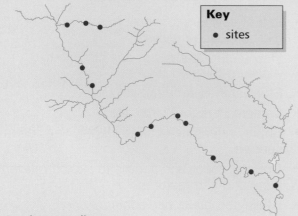

Key
● sites

Random sampling

Use a random number table to locate the 12 sites along the chosen stretch of river, or select numbers (distances along a section) from a hat.

* *Advantages* – this is rapid, simple and gives all points of the river an equal chance of selection.
* *Disadvantages* – the sites selected may not include important sections of the channel (a meander or weir), and may duplicate others. Sites at the start and end of the section may not be included. Access may be a problem.

b Systematic sampling

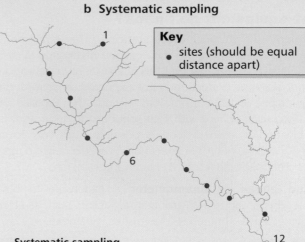

Key
● sites (should be equal distance apart)

Systematic sampling

Locate the 12 sites at equal intervals along the river section.

* *Advantages* – this gives an equal spread of sites and will accurately reflect continuous changes in river variables; relatively simple to carry out.
* *Disadvantages* – may exclude key sites (meanders and human modifications) and cause problems for access.

c Stratified sampling

Key
● sites

Lias group
(75%)
9 sites

Oolitic group
(20%)
2 sites

Corallian group
(5%)
1 site

Stratified sampling

Conduct an initial study of the channel, to divide it into straight, meandering, braided and artificially channelled sections. Calculate the total length in each category and their percentage importance. Allocate the 12 sites to these categories according to the weighting. For example, if 50 per cent of the channel length is meandering, then 6 sites should be selected within these sections by random numbers. In this example, changes in rock type have been taken into account. The sites are located according to the percentage occurrence of different rock types.

* *Advantages* – ensures that all types of channel are included in the survey. The sites do accurately reflect the whole of the river.
* *Disadvantages* – time and effort as well as problems in classifying the channel. Access may not be possible to some of the sites.

d Pragmatic sampling

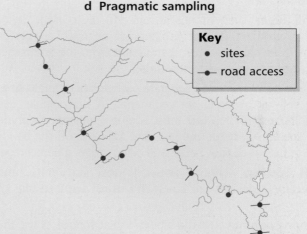

Key
● sites
●— road access

Pragmatic sampling

Suppose the site selected has poor or illegal access to the river channel. Rather than beating down a large bed of stinging nettles, penetrating a thicket of brambles or upsetting the local farmer, go to the next location downstream where there is a natural access to the channel. Fallen trees and particularly fast or deep sections of water are potentially dangerous and should be avoided.

'ideal' that often cannot be totally applied. It is important to try to be as objective as possible, but flexibility is required as, for example, not selecting a site with a large rock outcrop on a soil survey, or an obviously busy person during a shopping survey.

What equipment do I need?

Particularly in physical investigations, the equipment required to collect accurate data can be highly specialised and expensive. Remember that this is a student investigation and that absolute perfection is not expected. Improvisation of equipment is usually possible without the expense of specialised tools. Check with your school to find out what is available and/or see if a loan is possible from your local university, college of further education or even a neighbouring school. The following are some common problems.

- **Wind gauge or anemometer** – expensive to buy. Instead you can use environmental factors such as litter moving, leaves, branches, tree trunks, or smoke.

- **Flow meter** – very expensive, and demanding to use: try floating an orange or shaped dog biscuit over a measured stretch of the river.

- **Thermometer** – bulb thermometers may be too inaccurate to discriminate between sites. Use an electronic model sensitive to 0.1°C.

- **Relative humidity** – electronic devices are expensive and wet-and-dry bulb thermometers are slow and cumbersome.

- **Soil auger** – very expensive, and difficult to use in dry or stony ground. Pits dug with a spade or trowel are better, but make sure you have permission from the landowner first. Always clear up afterwards.

- **Sieve nests** – used to separate particles of different sizes in soil or beach/fluvial sediments. If unavailable, use sedimentation in a test tube, adding aluminium sulphate at the end to settle the fine clay. Measure off the bands of sediment as they settle and calculate as a percentage.

- **pH testing** – litmus paper is too inaccurate, and electronic devices are costly. Try using a soil-testing kit from a garden centre/nursery.

- **Drying cabinet** – try the biology lab at school, or use the oven at home (ask permission first).

- **Clinometer** – you can make your own clinometer using a protractor, but cheap spirit-level clinometers can be purchased at hardware shops.

Be aware of the potential limitations of your equipment, and try to minimise irregular results and inaccuracies.

River Study: equipment

- Tape long enough to extend from one bank to the other.
- Metre rule or piece of wood long enough to reach the bed from the surface.
- Pair of callipers to measure the diameter of the bed load particles.
- Flow meter, oranges or dog biscuits to measure the velocity.
- Stopwatch to time the velocity.
- Camera to record each site.
- Botany reference book to identify vegetation species.
- Drawing paper and pencils to sketch relevant bank and bed features.
- Wellington boots or waders.

Questionnaires – how and why to use them

In human topics, the questionnaire is the most common source of primary data but it can also be used in physical topics, particularly when management or conflict is an issue. The collection of questionnaires is time-consuming and it is essential that the design is right before you commit time and energy with the public. Careful thought and planning about the number and type of questions can save considerable time and energy at later stages of the investigation.

- Think about what data you wish to collect – work from the hypotheses.

- Try to isolate the main variables and make sure that these are covered by the questions. Refer to the background research to establish the relevant factors.

- If in doubt include the question, as the results can always be discarded if unnecessary.

- Always complete a draft and try it out as a pilot before the full collection. If questions are not working, you can change them.

- Be sure that there is a definite reason for each question being asked.

The design of the questionnaire is important because it must be completed rapidly (often under difficult conditions) and accurately. Tick-boxes, or 'forced choice format' are always preferable to writing and should be used whenever possible unless seeking individual perceptions or opinions.

River Study: questionnaire

A questionnaire is needed to determine the views of interested parties on river management, flooding, river use, water quality, or proposals for future change. It should include:

- status – local or non-local, gender, age
- socio-economic group/occupation
- use – how does the subject use the river: occupation, leisure, sport, view?
- awareness – is the subject aware of the issues or problems you are investigating?

- variables – questions on specific aspects such as water quality, change over time, flood frequency
- solutions – clear questions on what could be done.

These questions could be used to construct a picture of what the interested parties think about the river, its management, the issues and the solutions.

A clear and professional presentation helps to show subjects that you are serious.

Try to keep the questions on one side of the paper. This prevents confusion/loss and makes processing much easier.

Be prepared. Some people will be experts on your topic and may well reverse roles (question you, rather than be questioned). Know about the issues yourself.

Be sensitive when asking personal questions regarding gender, age, income or status.

Use clear tick-boxes to allow rapid and accurate completion. When necessary force a choice to avoid subjective answers.

Ask questions that directly relate to your hypothesis. Think about the kind of data you need, and focus on it.

Simple yes/no answers allow easier processing of data. This is particularly important if statistical testing will be used.

Do not ask too few or too many questions. People do not like to be stopped for no good reason when they are busy. Aim for 10–15 questions.

RIVER STUDY QUESTIONNAIRE

1	Gender	male ☐ female ☐
2	Age	under 16 ☐ 16 – 40 ☐ 41 – 60 ☐ 60+ ☐
3	Do you live locally?	yes ☐ no ☐
4	Do you use the river area for:	walking ☐ fishing ☐ boating ☐ swimming ☐ other (please specify) ☐
5	How frequently is this used?	daily ☐ weekly ☐ monthly ☐ rarely ☐
6	The river floods 2–3 times a year. Do you think it should be:	left as it is ☐ some reduction ☐ stop flooding ☐
7	What do you think of the water quality?	very good ☐
8	On what evidence do you base your answer?	good ☐ average ☐ poor ☐ very poor ☐
9	Would you like the river and its banks to be improved environmentally? If yes, how?	yes ☐ no ☐
10	Do you think that more use should be made of the river for recreation? If yes, how?	yes ☐ no ☐
11	Do you think that farmers should be restricted in using chemicals or fertilisers that enter the river?	yes ☐ no ☐
12	Would you like to see more wildlife?	yes ☐ no ☐ don't care ☐

Thank you for helping me in my research.

11.4 *A questionnaire*

2 Secondary data

Secondary data is the material that you use from other sources – that is, data that you have not personally collected. This supplements the primary data by including other variables, placing your investigation in an historical setting or in a wider framework. Secondary sources are numerous and diverse, and involve you in ferreting-out as much as you can. The most obvious place to begin is the library. Good local libraries hold more local information than many people realise, and librarians are generally extremely helpful. If you are not sure about where to find any particular element of secondary information, start with the librarian. Even if they haven't got it, they will often suggest avenues to follow which you may not have thought of.

Secondary sources include:

- maps – Ordnance Survey (1:50 000, 1:25 000, 1:10 000 and 1:2500), historical maps, soil and geology maps

- road maps for urban centres

- books and articles – local and central libraries, *Geofile*, *Geography Review*, *National Geographic Magazine*

- internet sites – particularly useful for hydrology, meteorology, urban functions and photographs; see also the Stanley Thornes website

- local government offices for reports and plans – particularly useful for management issues and historical data

- agencies and national government departments

- CD-ROM packages – British census data for 1981 and 1991 is available on Scamp-2 CD-ROM, and this includes a good map base, data at ward and enumeration district levels on the main social and economic variables, as well as a facility allowing your own data to be mapped and printed.

River Study: some secondary sources

- Maps at 1:50 000 and 1:25 000 to show location of sites and general drainage pattern.
- Older versions of OS maps to show changes in land use and settlement/routes.
- Maps to suggest past course of river, e.g. district boundaries often run along river courses.
- Geology maps and books to determine lithology and structure of drainage basin.
- Land use maps to determine vegetation and cropping in drainage basin.
- Local/national newspaper archives for incidences of flooding or channel modifications.
- National Rivers Authority (now the Environment Agency) and the Institute of Hydrology for any hydrological data available on the basin.
- Local meteorological offices for data on precipitation, evaporation and storm events.
- Local history section in local library for any reports, articles or books on your river.
- Would the local angling club be able to help? Try it.

C Processing the data

How should I present my data?

The primary data you have collected and the secondary data you have researched should be included in the investigation. This allows:

- the reader/examiner to assess the quantity and quality of the data
- the reader/examiner to draw their own conclusions from the material (thereby checking your inferences made in the same manner)
- basic patterns to be made evident
- the reasons for analysis and the use of statistics to be established.

The presentation of results generally takes the form of tables, visual representation of data and basic (descriptive) statistics, including mean, median and mode. This section should be separate from the results analysis.

These tasks are made simpler and more effective by the use of spreadsheet packages (such as Excel or Lotus 1-2-3), but remember:

- *do not* churn out an endless sequence of large graphs and pie charts. Try to select meaningful and *appropriate* graphs and diagrams which can be annotated and that are referred to in the written text.

- Photographs (well composed, in focus and illustrative) can be a useful tool both to show the reader elements of the study area, and to form the basis of annotative diagrams.
- Some graphs and diagrams should be hand-drawn. This is particularly relevant when composing complex composite diagrams, showing several different data sets on the same diagram. It is also an appropriate means of showing spatial (map) information.

Tables

Most data that is collected can be presented in the form of a table. This allows the comparison of variables and shows the whole set of data, giving a general picture. Tables also show the amount of time and effort you have made in primary research, so include all of the material you have gathered. Try to use separate tables for different aspects of the topic, and include a written summary of the main points either as annotations or below the diagram. This should include:

- maximum value
- minimum value
- the range
- the average, median or mode.

Take care over the presentation and layout so that items are presented in sequence and the results can be directly related to the results analysis. Also, draw the reader's attention to the important data.

River Study: tables

Clear identification of the variables included. Plotted as columns.

Rows represent sites, arranged from source to mouth.

All data collected is included, to allow selection of material as required. This also allows the reader to see the complete database.

Summary of data collected at 12 sites

Site	GR	Distance	Altitude	Width (m)	Interval (m)	Velocity (m/sec)	Bedload (cm length)				
1	209325	6.25	125	1.84	0.25	0.91	4	4.2	3.8	2.1	4.6
2	221282	9	118	4.1	0.25	1.37	3.5	3.4	2.5	4.1	3.7
3	236263	11.8	111	5.8	0.25	1.02	3.1	3.5	2.6	3.6	3.2
4	252223	17	106	4.4	0.25	1.57	2.7	3.7	3.5	2.1	3.5
5	268204	20	103	5.1	0.25	1.43	2.1	4.1	3.7	3.9	2.5
6	295182	23.3	98	7.1	0.25	1.51	2.8	2.7	3.6	3.2	3.1
7	299190	25.8	96	7.1	0.25	1.58	2.7	3.1	2.2	1.9	3.8
8	315203	28.8	90	7.22	0.25	1.67	2.6	1.8	3.4	1.9	2.5
9	331295	30.8	85	9.32	0.25	1.56	1.8	2.5	2.4	3.1	2.9
10	367172	37.8	78	8.62	0.25	1.72	3.1	2	1.8	2.1	2.8
11	385156	41	73	9.28	0.25	1.91	1.5	2.2	1.6	2.3	2.1
12	448102	57	64	10.3	0.25	2.1	1.6	2.2	1.8	1.6	2.5

summary of depth of channel at the 12 sites

site — depth

```
1  0.03 0.07 0.087 0.089 0.091 0.089
2  0.053 0.14 0.2 0.155 0.21 0.235 0.252 0.235 0.22 0.23 0.24
3  0.11 0.22 0.335 0.45 0.415 0.4 0.354 0.354 0.4 0.43 0.44 0.44 0.425 0.356 0.353 0.353 0.354 0.405
4  0.13 0.25 0.351 0.353 0.355 0.403 0.43 0.44 0.445 0.43 0.4 0.354 0.354 0.354
5  0.13 0.28 0.42 0.49 0.53 0.57 0.59 0.62 0.64 0.65 0.605 0.58 0.53 0.49 0.46
6  0.14 0.29 0.43 0.58 0.69 0.7 0.715 0.75 0.75 0.74 0.73 0.72 0.72 0.72 0.72 0.71 0.7 0.705 0.7
7  0.18 0.35 0.53 0.7 0.8 0.81 0.815 0.82 0.83 0.84 0.86 0.88 0.89 0.91 0.92 0.91 0.89 0.88 0.87 0.86 0.85 0.84 0.83
8  0.2 0.4 0.6 0.8 0.97 1.02 1.08 1.13 1.18 1.2 1.19 1.18 1.17 1.16 1.15 1.1 1.09 1.04 1 0.99 0.99 0.99
9  0.15 0.3 0.48 0.63 0.8 0.97 1.1 1.2 1.35 1.4 1.7 1.8 1.7 1.6 1.5 1.4 1.3 1.2 1.1 1.05 1 0.9 0.8 0.7 0.7 0.8 0.8 0.9
10 0.18 0.33 0.5 0.69 0.83 0.98 0.98 0.99 0.99 1 1.01 1.05 1.1 1.1 1.2 1.38 1.04 1.6 1.77 1.68 1.53 1.39 1.24 1.14 0.99 0.98 0.97 0.96
11 0.02 0.19 0.38 0.5 0.67 0.8 0.9 1.09 1.19 1.29 1.39 1.5 1.62 1.6 1.58 1.55 1.54 1.52 1.48 1.39 1.3 1.2 1.16 1.07 0.99 0.99 0.98 0.98 0.97
12 0.14 0.28 0.4 0.55 0.68 0.82 0.96 0.97 0.98 0.98 0.99 0.99 1.14 1.16 1.17 1.19 1.24 1.27 1.3 1.28 1.27 1.25 1.24 1.23 1.22 1.21 1.18 1.17 1.16 1.14 1.1
```

example
questionnaire results

Site 1

	1 gender	2 age	3 residence	4 use	5 frequency	6 flooding	7 quality	8 improved	9 environment	10 recreation
1	1 0	1 0	1 0	1 0	0 1	1 0	0 1	0 1	0 1	0 1
2	1 0	0 1	0 1	1 0	0 1	0 1	0 1	1 0	1 0	1 0
3	0 1	0 1	1 0	1 0	0 1	1 0	1 0	0 1	1 0	1 0
4	0 1	1 0	1 0	0 1	1 0	0 1	0 1	1 0	1 0	1 0
5	1 0	1 0	0 1	1 0	1 0	0 1	0 1	1 0	1 0	0 1
6	0 1	1 0	0 1	0 1	0 1	0 1	1 0	0 1	0 1	1 0
7	1 0	1 0	1 0	0 1	0 1	1 0	0 1	1 0	1 0	1 0
8	1 0	0 1	0 1	1 0	1 0	0 1	1 0	0 1	1 0	1 0
9	1 0	1 0	1 0	0 1	1 0	0 1	1 0	0 1	0 1	1 0
10	0 1	0 1	1 0	0 1	0 1	1 0	1 0	0 1	0 1	1 0

When plotting data on a spreadsheet, use 1 and 0 rather than 'yes' or 'no'. This allows rapid calculations during data analysis.

Totals and averages for each variable can be calculated using 'sum' and 'average' functions with 'fill, down'.

The questionnaire table includes all the data items. This allows the later analysis of any variable, and data can be rearranged by using 'data, sort' functions.

11.5 *Summary of data collected at 12 sites*

Visual representation of data

A wide range of techniques is available to show your data in visual form. These should be used to complement the tables, and should include:

- line graphs – useful in showing changes over time or distance
- bar graphs or histograms – to show variations between sub-totals
- pie charts – to show the composition of a total

- rose charts – to show changes with direction
- composite charts – a combination of other techniques (for example, line and bar)
- triangular graphs – when the total is made up from 3 components (soil texture shown as sand, silt and clay)
- Lorenz curves – showing the relationship between a population and area
- kite graphs – to show changes in quantity over distance or time.

NOTING ACTIVITIES

1   From the list above, describe data presentation situations where each of these methods might be appropriate.

2   Look at Figure 11.6. Each diagram displays the same data but in a different way.

a   What are the benefits of each of these methods?

b   What disadvantages might you experience with each of them?

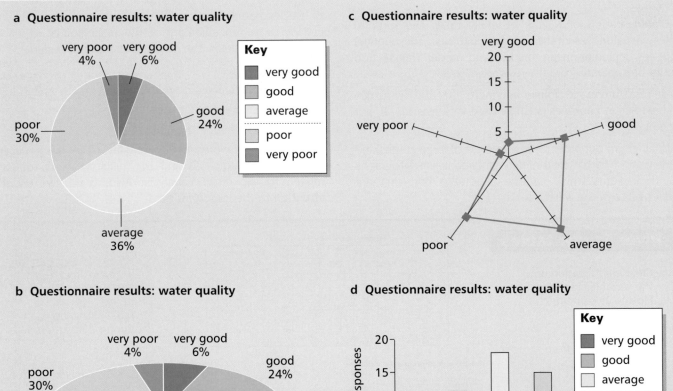

a Questionnaire results: water quality

b Questionnaire results: water quality

c Questionnaire results: water quality

d Questionnaire results: water quality

11.6   *Questionnaire results*

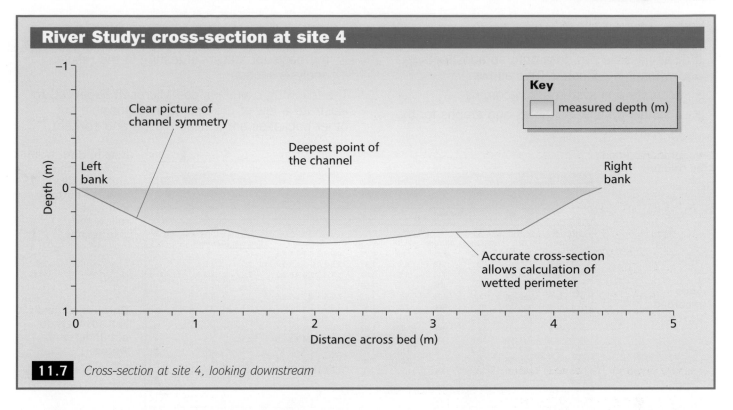

11.7   *Cross-section at site 4, looking downstream*

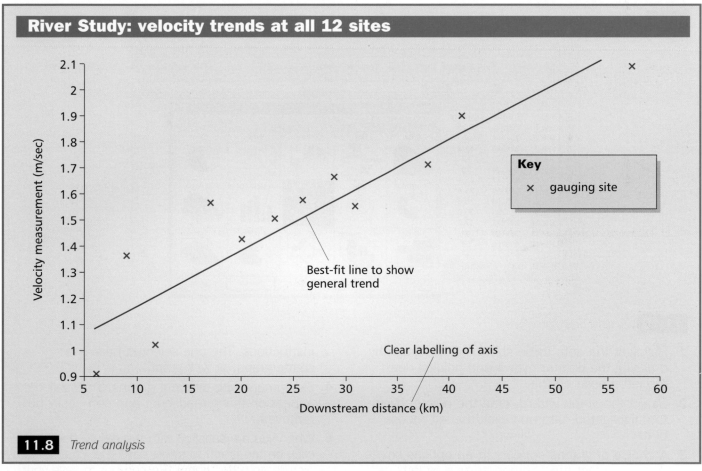

11.8   *Trend analysis*

Using spreadsheets

If at all possible put data onto spreadsheets as and when it is collected. This allows:

- accurate and systematic recording
- a simple transfer to tables and graphs for the results section

- the basis for statistical testing in the results analysis section.

The following examples use Microsoft Excel 5.0 to illustrate a few of the approaches possible but other packages are similar in style and content.

Variables recorded as columns

Help button. Click onto this and then click over the button in doubt – follow the instructions

Sites recorded as rows

Only use numerals. Spreadsheets do not understand words.

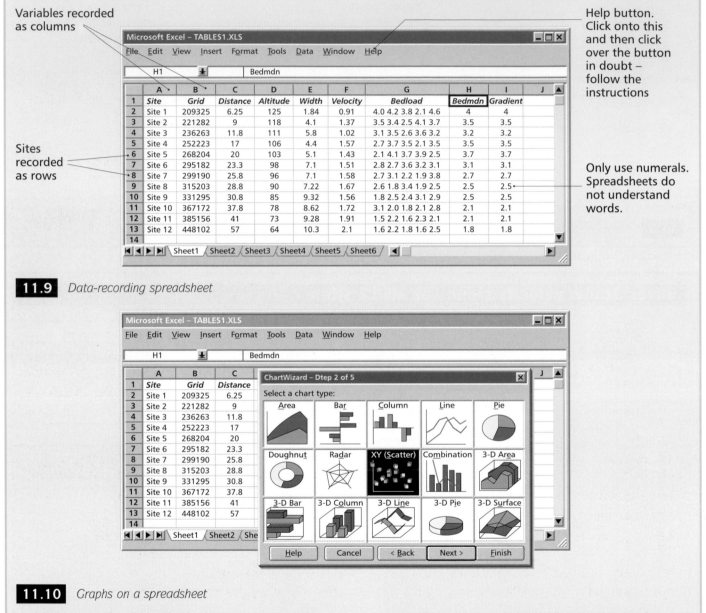

11.9   *Data-recording spreadsheet*

11.10   *Graphs on a spreadsheet*

1 Highlight the data to be shown on the graph by dragging the pointer over it and holding down the left mouse button.

2 Select the graph wizard, drag the pointer over the highlighted data and click the left mouse button.

3 A choice of graphs appears in an options box. You can select the appropriate one by left-clicking onto it. Then follow the on-screen

instructions. The one selected here is a scattergraph, or X/Y graph.

4 Experiment with different graphs. They will appear on the spreadsheet and can easily be removed.

5 When you are satisfied with the result, double-click on the graph to either save it or copy it, as a picture, to a document. Remember to save your work, as you may wish to make changes later.

How do I analyse my data?

Don't believe everything you measure! Analysis involves the manipulation and comparison of the data you have assembled and includes the use of formulae and statistics. More specifically it examines the patterns you have found and determines whether they are **significant** – that is, can they be trusted? The ultimate purpose of your analysis is to test your hypotheses or answer the questions you posed at the beginning of your study. If you can't trust your data, how can you have tested your question/hypothesis accurately? When we talk of statistics we are looking for proof of pattern and the ability to trust our findings. There are two broad types of statistics.

1 Descriptive statistics

These may be the average (mean), median or mode. Some simple calculations are normally possible, including conversion into percentages, standard deviations, calculation of averages, medians, modes, and inter-quartile ranges. The use of these varies with the type of data you have collected. Some examples are set out below.

- **Average or mean** – the total of all the sample divided by the number in the sample.

River Study: mean bed load size at site 4

The values of the 5 random particles are added together:

$$2.7 + 3.7 + 3.5 + 2.1 + 3.5 = 15.5$$

Divide the total by the number in the sample (5); $\frac{15.5}{5} = 3.1$ **Average** = 3.1

- **Median** – the central item in a sequence, calculated by ranking the data, counting the number in the sample and identifying the central value.

River Study: median bed load at site 4

Rank the data in sequence:

$$3.7, 3.5, 3.5, 2.7, 2.1$$

In a sample of 5 the median value is either 3rd from the top or 3rd from the bottom. **Median value** = 3.5

- **Mode** – the value with the greatest frequency; that is, the value that is listed the greatest number of times.

River Study: mode bed load at site 4

The most frequent value is 3.5, as it is mentioned twice. All other values are mentioned once. **Mode** = 3.5.

River Study: additional calculations

In the river survey more specialised calculations can also be used, including:

- cross-sectional area – calculated by multiplying width by depth
- wetted perimeter – the length of the channel cross-section that is in contact with the water; this can be calculated from the graph (see results in Figure 11.7)
- discharge – calculated by multiplying the cross-sectional area by velocity
- hydraulic radius – calculated by dividing the cross-sectional area by the wetted perimeter.

Site	Grid	Distance	Altitude	Width	Cross-sectional area	Velocity	Discharge (cumecs)	Median bedload	Wetted perimeter	Hydraulic radius
1	209325	6.25	125	1.84	0.13	0.91	0.12	4.0	1.88	0.07
2	221282	9.00	118	4.10	0.74	1.37	1.01	3.5	4.18	0.18
3	236263	11.80	111	5.80	2.06	1.02	2.10	3.2	6.15	0.33
4	252223	17.00	106	4.40	1.41	1.57	2.21	3.5	4.58	0.31
5	268204	20.00	103	5.10	2.30	1.43	3.29	3.7	5.36	0.43
6	295182	23.30	98	7.10	4.03	1.51	6.09	3.1	7.48	0.54
7	299190	25.80	96	7.10	5.09	1.58	8.04	2.7	7.70	0.66
8	315203	28.80	90	7.22	6.40	1.67	10.70	2.5	7.94	0.81
9	331295	30.80	85	9.32	8.81	1.56	13.70	2.5	10.40	0.85
10	367172	37.80	78	8.62	8.31	1.72	14.30	2.1	10.10	0.82
11	385156	41.00	73	9.28	9.03	1.91	17.20	2.1	10.10	0.90
12	448102	57.00	64	10.3	9.74	2.10	20.50	1.8	11.00	0.89

11.11 *River Study data*

2 Inferential statistics

This is an area of confusion and worry, particularly for students with a limited science background. There is an expectation that statistics will be included in a geography investigation, and they are important for the conclusion. In real patterns there are generally exceptions, anomalies or deviant samples that prevent a perfect result. Patterns do usually emerge but they are imperfect, and statistics are used to test whether or not they are strong enough to be trusted. Quite simply, the use of statistics replaces 'I think that there is a pattern' with 'There is a pattern with a 1 in 20 chance of error'.

The level of confidence

This expresses the level of trust that you can have in your results (it is similar to odds in gambling). The level of confidence is expressed in two main ways – **probability**, and **odds**.

Level of confidence	Probability	Odds
90%	0.1	a 1 in 10 chance of being wrong
95%	0.05	a 1 in 20 chance of being wrong
99%	0.01	a 1 in 100 chance of being wrong
99.9%	0.001	a 1 in 1000 chance of being wrong

Your research is not a matter of life and death, and millions of pounds will not be spent based on your outcome, so you do not have to set a very high level of proof. Generally the level of confidence will be 95% (±.05), as this is severe enough to reject weak patterns but not too stringent to reject sound patterns.

Statistics on spreadsheets

Many of the most commonly used statistics can be found on the spreadsheet program of your PC. Providing you have entered your data in a numerical form on your PC, you can use these programs to carry out much of the testing. To use these programs:

1 Highlight the data to be analysed. This will appear in the formula bar.

5 Remember, if you get stuck, click onto Help and hold the question mark over the item to be queried, and click again.

2 Click on to Function Wizard. This will open the options page shown.

3 Select the statistical option required.

4 A list of statistics functions will appear. To use, double click on a selected item and follow the instructions on the subsequent pages.

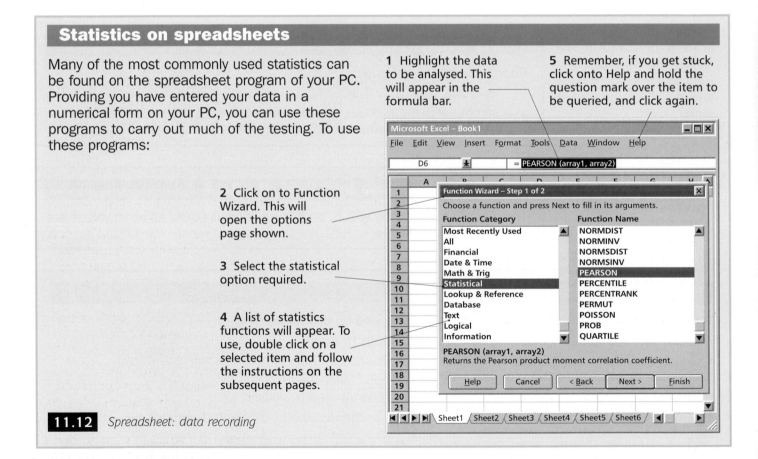

11.12 *Spreadsheet: data recording*

Which statistical test should I use?

In geography the **nominal** and the **correlation** tests are the most frequently used, but it is worth considering the application of tests of difference if the data allows.

Nominal data

This covers the data you may have collected in a questionnaire. Data collected in categories is 'weak' because the numerical information does not have the same value. For example, in questionnaire results the answer 'yes' may include responses ranging from 'absolute agreement' to 'I think I agree'. In contrast, when measuring the length of a pebble, a sample of 4 cm is twice as long as one of 2 cm and it has an absolute value. In terms of statistics this means that a nominal test requires a large difference to be trusted at a 95% confidence level, whilst absolute numerical data can be accepted with a smaller difference. Whenever possible numerical data is preferable to nominal.

The chi-squared test

- This is the only test for nominal data.
- It compares the pattern you have found (your data or observed data O) with what could be expected at random (called the expected pattern or E).

River Study: using chi-squared as a spreadsheet

- The first set of categories form the columns. In this example there are three categories, each one a different view of management. These values are counted from the questionnaire responses.

- The second subdivision is between locals and visitors. The sub-totals are shown in the rows, and are calculated by entering the following formula into cell I5:

  =C5+E5+G5 then Enter.

- The grand total in cell I11 can be calculated by entering the formula:

  =IS+I8 then Enter

- The expected values are calculated by multiplying the relevant sub-totals and dividing by the total. The expected value for cell 6 (H9)

can be calculated by the formula:

  =G11*I8/I11 then Enter.

- The x^2 figure must be calculated for each of the six cells in this example. Cell 1 can be calculated by clicking on cell D14 and entering the formula:

  =(C5-D6)*(C5-D6)/C6 then Enter.

  For cell 6 the formula entered in cell D19 is:

  =(G8-H9)*(G8-H9)/H9 then Enter.

- df or degrees of freedom is used when looking at the table of chi-squared values. It is calculated by:

  $(c - 1) \times (r - 1)$

  c = number of columns

  r = number of rows

  In this example $df = 2$

as $(2 - 1) \times (3 - 1) = 2$

- The critical value of x^2 is obtained from the table of values using the level of confidence and the degrees of freedom. At a 95% confidence level and with $df = 2$ the critical value of x^2 is 5.99.

- The total x^2 is calculated by inserting in cell G14 the formula:

  = sum (D14:D19) then Enter.

In this example the x^2 value of 7.19 is larger than the critical value of 5.99, indicating that the difference in views of river management by locals and visitors is different. The difference is strong enough to be trusted, with a greater than a 1 in 20 chance of being due to random error. If the x^2 calculation had been less than the critical value of 5.99 then any difference could not be accepted but rather seen as random error. As a result of the test in this example it is possible to:

- **reject the null hypothesis** – that there is no difference in view with residence, and

- **accept the working hypothesis** – that locals and visitors have different views of river management.

Hints: Construct the spreadsheet by inserting the formula, and save it without data. This will allow you to add various combinations of observed results to see what is significant. One matrix or table can be used over and over again.

Using Excel:
* means 'multiply'
= indicates 'start calculation'.

	A	B	C	D	E	F	G	H	I	J	K	L
		Microsoft Excel – CH13X2.XLS										
	File	Edit	View	Insert	Format	Tools	Data	Window	Help			
	F17			df = (2 – 1) (3 – 1) = 2								
1			*Comparison of views of flooding between residents and visitors*									
2												
3			leave as it is		some control		stop the flooding					
4		*locals*	cell 1		cell 2		cell 3					
5			5		8		17		30			
6				9		7.8		13.2				
7			cell 5		cell 6		cell 7					
8		*visitors*	10		5		5		20			
9				6		5.2		8.8				
10												
11			15		13		22		50			
12												
13			x2 calculations		sum of cells 1 – 7							
14			cell 1	1.777778	x2=		7.19					
15			cell 2	0.005128								
16			cell 3	1.093939	df = (2 – 1)	(3 – 1) = 2						
17			cell 4	2.666667								
18			cell 5	0.007692	*critical value = 5.99*							
19			cell 6	1.640909								
20												
21												

Sheet1 / Sheet2 / Sheet3 / Sheet4 / Sheet5 / Sheet6 /

11.13   *Using chi-squared as a spreadsheet*

- When the difference between the observed (O) and the expected (E) reaches a critical level the pattern can be trusted at the given level of significance.

- Chi-squared is calculated using the formula:

$$\chi^2 = \sum \frac{(O - E)^2}{E}$$

O = observed values

E = expected values (if at random)

Σ = sum of

- The test can either be worked through using a calculator, or by using the formula on a spreadsheet

- The spreadsheet has the advantage of making it possible to try out different categories to discover patterns that do and do not give significant results.

Correlations

A correlation is a comparison between two sets of data collected at a series of points to see if there is any relationship between them. Correlations can be shown on scattergraphs, or X/Y graphs. In geography, correlations are a basic approach, as they can compare frequently used variables, as long as they are numerical.

Frequently used variables include:

- distance
- time
- speed
- quantity or number
- height
- depth and width.

Spearman's rank correlation coefficient

Spearman's rank correlation coefficient (Rs) is used to determine how well two sets of data are related (correlated).

- Spearman's rank is an accurate test of relationship.

- It examines how variables behave in relation to each other.

- It does not mean that one variable is necessarily causing the change in the other.

- Spearman's rank is only a tool for preliminary analysis.

- It establishes patterns and anomalies for further research and explanation.

It uses the following formula:

$$Rs = 1 - \frac{6 \sum d^2}{n^3 - n}$$

The formula can either be worked through using a calculator or by constructing your own spreadsheet with inserted formula. As with chi-squared, the latter is preferable as it allows numerous calculations to be made by pasting new data to the same template and recording the results.

Significance of results

The results of multiple testing using Spearman's rank can be shown as a matrix which allows the identification of significant results.

Each of the Rs values represents a correlation between two variables. The critical value of Rs is obtained from a **table of significance** – an extract is shown in Figure 11.14.

- 0.591 is the critical value for a sample of 12 tested at a 95% confidence level.

- A value of +0.591 or greater represents a significant **positive** correlation. At worst there is a 1 in 20 chance of the relationship between the two variables being due to random error.

- A value of −0.591 or greater represents a **negative** correlation. At worst there is a 1 in 20 chance of the relationship between the variables being due to random error.

Interpreting correlations

Correlations provide a powerful tool to establish whether there is any relationship between variables. Three main conclusions can be drawn from the Rs statistic.

- **Positive correlation** – this means that as one variable increases so does the second variable. Equally, as one variable decreases so does the second (Figure 11.15).

- **Negative correlation** – this means that as the values of one variable increase so the values of the second variable decrease (Figure 11.16).

- **No correlation** – this means that as one variable increases or decreases, the second variable does not respond.

The river example has 12 sites so that n = 12.

The level of significance chosen for this example is 95% and the critical value is found in this column.

n (number in sample)	Level of significance		
	.1 or 90%	.05 or 95%	.01 or 99%
10	.564	.648	.794
12	.506	.591	.777
14	.456	.544	.715

The critical Rs value is found in this row.

The critical value must be exceeded for a relationship to be accepted.

Any values falling between +0.591 and −0.591 would not be significant.

11.14 *Extract from a table of significance*

River Study: does velocity increase with distance downstream?

- Overall there is a positive trend. With increasing distance downstream the velocity does increase. This is shown by the **best-fit line**.
- **A positive anomaly** (or deviation from the average)

  Velocity is greater than expected at site 2. This focuses research on channel characteristics at site 2 or on measurement error.
- **A negative anomaly**

  The velocity at site 7 is lower than expected. This focuses further research either on local channel characteristics or on measurement error.
- $Rs = -0.68$.

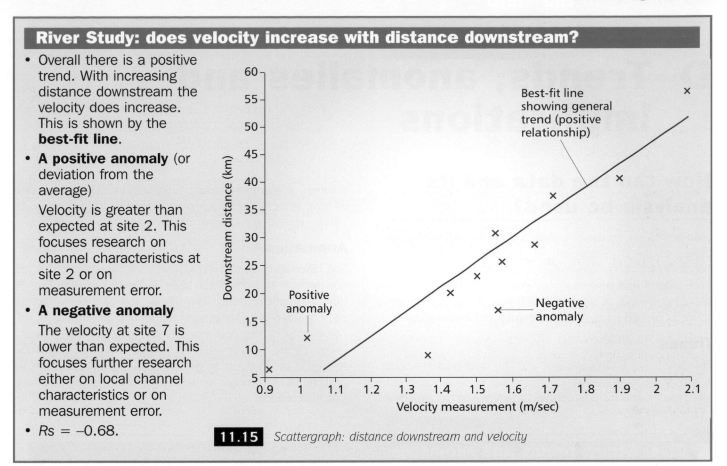

11.15   *Scattergraph: distance downstream and velocity*

River Study: does hydraulic radius decrease with height above sea level?

- Overall the hydraulic radius increases as height of the channel above sea level decreases. This is a negative correlation shown by the **best-fit line**.
- **A positive anomaly**

  The hydraulic radius at site 8 is higher than expected. This focuses attention on possible reasons why.
- **A negative anomaly**

  The hydraulic radius at sites 10 and 11 is lower than expected by the general trend. This focuses attention on possible reasons why.
- $Rs = -0.67$.

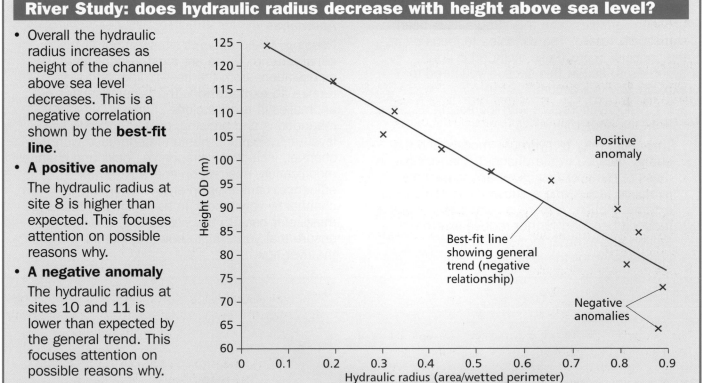

11.16   *Scattergraph: height and hydraulic radius*

D Trends, anomalies and implications

How can the data and its analysis be used?

Data collection, research and data analysis form the building blocks of the investigation. At the end of the work it is necessary for you to arrive at conclusions and to discuss the implications and limitations of your work. This discussion should come out of your data analysis. A useful strategy is to focus on two main themes: **trends** and **anomalies**.

Trends

This comprises a discussion of why the main patterns occur. It should generally involve a synthesis of theory and your own results, and include detailed reference to the processes involved. In the river example this could include a discussion of why velocity increases downstream, with reference to mass, external and internal frictions and hydraulic radius.

Anomalies

The discussion should explain why the general trends are not perfect, and why some sites and data vary significantly from the general trend. This allows you to include factors not considered in the general model and to emphasise the complexity of the real world – the main problem facing geographers. In the river example this could include reference to changes in geology, channel modification by artificial banks, bridges or weirs, vegetation in the channel, and meandering.

River Study: trends and anomalies

Trends

Your analysis should have identified relationships between variables, and this should have been statistically tested. The variables to focus on in the conclusion are those included in your hypotheses, and in this section you need to **explain** why this occurs.

- H1 (velocity increases with increasing distance from the source)

  This could refer to hydraulic models and the energy balance in the channel. The increase in mass of the river flow gives an increased hydraulic radius and a subsequent decrease in external friction. A more regular channel and a reduced bed load further reduce external friction and internal friction and these compensate for the loss of gradient.

Trends should be supported by theory whenever possible.

Anomalies

Readings taken at some sites will not fit the general trend. These are most evident on graphs as deviations from the best-fit line, and they require an **explanation**. The general causes of anomalies in rivers include changes in geology, meandering and braiding, past changes in sea level (rejuvenation), human interference with the channel, including weirs, dams, locks and channel modification, abstraction and wastewater effluence, bank erosion and invasion of the channel by vegetation. To establish the cause you must refer back to the site details, and you must ensure that you explain **how** the factors influence the river flow.

Limitations of your investigation

Reference should be made to the weaknesses in your research at all its stages. Reference could be made to:

- size of sample
- equipment
- conditions prevailing at the time
- limitations on time
- inaccuracies of measurement
- absence of important data.

Implications of your research

It is usually possible to make some reference to the wider implications of your research. Your conclusions should throw some light on the issue under consideration, and this should have some practical implications in the wider world.

You may also suggest further possible lines of research – this also shows an awareness of the limitations in your own research.

Main points

- Think through the whole investigation before you start.
- Check that your topic and data will meet the syllabus requirements.
- Frame precise hypotheses.
- When you are out in the field, be smart and keep safe.
- Try to use IT from the start.
- Enjoy your fieldwork enquiry, but don't become a martyr to it – if you couldn't complete a high-scoring project in the time allowed for it by the Exam Boards, they wouldn't suggest you do one.
- Keep coming back to this section, and also look up the tips on the Stanley Thornes website:

Good luck!

Semi-log graph paper

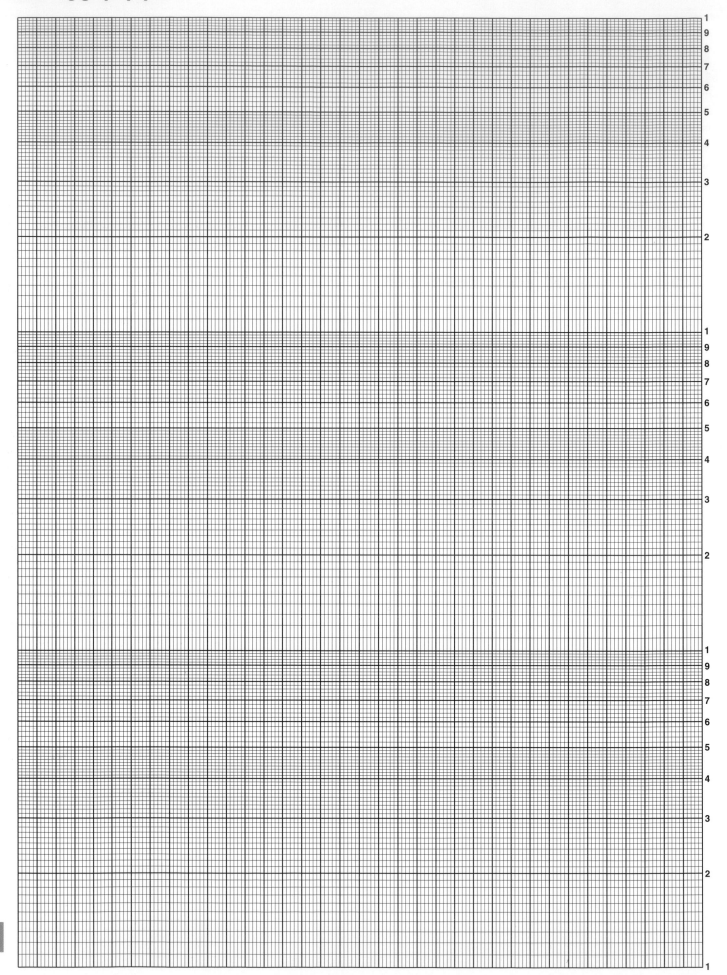

Semi-log graph paper

Acknowledgements

The authors and publishers would like to thank the following for permission to reproduce copyright photographs and other illustrations in this book:

Aerofilms Ltd, 4.39; Art Directors/Trip Photography, 4.42, 8.72, 10.9, 10.9, 10.26; Associated Press Ltd, 2.28, 2.32, 2.50, 4.47, 6.45; British Geological Survey, 9.17; Bettmann Corbis, 4.33b; Corbis /Joseph Sohm, Chromosolm, 10.22; Environmental Images, 7.50; Eye Ubiquitous, 2.14, 2.65, 5.2, 7.28, 8.35, 10.3(1),(2),(3); Frank Lane Picture Agency, 3.17, 3.21, 3.22, 5.2(b),(c), 5.8, 6.26 (a),(b),(c), 6.29, 6.36; Geophotos/Tony Waltham, 2.56, 2.63, 2.68, 2.71, 3.32, 5.26(a), 9.45; Geoscience Features, 2.21, 3.13, 3.42, 7.44; Hulton Getty Picture Library, 8.11, 8.49; Hutchison Library, 4.24(a), 10.19; ICCE/Joe Blossom, 7.61; Impact Photos, 10.11; J Allan Cash, 2.64, 3.6(a),(b),(c), 7.3; James Davis Travel Photography, 2.25(b), 2.45, 5.7, 8.6, 8.25, 8.56, 9.30; John Chaffey, 9.39, 9.43; Katherine James, 5.32; Landform Slides, 2.22, 2.70, 2.72, 5.20, 9.2; Magnum Photos, 4.42; Mary Evans Picture Library, 4.2, 8.54; NERC/University of Dundee, 6.17(2); Network Photographers, (p 112); Northern News, 10.1; Northumbrian Water, 7.54(both); OCR GCE A Level Geography 8/6/99; OCR Oxford and Cambridge Paper 2/92; Penni Bickle, 3.11; Rex Features, 4.6, 4.21, 4.24(b), 4.33(a), 4.48, 5.27, 6.43, 6.44, 7.56, 8.16, 8.27, 8.33(b), 8.38, 8.42, 8.46, 10.3(4) Richard Stanton, 7.58; Sealand, 9.11; Science Photo Library, 6.16(1), 6.27, 6.34, 7.47; Simon Ross, 2.36, 2.38, 2.39, 2.41, 2.76, 9.12, 9.13, 9.18, 9.19, 9.47; Simon Warner, 10.9(a); South Florida Water Management, 7.52(both) Still Pictures, 3.20, 3.27, 4.20, 5.24, 5.29, 5.39, 7.49, 7.63, 8.2, 8.4, 8.61, 8.63, 8.65, 8.68; The Evening Press, York, 6.19; The Fotomas Index, 4.37; Topham Picturepoint, 2.25(a), 4.18, 9.6; UK Perspectives Ltd, 5.35.

Map extracts 5.21, 7.59, 8.44 and 11.2 are reproduced from 1:50,000 Ordnance Survey mapping with the permission of the Controller of Her Majesty's Stationery Office © Crown Copyright. Licence No. 07000U.

The authors and publishers are grateful to the following for permission to use copyright text material in this book:

Assessment and Qualifications Alliance, Figure 1.11; Blackwell Publishers, The Urban Order, J Short, 1996, Figure 10.30; Cambridge University Press, Slopes and Weathering, M Clarke and J Small, 1982, Figures 2.46, 2.48; John Chaffey, Coastal Management in Western Dorset, 1999, Figure 9.38; Countryside Agency, Figures 5.28, 5.33; C Richardson, The Geography of Bradford, 1975, Figure 8.19; Crown Copyright, Source 1991 Census, ONS, Figure 4.41; Doonesbury ©, G B Trudeau. Reprinted with permission of Universal Press Syndicate. All rights reserved, Figure 8.36; Dorset Coast Forum, Dorset Coast Strategy, May 1999, Figure 9.10; Dorset County Council, Figures 9.8, 9.9; Financial Times, 30/3/99, Figure 5.25: 19–20/6/99, Figure 8.34; Geography Review, 'Patterns of nitrate concentration', T Burt, May 1989, Figure 7.65; Geography Review, 1994, Vol. 7, No. 4, Figure 8.39; Geologists Association, The Coastal Landforms of West Dorset, R J Allison, 1992, Figure 9.40; The Guardian, 14/8/99, John Vidal, Figure 4.1: 2/4/99, John Vidal, Figure 4.50: 5/9/98, Peter Hetherington, Figure 10.1: 12/7/99, C Denny and V Brittan, Figure 10.21; Hodder & Stoughton, Advanced Geography: Concepts and Cases, G Nagle and P Guiness, 1999, Figure 8.20: Natural Hazards, Frampton, Chaffey, Hardwick and McNaught, 1996, Figure 9.32; Oxford Cambridge and RSA Examinations, Figures 6.1, 6.2, 6.4, 6.5; Pearson Education Limited, Geographies of Development, Potter et al., 1999, Figure 8.3; River Restoration Centre/Environment Agency, Figure 7.53; Taylor & Francis, Environmental Hazards, K Smith, 1996, Figure 2.33: Whatever Happened To Planning?, P Ambrose, 1986, Figure 8.58; Times Newspapers Ltd, 26/4/87, Figure 2.40: 12/1/99, Figure 9.3; J Wiley & Sons Limited, 'Ecological Changes of the French Upper Rhône River since 1750', A L Roux et al., in Historical Change of Large Alluvial Rivers: Western Europe, G E Petts, 1989, Figures 3.16, Figure 8.26; N Woodcock, Geology and Environment in Britain and Ireland, 1994, Figure 2.60;

Every effort has been made to contact copyright holders. The publishers apologise to anyone whose rights have been inadvertently overlooked, and would be pleased to be advised of any errors or omissions so that these can be rectified.

Index

Page numbers refer to both text and figures. Page numbers in **bold** refer to a main reference or explanation of the subject.

Index

Index